T0211718

Lecture Notes in Computer Science   9282

Commenced Publication in 1973
Founding and Former Series Editors:
Gerhard Goos, Juris Hartmanis, and Jan van Leeuwen

Editorial Board

David Hutchison
  Lancaster University, Lancaster, UK
Takeo Kanade
  Carnegie Mellon University, Pittsburgh, PA, USA
Josef Kittler
  University of Surrey, Guildford, UK
Jon M. Kleinberg
  Cornell University, Ithaca, NY, USA
Friedemann Mattern
  ETH Zurich, Zürich, Switzerland
John C. Mitchell
  Stanford University, Stanford, CA, USA
Moni Naor
  Weizmann Institute of Science, Rehovot, Israel
C. Pandu Rangan
  Indian Institute of Technology, Madras, India
Bernhard Steffen
  TU Dortmund University, Dortmund, Germany
Demetri Terzopoulos
  University of California, Los Angeles, CA, USA
Doug Tygar
  University of California, Berkeley, CA, USA
Gerhard Weikum
  Max Planck Institute for Informatics, Saarbrücken, Germany

More information about this series at http://www.springer.com/series/7409

Tadeusz Morzy · Patrick Valduriez
Ladjel Bellatreche (Eds.)

Advances in Databases and Information Systems

19th East European Conference, ADBIS 2015
Poitiers, France, September 8–11, 2015
Proceedings

 Springer

Editors
Tadeusz Morzy
Poznan University of Technology
Poznán
Poland

Ladjel Bellatreche
LIAS/ISAE-ENSMA
Poitiers
France

Patrick Valduriez
INRIA
Montpellier
France

ISSN 0302-9743   ISSN 1611-3349   (electronic)
Lecture Notes in Computer Science
ISBN 978-3-319-23134-1   ISBN 978-3-319-23135-8   (eBook)
DOI 10.1007/978-3-319-23135-8

Library of Congress Control Number: 2015946766

LNCS Sublibrary: SL3 – Information Systems and Applications, incl. Internet/Web, and HCI

Springer Cham Heidelberg New York Dordrecht London
© Springer International Publishing Switzerland 2015
This work is subject to copyright. All rights are reserved by the Publisher, whether the whole or part of the material is concerned, specifically the rights of translation, reprinting, reuse of illustrations, recitation, broadcasting, reproduction on microfilms or in any other physical way, and transmission or information storage and retrieval, electronic adaptation, computer software, or by similar or dissimilar methodology now known or hereafter developed.
The use of general descriptive names, registered names, trademarks, service marks, etc. in this publication does not imply, even in the absence of a specific statement, that such names are exempt from the relevant protective laws and regulations and therefore free for general use.
The publisher, the authors and the editors are safe to assume that the advice and information in this book are believed to be true and accurate at the date of publication. Neither the publisher nor the authors or the editors give a warranty, express or implied, with respect to the material contained herein or for any errors or omissions that may have been made.

Printed on acid-free paper

Springer International Publishing AG Switzerland is part of Springer Science+Business Media
(www.springer.com)

Preface

This volume contains a selection of the papers presented at the 19th East-European Conference on Advances in Databases and Information Systems (ADBIS 2015), held during September 8–11, 2015, at Futuroscope, Poitiers, France.

The ADBIS series of conferences aims at providing a forum for the presentation and dissemination of research on database theory, development of advanced DBMS technologies, and their advanced applications. ADBIS 2015 in Poitiers continued the series after St. Petersburg (1997), Poznań (1998), Maribor (1999), Prague (2000), Vilnius (2001), Bratislava (2002), Dresden (2003), Budapest (2004), Tallinn (2005), Thessaloniki (2006), Varna (20007), Pori (2008), Riga (2009), Novi Sad (2010), Vienna (2011), Poznań (2012), Genoa (2013), and Ohrid (2014). This edition was special, as it was the first time that ADBIS took place in France. The conferences are initiated and supervised by an International Steering Committee consisting of representatives from Armenia, Austria, Bulgaria, Czech Republic, Estonia, Finland, Germany, Greece, Hungary, Israel, Italy, Latvia, Lithuania, Poland, Russia, Serbia, Slovakia, Slovenia, and the Ukraine.

The program of ADBIS 2015 included keynotes, research papers, two tutorials, and thematic workshops. The conference attracted 135 paper submissions from 39 countries from all continents with 330 authors. After rigorous reviewing by the Program Committee (77 reviewers from 22 countries), the 31 papers included in this LNCS proceedings volume were accepted as full contributions, making an acceptance rate of 23 %.

Furthermore, the Program Committee selected 18 more papers as short contributions and 30 papers from seven workshops that are published in a companion volume entitled *New Trends in Databases and Information Systems* in the Springer series *Communications in Computer and Information Science*. All papers were evaluated by at least three reviewers and most of them by four to five reviewers. The selected papers span a wide spectrum of topics in databases and related technologies, tackling challenging problems and presenting inventive and efficient solutions. In this volume, these papers are organized according to the 15 sessions: (1) Database Theory and Access Methods, (2) User Requirements and Database Evolution, (3) Multidimensional Modeling and OLAP, (4) ETL, (5) Transformation, Extraction and Archiving, (6) Modeling and Ontologies, (7) Time Series Processing, (8) Performance and Tuning, (9) Advanced Query Processing, (10) Approximation and Skyline, (11) Confidentiality and Trust.

For this edition of ADBIS 2015, we had two keynote talks: the first one from Serge Abiteboul from Inria and ENS Cachan, France, on "The Story of Webdamlog" and the second one by Jens Dittrich, from Saarland University, Germany, on "The Case for Small Data Management." In addition, we had two tutorials: the first by Nicolas Anciaux, Benjamin Nguyen, and Iulian Sandu Popa from Inria Paris-Rocquencourt and INSA Centre-Val de Loire, France, on "Towards an Era of Trust in Personal Data Management" and the second one by Boris Novikov, from St. Petersburg University, Russia, on "Query Processing: Beyond SQL and Relations."

ADBIS 2015 strived to create conditions for more experienced researchers to share their knowledge and expertises with the young researchers. In addition, the following seven workshops associated with the ADBIS were co-allocated with the main conference:

- Second International Workshop on Big Data Applications and Principles (BigDap 2015), organized by Elena Baralis (Politecnico di Torino, Italy), Tania Cerquitelli (Politecnico di Torino, Italy) and Pietro Michiardi (EURECOM, France).
- Workshop on Data Centered Smart Applications (DCSA 2015), organized by Ajantha Dahanayake (Prince Sultan University, Saudi Arabia) and Bernhard Thalheim (Christian Albrechts University, Germany).
- 4th International Workshop on GPUs in Databases (GID 2015), organized by Witold Andrzejewski (Poznan University of Technology, Poland), Krzysztof Kaczmarski (Warsaw University of Technology, Poland), and Tobias Lauer (Offenburg University of Applied Sciences, Germany).
- Workshop on Managing Evolving Business Intelligence Systems (MEBIS 2015), organized by Selma Khouri (National Engineering School for Mechanics and Aerotechnics (ISAE-ENSMA), France and National High School of Computer Science (ESI, Algeria), and Robert Wrembel (Poznan University of Technology, Poland).
- 4th International Workshop on Ontologies Meet Advanced Information Systems (OAIS 2015), organized by Ladjel Bellatreche (LIAS/ISAE-ENSMA, France), and Yamine Ait Ameur (IRIT-ENSEIHT, France).
- First International Workshop on Semantic Web for Cultural Heritage (SW4CH 2015), organizd by Béatrice Bouchou Markhoff (LI, University François Rabelais de Tours, France) and Stéphane Jean (LIAS/ISAE-ENSMA and University of Poitiers, France).
- Workshop on Information Systems for AlaRm Diffusion (WISARD 2015), organized by Rémi Delmas (ONERA, Toulouse, France), Thomas Polacsek (ONERA, Toulouse, France), Florence Sèdes (IRIT, Toulouse, France).

Each workshop has its own international Program Committee. The accepted papers were published by Springer in the series *Communications in Computer and Information Science* (CCIS).

The best papers of the main conference and workshop were invited to be submitted to special issues of the following journals: *Information Systems* - Elsevier, *Information Systems Frontiers* - Springer, and *International Journal on Semantic Web and Information Systems* - IGI.

We would like to express our gratitude to every individual who contributed to the success of ADBIS 2015. First, we thank all authors for submitting their research papers to the conference. We are also indebted to the members of the community who offered their precious time and expertise in performing various roles ranging from organizational to reviewing - their efforts, energy, and degree of professionalisms deserve the highest commendations. Special thanks to the Program Committee members and the external reviewers for evaluating papers submitted to ADBIS 2015, thereby ensuring the quality of the scientific program. Thanks also to all the colleagues, secretaries, and engineers involved in the conference organization, as well as the workshop organizers. We would like to thank Dr. Mickaël Baron, from LIAS/ISAE-ENSMA, for his endless help and support. Special thanks are due to the members of the Steering Committee, in particular, its chair Leonid Kalinichenko and his vice-chair Yannis Manolopoulos for all their help and guidance.

Finally, we thank Springer for publishing the proceedings containing invited and research papers in the LNCS series. The Program Committee work relied on Easy-Chair, and we thank its development team for creating and maintaining it; it offered a great support throughout the different phases of the reviewing process. The conference would not have been possible without our supporters and sponsors:

– Région Poitou Charentes
– ISAE-ENSMA
– Poitiers University
– INFORSID Association
– CRITT Informatique, Futuroscope
– LIAS laboratory

Last, but not least, we thank the participants of ADBIS 2015 for sharing their works and presenting their achievements, thus providing a lively, fruitful, and constructive forum, and giving us the pleasure of knowing that our work was purposeful.

September 2015

Ladjel Bellatreche
Tadeusz Morzy
Patrick Valduriez

Organization

General Chair

Ladjel Bellatreche   LIAS/ISAE-ENSMA, Poitiers, France

Program Committee Co-chairs

Patrick Valduriez   Inria of Montpellier, France
Tadeusz Morzy   Poznan University, Poland

Workshop Co-chairs

Athena Vakali   Aristotle University of Thessaloniki, Greece
Bernhard Thalheim   Kiel University, Germany

Doctoral Consortium Co-chairs

Sofian Maabout   Labri/Bordeaux, France
Boris Novikov   St. Petersburg University, Russia

Publicity Chair

Selma Khouri   LIAS/ISAE-ENSMA, France

Website Chair

Mickaël Baron   LIAS/ISAE-ENSMA, Poitiers, France

Proceedings Technical Editor

Stéphane Jean   LIAS/ISAE-ENSMA, Poitiers, France

Local Organizing Committee Chair

Patrick Girard   LIAS/ISAE-ENSMA, France

Local Organizing Committee

Mickaël Baron   LIAS/ISAE-ENSMA, Poitiers, France
Frédéric Carreau   LIAS/ISAE-ENSMA, Poitiers, France
Brice Chardin   LIAS/ISAE-ENSMA, Poitiers, France

Zoé Faget	LIAS/ISAE-ENSMA, Poitiers, France
Patrick Girard	LIAS/ISAE-ENSMA, Poitiers, France
Laurent Guittet	LIAS/ISAE-ENSMA, Poitiers, France
Stéphane Jean	LIAS/ISAE-ENSMA, Poitiers, France
Yassine Ouhammou	LIAS/ISAE-ENSMA, Poitiers, France
Claudine Rault	LIAS/ISAE-ENSMA, Poitiers, France
Okba Barkat	LIAS/ISAE-ENSMA, Poitiers, France
Selma Bouarar	LIAS/ISAE-ENSMA, Poitiers, France
Ahcène Boukorca	LIAS/ISAE-ENSMA, Poitiers, France
Lahcène Brahimi	LIAS/ISAE-ENSMA, Poitiers, France
Zouhir Djilani	LIAS/ISAE-ENSMA, Poitiers, France
Géraud Fokou	LIAS/ISAE-ENSMA, Poitiers, France
Nadir Guetmi	LIAS/ISAE-ENSMA, Poitiers, France
Yves Mouafo	LIAS/ISAE-ENSMA, Poitiers, France
Guillaume Phavorin	LIAS/ISAE-ENSMA, Poitiers, France

Supporters

Région Poitou Charentes
ISAE-ENSMA
Poitiers University
INFORSID Association
CRITT Informatique, Futuroscope
LIAS laboratory

Steering Committee

Paolo Atzeni	Italy
Andras Benczur	Hungary
Albertas Caplinskas	Lithuania
Barbara Catania	Italy
Johann Eder	Austria
Theo Haerder	Germany
Marite Kirikova	Latvia
Hele-Mai Haav	Estonia
Mirjana Ivanovic	Serbia
Hannu Jaakkola	Finland
Mikhail Kogalovsky	Russia
Yannis Manolopoulos	Greece
Rainer Manthey	Germany
Manuk Manukyan	Armenia
Joris Mihaeli	Israel
Tadeusz Morzy	Poland
Pavol Navrat	Slovakia
Boris Novikov	Russia

Mykola Nikitchenko	Ukraine
Jaroslav Pokornyv	Czech Republic
Boris Rachev	Bulgaria
Bernhard Thalheim	Germany
Gottfried Vossen	Germany
Tatjana Welzer	Slovenia
Viacheslav Wolfengagen	Russia
Robert Wrembel	Poland
Ester Zumpano	Italy

Program Committee

Reza Akbarinia	Inria, France
Paolo Atzeni	Università Roma Tre, Italy
Andreas Behrend	University of Bonn, Germany
Ladjel Bellatreche	ISAE-ENSMA, France
Omar Boucelma	Aix-Marseille University, France
Mahdi Bohlouli	University of Siegen, Germany
Albertas Caplinskas	Institute of Mathematics and Informatics, Italy
Barbara Catania	DISI-University of Genoa, Italy
Wojciech Cellary	Poznan School of Economy, Poland
Ricardo Rodrigues Ciferri	Federal University of São Carlos, Brazil
Alfredo Cuzzocrea	University of Trieste, Italy
Todd Eavis	Concordia University, Canada
Johann Eder	Alpen-Adria-Universität Klagenfurt, Austria
Markus Endres	University of Augsburg, Germany
Pedro Furtado	University of Coimbra/CISUC, Portugal
Johann Gamper	Free University of Bozen-Bolzano, Italy
Jérôme Gensel	Grenoble University, France
Shahram Ghandeharizadeh	University of Southern California, USA
Matteo Golfarelli	DISI - University of Bologna, Italy
Goetz Graefe	Hewlett-Packard Laboratories, USA
Dawid Gross-amblard	IRISA, Rennes University, France
Jarek Gryz	York University, Canada
Mohand-Said Hacid	University of Claude Bernard Lyon 1 - UCBL, France
Theo Härder	TU Kaiserslautern, Germany
Mirjana Ivanovic	University of Novi Sad, Serbia
Hannu Jaakkola	Tampere University of Technology, Finland
Leonid Kalinichenko	Russian Academy of Science, Russia
Ahto Kalja	Küberneetika Instituut, Estonia
Kalinka Kaloyanova	University of Sofia - FMI, Bulgaria
Mehmed Kantardzic	University of Louisville, USA
Marite Kirikova	Riga Technical University, Latvia
Mikhail Kogalovsky	Market Economy Institute of the Russian Academy of Sciences, Russia
Christian Koncilia	Alpen-Adria University of Klagenfurt, Austria

Margita Kon-popovska	Ss Cyril and Methodius University, Macedonia
Harald Kosch	University of Passau, Germany
Georgia Koutrika	HP Labs, USA
Regine Laleau	Paris Est Creteil University, France
Wolfgang Lehner	TU Dresden, Germany
Pericles Loucopoulos	University of Manchester, UK
Ivan Lukovic	University of Novi Sad, Serbia
Yannis Manolopoulos	Aristotle University of Thessaloniki, Greece
Rainer Manthey	University of Bonn, Germany
Pascal Molli	Nantes University, France
Tadeusz Morzy	Poznan University of Technology, Poland
Pavol Navrat	Slovak University of Technology, Slovakia
Kjetil Nørvåg	Norwegian University of Science and Technology, Norway
Gultekin Ozsoyoglu	Case Western Reserve University, USA
M. Tamer Ozsu	University of Waterloo, Canada
Oscar Pastor	Valencia University of Technology, Spain
Dana Petcu	Institute e-Austria Timisoara, Romania
Jean-Marc Petit	Université de Lyon, INSA Lyon, France
Olivier Pivert	IRISA, Rennes University, France
Neoklis Polyzotis	University of California Santa Cruz, USA
Boris Rachev	Technical University of Varna, Bulgaria
Peter Revesz	University of Nebraska, USA
Stefano Rizzi	DEIS - University of Bologna, Italy
Viera Rozinajova	Slovak University of Technology in Bratislava, Slovakia
Henryk Rybinski	Warsaw University of Technology, Poland
Gunter Saake	University of Magdeburg, Germany
Klaus-Dieter Schewe	Software Competence Center Hagenberg, Germany
Timos Sellis	RMIT University, Australia
Bela Stantic	Griffith University, Australia
Manolis Terrovitis	Institute for the Management of Information Systems, RC Athena, Greece
Martin Theobald	University of Antwerp, Belgium
Farouk Toumani	LIMOS, Blaise Pascal University, Clermont-Ferrand, France
Patrick Valduriez	Inria, France
Panos Vassiliadis	University of Ioannina, Greece
Jari Veijalainen	University of Jyvaskyla, Finland
Goran Velinov	UKIM, Skopje, Macedonia
Krishnamurthy Vidyasankar	Memorial University, Canada
Stratis Viglas	University of Edinburgh, UK
Peter Vojtas	Charles University of Prague, Czech Republic
Gerhard Weikum	Max Planck Institute for Informatics, Germany
Tatjana Welzer	University of Maribor, Slovenia

Robert Wrembel   Poznan Unviersity of Technology, Institute of
     Computing Science, Poland
Vladimir Zadorozhny   University of Pittsburgh, USA

Additional Reviewers

Fabian Benduhn   Magdeburg University, Germany
Jevgeni Marenkov   Tallinn University of Technology, Estonia
Sonja Ristic   University of Novi Sad, Serbia
Giorgos Giannopoulos   National Technical University of Athens, Greece
Karoly Bosa   Johannes Kepler University Linz, Austria
Grégory Smits   IRISA, France
Fatma Slaimi   LSIS, Marseille, France
Olga Gkountouna   National Technical University of Athens (NTUA),
     Athens, Greece
John Liagouris   University of Hong Kong SAR China
Panagiotis Symeonidis   Aristotle University, Thessaloniki, Greece
Konstantinos Theocharidis   IMIS, Research Center Athena, Greece
Mustafa Al-Hajjaji   University of Magdeburg, Germany
Sebastian Dorok   University of Magdeburg, Germany
Loredana Tec   AIT Austrian Institute of Technology GmbH, Vienna,
     Austria
Anton Dignos   University of Zürich, Switzerland
Felix Kossak   Software Competence Center Hagenberg GmbH,
     Hagenberg, Austria
Amel Mammar   Telecom/Telecom SudParis, France
Sahar Vahdati   University of Bonn, Germany
Nabil Hameurlain   University of Pau, France
Tarmo Robal   Tallinn University of Technology, Estonia
Hala Skaf-Molli   LINA, Nantes University, France
Zoltan Miklos   Inria, Rennes, France
Farida Semmak   Université Paris-Est, France
Christophe Gnaho   Université Paris Est, France
Lorena Paoletti   Universidad de Santiago de Chile, Chile
Gilles Nachouki   LINA, Nantes University, France
Irina Astrova   Tallinn University of Technology, Estonia
Shuaiqiang Wang   University of Jyvaskyla, Finland
Zoé Faget   LIAS/ISAE-ENSMA, France
Vladimir Ivančević   University of Novi Sad, Serbia
Saulius Gudas   Vilnius University, Lithuania
Dirk Habich   Technische Universität Dresden, Germany
Slavica Kordić   University of Novi Sad, Serbia
Eike Schallehn   Otto von Guericke University of Magdeburg, Germany
Vladimir Dimitrieski   University of Novi Sad, Serbia
Christian Koncilia   Alpen-Adria-Universität Klagenfurt, Austria
Ioannis N. Athanasiadis   Hellenic Open University, Kozani, Greece

Keynotes

The Story of Webdamlog

Serge Abiteboul

INRIA Saclay and ENS Cachan

Abstarct. We summarize in this paper works about the management of data in a distributed manner based on Webdamlog, a datalog-extension. We point to relevant articles on these works. More references may be found there.

1 The Webdamlog Approach

Information of interest may be found on the Web in a variety of forms, in many systems, and with different access protocols. Today, the control and management of the diversity of data and tasks in this setting are beyond the skills of casual users [1]. Facing similar issues, companies see the cost of managing and integrating information skyrocketing. We are concerned with the management of Web data in place in a distributed manner, with a possibly large number of autonomous, heterogeneous systems collaborating to support certain tasks. We summarize in this paper works in this setting around Webdamlog and point to the relevant articles on it.

The thesis is that managing the richness and diversity of data residing on the Web can be tamed using a holistic approach based on a distributed knowledge base. Our approach is to represent all Web information as logical facts, and Web data management tasks as logical rules. A variety of complex data management tasks that currently require intense work and deep expertise may then greatly benefit from the automatic reasoning provided by inference engines, operating over the distributed Web knowledge base: for instance, information access, access control, knowledge acquisition and dissemination.

We propose to express the peers logic in Webdamlog, a datalog-style rule-based language. In Webdamlog, peers exchange facts (for information) and rules (in place of code). The use of declarative rules provides the following advantages. Peers may perform automatic reasoning using the available knowledge. Because the model is formally defined, it becomes possible to prove (or disprove) desirable properties. Because the model is based on a datalog-style language, query processing can benefit from optimization techniques. Because the model represents provenance and time, the quality of data can be better controlled. Because the model is general, a wide variety of scenarios and protocoles may be captured, which is a requirement for todays Web.

This work was realized in the context of the European Research Council grant Webdam [6, 13]. The system is available in opensource at [8]. The work on Webdamlog was inspired by previous works on ActiveXML [3] at INRIA, as well as Bud [7, 12] at Berkeley University. The system has been demonstrated in [2]. An extensive experimental evaluation of the implementation (showing notably that the computational cost of access control is modest) is presented in [11].

In the remaining of this paper, we briefly mention three main contributions: (i) The Webdamlog language that facilitates the exchange of data and rules between distributed peers; (ii) A collaborative access control mechanism for Webdamlog that enables controlling the dissemination of data in a network; and (iii) A probabilistic semantics for datalog with functional dependencies that can serve as the basis for managing uncertain, noisy, possibly contradicting data.

2 Three Main Contributions

Webdamlog. There is a new trend to use datalog-style rule-based languages to specify modern distributed applications, notably on the Web [9, 10]. The Webdamlog language was first formally described in [4]. It is a version of distributed datalog that allows specifying distributed applications where peers exchange messages (i.e. logical facts) as well as rules (i.e., the analog of code). An example of rule is as follows:

```
[at alice] album@alice($photoId,$photo,$f) :- friend@alice($f),
            album@$f($photoId,$photo,$source), tags@f($photoId,"Alice")
```

Ignore the details of the syntax. With this rule, Alice deploys, at each peer corresponding to one of her friends, a rule that sends her all photos this friend owns that is tagged by her name. The main originality of the language is the use of *delegation* that allows delegating rules to other peers. Distributed computating is realized by delegating some rules to perform some tasks to other peers. Knowledge acquisition, i.e., the Webdamlog analog to "downloading apps", is also performed using rule delagations. The main contribution of [4] is the presentation of the language. A study of the impact on expressiveness of "delegations" is also provided.

Access control. Users wish to share data using these systems, but avoiding the risks of unintended disclosures or unauthorized access by applications has become a major challenge. An important issue for users in a distributed setting is thus the control of the access to their data by others. In [11], we introduce a collaborative access control mechanism for Webdamlog. Using this model, users can specify access control policies providing flexible tuple-level control derived using provenance information.

Inconsistency and imprecision. In [5], we study deduction in the presence of inconsistencies and probabilites for datalog programs. (The results can be extended to Webdamlog in a straightward manner). Inconsistencies are captured through violations of functional dependencies (FDs). We propose nondeterministic semantics for datalog with FDs. We introduce a PTIME (in the size of the extensional data) algorithm, that given a datalog program, a set of FDs and an input instance, produces a c-table representation of the set of possible resulting worlds.

We then propose to quantify nondeterminism with probabilities, by means of a probabilistic semantics. We consider the problem of capturing possible worlds along with their probabilities via probabilistic c-tables. We then study classical computational problems in this novel context. We consider the problems of computing the probabilities of answers, of identifying most likely supports for answers, and of determining the extensional facts that are most influential for deriving a particular fact. We show that the interplay of recursion and FDs leads to novel technical challenges in the context of these problems.

Acknowlegements. We thank all the researchers who participated in the Webdamlog project and in particular, Meghyn Bienvenu, Pierre Bourhis, Daniel Deutch, Alban Galland, Gerome Miklau, Vera Zaychik Moffitt, Marie-Christine Rousset, Julia Stoyanovich, Jules Testard, and Victor Vianu.

References

1. Abiteboul, S., André, B., Kaplan, D.: Managing your digital life. Commun. ACM **58**(5), 32–35 (2015)
2. Abiteboul, S., Antoine, E., Miklau, G., Stoyanovich, J., Testard J.: [Demo] rule-based application development using WebdamLog. In: SIGMOD (2013)
3. Abiteboul, S., Benjelloun, O., Milo, T.: The active XML project: an overview. VLDB J. **17** (5), 1019–1040 (2008)
4. Abiteboul, S., Bienvenu, M., Galland, A., Antoine, E.: A rule-based language for Web data management. In: PODS (2011)
5. Abiteboul, S., Deutch, D., Vianu, V.: Deduction with contradictions in datalog. In: International Conference on Database Theory (2014)
6. Abiteboul, S., Senellart, P., Vianu, V.: The ERC webdam on foundations of web data management. In: Proceedings of the 21st World Wide Web Conference, WWW 2012, Lyon, France, 16–20 April 2012 (Companion Volume), pp. 211–214 (2012)
7. Alvaro, P., Conway, N., Hellerstein, J., Marczak W.R.: Consistency analysis in bloom: a calm and collected approach. In: CIDR, pp. 249–260 (2011)
8. Antoine, E.: The webdamlog system on github (2013). https://github.com/Emilien-Antoine/webdamlog-engine
9. Hellerstein, J.M.: Datalog redux: experience and conjecture. In: Proceedings of the Twenty-Ninth ACM SIGMOD-SIGACT-SIGART Symposium on Principles of Database Systems, pp. 1–2. ACM (2010)
10. Huang, S.S., Green, T.J., Loo, B.T.: Datalog and emerging applications: an interactive tutorial. In: Proceedings of the 2011 ACM SIGMOD International Conference on Management of Data, pp. 1213–1216. ACM (2011)
11. Moffitt, V.Z., Stoyanovich, J., Abiteboul, S., Miklau G.: Collaborative access control in webdamlog. In: Proceedings of the 2015 ACM SIGMOD International Conference on Management of Data, Melbourne, Victoria, Australia, 31 May – 4 June 2015, pp. 197–211 (2015)
12. B. O. O. M. project. Bloom programming language. http://www.bloom-lang.net/
13. The Webdam ERC Project. http://webdam.inria.fr/

The Case for Small Data Management

Jens Dittrich

Saarland University
http://infosys.uni-saarland.de

Abstract. Exabytes of data; several hundred thousand TPC-C transactions per second on a single computing core; scale-up to hundreds of cores and a dozen Terabytes of main memory; scale-out to thousands of nodes with close to Petabyte-sized main memories; and massively parallel query processing are a reality in data management. But, hold on a second: for how many users exactly? How many users do you know that really have to handle these kinds of massive datasets and extreme query workloads? On the other hand: how many users do you know that are fighting to handle relatively small datasets, say in the range of a few thousand to a few million rows per table? How come some of the most popular open source DBMS have hopelessly outdated optimizers producing inefficient query plans? How come people don't care and love it anyway? Could it be that most of the worlds data management problems are actually quite small? How can we increase the impact of database research in areas when datasets are small? What are the typical problems? What does this mean for database research? We discuss research challenges, directions, and a concrete technical solution coined PDbF: Portable Database Files (open source at https://github.com/uds-datalab/PDBF). See also our VLDB 2015 demo (https://infosys.uni-saarland.de/publications/p2199-dittrich.pdf).

CV. Jens Dittrich is a Full Professor of Computer Science in the area of Databases, Data Management, and Big Data at Saarland University, Germany. Previous affiliations include U Marburg, SAP AG, and ETH Zurich. He is also associated to CISPA (Center for IT-Security, Privacy and Accountability). He received an Outrageous Ideas and Vision Paper Award at CIDR 2011, a BMBF VIP Grant, a best paper award at VLDB 2014, two CS teaching awards in 2011 and 2013, as well as several presentation awards including a qualification for the interdisciplinary German science slam finals in 2012 and three presentation awards at CIDR (2011, 2013, and 2015). His research focuses on fast access to big data including in particular: data analytics on large datasets, Hadoop MapReduce, main-memory databases, and database indexing. He has been a PC member and/or area chair of prestigious international database conferences such as PVLDB, SIGMOD, and ICDE. Since 2013 he has been teaching his classes on data management as flipped classrooms. See http://datenbankenlernen.de or http://youtube.com/jensdit for a list of freely available videos on database technology in German and English (about 80 videos in German and 80 in English so far).

Tutorials

Towards an Era of Trust in Personal Data Management

Nicolas Anciaux[1], Benjamin Nguyen[2], and Iulian Sandu Popa[1]

[1] INRIA Paris-Rocquencourt, Domaine du Voluceau, 78153 Le Chesnay, France
{Nicolas.Anciaux, Iulian.Sandu_Popa}@inria.fr
[2] INSA Centre-Val de Loire, 88 boulevard Lahitolle, 18022 BOURGES, France
Benjamin.Nguyen@insa-cvl.fr

Managing personal data with strong privacy guarantees has become an important topic in an age where your glasses record and share everything you see, your wallet records and shares your financial transactions, and your set-top box records and shares your energy consumption, while several recent affairs have unveiled the severe consequences of the loss of privacy. In this context, more and more alternatives are proposed based on user centric and decentralized solutions, capitalizing on the use of trusted personal devices controlling the data at the edges of the Internet. Decentralized solutions are promising because they do not exhibit the intrinsic limitations of classical centralized solutions, e.g., sudden changes in privacy policies of companies holding the data, data exposures by negligence or because it is regulated by too weak policies, exposure to sophisticated attacks whose benefit/cost ratio is high for centralized databases. Hence, such solutions appear as a sea change for personal data management, where the control over personal data is pushed to the edges of the Internet, within sensors acquiring the data and in a variety of user devices endowed with a form of trust, e.g., tamper-resistant secure hardware-based devices.

This tutorial reviews several existing solutions going in this direction, presents a functional architecture encompassing these alternatives, and exposes the underlying techniques and open issues dealing with user centric and decentralized data management platforms. In a first part, we review the recent initiatives pursuing the objective of reestablishing user control over their data by decentralizing this control in personal secure or trusted devices. We discuss an abstract distributed architecture focusing on secure storing, managing and sharing of personal data, i.e., the asymmetric architecture, and indicate the main challenges inherent to decentralized data management. In a second part, we explore data management techniques exercised within a trusted device at the client side. We review the main attempts proposed in the literature and concentrate on those addressing the specific context of microcontrollers equipping sensors and mobile phones (SIM cards). In a third part, we investigate the problem of performing global processing without any compromise on data privacy. We present the difficulties to overcome to execute privacy preserving computations on populations of personal devices, and illustrate it by focusing on Group By SQL queries and Privacy Preserving Data Publishing. In a fourth part, we conclude the tutorial by presenting existing and future instances of decentralized privacy preserving data management architectures. We mainly focus on attempts and proposals targeting social-medical, smart houses, and rural areas contexts.

Query Processing: Beyond SQL and Relations

Boris Novikov

Saint-Petersburg University
b.novikov@spbu.ru

Query processing and optimization are essential for any data processing system since introduction of high-level declarative query languages in early 80-ies. During the last decade several new techniques were introduced in order to address requirements of new classes of applications, data models, storage and indexing, and querying paradigms.

Modern query processing and optimization extends far beyond relational queries. Several techniques were revised and a number of new techniques have been introduced to make the query processing efficient. Several systems that were originally designed as low-level storage facilities implementing persistence layer, were augmented with high level declarative features. The declarative scripting languages provide a technique for easy-to-understand specification of complex analytical scenarios that look like sequential but are executed on massively parallel systems.

The main focus of this tutorial is on the query optimization and processing in new environments and for new classes of applications.

Although many of declarative languages are designed as extensions to SQL, the internals of the implementations usually have significant differences with well-known optimization and processing techniques developed for relational systems using row-based storage structures.

Column stores are considered to be the most efficient for analytical processing on modern hardware. The physical algebraic operations for column stores differ from those used in row-based ones, and optimization strategies and heuristics are different.

Distributed data processing systems such as Hadoop weren't originally intended for declarative query processing. However, several query languages are implemented on top, bringing back the need for optimization. Examples of these languages and systems include ASTERIX, SCOPE, and Apache Hive.

Processing of semi-structured and unstructured data ultimately requires fuzzy (e.g. similarity) queries resulting in several obstacles for relational optimizers that are mostly oriented on re-ordering of join operations. Although some of recently introduced techniques, such as efficient top-down enumeration algorithms might be helpful, many issues are still open.

Parametric and dynamic optimization techniques seem to be especially useful for distributed heterogeneous environments where availability of data statistics is often severely limited and cost estimations are unreliable.

Finally, holistic optimization is an emerging technology that optimizes the database queries and application together with the goal to improve the overall application performance.

Contents

Transformation, Extraction and Archiving

Modeling and Ontologies

Time Series Processing

Database Theory and Access Methods

Conditional Differential Dependencies (CDDs)

Selasi Kwashie[1]([✉]), Jixue Liu[1], Jiuyong Li[1], and Feiyue Ye[2]

[1] ITMS, University of South Australia, Adelaide, Australia
`Selasi.Kwashie@mymail.unisa.edu.au`
[2] CSE, Jiangsu University of Technology, Changzhou, China

Abstract. Differential dependency (DD) is a newly proposed data dependency theory that captures the relationships amongst data values. Like the classical functional dependency (FD) theory, DDs are defined to hold over entire instances of relations. This paper proposes a novel extension of the DD theory to hold over subsets of relations, called conditional DD (CDD), similar to the relaxations of FD to conditional FD (CFD) [4] and conditional FD with predicates (CFDPs) [6]. In this work, we present: the formal definitions; the consistency and implication analysis; and a set of axioms to infer CDDs. Furthermore, we study the discovery problem of CDDs and present an algorithm for mining a minimal cover set Σ_c of *constant* CDDs from a given instance of a relation. And, we propose an interestingness measure for ranking discovered CDDs and reducing the size $|\Sigma_c|$ of Σ_c. We demonstrate the efficiency, effectiveness and scalability of the discovery algorithm through experiments on both real and synthetic datasets.

1 Introduction

Functional dependencies (FDs) have recently been extended to capture the semantics of distance in data, namely, differential dependencies (DDs) in [13]. DDs relax the strict equality constraint in FDs to distance constraints. A DD $B[c, d] \rightarrow A[e, f]$ holds on an instance r of a relation R if for any two distinct tuples $t_1, t_2 \in r$, if the distance between $t_1[B]$ and $t_2[B]$ is within c and d, then the distance between $t_1[A]$ and $t_2[A]$ is within e and f. This definition allows the declaration of dependencies based on *similarities* of attribute values, unlike an FD which uses exact match of attribute values. DDs are useful in data management applications involving the semantics of distance.

From a different perspective, new dependencies of conditional FDs (CFDs) [4], conditional FDs with predicates (CFDPs) [6], and conditional inclusion dependencies (CINDs) [5] have been proposed to represent dependencies that hold on subsets of a dataset (hold locally), unlike FDs which hold on the whole dataset (hold globally). These new dependencies have been shown useful in data quality management and cleaning practices in these proposals.

In this study, we propose a novel type of dependencies, namely, *Conditional Differential Dependencies*, (CDDs) which extends DDs by specifying constraints enabling DDs to apply to subsets of data. Examples of uses of the new dependencies in knowledge discovery and data quality management are as follows.

© Springer International Publishing Switzerland 2015
T. Morzy et al. (Eds.): ADBIS 2015, LNCS 9282, pp. 3–17, 2015.
DOI: 10.1007/978-3-319-23135-8_1

In knowledge discovery. Different types of dependencies represent knowledge in different ways and have different expressive power. The CDDs of our extension enables locally satisfied dependencies defined on value similarity to be represented. For example, from the Iris[1] dataset [3], the global DD $\phi : PW[0] \rightarrow SL[0, 2.3]$ can be discovered which indicates that for any two iris plants, if they have the same petal width ($PW[0]$), the distance in their sepal length is within 2.3 cm ($SL[0, 2.3]$). However, from the same dataset, three local DDs (our CDDs) can be found:

  (i) if $CL = setosa$, $PW[0] \rightarrow SL[0, 1.4]$;
 (ii) if $CL = versicolor$, $PW[0] \rightarrow SL[0, 1.8]$; and
(iii) if $CL = virginica$, $PW[0] \rightarrow SL[0, 2.3]$.

Although the global DD ϕ covers all the local ones (i) - (iii), the knowledge represented in (i) and (ii) differs significantly from ϕ. Thus, (i) - (iii) can be used to classify iris plants much more accurately than ϕ. This result confirms Simpson's paradox [12] that local rules (dependencies) can be completely different from global rules. Our extension addresses the representation problem of local dependencies based on the semantics of distance.

In Data Quality Management. Given a dataset shown in Table 1 where the attributes mean employee Gender, Educational Qualification, Professional Category (academic or administrative), Years of Service, Staff Position and Salary in order, the application has the requirement that for administrative staff, when their position is similar (difference ≤ 1), their number of years in service is similar (difference ≤ 5), then their salary should be similar (difference ≤ 5). The representation of the requirement in our CDD is:

Table 1. A snippet of employees' data

tid	Gen	Qua	Cat	Yrs	Pos	Sal
1	1	4	1	12	24	136
2	0	4	1	12	24	124
3	0	3	0	4	13	62
4	1	4	1	15	25	159
5	1	1	0	11	12	69
6	0	2	1	5	21	42
7	0	3	0	6	12	59
8	1	2	0	2	11	15
9	1	3	1	3	22	20
10	0	1	1	2	21	19

Gen: 1=Male; 0=Female
Qua: 1=Dip.; 2=BSC.; 3=MSc.; 4=PhD. **Cat**: 0=Adm.; 1=Aca.

  if $Cat = 0$, $Pos[0, 1]Yrs[0, 5] \rightarrow Sal[0, 5]$.
This requirement cannot be represented by a global DD as DDs do not allow the specification of the 'if' condition. We note that the global DD $Cat[0]Pos[0, 1]Yrs[0, 5] \rightarrow Sal[0, 5]$ will also apply to the academic category, which is against the requirement. The requirement cannot be represented by a CFD either as CFD uses exact value match, and not defined with value similarity.

Contributions. The foregoing discussions highlight the need for extending DDs to conditionally hold on subsets of data to enable the capturing of some latent knowledge and inconsistencies. To achieve this, we: (a) propose a new class of data dependencies, CDDs, an extension of DDs, that are capable of capturing

---

[1] PW is petal width; SL is sepal length; CL is Iris class.

the distance relationships amongst subsets of tuples in a relation; (b) investigate the static analyses of CDDs – study two reasoning problems for CDDs and present their complexity bounds, and design a set of axioms to infer CDDs; (c) investigate the discovery problem of CDDs and design an efficient algorithm to mine *constant* CDDs in data; and (d) define an interestingness measure for CDDs to allow the return of a smaller set of CDDs that captures interesting patterns in data. We empirically evaluate the feasibility, efficiency and effectiveness of our discovery algorithm on both real-world and synthetic data sets.

2   Conditional Differential Dependencies (CDDs)

Here, we first recall the notions and definition of DDs. Then, we present the formal definition of CDDs; and show the relationship between CDDs and DDs, as well as CDDs and other conditional dependencies.

Let $R(A_1, \cdots, A_n)$ be a relation schema with instance r; $A_i \in R$ is an attribute with domain $dom(A_i)$; and $X, Y, Z \subset R$ are subsets of attributes in R.

A **distance metric**, $d_A(a_1, a_2)$, is a function defined over attribute A that returns the distance between any two values a_1, a_2 of A in r. $d_A(a_1, a_2)$ is assumed to exhibit: (a) non-negativity: $d_A(a_1, a_2) \geq 0$; (b) identity of indiscernible: $d_A(a_1, a_2) = 0$ iff. $a_1 = a_2$; and (c) symmetry: $d_A(a_1, a_2) = d_A(a_2, a_1)$; where $a_1, a_2 \in dom(A)$. Examples of distance metrics are: edit distance, cosine similarity (for textual values); and absolute value of difference (for numeric values).

A **differential function** (DF) of A w.r.t. the *distance interval* $w = [x, y]$ is denoted by $A[w]$, and returns a boolean value indicating whether or not for any two values a_1, a_2 of A in r, $x \leq d_A(a_1, a_2) \leq y$. $A[w]$ is written as $A[x]$ if $x = y$. A DF of $X = \{A_1, \cdots, A_m\}$ on $W_X = w_1 \times \cdots \times w_m$ is denoted by $X[W_X]$. A tuple pair $t_1, t_2 \in r$ is said to *agree on* (*satisfy*) a DF $X[W_X]$ if for all $A_i[w_i] \in X[W_X]$, $d_{A_i}(t_1[A_i], t_2[A_i])$ is within/on w_i. $\mathcal{T}(X[W])$ represents *the set of all tuple pairs* in r of R that agree on $X[W]$.

Given two differential functions (DFs) $X[W_X]$ and $Y[W_Y]$: $X[W_X]$ is said to *subsume* $Y[W_Y]$, denoted by $X[W_X] \succeq Y[W_Y]$, if and only if any tuple pair t_1, t_2 satisfying $Y[W_Y]$ also satisfies $X[W_X]$, i.e., for all $A_i[w_i] \in X[W_X]$, there exists $A_i[w_i'] \in Y[W_Y]$ such that $w_i' \subseteq w_i$.

A **differential dependency** (DD) [13], ϕ, is a statement $\phi : X[W_L] \to Y[W_R]$ between two DFs $X[W_L], Y[W_R]$. ϕ holds over r of R iff. for any two distinct tuples $t_1, t_2 \in r$, if $X[W_L]$ returns true, $Y[W_R]$ returns true.

For example, assume $d_A(a_1, a_2) = |a_1 - a_2|$ for all $A \in R$ in Table 1. Then the following are true: (a) $Pos[0, 3] \succeq Pos[0, 2]Cat[0]$ since any pair of tuple that agrees on $Pos[0, 2]Cat[0]$ also agrees on $Pos[0, 3]$; (b) $Cat[0]Yrs[0, 3] \to Sal[1, 47]$ since $\mathcal{T}(Cat[0]Yrs[0, 3]) \subseteq \mathcal{T}(Sal[1, 47])$.

We now extend DDs with conditions to hold over subsets of (instead of the entire) r as follows.

Let $\psi(A) = A \text{ op } a$ denote a *predicate* on $A \in R$, where 'op' is one of the relation symbols in the set $\mathsf{S} = \{=, <, >, \leq, \geq, \in\}$ and 'a' is either:

- A set of values in $dom(A)$ when op is $\in$ and $dom(A)$ is categorical; or
- An interval in $dom(A)$ when op is $\in$ and $dom(A)$ is continuous; or
- A value in $dom(A)$ when op is $=, <, >, \leq$ or $\geq$.

To represent any value in $dom(A)$, we use the wild-card character '_' (i.e. a is '_'). Examples of predicates on some attributes in the relation in Table 1 are: $\psi_1(Qua) = Qua \in \{2,3\}$; $\psi_2(Yrs) = Yrs \in [5-10]$; $\psi_3(Pos) = Pos \geq 24$.

A **conditional statement (CS)**, $\zeta(Z)$, on a set $Z = \{A_1, \cdots, A_m\}$ of attributes is a conjunction of predicates on the attributes in Z. That is, $\zeta(Z) = \psi_1(A_1) \cdots \psi_m(A_m)$. For any tuple $t \in r$, $\zeta(Z)$ returns *true* on t, denoted by $t \asymp \zeta(Z)$ iff: for each $\psi_i(A_i) \in \zeta(Z)$, $t[A_i]$ *agrees* on the constraint specified by $\psi_i(A_i)$. The set of all tuples $t \in r$ that agree on $\zeta(Z)$ is denoted by $\mathsf{supp}(\zeta(Z), r)$. A CS $\zeta(Z)$ is *constant* if for all $\psi(A) \in \zeta(Z)$, $\psi(A) = A$ op a is such that a is a single value in $dom(A)$ and op is '=' and we simply denote $\psi(A)$ as $A\langle a \rangle^2$ for brevity; $\zeta(Z)$ is otherwise termed *variable*.

Given two CSs $\zeta_1(Z), \zeta_2(Y)$ on the sets Z, Y respectively, $\zeta_1(Z)$ is said to *dominate* $\zeta_2(Y)$, denoted by $\zeta_1(Z) \geq_d \zeta_2(Y)$, if $\zeta_1(Z)$ is *more general* than $\zeta_2(Y)$. That is, (a) $Z \subseteq Y$ and (b) for all $\psi_i(A) \in \zeta_1(Z)$ there exits $\psi_j(A) \in \zeta_2(Y)$ such that for any tuple $t \in r$, if t agrees on $\psi_j(A)$, then it agrees on $\psi_i(A)$. In other words, $\zeta_1(Z)$ dominates $\zeta_2(Y)$ if $Z \subseteq Y$ and the predicates in $\zeta_2(Y)$ are the same as or more restrictive than those in $\zeta_1(Z)$. Examples of some CSs in Table 1 are: $Qua \geq 2$; $Gen\langle 0 \rangle Cat\langle 1 \rangle$; $Sal \in [100-160] \wedge Yrs \in [10-15] \wedge Qua\langle 4 \rangle$. Let $\zeta_1(Cat) = Cat\langle 1 \rangle$, and $\zeta_2(Sal, Yrs, Cat) = Sal \leq 50 \wedge Yrs \in [2-6] \wedge Cat\langle 1 \rangle$: the relation $\zeta_1(Cat) \geq_d \zeta_2(Sal, Yrs, Cat)$ is true.

A **conditional differential function (cDF)**, η, on the sets $Z \supseteq X$ of attributes is a pair $\eta = (\zeta(Z), X[W_X])$: where $\zeta(Z)$ is a CS on Z and $X[W_X]$ is a DF on X. A cDF, $\eta = (\zeta(Z), X[W_X])$ returns true if the DF $X[W_X]$ returns true on a tuple pair drawn from the set of all tuples that agree on the CS $\zeta(Z)$. That is, $\eta = (\zeta(Z), X[W_X])$ returns true on a pair of tuples $t_1, t_2 \in r$ if and only if: (a) $t_1, t_2 \in \mathsf{supp}(\zeta(Z), r)$; and (b) t_1, t_2 agrees on $X[W_X]$. We use $\mathsf{agrT}(\eta)$ to denote the set of all tuple pairs in $\mathsf{supp}(\zeta(Z), r)$ that agree on $X[W_X]$.

Given two cDFs $\eta_1 = (\zeta(Z), X[W_X])$ and $\eta_2 = (\zeta(U), M[W_M])$, we say η_1 *subdues* η_2, denoted by $\eta_1 \succcurlyeq \eta_2$ if and only if: (a) $\zeta(Z) \geq_d \zeta(U)$ and (b) $X[W_X] \succeq M[W_M]$. That is, the CS $\zeta(Z)$ and DF $X[W_X]$ of η_1 are respectively less restrictive and more general than the CS $\zeta(U)$ and DF $M[W_M]$ of η_2. For example, if $\eta_1 = (Yrs \leq 10 \wedge Qua \in \{1,2,3\}, Yrs[0,3]Qua[0])$ and $\eta_2 = (Yrs \in [5-10] \wedge Qua \in \{2,3\}, Yrs[0,2]Qua[0])$, then $\eta_1 \succcurlyeq \eta_2$.

cDFs serve as the constraints in CDDs, formally defined as follows.

A **conditional differential dependency (CDD)**, σ, defined on a relation R is a constraint between two cDFs, $\eta = (\zeta(Z_L), X[W_X]), \gamma = (\zeta(Z_R), Y[W_Y])$, in the form $\sigma : \eta \to \gamma$. η is termed the LHS, $lhs(\sigma)$; and γ is the RHS, $rhs(\sigma)$.

A CDD σ is satisfied by an instance r of R if for the set $\mathsf{supp}(\zeta(Z_L), r)$ of all tuples satisfying constraint $\zeta(Z_L)$ individually, if the distance of any two distinct tuples $t_1, t_2 \in \mathsf{supp}(\zeta(Z_L), r)$ on attribute set X is within W_X, then the distance of t_1, t_2 on attribute set Y must be within W_Y, and t_1 and t_2 must also satisfy the constraint $\zeta(Z_R)$ individually. Formally, r *satisfies* the CDD σ, denoted by

---

[2] we note that we use $A\langle a \rangle$ for a constant CS of A; and $A[w]$ for a DF of A.

$r \models \sigma$, if for any pair of tuples $t_1, t_2 \in r$: if $\eta = (\zeta(Z_L), X[W_X])$ on (t_1, t_2) returns true, then $\gamma = (\zeta(Z_R), Y[W_Y])$ on (t_1, t_2) returns true.

A CDD $\sigma : (\zeta(Z_L), X[W_X]) \rightarrow (\zeta(Z_R), Y[W_Y])$ is said to be *constant* if for all $\psi(A) \in \zeta(Z_L) \cup \zeta(Z_R)$, $\psi(A) = A\langle a \rangle$ or $\psi(A) =$'_'; and *variable* otherwise. We refer to the function $\phi : X[W_X] \rightarrow Y[W_Y]$ as the *embedded DD* in σ. We say a CDD σ is in the *normal form* if its RHS has only one attribute, i.e. $\sigma : (\zeta(Z_L), X[W_X]) \rightarrow (\zeta(A), A[w_a])$. Unless otherwise specified, in the rest of the paper, we consider CDDs in the normal form.

CDDs and Other Dependencies. CDDs are different from DDs. DDs specify distance constraints on sets of attributes which hold on an entire instance of a relation whereas CDDs define distance constraints on attribute sets valid over *only* the subsets of instances that agree on CSs. Furthermore, CFDs and CFDPs can be considered as special cases of CDDs where the distance constraints on attributes are zero (0), like how FDs are seen as a special case of DDs with a distance of zero (0) on all attributes. Examples of some CDDs on the relation in Table 1 are as follows. $\sigma_1 : (Qua\langle 4 \rangle Cat\langle 1 \rangle Pos=$'_', $Pos[0]) \rightarrow (Sal \geq 80, Sal[0, 12])$; and $\sigma_2 : (Qua\langle 4 \rangle Cat\langle 1 \rangle Yrs=$'_', $Yrs[0]) \rightarrow (Pos=$'_', $Pos[0])$. σ_1 states that for any two academic staffs ($Cat\langle 1 \rangle$) with PhD. degrees ($Qua\langle 4 \rangle$), irrespective of their position ($Pos=$'_'), if there exists no difference in their position ($Pos[0]$), then their salary difference must be within \$12K ($Sal[0, 12]$) and their individual salaries should be at least \$80K ($Sal \geq 80$). The CDD σ_2 is equivalent to a CFD: the years of service (Yrs) of academics with PhD ($Qua\langle 4 \rangle Cat\langle 1 \rangle$) uniquely determine their position (Pos).

Static Analysis of CDDs. For two of the fundamental reasoning problems (i.e. consistency and implication analysis) that come with dependencies, we give here results with brief sketches of proof due to space limit; and present inference rules for CDDs.

1. Consistency Analysis of CDDs: A set Σ of CDDs is said to be *consistent* if there exists an instance r of R that has at least two tuples (ALTT) and that satisfies Σ. The *consistency problem* $\mathcal{C}$ is to determine the existence of an ALTT instance for a given set Σ of CDDs.

$\mathcal{C}$ is NP-complete for DDs [13]; and CDDs maintain the same complexity.

Theorem 1. *The consistency problem for CDDs is NP-complete.*

Proof. The NP-hard bound follows the NP-hardness of $\mathcal{C}$ for DDs, a special case of CDDs. NP bound is verified by an NP algorithm that can decide if $r \models \Sigma$. □

2. Implication Analysis of CDDs: Given two sets Σ_1, Σ_2 of CDDs, Σ_1 is said to be a *logical implication* of (or to *imply*) Σ_2, denoted by $\Sigma_1 \models \Sigma_2$, if for any instance r of R, if $r \models \Sigma_1$, then $r \models \Sigma_2$. The *implication problem*, $\mathcal{I}$, for CDDs is to determine, given a set Σ of CDDs and a single CDD σ over a relation R, whether Σ implies σ, denoted by $\Sigma \models \sigma$. An implication analysis ensures discovered rules are free from redundancy.

$\mathcal{I}$ is coNP-complete for DDs [13]; and CDDs maintain the same complexity.

Theorem 2. *The implication problem of CDDs is coNP-complete.*

Proof. The coNP-hard bound is verified by the coNP-hardness of DDs. The coNP bound is verified by an NP algorithm that can verify if $\Sigma \not\models \sigma$.   □

3. Inference System for CDDs: An inference system allows pruning of implied rules during discovery. For example, the well-known Armstrong's Axioms [2] form the bases of the finite axiomatizability of FDs and similar axioms have been derived f or CFDs and DDs. We present a few (because of space limit) of all the inference rules used in pruning in Sect. 4 below for CDDs.

Some inference rules for CDDs

$\mathcal{I}_1$. Reflexivity: If $(\zeta(Z_R), Y[W_Y]) \succcurlyeq (\zeta(Z_L), X[W_X])$, then $\Sigma \vdash_{\mathcal{I}} (\zeta(Z_L), X[W_X]) \rightarrow (\zeta(Z_R), Y[W_Y])$.

$\mathcal{I}_2$. Transitivity: If $\Sigma \vdash_{\mathcal{I}} (\zeta(Z_L), X[W_X]) \rightarrow (\zeta(U_R), N[W_N])$ and $\Sigma \vdash_{\mathcal{I}} (\zeta(U_L), M[W_M]) \rightarrow (\zeta(Z_R), Y[W_Y])$, where $(\zeta(U_L), M[W_M]) \succcurlyeq (\zeta(U_R), N[W_N])$, then $\Sigma \vdash_{\mathcal{I}} (\zeta(Z_L), X[W_X]) \rightarrow (\zeta(Z_R), Y[W_Y])$.

$\mathcal{I}_3$. Augmentation: If $\Sigma \vdash_{\mathcal{I}} (\zeta(Z_L), X[W_X]) \rightarrow (\zeta(Z_R), Y[W_Y])$, then $\Sigma \vdash_{\mathcal{I}} (\zeta(Z_L), X[W_X]) \wedge (\zeta(U), M[W_M]) \rightarrow (\zeta(Z_R), Y[W_Y]) \wedge (\zeta(U), M[W_M])$ for any $(\zeta(U), M[W_M])$.

$\mathcal{I}_4$. Finite domains: If (1) $\Sigma \vdash_{\mathcal{I}} (\zeta(Z_L), X[W_X]) \wedge (\zeta_i(U), B[w_b]) \rightarrow (\zeta(Z_R), Y[W_Y])$ for $b \in [1, k]; i \in [1, p]$ and $B \subseteq U$, (2) $dom(B), dom(U)$ are finite w.r.t. Σ and $(\zeta_1(U), B[w_1]) \vee \cdots \vee (\zeta_p(U), B[w_k])$ is consistent w.r.t. Σ, then $\Sigma \vdash_{\mathcal{I}} (\zeta(Z_L), X[W_X]) \rightarrow (\zeta(Z_R), Y[W_Y])$.

---

3   Discovery of CDDs

Given an instance r of a relation R, the discovery problem of CDDs is to find the set of all valid CDDs in r. The set of all valid CDDs in any given instance of relation can be very large. From discussions in Sect. 2, such a set may contain several implications, hence, redundant. It is, therefore, interesting to find a set of valid CDDs with no redundancy from which all other valid CDDs can be inferred. We want a cover of all valid CDDs in a given r of R with the least possible number of CDDs, namely, a *minimal cover*, formally defined as follows.

Definition 1 (Minimal CDD, Cover and Minimal Cover). *Let Σ be a set of valid CDDs in r of R. A CDD $\sigma : (\zeta(Z_L), X[W_X]) \rightarrow (\zeta(A), A[w_a]) \in \Sigma$ is said to be* **minimal** *iff. it is: (a) left-reduced – there does not exist another $\sigma_1 : (\zeta_1(U), Y[W_Y]) \rightarrow (\zeta(A), A[w_a]) \in \Sigma$ s.t. $(\zeta_1(U), Y[W_Y]) \succcurlyeq (\zeta(Z_L), X[W_X])$; (b) right-subdued – there does not exist any $\sigma_2 : (\zeta(Z_L), X[W_X]) \rightarrow (\zeta_2(A), A[w_2]) \in \Sigma$ s.t. $(\zeta(A), A[w_a]) \succcurlyeq (\zeta_2(A), A[w_2])$.*

The set Σ_1 is a **cover** *of the set Σ if every CDD in Σ is in or implied by DDs in Σ_1. Let Σ be a set of minimal CDDs. We say a cover Σ_c of Σ is* **minimal cover** *iff. there does not exist a cover Σ' of Σ s.t. $\Sigma' \subset \Sigma_c$.*

Finding a minimal cover Σ_c of CDDs in a given instance of a relation is non-trivial and important. This is because, it presents a concise set of valid CDDs, making its utilization less costly. For instance, in data management applications that may require the validation of CDDs in data, a smaller Σ_c reduces the cost of validation, resulting in more efficient and useful applications.

Given an instance r of a relation schema R, the discovery problem of CDDs is to find a minimal cover of all CDDs that hold in r.

Problem Statement. The search space of possible CDDs is large. To reduce the search space while discovering highly relevant CDDs, we restrict the maximum distance of DFs in the LHS cDFs of CDDs. Indeed, we incorporate the notion of ε-DFs [8] into CDD discovery and accordingly define ε-minimal CDDs. Let the function $up(w)$ on the distance interval ($w = [x, y]$) of a DF $A[w]$ return y.

A DF $X[W_X] = A_1[w_1] \wedge \cdots \wedge A_m[w_m]$ on the set X of attributes is said to be an **ε-DF** if for all $A_i[w_i] \in X[W]$, $up(w_i) \leq \varepsilon \times range(A_i)$, where $\varepsilon \in [0, 1]$ is user-specified; $range(A_i)$ is the range of distance values of A_i.

Definition 2 (ε-minimal CDD). *Given a set Σ^ε of valid CDDs in r, a CDD $\sigma : (\zeta(Z), X[W_X]) \rightarrow (\zeta(A), A[w_a]) \in \Sigma^\varepsilon$ is ε-minimal iff. $X[W_X]$ is an ε-DF and: (a) left-reduced – there does not exist another $\sigma_1 : (\zeta_1(U), Y[W_Y]) \rightarrow (\zeta(A), A[w_a]) \in \Sigma^\varepsilon$ s.t. $(\zeta_1(U), Y[W_Y]) \succcurlyeq (\zeta(Z_L), X[W_X])$ where $Y[W_Y]$ is an ε-DF; (b) right-subdued – there does not exist any $\sigma_2 : (\zeta(Z_L), X[W_X]) \rightarrow (\zeta_2(A), A[w_2]) \in \Sigma^\varepsilon$ s.t. $(\zeta(A), A[w_a]) \succcurlyeq (\zeta_2(A), A[w_2])$.*

In this paper, we study the following discovery problem: given an instance r of a relation R, we find a minimal cover Σ_c^ε of ε-minimal CDDs that hold in r.

4   The Discovery Algorithm

This section introduces our algorithm, MineCDD, for *constant* CDD discovery. A sketch of the pseudo-code for MineCDD is presented in Algorithm 1. It consists of four major procedures viz: generating constant CSs; forming candidate LHS cDFs; finding valid RHS cDFs to form CDDs; and pruning the set of valid CDDs. We elaborate on each step in the following.

Generating Constant CSs. The first step in our discovery process is generating all constant CSs using the *closed patterns* (CPs) of *equivalence classes* (ECs). This approach is based on a connection between constant CSs and ECs in a relation. We recall the notions of ECs and CPs; and present a link between CSs and ECs below.

Let $R(A_1, \cdots, A_n)$ be a relation schema, and r be an instance of R. An *item* is a predicate $\psi(A) : A$ **op a** on an attribute $A \in R$ where op is '=' and **a** is a constant in

---

Algorithm 1 MineCDD

Data: An instance r of a relation R
Input parameters: k, ε, mi.
Result: minimal cover Σ_c^ε of constant ε-minimal CDDs.

1: mine all k-frequent CSs in r
2: **for each** k-frequent CS in r **do**
3:    form candidate LHS cDFs
4:    **for each** candidate LHS **do**
5:       find set of valid RHS cDFs;
6:       generate valid CDDs
7: prune the set Σ of valid CDDs
8: **return** Σ_c^ε (Pruned Σ)

the domain $dom(A)$. For brevity, we denote an item by $A\langle a\rangle$. An *itemset* $\zeta(Z)$ on the set Z of attributes is a set of items with distinct attributes. Let $\mathsf{supp}(\zeta(Z), r)$ be the set of tuples in r that contain (support) $\zeta(Z)$. Two itemsets $\zeta_1(Z)$ and $\zeta_2(Y)$ are said to be *equivalent* if they co-occur in r.

An equivalence class (EC), $\mathcal{E}_r[\zeta(Z)]$, of an itemset $\zeta(Z)$ in r of R is the set $\mathcal{E}_r[\zeta(Z)] = \{\zeta(Y) \mid \mathsf{supp}(\zeta(Z), r) = \mathsf{supp}(\zeta(Y), r)\}$. That is, an EC consists of a set of itemsets with a common agree (supporting) tuples set. Hence, to avoid redundancy, we represent each EC with the most maximal itemset within the class. And, as show in [9], an EC can be uniquely and concisely represented by a closed pattern (CP) and a set of generators (free patterns): and CPs are the maximal itemsets in ECs. We, therefore, mine CPs as representatives of ECs.

An itemset $\zeta(Z)$ is said to be *closed* iff. there does not exist another itemset $\zeta(Y)$ such that: (a) $\zeta(Z) \subset \zeta(Y)$; and (b) $\mathsf{supp}(\zeta(Z), r) = \mathsf{supp}(\zeta(Y), r)$. A CP $\zeta(Z)$ is *k-frequent* if $|\mathsf{supp}(\zeta(Z), r)| \geq k$, where k is a user-specified natural number. We adopt and adapt the CPs mining algorithm in [15] to discover all k-frequent CPs and their agree tuples set in r (line 1 of Algorithm 1), and transform them to constant CSs by Lemma 1 below.

Lemma 1 (ECs, CPs and Constant CSs). *Given an instance r of a relation R, a CP $\zeta(Z) = A_1\langle a_1\rangle \cdots A_m\langle a_m\rangle$ is the maximal itemset in the EC $\mathcal{E}_r[\zeta(Z)] = \{\zeta(Y) \mid \mathsf{supp}(\zeta(Z), r) = \mathsf{supp}(\zeta(Y), r)\}$; and $\zeta(Z)$ corresponds to a constant CS with an agree tuples set $\mathsf{supp}(\zeta(Z), r)$.*

Forming Candidate LHS cDFs. Let C be the set of all constant CSs generated in the previous step. Here, for each constant CS $\zeta(U) \in$ C, we form a set of candidate LHS cDFs. That is, we form all valid LHS cDFs for the tuples set $\mathsf{supp}(\zeta(U), r)$ in r that agrees on $\zeta(U)$.

The search space of possible LHS cDFs for even a single CS can be very large due the combinatorial distance intervals of DFs. Hence, the need for a user-specified constraint on the DFs in LHS cDFs and the definition of ε-minimal CDDs in Sect. 3. To ensure the discovery of ε-minimal CDDs, we mine CDDs with ε-DFs in their LHS cDFs. Naturally, we adapt and incorporate the technique in [8] for mining ε-DFs for this task (line 3 of Algorithm 1).

A LHS cDF η is a pair $\eta = (\zeta(Z_L), X[W_X])$. We form candidate LHSs with each constant CS $\zeta(U) \in$ C as follows. First, we find all left-reduced ε-DFs $X[W_X]$ valid over $\mathsf{supp}(\zeta(U), r)$; then form their respective CS $\zeta(Z_L)$.

The generation of candidate LHS ε-DFs uses the building of an attribute lattice [1]. Given a constant CS $\zeta(U)$, an attribute lattice $\mathcal{L}$ is built using attributes in the set $R' = \{R \setminus U\}$. Each node N in the lattice is a triplet $N = (X, W_X, \mathcal{T})$ where $X \subset R'$; W_X is the set of distance intervals of the ε-DF $X[W_X]$; and $\mathcal{T}$ represents the set, $\mathsf{agrT}(\eta)$, of all tuple pairs that agree on $X[W_X]$ in $\mathsf{supp}(\zeta(U), r)$. The discovery of all candidate ε-DFs in the lattice is based on the established relationship between δ-nClusters [10] and ε-DFs in [8]. We extend this relationship for subsets of relations below.

Lemma 2 (δ-nClusters and ε-DFs). *Given are the subset, $\mathsf{supp}(\zeta(U), r)$, of r and the subspace (set of attributes) $X \subset R$. If there exists a free set $\Psi_X = \{I_1, I_2, \cdots, I_l\}$ of all maximal δ-nClusters valid in X over $\mathsf{supp}(\zeta(U), r)$, then, an ε-DF $X[W_X] = B_1[w_1] \wedge \cdots \wedge B_m[w_m]$ holds in $\mathsf{supp}(\zeta(U), r)$ s.t.: $\mathcal{T}(X[W]) = pr(T_1) \cup pr(T_2) \cup \cdots \cup pr(T_l)$, where $pr(T_\alpha)$ is the set of all tuple pairs in the tuple set T_α of a maximal δ-nCluster $I_\alpha = (T_\alpha, X)$.*

Lemma 2 allows us to adopt and adapt techniques for mining δ-nClusters to efficiently find valid ε-DFs. More precisely, given the free sets of all maximal δ-nClusters, we generate the set of candidate LHS ε-DFs by transforming each free set of maximal δ-nClusters into a corresponding valid ε-DF.

The lattice is built using the breadth-first approach, starting with single attributes. At the first level $\mathcal{L}_1$ of the lattice $\mathcal{L}$, for every $A \in R'$, if there exists a free set Ψ_A of maximal δ-nClusters in the subspace A then $\mathcal{T}(A[w_a])$ and $A[w_a]^3$ are generated according to Lemma 2 to form the node $N_A(A, w_a, \mathcal{T})$. A node is included in the lattice if the support of its ε-DF is greater than the minimum. The nodes at $\mathcal{L}_1$ form the first level of $\mathcal{L}$.

Nodes at other levels of $\mathcal{L}$ are formed as follows. For any level i such that ($2 \leq i < m$) and $m = |R'|$, $\mathcal{L}_i$, are formed from $\mathcal{L}_{i-1}$ nodes. Any two nodes $N_l, N_k \in \mathcal{L}_{i-1}$ are parent to a node N_c of $\mathcal{L}_i$ if and only if: N_l, N_k have up to $(i-2)$ preceding single-attributes in common on their first-triplet X (set of attribute, sorted in lexicographical order) and their remaining attributes in X are different. If this condition is satisfied, then $(N_c).X = X_c = \{(N_l).X_l\} \cup \{(N_k).X_k\}$. If there exists $\Psi_{X_c} \in \mathcal{C}$, then node $N_c(X_c, W_{X_c}, \mathcal{T})$ of $\mathcal{L}_i$ is formed (by Lemma 2) with $N_l, N_k \in \mathcal{L}_{i-1}$ as parents.

For every ε-DF $X[W_X]$, we form $\zeta(Z_L) = \zeta(U) \wedge \zeta(X)$, where for every $\psi(A) \in \zeta(X)$, $\psi(A) = '\_'$. The pair $(\zeta(Z_L), X[W_X])$ then forms a LHS cDF η. That is, the LHS cDF, $\eta(N)$, of a node N is thus $\eta(N) = (\zeta(Z_L), X[W_X])$.

Finding Valid RHS cDFs. Next, we find valid RHS cDFs, $\gamma = (\zeta(A), A[w_a])$ for each LHS cDF $\eta = (\zeta(Z_L), X[W_L])$ (line 4, 5 of Algorithm 1). In this discovery, we set all $\zeta(A) = '\_'$, to avoid missing any similarity among the values of A. This, thus allows the mining of all valid RHS DFs on A. A RHS cDF $\gamma = (\zeta(A), A[w_a])$ is *valid* iff: (1) $Z_L \cap \{A\} = \emptyset$; (2) $\mathsf{agrT}(\zeta(Z_L), X[W_L]) \subseteq \mathsf{agrT}(\zeta(A), A[w_a])$; (3) there does not exist γ' such that $\gamma \succcurlyeq \gamma'$ and $\mathsf{agrT}(\zeta(Z_L), X[W_L]) \subseteq \mathsf{agrT}(\gamma')$. The first requirement ensures that the RHS has only one attribute A, and $A \notin Z_L$ to avoid the generation of redundant and implied CDDs. Conditions 2 and 3 require that γ forms a valid CDD with η and $A[w_a] \in \gamma$ has the smallest valid distance interval respectively.

The lattice $\mathcal{L}$ is traversed level by level to find a set of valid RHS cDFs for each candidate LHS cDF, $\eta(N) \in \mathcal{L}$. For each level i ($1 \leq i \leq m$) in $\mathcal{L}$, for every node $N \in \mathcal{L}_i$, if there exists no DD amongst the ε-DF, $N.W$, of N, then we find the set $\mathcal{R}$ of valid RHS cDFs for $\eta(N)$. Let Y be the set of candidate attributes to form a valid RHS cDF with $\eta(N)$ (i.e. $Y = \{R' \setminus N.Z_L\}$). For each

---

^{3} $w_a = [x, y]$ where x, y are the min. and max. distance of values in Ψ_A respectively.

$A \in Y$, the projection of values of A with respect to the set of agree tuples $N.\mathcal{T}$ of the node N is generated. The minimum and maximum distance values in the projection forms the distance interval w_a of A. If w_a does not cover the entire distance space of A, then pair $(\zeta(A), A[w_a])$ is added to the set $\mathcal{R}$. Next, valid CDDs are formed with the LHS cDF (line 6 of Algorithm 1).

Pruning and Generating a Minimal Cover. To ensure that the set of CDDs discovered for a given CS is (ε)-minimal and devoid of all implications, we adopt and extend the DD-Tree structure proposed in [11] for pruning implied CDDs.

A DD-Tree, $\mathtt{tr}(A[w_a])$, stores all DDs with the same RHS DF $A[w_a]$. Thus, the root of each DD-Tree is the common RHS $A[w_a]$. Other nodes in the tree are single-attribute DF of the LHS DFs of the DDs in the tree. Hence, a tree-path $\mathtt{p}(\phi) = A[w_a]/B_1[w_1]/\cdots/B[w_m]$ represents a DD $\phi: B[w_1]\cdots B[w_m] \rightarrow A[w_a]$. Child nodes in the tree are sorted by their attribute and the distance intervals. The implication of DDs is detected via the notion of *path-prefix* [11]. A path $\mathtt{p}(\phi_1) = A[w_a]/B_1[w_1]/B_2[w_2]/\cdots/A_m[w_m]$ is a prefix of another path $\mathtt{p}(\phi_2) = A[w_a]/B_1[\bar{w}_1]/B_2[\bar{w}_2]/\cdots/B_k[\bar{w}_k]$ if $m < k$ and for each $i \in [1 \cdots m]$, we have $B_i[w_i] \succeq B_i[\bar{w}_i]$. Given two DDs ϕ_1, ϕ_2: ϕ_1 implies ϕ_2 if $\mathtt{p}(\phi_1)$ is a prefix of $\mathtt{p}(\phi_2)$.

To eliminate implications in the set Φ of embedded DDs of a given constant CS, $\zeta(U)$, we build a hash-table $\mathcal{H}_\Phi = \langle A[w_a], \mathtt{tr}(A[w_a]) \rangle$ of all DD-Trees for each RHS cDF $\gamma = (\zeta(A), A[w_a])$. The implication of a CDD σ can be checked with $\mathcal{H}_\Phi$ as follows. If there exists $\mathtt{tr}(A[w_a])$ for σ with embedded DD ϕ, and $\exists\, \mathtt{p}(\phi_i) \in \mathtt{tr}(A[w_a])$ such that $\mathtt{p}(\phi_i)$ is a prefix of $\mathtt{p}(\phi)$, then ϕ is implied, hence not added to $\mathtt{tr}(A[w_a])$ (i.e. σ is eliminated). Otherwise, ϕ is added in $\mathtt{tr}(A[w_a])$.

Up to this point, for every valid constant CS $\zeta(U) \in \mathsf{C}$, we find a set of valid CDDs. These sets of CDDs are non-redundant w.r.t their CSs. However, there may exist implications among the CDDs mined for different CSs. Therefore, further pruning is required on the set of all CDDs discovered to produce a minimal cover Σ_c^ε (line 7 of Algorithm 1). To eliminate these implied CDDs, we utilize the dominance relation among CSs.

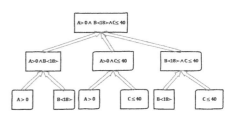

Fig 1. An example of a d-tree

We introduce here, the concept of d-tree (dominance tree) for CSs: to capture the dominance relation among valid LHS CSs. An example of a d-tree is shown in Fig. 1: the directions of dominance among the CSs are shown by the arrows. Let $\mathtt{n}_i$ be a node in the d-tree D, $\mathtt{N}_i$ be the set $\mathtt{N}_i = \{\mathtt{n}_j \in D \mid \mathtt{n}_j \geq_d \mathtt{n}_i\}$. Let Φ_i be the set of embedded DDs of CDDs of $\mathtt{n}_i$. Pruning of implied CDDs is done top-down the d-tree as follows. Given $\mathtt{n}_i$ with Φ_i, for every $\phi_u \in \Phi_i$ and all $\mathtt{n}_j \in \mathtt{N}_i$, if $\exists\, \phi_v \in \Phi_j$ s.t. $\phi_u = \phi_v$ and both ϕ_u, ϕ_v have the same set of agreeing tuple pairs, where Φ_j is the set of embedded DDs of $\mathtt{n}_j$: then prune ϕ_v (rsptly. the corresponding CDD of ϕ_v). For example, let Φ be the set of embedded DDs of CDDs that hold over the root CS in the d-tree in Fig. 1. Then for any $\phi \in \Phi$, if there exist an embedded DD ϕ_i at any other node such that the above conditions are satisfied, then we eliminate the CDDs containing ϕ_i at those nodes except at the root, since the root node is dominated by all other nodes.

5   Ranking CDDs

Although a minimal cover of CDDs is concise and non-redundant, its size can still be large. In this section, we propose an interestingness measure for CDDs to score minimal (or ε-minimal) CDDs in r. This enables us to prune a minimal cover of CDDs further to a smaller set of interesting CDDs.

Our definition of interestingness is based on the semantics of CDDs. A CDD is a simply a DD that holds on a pattern in data. In other words, a CDD is a DD that hold over a subset of (tuples that agree on a CS in) a relation instance. Hence, two intuitive, yet significant factors that determine the importance of a minimal CDD is: the *coverage* of the pattern of data (CSs); and the *relevance* of its embedded DD. We define the interestingness $\mathtt{intr}(\sigma)$ of a CDD $\sigma : (\zeta(Z), X[W_X]) \to (\zeta(A), A[w_a])$ as a linear combination of these two factors as follows:

$$\mathtt{intr}(\sigma) = \alpha \cdot cov(\zeta(Z) \cup \zeta(A)) + (1 - \alpha) \cdot rel(\phi), \tag{1}$$

where (a) $cov(\zeta(Z)) = \frac{|\mathtt{supp}(\zeta(Z) \cup \zeta(A), r)|}{|r|} = \frac{|\mathtt{supp}(\zeta(Z), r)|}{|r|}$;

(b) $rel(\phi) = \frac{1}{3}(s(\phi) + \frac{\sum_{i=1}^{m} \lambda(B_i[w_i])}{m} + \mu(A[w_a]))$ – for the embedded DD $\phi = X[W_X] \to A[w_a]$ s.t. $X[W_X] = B_1[w_1] \cdots B_m[w_m]$:

- $s(\phi) = \frac{|\mathcal{T}(X[W_X]) \cap \mathcal{T}(A[w_a])|}{|\mathcal{T}(\mathtt{supp}(\zeta(Z), r)|} = \frac{|\mathcal{T}(X[W_X])|}{|\mathcal{T}(\mathtt{supp}(\zeta(Z), r)|}$ is the probability of occurrence of the embedded DD in $\mathtt{supp}(\zeta(Z), r)$;

- $\lambda(B_i[w_i]) = \frac{|w_i^d|}{|w_i| + |w_i^d|}$ reflects of the closeness of the LHS DF $B_i[w_i]$ of ϕ;

- $\mu(A[w_a])^4 = \frac{|w_a^d| - |w_a|}{|w_a^d|}$ reflects the similarity amongst the values of the RHS attribute A in ϕ. w_i^d, w_a^d are the maximum distance intervals of B_i and A rsptly.; $|\mathcal{T}(\mathtt{supp}(\zeta(Z), r)|$ is the total number of tuple pairs in $\mathtt{supp}(\zeta(Z), r)$.

$cov(\zeta(Z) \cup \zeta(A))$ gives a score (in the range [0,1]) that indicates the statistical significance of patterns of data. Its definition is straight-forward and self explanatory. $rel(\phi)$, other the other hand, scores (between 0 and 1) the 'informativeness' of an embedded DD ϕ in a CDD. The definition of $rel(\phi)$ is a variant of the interestingness measure of DDs presented in [8]. It is influenced by the support of ϕ; the closeness conveyed by the LHS DF of ϕ; and the degree of similarity revealed by the RHS DF of ϕ. And, $\alpha \in [0, 1]$ is a scaling factor.

Rank-Aware Pruning of CDDs. The minimal cover of CDDs Σ_c can be pruned based on the above definition of interestingness. Indeed, given a minimum interestingness $mi \in [0, 1]$ value, we reduce Σ_c as follows. First, in every DD-Tree

---

[4] similar to the dependent quality measure in [14].

of a valid CS, we lower bound the $rel(\phi)$ value of every embedded DD in the tree to $\frac{mi-c}{1-\alpha}$ where $c = \alpha \cdot cov(\zeta(Z) \cup \zeta(A))$ – a constant value for all embedded DDs in the tree since they share a common constant CS $\zeta(U)$. In this case, any embedded DD ϕ in the DD-tree with $rel(\phi) < \frac{mi-c}{1-\alpha}$ is removed during stage 3 of finding CDDs in Sect. 4. Furthermore, when pruning CDDs of implications across different CSs using the dominance relation, we eliminate all those that have a lower interestingness value than mi.

6   Empirical Evaluation

In this section, we present our experimental set-up, datasets and a discussion of results obtained from mining constant CDDs.

Experimental Set-Up. The proposed algorithm is implemented in Java. The experiments were conducted on an Intel Core i5-2520M CPU @ 2.5 GHz processor computer with 4.0 GB of memory running Windows 8 OS.

We used both real-world and synthetic data sets to evaluate the proposed discovery algorithm. Table 2 briefly describes the real-world data set from the UCI Machine Learning data repository [3]. We generate a set of synthetic data sets with varying: arity ($|R|$); size ($|r|$); and correlation (CF) amongst attributes and tuples to further evaluate the scalability of our algorithm. For any dataset, various distance functions can be defined for each attribute based on domain-knowledge. For our experiments, we use: absolute value of difference as distance metric for numeric attributes and equality function for categoric attributes.

Table 2. Description of data sets

Data sets	Size	No. of Attributes
Chess (KRK)	28,056	7 (6 N; 1 C)
Mammographic Mass*	830	6 (1 N; 5 C)
Iris	150	5 (4 N; 1 C)

N = numeric, C = categorical
* – this version has no missing value.

Results and Analysis. The results of the experiments are discussed below.

1. Time performance: We evaluate the time performance of our algorithm (MineCDD) on the generated synthetic datasets for varying $|r|$, $|R|$ and CF. In these experiments, we set: the minimum support for CSs to $k = 0.1$; $\varepsilon = 0$ for DFs of the LHS cDFs; and the minimum interestingness value for all CDDs to $mi > 0$.

(a). Scalability w.r.t. $|r|$:– Part (a) of Fig. 2 show how MineCDD performs for vary $|r|$ sizes for different CF values. In this experiment, $|R|$ is fixed to 7 attributes. The graph shows that the runtime (in sec. on the y-axis) of MineCDD for increasing $|r|$ sizes (on the x-axis) for different CF values. The plots show that the runtime of MineCDD is dependent on both $|r|$ and CF. However, the runtime is, clearly, affected more by higher CF values as compared to larger $|r|$ sizes. The runtime increases, generally, for increasing $|r|$. For highly correlated

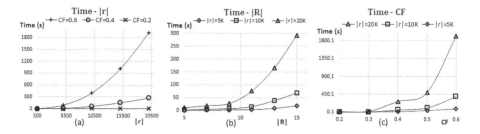

Fig 2. Runtime

datasets (high CF values), the chances of finding more persistent patterns (CSs) and valid relationships (embedded DDs) is higher. In other words, the search space of possible CDDs is larger for datasets with high CF values. Hence, a much longer discovery time.

(b). Scalability w.r.t. $|R|$:– Fig. 2 (b) shows the runtime (in sec. on the y-axis) of MineCDD for increasing $|R|$ sizes (on the x-axis), for different sizes of r with a constant CF of 0.5. In general, for any instance size, more time is required to mine CDDs in datasets with higher arity. This is because, relations with more attributes have more combinatorial possibilities, hence, larger search spaces. This is confirmed by the characteristics of the three plots in Fig. 2 (b).

(c). Scalability w.r.t. CF:– For a fixed $|R| = 7$, we investigate how MineCDD performs for increasing CF values for differently sized instance $|r|$ sizes of data. The results are presented in Part (c) of Fig. 2. On the y-axis is the runtime (in sec. – log scale) for varying CF values (on the x-axis). This plot shows that for all $|r|$ sizes, the runtime of MineCDD increases with increasing CF. Giving a clearer depiction of the influence of CF (high correlation in data) on runtime as discussed in (a) above.

2. Effect of the parameters: In this set of experiments, we show how the input parameters, k, ε, mi, affect the time performance of our algorithm. We demonstrate this on both the synthetic and real-world data sets.

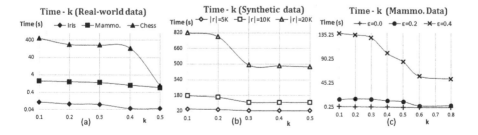

Fig 3. Effect of parameters on runtime

(a). Influence of k:– First, we show how increasing k value affects the runtime of MineCDD. Parts (a) and (b) of Fig. 3 show the effect of k on runtime for real-world and synthetic data sets respectively. In this experiment, we used all the instance sizes of the real-world data sets and set $|R|$ of the synthetic datasets to 7. In general, as expected, efficiency of MineCDD improves for high k vales.

(b). Influence of ε:– Next, we show how ε affects the efficiency of MineCDD. Figure 3 (c) presents a plot of runtime vs. k for different ε values. For lack of space, we present only the results on the Mammographic data set. The runtime of MineCDD increases for high ε values – more DFs are considered for the LHS cDFs, increasing the search space of CDDs. This is shown in the graph.

3. CDDs mined: Here, we present the effect of the interesting-ness measure (α=0.3 is best for $\texttt{intr}(\sigma)$) on pruning the minimal cover of CDDs. In Fig. 4 is a plot of the number of discovered CDDs (y-axis) versus the varying minimum interestingness values (x-axis). Our experiments on both categories of data sets show that with an $mi \geq 0.3$ a significant amounts of CDDs are pruned.

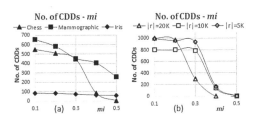

Fig 4. Effect of mi on size of Σ_c^ε

7   Conclusion and Future Works

This paper propose CDDs, a novel extension of DDs which allows DDs to be specified on patterns of data instead of the entire instance of a relation. CDDs present the opportunity to discover latent knowledge and inconsistencies in data. We show that, although CDDs have more expressivity than CFDs and DDs, their static analysis have the same complexity. Furthermore, we study the discovery problem of CDDs, and develop an efficient algorithm for mining constant CDDs in data. Also, an interestingness measure is designed to reduce the set of discovered CDDs. In our next studies, we shall extend the discover algorithm to mine variable CDDs. Furthermore, we shall investigate the use of CDDs in the detection of inconsistencies and repair of data along the lines of work in [7].

Acknowledgement. This work is partially supported by NSFC 61472166.

References

1. Agrawal, R., Srikant, R.: Fast algorithms for mining association rules in large databases. In: 20th International Conference on VLDB, pp. 487–499 (1994)
2. Armstrong, W.W.: Dependency structures of data base relationships. In: World Computer Congress - IFIP, pp. 580–583 (1974)

3. Bache, K., Lichman, M.: UCI Machine Learning repository (2013). http://archive. ics.uci.edu/ml
4. Bohannon, P., Fan, W., Geerts, F., Jia, X., Kementsietsidis, A.: Conditional functional dependencies for data cleaning. In: 23rd ICDE, pp. 746–755 (2007)
5. Bravo, L., Fan, W., Ma, S.: Extending dependencies with conditions. In: 33rd International Conference on VLDB, pp. 243–254 (2007)
6. Chen, W., Fan, W., Ma, S.: Analyses and validation of conditional dependencies with built-in predicates. In: Bhowmick, S.S., Küng, J., Wagner, R. (eds.) DEXA 2009. LNCS, vol. 5690, pp. 576–591. Springer, Heidelberg (2009)
7. Fan, W., Geerts, F.: Foundations of Data Quality Management. Synth. Lect. Data Manage. 4(5), 1–127 (2012). doi:10.2200/S00439ED1V01Y201207DTM030
8. Kwashie, S., Liu, J., Li, J., Ye, F.: Mining differential dependencies: a subspace clustering approach. In: Wang, H., Sharaf, M.A. (eds.) ADC 2014. LNCS, vol. 8506, pp. 50–61. Springer, Heidelberg (2014)
9. Li, J., Liu, G., Wong, L.: Mining statistically important equivalence classes and delta-discriminative emerging patterns. In: 13th ACM SIGKDD, pp. 430–439 (2007)
10. Liu, G., Li, J., Sim, K., Wong, L.: Distance based subspace clustering with flexible dimension partitioning. In: 23rd ICDE, pp. 1250–1254 (2007)
11. Liu, J., Kwashie, S., Li, J., Ye, F., Vincent, M.W.: Discovery of approximate differential dependencies. CoRR, abs/1309.3733 (2013)
12. Simpson, E.H.: The interpretation of interaction in contingency tables. J. Roy. Stat. Soc. Ser. B (Stat. Meth.) 13(2), 238–241 (1951)
13. Song, S., Chen, L.: Differential dependencies: reasoning and discovery. ACM Trans. Database Syst. 36(3), 16:1–16:41 (2011)
14. Song, S., Chen, L., Cheng, H.: Parameter-free determination of distance thresholds for metric distance constraints. In: 28th ICDE, pp. 846–857 (2012)
15. Uno, T., Kiyomi, M., Arimura, H.: LCM ver.3: collaboration of array, bitmap and prefix tree for frequent itemset mining. In: 1st International Workshop on Open Source Data Mining, pp. 77–86 (2005)

Improving the Pruning Ability of Dynamic Metric Access Methods with Local Additional Pivots and Anticipation of Information

Paulo H. Oliveira[1]($\boxtimes$), Caetano Traina Jr.[2], and Daniel S. Kaster[1]

[1] Department of Computer Science, University of Londrina (UEL), Londrina, Brazil
oliveiraph17@gmail.com, dskaster@uel.br
[2] Institute of Mathematics and Computer Science, University of São Paulo (USP),
São Paulo, Brazil
caetano@icmc.usp.br

Abstract. Metric Access Methods (MAMs) have been proved to allow performing similarity queries over complex data more efficiently than other access methods. They can be considered dynamic or static depending on the pivot type used in their construction. Global pivots tend to compromise the dynamicity of MAMs, as eventual pivot-related updates must be propagated through the entire structure, while local pivots allow this maintenance to occur locally. Several applications handle online complex data and, consequently, demand efficient dynamic indexes to be successful. In this context, this work presents two techniques for improving the pruning ability of dynamic MAMs: (i) using cutting local additional pivots to reduce distance calculations and (ii) anticipating information from child nodes to reduce unnecessary disk accesses. The experiments reveal significant improvements in a dynamic MAM, reducing execution time in more than 50 % for similarity queries posed on datasets ranging from moderate to high dimensionality and cardinality.

Keywords: Similarity queries · Metric access methods · Cutting local additional pivots · Anticipation of child information

1 Introduction

In recent years, it has been noticed a fast-growing volume of complex data. Multimedia data, georeferenced data and time series are examples of such data. Some reasons for the growth are: lower prices of digital cameras and other video capture devices, high-definition cameras embedded in mobile phones, user-friendly tools for processing and editing images and videos, acquisition of data from medical equipment, data capture through sensor networks and high-speed internet connections. The success of multimedia sharing services such as YouTube, Flickr and social media is another evidence of this growth.

This research has been supported by scholarship grants from the Brazilian Coordination for the Improvement of Higher Education Personnel (CAPES).

© Springer International Publishing Switzerland 2015
T. Morzy et al. (Eds.): ADBIS 2015, LNCS 9282, pp. 18–31, 2015.
DOI: 10.1007/978-3-319-23135-8_2

In this work, complex data are considered as information that are not represented by traditional types, such as numbers, characters, dates and short texts. A key observation is that the *order relation* does not apply to most complex domains [8]. The order relation is a property that allows identifying which element precedes the other, according to some criterion, in each pair of elements of the domain. Since traditional index structures are based on this property, they are not suitable for complex data. Nevertheless, there are structures well-suited for complex domains, such as the Metric Access Methods (MAMs).

There are several MAMs related in the literature, categorized in different ways depending on which factors are taken into account to structure the data. The factors *pivot type* and *structure dynamicity* are directly related to each other. Pivots are elements that act as representatives of certain regions of the search space and are used to prune irrelevant elements during the query execution. It is said that a pivot is global when all elements of the dataset are referenced to it, whereas a pivot is local when only a portion of the dataset is referenced to it. Because global pivots are referenced by the whole dataset, they have a high impact in the pruning process of irrelevant elements, once that a single global pivot can be used to discard a large amount of irrelevant elements. However, MAMs based on global pivots may have their dynamicity compromised by the fact that eventual pivot-related updates need to be propagated through the entire structure. Local pivots, on the other hand, allow the maintenance to occur locally at the price of a lower pruning ability. In this context, the challenge addressed in this work is to improve the pruning ability of dynamic MAMs without harming their dynamicity.

This paper presents two new techniques that significantly improve the performance in similarity queries of dynamic MAMs based on local pivots. The first technique is to employ local additional pivots to reduce the uncertainty area in the search space, i.e. the area that may contain elements that are not part of the answer but cannot be pruned without being analyzed. The second technique is to anticipate information from child nodes to their parents to enable pruning irrelevant elements before visiting the disk pages that actually store them. Differently from other approaches regarding multiple pivots to define a space region, our proposal allows reducing both the number of distance calculations and the number of disk accesses as well as it does not impose any constraint in the index dynamicity.

The new techniques presented in the paper have been applied to the dynamic MAM Slim-tree [15] and evaluated through an extensive set of experiments over real datasets, varying the number of elements, the dimensionality and employing distance functions with different computational costs. In the paper, we present results that confirm their efficiency, as they enabled gains of more than 50 % in execution time, number of distance calculations and number of disk accesses when compared to the original structure, regarding every evaluated dataset.

The paper is organized as follows: Sect. 2 covers essential concepts regarding similarity queries over complex data and MAMs, as well as it presents the related work; Sects. 3 and 4 describe the two new proposed techniques; Sect. 5 presents how the techniques were applied to Slim-tree; Sect. 6 describes the experiments and discusses the results; and Sect. 7 presents the conclusions and future work.

2   Background and Related Work

2.1   Similarity Queries

In order to allow performing queries over complex domains, the elements of a given complex dataset usually have features extracted from their content. The extracted features are used in place of the original data to execute the queries. The retrieval of complex data based on this process is known as *content-based retrieval* and the set of extracted features of an element is called its *feature vector* or *signature*. Feature vectors can be, using images for example, shape and texture attributes, color histograms and results from transformations applied to data [8]. Usually, complex data are compared by dissimilarity relations between pairs of feature vectors. This is performed by employing a *distance function* that calculates how dissimilar are the two feature vectors from each other. Those are known as *similarity queries*, as they retrieve the elements from the dataset that satisfy a given similarity-based criterion.

There are several types of similarity queries [18], which range from similarity selections and joins to aggregate similarity queries. The two most common types are the Range and the k-Nearest Neighbors queries. Given a maximum threshold ξ, the *Range query* (Rq) retrieves every tuple t_i from relation R that has a value s_i for attribute S_j, which represents the feature vector of the element, satisfying the condition $\delta(s_i, s_q) \leq \xi$, where s_q is the value of the query element for attribute S_j and δ is the distance function that returns the dissimilarity between s_i and s_q. Considering R as a relation of images, an example of Range query is: "Select the images which are similar to the image Q by up to 5 units". Given an integer value $k \geq 1$, the *k-Nearest Neighbor query* (k-NNq) retrieves k tuples t_i from relation R with values s_i for attribute S_j that have the lowest distance from s_q according to δ. An example of k-Nearest Neighbors query is: "Select the 4 most similar images to the image Q".

2.2   Metric Access Methods

There are two main categories of access methods for indexing complex data. The first category is a class of access methods that support data domains represented in dimensional spaces, especially spatial data, known as Spatial Access Methods (SAMs) [9]. However, many complex data types are high-dimensional and SAMs degrade quickly as the number of dimensions grows. Furthermore, some complex data types are *dimensionless*, that is, they cannot be represented by coordinates in orthogonal axes. The second category is given by the Metric Access Methods (MAMs). MAMs rely on the premise that data are immersed in a metric space. A metric space is defined by a pair $\langle \mathbb{S}, \delta \rangle$, where $\mathbb{S}$ is a complex data domain (i.e. feature vectors) and δ is a distance function $\delta : \mathbb{S} \times \mathbb{S} \mapsto \mathbb{R}^+$, known as *metric*. A metric has properties called *metric postulates* [18] $\forall x, y, z \in \mathbb{S}$: (i) $\delta(x,y) \geq 0$ (non-negativity), (ii) $\delta(x,y) = \delta(y,x)$ (symmetry), (iii) $x = y \iff \delta(x,y) = 0$ (identity) and (iv) $\delta(x,z) \leq \delta(x,y) + \delta(y,z)$ (triangular inequality). An important characteristic of metric spaces is that, in addition to comprehending vector spaces, they include dimensionless spaces. Therefore, almost any data type can be immersed into a metric space, including geographic coordinates, images, sounds, words and DNA sequences.

The general idea of most MAMs is to choose some elements as representatives of certain subsets of data. Such elements are also known as pivots. Whenever an element $s_i \in \mathbb{S}$ is inserted, the distance from its representative is calculated and stored in the structure. Afterwards, these distance values are used in similarity queries for pruning elements by using the triangular inequality property. More formally, an element can be pruned, i.e. discarded without calculating its distance from the query element, when one of the following conditions [3] is true, where s_{rep} is the representative of the subset, s_q is the query element, s_i is any of the remaining elements in the subset and r_i is the covering radius of s_i:

$$\delta(s_{rep}, s_i) + r_i < \delta(s_{rep}, s_q) - \xi \tag{1}$$

$$\delta(s_{rep}, s_i) - r_i > \delta(s_{rep}, s_q) + \xi \tag{2}$$

2.3 Related Work

The problem addressed in this work is about improving the pruning ability of dynamic MAMs without compromising their dynamicity. MAMs which rely on global pivots, such as the OMNI-family [14] and the permutation-based approximate index PP-Index [7], allow updates in the structure and thus are classified as dynamic. However, if it is necessary to perform a pivot-related update, the cost can be close to rebuilding the whole structure. Therefore, we consider truly dynamic the MAMs in which updates are managed locally, being at most propagated through the path between a descendant and its highest ancestor.

In this sense, the pioneer dynamic MAM is M-tree [5], a balanced hierarchic MAM based on local pivots, with ball partitioning and bottom-up construction. Due to the success of M-tree, several access methods sharing similar principles have been proposed with the goal of achieving a better performance, but keeping their structure dynamic. One of them is the Slim-tree [15], an evolution of M-tree with improvements such as the evaluation and minimization of the overlap level between nodes and a new split algorithm. Another example is the DBM-tree [17], which allows a controlled unbalance to better fit the dataset density variations. These structures have a single local pivot per node. There are MAMs that employ multiple pivots per node, such as MM-tree [11] and Onion-tree [4]. However, the primary goal of these structures is to index data in main memory avoiding overlaps among nodes, being subject to end up highly unbalanced after updates. Other example is the M*-tree [12], a variation of M-tree that stores, for every node, a nearest-neighbors graph containing the nearest neighbor of each element in the node and the distance between them. This turns every element of a node into a kind of local pivot, allowing saving distance calculations.

Another strategy employed in some works is to include additional global pivots in a local pivot-based MAM. An example of a method using this strategy is the DF-tree [16]. Its structure is similar to the Slim-tree, but it uses global pivots in the pruning process, embodying the idea of global pivots of the OMNI-family.

Other examples are PM-tree (*Pivoting M-tree*) [13] and PM*-tree [12]. The PM-tree is a ball-partitioned structure which restricts the uncertainty region by the intersection of hyper-rings defined by global pivots. The PM*-tree adds multiple local pivots to the PM-tree in the way M*-tree does. The main advantage of these MAMs is that the number of distance calculations is significantly reduced due to the improved pruning ability provided by global pivots. However, pivot-related changes require rebuilding the whole index.

Our proposal differs from the existing ones as we include local additional pivots in local pivot-based MAMs and anticipate information from child nodes in such a way that it improves the pruning ability of the structure while maintaining every update locally contained. The first technique consists of adding local pivots to each node and having the distances of their elements from each additional pivot calculated and stored. The target for this technique is to reduce the uncertainty region and, consequently, the number of distance calculations by using the triangular inequality property in each node accessed by a query. The second technique consists of anticipating information from child nodes, such as distance and radius values, and use them to avoid unnecessary disk accesses. Although M*-tree employs local additional pivots in each node too, the benefits obtained are only in terms of distance calculations. Our techniques, on the other hand, allow reducing both distance calculations and disk accesses. They are presented in the next sections.

3   The CLAP Technique

This section describes our new technique to improve the pruning ability of dynamic MAMs, named *Cutting Local Additional Pivots* (CLAP). It aims at reducing the uncertainty region of each node accessed in a similarity query. However, distinctly from other strategies that employ multiple local pivots, in which the pivots define the covering region of a node, as in the MM-tree and the Onion-tree, the CLAP technique is used to cut the region defined by a node representative applying the triangular inequality pruning mechanism using local additional pivots. In spite of changing the node structure, this technique does not change its partitioning, i.e. the covering radius of each node is still defined from s_{rep} and therefore it does not compromise the dynamicity of the structure.

As described in Sect. 2.2, pruning by the triangular inequality regarding the representative, known as the *main pivot*, consists of discarding every element which satisfies one of the Eqs. 1 and 2. Figure 1a illustrates this property, where the child node whose representative is also s_{rep} (the black dashed circumference, centered in s_{rep}) is pruned, as it does not touch the uncertainty region (the orange dashed ring). What the CLAP technique allows is to extend the use of this property for all local additional pivots. Thus, for every element s_i which has not been pruned by the triangular inequality involving the main pivot, each additional pivot p_j, $1 \leq j \leq n$, where n is the number of additional pivots, is used to prune the elements that satisfy the following additional conditions:

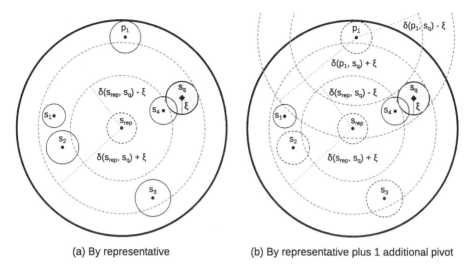

(a) By representative   (b) By representative plus 1 additional pivot

Fig. 1. Pruning by triangular inequality with 1 additional pivot. Dashed circumferences centered in s_{rep}, s_2, s_3 and p_1 represent pruned elements

$$\delta(p_j, s_i) + r_i < \delta(p_j, s_q) - \xi \qquad (3)$$

$$\delta(p_j, s_i) - r_i > \delta(p_j, s_q) + \xi \qquad (4)$$

Similarly to In Eqs. 1 and 2, these conditions involve two distance calculations. One of them, the value $\delta(p_j, s_i)$, was calculated and stored into the node when the MAM was built. The value $\delta(p_j, s_q)$, even though it must be calculated for each pivot p_j when the node is visited, allows reducing even more the number of distance calculations. Figure 1b shows the new uncertainty region as the intersection of two rings: the orange dashed one, centered in the representative s_{rep}, and the blue dashed one, centered in the cutting local additional pivot p_1 (this figure considers one cutting local additional pivot). It can be verified that the uncertainty region is much smaller than that shown in Fig. 1a, pruning the child nodes which have s_{rep}, s_2, s_3 and p_1 as their representatives.

4   Anticipation of Child Information

This section presents a new approach which aims at avoiding unnecessary disk accesses by modifying the structure of nodes in MAMs in order to get, in advance, information from child nodes that would only be accessed when these nodes were read from disk. Taking the Slim-tree for example, when a node is visited during a query, the first step is to use the triangular inequality pruning mechanism. For every element s_i not pruned in an index node, the corresponding node must be accessed from disk even if none of its elements touches the search region.

Our second proposed technique is ACIR (*Anticipation of Child Information regarding Representatives*), which consists of anticipating, for each child node of the current node, the array of distances from the representative s_i plus the array of covering radii (only the distances when the child nodes are leaf nodes, since their entries do not have radii). Consequently, the triangular inequality pruning mechanism regarding the representative is anticipated for each child node which intercepts the search region. With the ACIR technique, the sequence of steps for each node accessed during a similarity query is: (i) to use the triangular inequality pruning mechanism by the main pivot and by the additional pivots; (ii) to calculate the distances between s_q and all elements not pruned in the previous step; (iii) before reading from disk the elements which intercept the search region defined by s_q, execute additional steps involving the information anticipated from child nodes. In doing so, it is possible to avoid unnecessary disk accesses by identifying nodes that intercept the search region and, nevertheless, are irrelevant for the result.

This allows avoiding unnecessary disk accesses in situations like the one depicted in Fig. 2, where the child node centered in s_1 intercepts the search region defined by s_q, but none of its children — s_{11}, s_{12}, s_{13} and s_{14}, which are grandchild nodes of the current node (the biggest one, centered in s_{rep}) — intercepts the uncertainty region represented by the dashed ring within the child node centered in s_1. The evaluation of whether the grandchild nodes intercept or not this uncertainty region can happen in the current level only because their distances and radii have been anticipated. If they had not, the child node centered in s_1 would have to be read from disk (unnecessarily) for this evaluation to be done. Still in Fig. 2, the small circumferences centered in s_{11}, s_{12}, s_{13} and s_{14} are the covering radii of the grandchild nodes. Those regions appear in the figure just to illustrate the anticipated use of the triangular inequality pruning mechanism allowed by the strategy. They are not known when the current node (centered in s_{rep}) is in fact processed, once that the representatives of the grandchild nodes are not stored in the current node.

5   Application of CLAP and ACIR to Slim-Tree

Slim-tree has two types of node: index nodes and leaf nodes. A leaf node presents the following structure (the characters $\langle$ and $\rangle$ delimit an array):

$$leaf\ node[\langle OID_i, s_i, \delta(s_i, s_{rep})\rangle]$$

where OID_i is the identifier of the element; s_i is the element itself, stored as a feature vector; $\delta(s_i, s_{rep})$ is the distance of s_i from the representative. An index node presents the following structure:

$$index\ node[\langle s_i, r_i, \delta(s_i, s_{rep}), Ptr(T_{s_i}), \#Ent(T_{s_i})\rangle]$$

where s_i is the feature vector of the representative of the subtree T_{s_i}, pointed by $Ptr(T_{s_i})$; r_i is the covering radius of this subtree, determined by the distance of

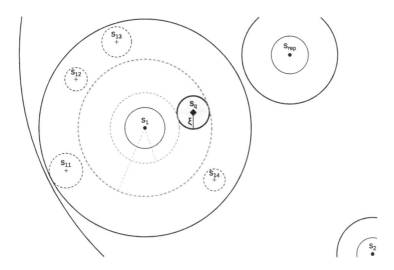

Fig. 2. Avoiding a disk access by anticipation of information. None of the grandchild nodes (s_{11}, s_{12}, s_{13}, s_{14} and the representative s_1 itself, which are children of the child node centered in s_1) touches the uncertainty region, represented by the dashed ring

s_i from the farthest element in the node of this subtree; $\delta(s_i, s_{rep})$ is the distance of s_i from the representative of the current node; $\#Ent(T_{s_i})$ is the number of entries in T_{s_i}.

After including the CLAP and ACIR techniques, both index and leaf nodes present new structures. In the following definitions, bold symbols represent the information added by CLAP technique and blue symbols represent the changes promoted by ACIR. A leaf node presents the following new structure:

$$leaf\ node[\langle \boldsymbol{Pos_j} \rangle, \langle OID_i, s_i, \langle \boldsymbol{\delta(s_i, p_j)} \rangle \rangle]$$

where Pos_j is the position of additional pivot p_j (e.g. Pos_j equals 2 if p_j is the second element stored in the node); $\delta(s_i, p_j)$ is the distance of s_i from pivot p_j; the remaining information are the same as the original structure. The criterion for choosing the local additional pivots used in this implementation is the greater sum of distances from the previous pivots. Nevertheless, other criteria can be analyzed. In the case of the first additional pivot, it is the element which has the greater distance from s_{rep}; in the case of the second additional pivot, it is the element which has the greater sum of the distance from s_{rep} plus the distance from p_1 and so on. In the ACIR technique strategy, once that the distance values from the representative are anticipated one level above in the MAM, these are removed from the structure of the leaf node.

The index nodes, on the other hand, are divided into two types: index nodes which are parents of leaf nodes, called *l-index node*, and index nodes which are parents of index nodes, called *i-index node*. Their structures are the following:

$$l\text{-}index\ node$$
$$[\langle \boldsymbol{Pos_j} \rangle, \langle s_i, r_i, \delta(s_i, s_{rep}), \langle \boldsymbol{\delta(s_{il}, s_i)} \rangle, Ptr(T_{s_i}), \#Ent(T_{s_i}), \langle \boldsymbol{\delta(s_i, p_j)} \rangle \rangle]$$

$$i\text{-}index\ node$$
$$[\langle \boldsymbol{Pos_j} \rangle, \langle s_i, r_i, \delta(s_i, s_{rep}), \langle \boldsymbol{\delta(s_{il}, s_i)} \rangle, \langle \boldsymbol{r_{il}} \rangle, Ptr(T_{s_i}), \#Ent(T_{s_i}), \langle \boldsymbol{\delta(s_i, p_j)} \rangle \rangle]$$

The difference between those nodes is that, in an l-index node, only the array of distances $\langle \delta(s_{il}, s_i) \rangle$ are added, where s_{il} is the l-th entry of the i-th child node and s_i is its representative. In an i-index node, the covering radius r_{il} of each s_{il} is also added.

6   Experimental Results

We performed extensive evaluations on both proposed techniques. The first technique has been implemented for only one local additional pivot in order to analyze its impact. We carried out experiments over datasets with different dimensionalities and cardinalities and employed metrics with different computational costs, so that we could evaluate our techniques in varied scenarios. In this section, we present the results achieved through combinations of three datasets and two metrics. For running the experiments, we used a machine with an Intel Core i5 2400@3.1 GHz processor, 4 GB of RAM@1333 MHz and HDD SATA III 6 Gb/s.

The datasets ALOI-T and ALOI-H belong to the *Amsterdam Library of Object Images*[1] (ALOI) [10]. These datasets are based on feature vectors extracted from 108,000 images of objects photographed several times, varying the position, the illumination and the combination of colors. ALOI-T consists of texture feature vectors with 140 dimensions, whereas ALOI-H consists of color histograms with 256 dimensions.

The test-collection CoPhIR[2] [2] contains 106 million images processed from Flickr. For all the images, the standard MPEG-7[3] features have been extracted: Scalable Color, Color Structure, Color Layout, Edge Histogram, Homogeneous Texture. In the experiments, the full feature vector of 282 dimensions was used in datasets of cardinality ranging from 10k to 10M elements, generating the datasets CoPhIR-10k-WL2, CoPhIR-100k-WL2, CoPhIR-1M-WL2, CoPhIR-10M-WL2, in order to evaluate the scalability of the techniques using a weighted euclidean distance. We also built the dataset CoPhIR-1M-M with 1M elements consisting of the Color Structure feature, with 64 dimensions, employing the Mahalanobis metric, also known as *histogram quadratic distance* [6]. This metric is expensive because it considers the correlation between bins of color histograms, which leads to more desirable results.

Since ALOI-H and CoPhIR-1M-M consist of color histograms, they were the chosen datasets for employing the Mahalanobis metric. The L2 metric, which is the euclidean distance, was employed on the rest of the datasets. On the CoPhIR datasets, the weights suggested in [1] were used for the Weighted L2 metric.

---

[1] Available at: http://aloi.science.uva.nl.

[2] Available at: http://cophir.isti.cnr.it.

[3] http://mpeg.chiariglione.org/standards/mpeg-7.

6.1   Performance in Similarity Queries

This subsection presents the results comparing the performance of Slim-tree + CLAP and ACIR with the original Slim-tree to execute similarity queries. In these experiments, k-NN queries and Range queries were performed varying the k value (1, 10, 25, 50, 100, 150, 200, 250 and 300) and using the corresponding radius values to retrieve k elements in Range queries. For each k, the results were obtained by performing queries multiple times (500 when using the L2 metric and 100 when using the Mahalanobis metric), each time with a random query element, and taking the average value.

Figure 3 shows the obtained results. The graphs show that our techniques lead to notable gains when compared to the original Slim-tree. For low selectivities (e.g. $k = 1$ for k-NN queries and Range queries returning 1 element) the improvement was very high, being up to 62.51 % regarding execution time, 62.58 % in distance calculations and 96.93 % in disk accesses. When the selectivity was 50 or more, the gains were less expressive. Nevertheless, our techniques consistently outperformed the original structure, regarding every dataset, in execution time, number of distance calculations and number of disk accesses.

The first row of graphs in Fig. 3 corresponds to results over ALOI-T. In k-NN queries, the gain ranged from 21.59 % to 46.61 % in execution time, from 33.12 % to 62.58 % in distance calculations and from 9.57 % to 51.83 % in disk accesses. In Range queries, the gain ranged from 37.02 % to 62.51 %, from 40.48 % to 58.24 % and from 10.55 % to 82 %, respectively for the same variables.

The second row of graphs in Fig. 3 refers to ALOI-H, which employs the costly Mahalanobis metric. Our proposal presented a noticeable speedup, although the gains, especially in disk accesses, were lower if compared to the previous dataset. In k-NN queries, the gain ranged from 18 % to 53.2 % in execution time, from 17.96 % to 53.2 % in distance calculations and from 5.8 % to 56.5 % in disk accesses. In Range queries, the gain ranged from 19.77 % to 32.14 %, from 19.68 % to 32.12 % and from 6.85 % to 89.43 %, respectively for the same variables.

Regarding the CoPhIR datasets, the third row in Fig. 3 shows the results of experiments carried out over 1M elements with 282 dimensions by varying k (dataset CoPhIR-1M-WL2). This is the dataset for which our techniques presented the lowest gains for high values of k. Nonetheless, its use allowed improving the performance of every evaluated aspect. Regarding execution time, distance calculations and disk accesses, respectively, the gain in k-NN queries ranged from 5.12 % to 53.05 %, from 7.29 % to 55.09 % and from 3.73 % to 66.82 %, while the gain in Range queries ranged from 6.92 % to 60.36 %, from 5.17 % to 13.05 % and from 3.80 % to 96.93 %. In the set of experiments using CoPhIR-1M-M, the proposed techniques achieved even better results, as the main improvement of the techniques regards distance calculations and the cost of the Mahalanobis metric is much higher than the cost of the L2 metric. In this dataset, the gain in k-NN queries ranged from 11.25 % to 48.51 % in execution time, from 11.53 % to 49.09 % in distance calculations and from 6.59 % to 56.65 % in disk accesses. Finally, the gain in Range queries ranged from 12.72 % to 20.37 %, from 9.65 % to 20.39 % and from 6.49 % to 87.6 %, respectively for the same variables.

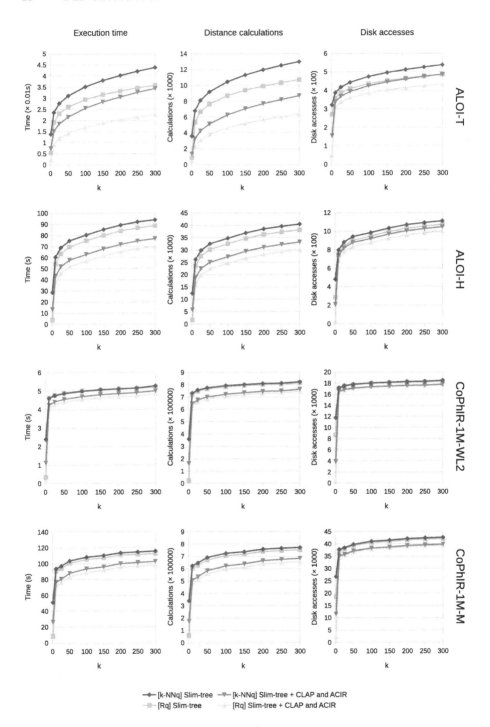

Fig. 3. Results of the experiments varying k

6.2   Evaluation of Construction Issues and Scalability of Gain

This section evaluates the impact of the proposed techniques when compared to the original structure in terms of building time, page size and resulting data file size, as well as how the gain promoted by the techniques behaves with the size of the dataset. The information of both structures regarding their construction are presented in Table 1. Note that the Slim-tree with CLAP and ACIR required two page sizes, one for index nodes and one for leaf nodes. This is because we wanted to minimize the overhead of information in index nodes by increasing their page size. The page size of index nodes will be usually a multiple of the page size of leaf nodes, to allow using the same buffer pool for both node types. Although the index nodes in Slim-trees + CLAP and ACIR are the double of the size of the original Slim-trees, taking a longer time to read them from disk, the higher pruning ability of CLAP and ACIR allowed a better performance in our experiments. Also, note that both structures have similar file sizes. Slim-tree + CLAP and ACIR presented files from 2.4 % to 6.91 % smaller than Slim-tree did. It can be explained by the fact that, with the anticipation of information, leaf nodes store less information than before. Since there are many more leaf nodes than index nodes, storing less information within leaf nodes leads to a little smaller file sizes.

The building time for Slim-trees with CLAP and ACIR was from 12.32 % to 34.32 % higher, as expected, because it involves additional computations such as distance calculations regarding the cutting local additional pivots. However, the worst case was on the smallest dataset, CoPhIR-10k-WL2, which resulted in a difference of only 1.528s. Thus, considering the performance gain in queries, the proposed techniques are worth the higher building time.

Table 1. Construction information of both structures for all datasets

Dataset	Slim-tree	Time (s)	Page size (KB)	File size
ALOI-T	Original	37.296	32	103 MB
	CLAP and ACIR	43.650	64 (index) — 32 (leaf)	101.2 MB
ALOI-H	Original	29637.4	64	162 MB
	CLAP and ACIR	33289.4	128 (index) — 64 (leaf)	155.9 MB
CoPhIR-10k-WL2	Original	4.452	64	15 MB
	CLAP and ACIR	5.98	128 (index) — 64 (leaf)	14.640 MB
CoPhIR-100k-WL2	Original	58.967	64	139 MB
	CLAP and ACIR	71.684	128 (index) — 64 (leaf)	129.4 MB
CoPhIR-1M-WL2	Original	740.946	64	1.4 GB
	CLAP and ACIR	871.27	128 (index) — 64 (leaf)	1.334 GB
CoPhIR-10M-WL2	Original	9481.47	64	14 GB
	CLAP and ACIR	10954.6	128 (index) — 64 (leaf)	13.329 GB
CoPhIR-1M-M	Original	11763.2	8	449 MB
	CLAP and ACIR	13385.5	16 (index) — 8 (leaf)	434 MB

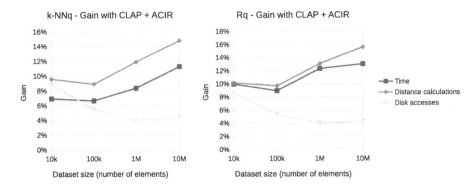

Fig. 4. Results of the scalability experiments over CoPhIR

Finally, we evaluated how the gain promoted by CLAP and ACIR techniques behaves according to the dataset size. In this experiment, we fixed k to 10 and varied the dataset size from 10k to 10M elements with 282 dimensions, using the Weighted L2 metric and random query elements. Figure 4 presents the obtained results. It can be seen that the improvement raises with the dataset size regarding distance calculations and execution time. The gain in disk accesses drops until 1M elements and afterwards presents a sensible increase, being always positive. These results show that the proposed techniques scale well with increasing dataset size.

7   Conclusions

We have proposed new techniques based on local additional pivots and anticipation of information for improving the pruning ability of dynamic MAMs: CLAP and ACIR. Our techniques were extensively tested and achieved better results in all evaluated scenarios, for both k-NN and Range queries. We also showed that the gain promoted by the techniques scales well with the dataset size. Moreover, the CLAP and ACIR techniques do not affect the dynamicity of the underlying MAM and, just like they were implemented over Slim-tree, other dynamic hierarchic MAMs could be improved by using them as well.

Both contributions of this work opens possibilities for future work. Regarding the CLAP technique, the use of more than one additional pivot per node could be explored, as well as strategies for selecting cutting local additional pivots. In ACIR, the information considered in this work to be anticipated from child nodes are the distances of each element from their representative and the radius values. However, other information could be anticipated due to the CLAP technique, such as the feature vectors of each additional pivot and the distances of each element from the additional pivots. Our insight is that, by having more information anticipated, the improvements can be even better. We are working on these extensions to present them in a next work.

References

1. Batko, M., Kohoutkova, P., Novak, D.: CoPhIR image collection under the microscope. In: 2nd International Workshop on Similarity Search and Applications, pp. 47–54. IEEE Computer Society, Washington, DC (2009)
2. Bolettieri, P., Esuli, A., Falchi, F., Lucchese, C., Perego, R., Piccioli, T., Rabitti, F.: CoPhIR: A Test Collection for Content-Based Image Retrieval. Computing Research Repository abs/0905.4627v2 (2009)
3. Burkhard, W.A., Keller, R.M.: Some approaches to best-match file searching. Commun. ACM **16**(4), 230–236 (1973)
4. Carélo, C.C.M., Pola, I.R.V., Ciferri, R.R., Traina, A.J.M., Traina Jr., C., Ciferri, C.D.A.: Slicing the metric space to provide quick indexing of complex data in the main memory. Inf. Syst. **36**(1), 79–98 (2011)
5. Ciaccia, P., Patella, M., Zezula, P.: M-Tree: An efficient access method for similarity search in metric spaces. In: 23rd International Conference on Very Large Data Bases, pp. 426–435. Morgan Kaufmann, San Francisco (1997)
6. Deza, M.M., Deza, E.: Encyclopedia of Distances. Springer, Heidelberg (2009)
7. Esuli, A.: Use of permutation prefixes for efficient and scalable approximate similarity search. Inf. Process. Manage. **48**(5), 889–902 (2012)
8. Faloutsos, C.: Searching Multimedia Databases by Content. Advances in Database Systems, vol. 3. Springer, New York (1996)
9. Gaede, V., Gunther, O.: Multidimensional access methods. ACM Comput. Surv. **30**(2), 170–231 (1998)
10. Geusebroek, J.M., Burghouts, G.J., Smeulders, A.W.M.: The amsterdam library of object images. Int. J. Comput. Vis. **61**(1), 103–112 (2005)
11. Pola, I.R.V., Traina Jr., C., Traina, A.J.M.: The MM-tree: a memory-based metric tree without overlap between nodes. In: Ioannidis, Y., Novikov, B., Rachev, B. (eds.) ADBIS 2007. LNCS, vol. 4690, pp. 157–171. Springer, Heidelberg (2007)
12. Skopal, T., Hoksza, D.: Improving the performance of M-Tree family by nearest-neighbor graphs. In: Ioannidis, Y., Novikov, B., Rachev, B. (eds.) ADBIS 2007. LNCS, vol. 4690, pp. 172–188. Springer, Heidelberg (2007)
13. Skopal, T., Pokorný, J., Snášel, V.: Nearest neighbours search using the PM-tree. In: Zhou, L., Ooi, B.-C., Meng, X. (eds.) DASFAA 2005. LNCS, vol. 3453, pp. 803–815. Springer, Heidelberg (2005)
14. Traina Jr., C., Filho, R.F.S., Traina, A.J.M., Vieira, M.R., Faloutsos, C.: The omni-family of all-purpose access methods: a simple and effective way to make similarity search more efficient. VLDB J. **16**(4), 483–505 (2007)
15. Traina Jr., C., Traina, A.J.M., Faloutsos, C., Seeger, B.: Fast indexing and visualization of metric data sets using slim-trees. IEEE Trans. Knowl. Data Eng. **14**(2), 244–260 (2002)
16. Traina Jr., C., Traina, A.J.M., Filho, R.F.S., Faloutsos, C.: How to improve the pruning ability of dynamic metric access methods. In: 11th International Conference on Information and Knowledge Management, pp. 219–226. ACM, New York (2002)
17. Vieira, M.R., Traina Jr., C., Chino, F.J.T., Traina, A.J.M.: DBM-Tree: trading height-balancing for performance in metric access methods. J. Braz. Comput. Soc. **11**(3), 37–51 (2005)
18. Zezula, P., Amato, G., Dohnal, V., Batko, M.: Similarity Search: The Metric Space Approach. Advances in Database Systems, vol. 32. Springer, New York (2006)

The Structure of Preference Orders

Markus Endres [✉]

Department of Computer Science, University of Augsburg,
86135 Augsburg, Germany
endres@informatik.uni-augsburg.de
http://www.informatik.uni-augsburg.de/dbis

Abstract. Preferences are an important natural concept in real life and
are well-known in the database and artificial intelligence community.
Modeling preferences as strict partial orders closely matches people's
intuition. There are many algorithms for the evaluation of these strict
partial orders. In particular some algorithms rely on the total order or the
lattice structure constructed by a preference query. This paper provides
an overview of the structure of preference orders. We present several
measures of the different "better-than graphs" and give a deep insight
into the structure of preferences. In fact, a careful analysis of the underlying "better-than graph" enables one to develop efficient algorithms for
preference computation.

Keywords: Preference · Better-than graph · Lattice

1 Introduction

Preferences have always been an important natural concept in real life. Preferences in computer science form popular research topics not only in databases,
but also in fields as AI, constraint logic programming, or decision making [1].

A preference is often modeled as *strict partial order* and therefore *transitivity* holds [2–4]. Figure 1 expresses a simple user preference on the domain of colors
dom(*color*), where {red, blue} is preferred over all other colors, except {purple},
which is the least preferred value. A *Better-Than graph* (BTG) is a visualization of
the domination of domain elements for a preference, cp. Figure 1. The *nodes* in the
BTG represent *equivalence classes*. Each equivalence class contains objects which
are mapped to the same level by a scoring function. All values in the same equivalence class are considered substitutable. The *edges* in the BTG state dominance.

For the complex Pareto preference, the BTG constitutes a *lattice* [5], cp.
Figure 3. A Pareto preference (also known as Skyline query [6]) selects those
objects from a dataset R that are not dominated by any others. An object p
having m attributes (dimensions) dominates an object q, if p is better than q
in at least one dimension and not worse than q in all other dimensions, for
a defined comparison function. There are many algorithms which exploit the
lattice structure of Pareto for efficient preference evaluation, cp. [7–13].

In this paper we provide a deep insight into the structure of preference orders.
We will discuss the *visualization* of simple preferences as well as complex preferences like Pareto or Prioritization. For this we present several measures of the

© Springer International Publishing Switzerland 2015
T. Morzy et al. (Eds.): ADBIS 2015, LNCS 9282, pp. 32–45, 2015.
DOI: 10.1007/978-3-319-23135-8_3

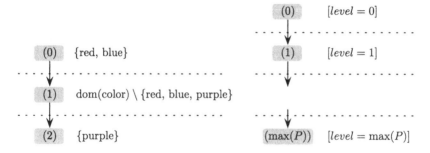

Fig. 1. Sample preference on colors.   **Fig. 2.** BTG for a WOP.

different better-than graphs, e.g., height, width, number of nodes, or number of edges. In addition we show how to integrate general strict partial orders into lattices, even if the values in the same equivalence class are not considered as substitutable. Understanding the measures of a BTG for a preference is essential in understanding the BTG itself. These measures can then be used to design new algorithms for preference computation or just to apply existing lattice based algorithms like [7–12].

The rest of this paper is organized as follows: Sect. 2 contains the formal background. Section 3 presents the visualization of preference orders, Sect. 4 provides a method for the integration of preferences with trivial SV-semantics. Section 5 contains our concluding remarks.

2   Background

Following [2,3] a database preference $P = (A, <_P)$ is a *strict partial order* on the domain of the attribute set A. The term $\mathbf{x} <_\mathbf{P} \mathbf{y}$ is interpreted as "*I like y more than x*". As strict partial orders are *transitive*, better-than relations in this preference model are, too. Given $<_P$, the *indifference relation* $\sim_P$ is defined as: $x \sim_P y \iff \neg(x <_P y) \land \neg(y <_P x)$.

An important subclass of strict partial orders are *weak order preferences* (WOP). Following [14], a weak order preference is a strict partial order, in which indifference is transitive. For each WOP $P = (A, <_P)$ we can define an utility function $level_P$ that can be used to determine dominance between two values.

$$level_P : \mathrm{dom}(A) \to \mathbb{R}_0^+ \qquad x <_P y \iff level_P(x) > level_P(y) \qquad (1)$$

Note that for WOPs the domain values x and y with the same level are either *equal* or *indifferent*. Two values with the same level belong to the same *equivalence class*.

Definition 1 (max(P)). $\max(P) \in \mathbb{N}_0$ is the *maximum level* for a preference P.

To specify a database preference, a variety of intuitive preference constructors have been defined, cp. [15].

2.1   Base Preference Constructors

Preferences on single attributes like *discrete (categorical)* or *continuous (numerical)* domains are called *base preferences*. Usually they can be defined as WOPs. Most of the base preferences can be specified by a score function $f : \text{dom}(A) \rightarrow \mathbb{R}_0^+$, such that

$$level_P(v) := \begin{cases} f(v) & \text{if } d = 0 \\ \left\lceil \frac{f(v)}{d} \right\rceil & \text{if } d > 0 \end{cases} \qquad (2)$$

In the case of $d = 0$ the function $f(v)$ models the *distance* to the best value. A d-parameter $d > 0$ represents a discretization, which is used to group ranges of scores together. The d-parameter maps different function values to a single integer number. Choosing $d > 0$ effects that attribute values with identical $level_P(v)$ value become *indifferent* and stay in the same *equivalence class*.

Note that the definition of the function f depends on the type of preference. The $\text{BETWEEN}_d(A, [low, up])$ preference for example expresses the wish for a value between a *lower* and an *upper* bound. The scoring function is $f(v) = \max\{low - v, 0, v - up\}$. The $\text{AROUND}_d(A, v)$ is a special case of the former, where $low = up =: v$. In a categorical domain the $\text{LAYERED}_m(A, \{L_1, \ldots, L_m\})$ preference expresses that a user has a set of preferred values given by the disjoint sets L_i, which form a partition of $\text{dom}(A)$. Thereby the values in L_1 are the most preferred values. The scoring function equals $f(v) = i - 1 \iff x \in L_i$.

2.2   Complex Preference Constructors

A *Pareto* preference models "equal importance of preferences" whereas a *Prioritization* expresses that "a preference is more important than the other". Such *complex preferences* are built of constructs like "Better w.r.t. P_1, equal w.r.t. P_2", where P_1 and P_2 are preferences. A simple approach for the notion of equality w.r.t. a preference is to use strict equality of the domain values. But often we have base preferences where values x, x' are equally good in the sense that $x <_P y \Leftrightarrow x' <_P y$ for all y. For example, this is the case if $level_P(x) = level_P(x')$, i.e., the tuples have the same level value; they belong to the same *equivalence class* (later denoted as *node*). This behavior is called *regular substitutable values semantics* (SV semantics) [16], denoted by $\sim_P$. Requiring strict equality leads to the *trivial SV-semantics*, denoted by $=_P$. Note that a general SV-relation $(\cong_P, A)$ on an attribute set A is an equivalence relation on $\text{dom}(A)$ [17].

Definition 2 (Pareto). *Let $P_i = (A_i, <_{P_i})$ be m weak order preferences and $x = (x_1, \ldots, x_m), y = (y_1, \ldots, y_m) \in \text{dom}(A_1 \times \cdots \times A_m)$. A Pareto preference $P := P_1 \otimes \ldots \otimes P_m$ is defined as:*

$$x <_P y \Leftrightarrow \exists i : x_i <_{P_i} y_i \wedge \left(\forall i, j \in \{1, \ldots, m\}, j \neq i : (x_j <_{P_j} y_j \vee x_j \cong_{P_j} y_j)\right)$$

Definition 3 (Prioritization). *In a Prioritization $P := P_1 \& \ldots \& P_m$ the preference $P_1 = (A_1, <_{P_1})$ is more important than $P_2 = (A_2, <_{P_2})$, and so on.*

$$x <_P y \iff \exists k \in \{1, \ldots, m\} : \forall i \in \{1, \ldots, k-1\} : x_i \cong_{P_i} y_i \ \wedge \ x_k <_{P_k} y_k$$

The Semi-Pareto preference has no intuitive interpretation, but is useful for algebraic query optimization, cp. [10]. Comparing the definition of Semi-Pareto to Pareto, it is evident that Semi-Pareto is half of a Pareto preference.

Definition 4 (Semi-Pareto). *Given preferences* $P_1 = (A_1, <_{P_1})$ *and* $P_2 = (A_2, <_{P_2})$. *Then we define Left-Semi-Pareto (LSP)* $P := P_1 \otimes P_2$ *and Right-Semi-Pareto (RSP)* $P := P_1 \otimes P_2$ *as*

(a) $P := P_1 \otimes P_2$ *iff* $x <_P y \iff x <_{P_1} y \ \wedge \ (x <_{P_2} y \ \vee \ x \cong_{P_2} y)$
(b) $P := P_1 \otimes P_2$ *iff* $x <_P y \iff (x <_{P_1} y \ \vee \ x \cong_{P_1} y) \ \wedge \ x <_{P_2} y$

Another form of preference combination is by associating numerical scores to each individual preference and then applying a combining function to decide the "better-than" relation.

Definition 5 (Rank). *Given some preference* P_i *with scoring functions* f_i. *The numerical ranking preference* $rank_{F,d}$ *(d > 0) with an m-ary combining function* $F : \mathbb{R}^m \to \mathbb{N}_0$ *is defined as:*

$$x <_P y \iff \left\lceil \frac{F(f_1(x_1), \ldots, f_m(x_m))}{d} \right\rceil > \left\lceil \frac{F(f_1(y_1), \ldots, f_m(y_m))}{d} \right\rceil$$

2.3 Better-Than Graph

Visualization of strict partial orders is often done using Hasse diagrams, graphs (directed and acyclic) in which edges state dominance [5], also known as *better-than graphs*.

Definition 6 (Better-Than Graph (BTG)). *The* better-than graph *(BTG) for a preference* $P = (A, <_P)$ *is the Hasse diagram of* $<_P$ *where*

(a) each equivalence class in dom(A) *is represented by one node in the BTG.*
(b) a directed edge (a_1, a_2) *is drawn from* a_1 *to* a_2 *for each pair of nodes* a_1, a_2 *for which holds:* $a_2 <_P a_1 \ \wedge \ (\neg \exists a_3 \in$ dom$(A) : a_2 <_P a_3 <_P a_1)$.
Note that edges following from transitivity *of domination are omitted.*
(c) the level value *level$_P$ of a node is the length of a longest-path leading to it.*

For simplicity, we use the terms *node* and *equivalence class* synonymosuly, as an equivalence class for a preference P is represented by exactly one node in the BTG and vice versa.

3 Analysis of BTGs with Regular SV-Semantics

In this section we consider BTGs for preferences with regular SV-semantics.

3.1   BTGs for WOPs

For base preferences being WOPs, each level value represents exactly one *equivalence class* (node) in the input domain, i.e., in the interval $[0, \max(P)]$. Since domination can be directly seen from the level value, this leads to BTGs as shown in Fig. 2. Best values have a level value of zero and therefore belong to the top node. The worst node of the BTG has the largest possible level, $\max(P)$.

The very simple chain structure of BTGs for WOPs makes their analysis very simple. The proofs can be found in [18].

Theorem 1 (Properties of BTGs for WOPs). *Let $P = (A, <_P)$ be a weak order preference, then:*

(a) $\text{width}(BTG_P) = 1$
(b) $\text{height}(BTG_P) = \max(P) + 1$
(c) $\text{nodes}(BTG_P) = \text{height}(BTG_P)$

Example 1. For the LAYERED$_m$ preference in Fig. 1 we have 3 level values, 0 for {red, blue}, 2 for {purple}, and 1 for all other colors in dom(*color*). The width of the BTG is 1, the height = 2+1 = 3, which is also the number of nodes.

3.2   BTGs for Pareto Preferences

The key point when drawing Pareto preferences $P := P_1 \otimes \ldots \otimes P_m$ is that they do *not* create weak orders, even if they are build up by WOPs P_i, cp. [3]. Therefore, dominance can not be decided by the value of the level function $level_P$. The partial order induced by a Pareto preference with only WOPs as input preferences constitutes a *complete distributive lattice* [5]. This means if $a, b \in \text{dom}(A)$, the set $\{a, b\}$ has a least upper bound and a greatest lower bound in $\text{dom}(A)$.

Example 2. In Fig. 3, we see a BTG for a Pareto preference consisting of two WOPs with maximum level value of 2 and 4. As each value in the domain of each P_i can be represented by its level value, a node is a combination of level values for the P_i.

Theorem 2 (Properties of BTGs for Pareto Preferences). *Let $P := P_1 \otimes \ldots \otimes P_m$ be a Pareto preference on $\text{dom}(A) := \text{dom}(A_1) \times \cdots \times \text{dom}(A_m)$ and $P_i, i = 1, \ldots, m$ be WOPs. Then:*

(a) $\max(P) = \sum_{i=1}^{m} \max(P_i)$
(b) $\text{nodes}(BTG_P) = \prod_{i=1}^{m} (\max(P_i) + 1)$
(c) $level_P(a) = \sum_{i=1}^{m} level_{P_i}(a_i)$ *for* $a = (a_1, \ldots, a_m) \in \text{dom}(A)$
(d) $\text{height}(BTG_P) = 1 + \sum_{i=1}^{m} \max(P_i)$
(e) $\text{edges}(BTG_P) = \prod_{i=1}^{m} (\max(P_i) + 1) \cdot \sum_{i=1}^{m} \frac{\max(P_i)}{\max(P_i)+1}$

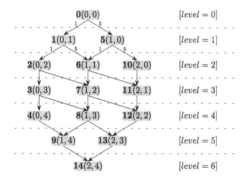

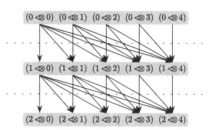

Fig. 3. BTG for Pareto.   **Fig. 4.** BTG for a 2d LSP preference.

Proof. We prove e); a)–d) can be found in [18].

Consider those edges in BTG_P which result from the domination with respect to P_i. All nodes with a level value smaller than $\max(P_i)$ have such an outgoing edge. The number of such nodes is given by $\frac{nodes(BTG_P)}{\max(P_i)+1} \cdot \max(P_i)$. These are all nodes with any possible level values for all $P_j \in P$ with $i \neq j$ and a level value in the interval $[0, \max(P_i) - 1]$ for P_i. This leads to the following sum of edges for all preferences in P: $\sum_{i=1}^{m} \left(\frac{nodes(BTG_P)}{\max(P_i)+1} \cdot \max(P_i) \right)$ $\square$

The graph of a Pareto preference is *symmetric* with respect to its middle axis (duality principle in lattices [5]). The top level contains only one node: $(0, \ldots, 0)$, level 1 contains m nodes. The levels grow wider until a maximum width is reached before or at level $\frac{1}{2} \cdot \max(P)$. It starts to get "thinner" while increasing the overall level value and finally reaches the bottom node. This leads to the characteristic *hexagon shape* depicted in Fig. 3. The method to find the width of a BTG is based on products of lattices as described in [5,18].

Theorem 3 (Width of the BTG of a Pareto Preference). *Consider* $P_K := P_{K_1} \otimes \ldots \otimes P_{K_m}$ *and* $P_L := P_{L_1} \otimes \ldots \otimes P_{L_n}$, *all* P_{i_j} *WOPs. We construct a preference* $P := P_K \otimes P_L$. *Then, the* width *for* P *in level* v *is*

$$width(P, v) = \sum_{i=0}^{v} \left(width(P_K, i) \cdot width(P_L, v - i) \right)$$

In some cases, there are shortcuts for this calculation.

Lemma 1. *For* $m = 2$, *i.e.,* $P = P_1 \otimes P_2$ *the width can be computed as:*

(a) $v \leq \min(\max(P_1), \max(P_2)) : width(BTG_P, v) = v + 1$
(b) $\min(\max(P_1), \max(P_2)) < v \leq \frac{1}{2}(\max(P_1) + \max(P_2)) :$

$$width(BTG_P, v) = \min(\max(P_1), \max(P_2)) + 1$$

(c) $\frac{1}{2}(\max(P_1) + \max(P_2)) < v$:

$$width(BTG_P, v) = width(BTG_P, height(BTG_P - v)$$

Proof. (a) Consider a BTG node $x = (x_1, x_2)$ in level $v \leq \min(\max(P_1), \max(P_2))$. So we get $v + 1$ possible level combinations for x: $(0, v)$, $(1, v - 1)$, ..., $(v, 0)$. (b) Consider $\max(P_1) < \max(P_2)$ and a node $x = (x_1, x_2)$ in the BTG so that $\max(P_1) < x_1 + x_2 \leq \max(P_2)$. Then, we have $\max(P_1)$ possible combinations in a given level $v \geq \max(P_1)$: $(0, v)$, $(1, v - 1)$, ..., $(\max(P_1), v - \max(P_1))$. (c) This follows from the symmetry of BTGs. □

For each node in the BTG of a weak order preference, the level function value can also be used as a *unique node identifier* (ID), because the BTG forms a total order. For the nodes in the BTG of a Pareto preference this is not possible, because in each level there is more than one node representing an equivalence class (except the top and the bottom level). Nevertheless, it is possible to define unique node IDs. For this, we need the definition of *edge weights*.

Definition 7 (Edge weights). *Let $P := P_1 \otimes \ldots \otimes P_m$ be a Pareto preference and P_i be WOPs. The* weight *of an edge in the BTG expressing dominance between two direct connected nodes with respect to any P_i is characterized by*

$$weight(P_i) = \prod_{j=i+1}^{m} (\max(P_j) + 1)$$

For $j > m$ we set $weight(P_i) = 1$.

These edge weights can be used to define unique identifiers, cp. [8].

Theorem 4 (Unique node IDs). *Let $a = (a_1, \ldots, a_m) \in dom(A)$ be a node in BTG_P. Let ID: $(\mathbb{N}_0)^m \rightarrow \{0, 1, 2, \ldots |BTG| - 1\}$ be a mapping such that*

$$\text{ID}(a) = \sum_{i=1}^{m} (weight(P_i) \cdot a_i)$$

Then the following properties hold:

(a) *ID is unique for every node in the BTG.*
(b) *Every value in the set $\{0, 1, 2, \ldots, |BTG - 1\}$ is a valid ID for one node in the BTG.*
(c) *ID is a bijective mapping. $\text{ID}^{-1}(n) = (a_1, \ldots, a_m)$ maps a unique integer ID n to the corresponding level value combination $(a_1, \ldots, a_m)$, where the a_i are found the following way*

$$a_i = \left\lfloor \frac{\text{ID}(n) - \sum_{j=1}^{i-1} a_j \cdot weight(P_j)}{weight(P_i)} \right\rfloor$$

The *unique path* leading to every node in the lattice can be found by the following simple rule: in each node you visit, follow the edge with the highest weight leading to the target node. This path clearly is unambiguous.

Example 3. Consider Fig. 3. The BTG has $(2+1) \cdot (4+1) = 15$ nodes. The height is $1 + (2 + 4) = 7$ and the node $(0, 4)$ resides in level 4. The number of edges is $(3 \cdot 5) \cdot (2/3 + 4/5) = 22$. The maximum level value is $\max(P) = 2 + 4 = 6$. The width of level $k = 2$ for example can be computed as $w(P, k) = w(P_2, 2) + w(P_2, 1) + w(P_2, 0) = 3$. The edge weight for P_1 is $\text{weight}(P_1) = \max(P_2) + 1 = 5$ and for P_2 we have $\text{weight}(P_2) = 1$. For the node $n = (1, 1)$ we compute $\text{ID}(n) = 5 \cdot 1 + 1 \cdot 1 = 6$.

3.3 BTGs for Prioritization

Similar to Pareto, Prioritization is generally not restricted to contain weak order preferences. If Prioritization contains only WOPs, it forms a weak order preference, too [3]. Therefore, dominance can be decided by a $level_P$ function and the graphical representation corresponds to the BTG of WOPs, cp. Section 3.1.

Theorem 5 (Level Function for Prioritization). *Let $P := P_1 \& \ldots \& P_m$ be a Prioritization where all P_i are WOPs and $x = (x_1, \ldots, x_m) \in \text{dom}(A)$. Then*

(a) $level_P(x) = \sum_{i=1}^{m} \left(level_{P_i}(x_i) \cdot \prod_{j=i+1}^{m}(\max(P_j) + 1) \right)$
(b) $\max(P) = \prod_{i=1}^{m}(\max(P_i) + 1) - 1$

Proof. Consider $s, t \in \text{dom}(A)$. We assume $s <_P t$. Since $s <_P t$, there is a k for which $s_k <_{P_k} t_k$. Then, for all $i < k$ s_i and k_i are substitutable, i.e. $level_{P_i}(s_i) = level_{P_i}(t_i)$. Consider $level_P(t) - level_P(s)$, which is negative:
$$\sum_{i=k}^{m} \left((level_{P_i}(t) - level_{P_i}(s)) * \prod_{j=i+1}^{m}(\max(P_j) + 1) \right) < 0.$$
We know that $level_{P_k}(s_k) > level_{P_k}(t_k)$. So the above inequation is true, iff the amount the level value of s has to be higher than t's level value due to $s_k <_{P_k} t_k$ is bigger than the sum of products with the biggest possible level values for t for $P_{k+1}, \ldots, P_m$.

As $level_{P_k}(s_k)$ has to be at least bigger than $level_{P_k}(t_k)$ by 1, hence it is to be proved: $\prod_{i=k+1}^{m}(\max(P_i) + 1) > \sum_{i=k+1}^{m} \left(\max(P_i) \cdot \prod_{j=i+1}^{m}(\max(P_j) + 1) \right)$. For $k = m$ this leads to $1 > 0$ as base step for induction. Assuming the statement holds for some value of k, induction will prove it for $k - 1$. $\square$

3.4 BTGs for Semi-Pareto

The graphical representation of a Semi-Pareto preference can be used to develop efficient algorithms for Semi-Skyline computation, cp. [10]. However, the structure of these BTGs was never considered in detail.

Example 4. Consider $P = P_1 \otimes P_2$ with maximum level values 2 and 4. Figure 4 shows the BTG of P and its typical shape of a *rectangle* for 2dim Semi-Pareto. As in Pareto, a node is a combination of level values for the P_i. To distinguish the BTG for LSP and RSP, we "annotate" each node with the corresponding operator sign, e.g., $(x \otimes y)$ for a LSP preference.

In general, Semi-Pareto does not form a weak order preference, except for a two-dimensional Semi-Pareto containing an ANTICHAIN $A^{\leftrightarrow}$ preference. For $A^{\leftrightarrow}$ it holds that $<_P = \emptyset$, i.e., it returns all values without any ordering.

Lemma 2. *Let P_1, P_2, and $A^{\leftrightarrow}$ be preferences. If P_1 is a WOP $\Longrightarrow P_1 \otimes\!\!\!\!<\, A^{\leftrightarrow}$ and $A^{\leftrightarrow} \,>\!\!\!\!\otimes P_1$ are WOPs.*

Proof. In $A^{\leftrightarrow}$ each element in $\mathrm{dom}(A)$ is mapped to level 0, hence all $x \in \mathrm{dom}(A)$ are substitutable. Therefore, the domination of x is based on P_1. Since P_1 is a WOP, Left-Semi-Pareto and the Right-Semi-Pareto are WOPs, too.   □

Theorem 6 (Properties of BTGs for Semi-Pareto). *Let P_1 and P_2 be (complex) preferences containing WOPs and $a \in \mathrm{dom}(A)$ of $P := P_1 \otimes\!\!\!\!<\, P_2$ or $P := P_1 \,>\!\!\!\!\otimes P_2$. Then*

$P := P_1 \otimes\!\!\!\!<\, P_2$	$P := P_1 \,>\!\!\!\!\otimes P_2$
(a) $level_P(a) = level_{P_1}(a)$	$level_P(a) = level_{P_2}(a)$
(b) $\max(P) = \max(P_1)$	$\max(P) = \max(P_2)$
(c) $nodes(BTG_P) = nodes(BTG_{P_1}) \cdot nodes(BTG_{P_2})$	
(d) $height(BTG_P) = height(BTG_{P_1})$	$height(BTG_{P_2}) = height(BTG_{P_2})$

Proof. W.l.o.g. we prove the theorem only for LSP. (a) Consider $a \in \mathrm{dom}(A)$ having the node $v = (0, \ldots, 0 \otimes\!\!\!\!<\, 0, \ldots, 0)$. A node v' is worse than v only if it is worse concerning P_1, i.e., if on of the 0's left of "$\otimes\!\!\!\!<$" is a higher value, say 1. Then v' is a direct successor of v in the BTG_P. If $v' <_P v$ we know that v' must be worse or as good as v concerning P_2, leading to $level_P(v') = level_P(v) + 1$. Induction over the height of the BTG proves the theorem. (b) The level value of $\max(P)$ is computed by maximizing the level values of P_1 and P_2, respectively. Following a) we get $\max(P) = \max(P_1)$. (c) For P_1 and P_2 we have $nodes(BTG_{P_1})$ and $nodes(BTG_{P_2})$ nodes. Thus, the number of all possible combinations is $nodes(BTG_{P_1}) \cdot nods(BTG_{P_2})$. (d) The height of the BTG is defined by the number of different levels. All integer numbers in $[0, \max(P_1)]$ are valid level values, hence $height(BTG_P) = height(BTG_{P_1})$.   □

Theorem 7 (Width of the BTG of Semi-Pareto). *Let $P_1 = Q_1 \otimes \ldots \otimes Q_i$, $P_2 = Q_{i+1} \otimes \ldots \otimes Q_m$ be Pareto preferences containing only WOPs. The width of Semi-Pareto for a given level v is as follows:*

(a) $P := P_1 \otimes\!\!\!\!<\, P_2$: $width(P, v) = width(P_1, v) \cdot nodes(BTG_{P_2})$
(b) $P := P_1 \,>\!\!\!\!\otimes P_2$: $width(P, v) = nodes(BTG_{P_1}) \cdot width(P_2, v)$

Proof. W.l.o.g. we prove it for LSP.
Consider a node $a = (a_1, \ldots, a_i \otimes\!\!\!\!<\, a_{i+1}, \ldots, a_m)$ in level v of the BTG_P. Since $level_P(a) = level_{P_1}(a)$ there are $width(P_1, v)$ possible combinations for $(a_1, \ldots, a_i)$. Since all nodes in level v are indifferent due to P_1, each combination for $(a_{i+1}, \ldots, a_m)$ is allowed, hence we have $nodes(BTG_{P_2})$ possibilities for a. Together, we have $width(P_1, v) \cdot nodes(BTG_{P_2})$ nodes in level v.   □

A further notable property of Semi-Pareto is the fact that the BTG is always symmetric to its middle axis.

Lemma 3 (Duality Principle of Semi-Pareto). *The BTG of a Semi-Pareto preference is symmetric with respect to its middle axis. That means, in a* level v *of the BTG, there are exactly as many nodes as in* level $\max(P) - v$.

Proof. W.l.o.g. consider $P := P_1 \otimes P_2$, P_i WOPs. For each node $a = (a_1, a_2)$ in level v we can find a node $\bar{a} = (\bar{a}_1, \bar{a}_2)$ with $\bar{a}_k := \max(P_k) - a_k$. The level of $\bar{a}$ is $level_{P_1}(\bar{a}_1) = \max(P) - level_P(a)$. Since for each node a in level v a node $\bar{a}$ exists, the lemma is proven. □

3.5   BTG for Rank

A Rank preference uses a number of score functions as input and merges their (weighted) function values to an overall score. Similar to Prioritization, Rank forms a weak oder if it only consists of weak order score preferences, i.e., the BTG follows the structure of a chain.

Theorem 8 (Level Function for Rank). *A Rank preference is a weak order preference with the following level function:*

$$level_{rank_{F,d}} := \left\lceil \frac{F(f_1(x_1), \dots, f_m(x_m))}{d} \right\rceil$$

Proof. Using the mentioned level function in Definition 5 yields Eq. 2. □

4   Analysis of BTGs with Trivial SV-Semantics

Originally, in [2] all preferences were defined with *trivial* instead of *regular* SV-semantics, as described in Sect. 2. That means, the relation $x \cong_P y$ is substituted by $x = y$. Hence, one value may be *better than, equal to, worse than,* or *incomparable* to another value, but *not* substitutable.

For example, for categorical base preferences like LAYERED$_m$ using trivial SV-semantics, all different values in the same layer are *incomparable* to each other. Hence, the level value of this layer alone is not sufficient to determine domination. Considering only base preferences, this makes no semantical difference. The difference occurs when such preferences are combined to complex preferences, e.g., a Pareto preference.

In this section we will embed base preferences with trivial SV-semantics into lattice structures. For this we replace the single integer level values to determine domination by *two integers* which model the same order. This allows us to use efficient lattice based algorithms, especially when combining base preferences to Pareto preferences.

4.1   Numerical Base Preferences with Trivial SV-Semantics

In the case of numerical base preferences, we only consider $BETWEEN_d$ with trivial SV-semantics, because all other numerical base preferences can be derived from it.

Theorem 9. *Consider* $P := BETWEEN_d(A, [low, up])$ *and* P' *derived from* P *by replacing regular by trivial SV-semantics. We map* $x \in dom(A)$ *to the integer combination* (l_1, l_2) *in* P' *as follows:*

$$x \rightarrow (l_1, l_2) = \begin{cases} (level_P(x), level_P(x) - 1) \Leftrightarrow up < x \\ (level_P(x) - 1, level_P(x)) \Leftrightarrow x < low \end{cases}$$

Then, P' models the same order w.r.t. $dom(A)$ as P, but distinguishes between values lower and values higher than the interval borders.

Proof. Consider a value v mapped to (v_1, v_2), and a value w mapped to (w_1, w_2). The following cases may occur:

- $level_P(v) = level_P(w) + 1$:
  - $v < low \wedge w < low \Rightarrow (w_1 = v_1 + 1) \wedge (w_2 = v_2 + 1)$
  - $v < low \wedge up < w \Rightarrow (w_1 = v_1 + 2) \wedge (w_2 = v_2)$
- $level_P(v) = level_P(w)$:
  - $v < low \wedge w < low \Rightarrow (v_1, v_2) = (w_1, w_2) \Rightarrow v \cong_{P'} w$
  - $v < low \wedge up < w \Rightarrow \begin{matrix} (v_1, v_2) = (level_P(v), level_P(v) + 1) \\ (w_1, w_2) = (level_P(v) + 1, level_P(v)) \end{matrix} \Rightarrow v \sim_{P'} w$

All other possible cases can be derived from those above. So the level combination assigned to domain values fulfills the specification of the preference. $\square$

From a technical point of view, two WOPs are connected and used to model a strict partial order. A "virtual" Pareto preference is constructed by the numerical base preference.

Lemma 4. *The number of nodes in the BTG* P' *defined by a* $BETWEEN_d$ *preference* P *by replacing regular by trivial SV-semantics is given by:*

$$(\max(level_P(\min(dom(A))), level_P(\max(dom(A)))) + 1)^2 = (\max(P) + 1)^2$$

Proof. Looking at the computation of level combinations for values to be rated, the BTG that is constructed is identical to one for a Pareto preference containing two WOPs with maximum level values of $\max(P) + 1$.

Lemma 5. *Consider a preference* P' *which is defined as a* $BETWEEN_d$ *preference* P *with trivial instead of regular SV-semantics. The number of used nodes (i.e. the number of nodes that can be matched by values evaluated by* P'*) in the* $BTG_{P'}$ *is given by* $2 * \max(P) + 1$.

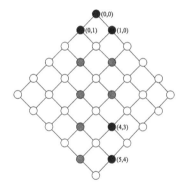

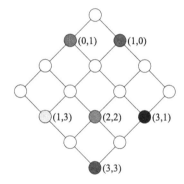

Fig. 5. BTG for AROUND$_d$ with trivial SV-semantics.

Fig. 6. BTG for LAYERED$_m$ with trivial SV-semantics.

Proof. The node $(0,0)$ is used for perfect matches. Other nodes used have level combinations of $(x, x+1)$ or $(x+1, x)$. The minimum value for x is 1, the maximum is $\max(P)$, leading to $2 * \max(P) + 1$ values in use.   □

Example 5. Let $P := \mathrm{AROUND}_5(A, 50)$ with $\mathrm{dom}(A) = \{45, 50, 55, 70, 75\}$. Then $\max(P) = 5$. We derive P' with trivial SV-semantics and create level pair mappings: A perfect value of 50 is mapped to $(0,0)$, 45 and 55 (with level 1) are mapped to incomparable value combinations $(0,1)$ and $(1,0)$, respectively.

Figure 5 shows $\mathrm{BTG}_{P'}$. The black nodes have tuples belonging to them, the gray nodes represent valid integers for l_1 and l_2, while the white nodes are unused dummy nodes given by the graph structure. The height/width of $BTG_{P'}$ is $\max(P)+1 = 6$. The number of nodes is $(5+1)^2 = 36$ from which $2 \cdot 5 + 1 = 11$ might be used.

4.2   Categorical Base Preferences with Trivial SV-Semantics

Using trivial SV-semantics in $\mathrm{LAYERED}_m(A, \{L_1, \ldots, L_m\})$, all values in one of the L_i are incomparable.

Theorem 10. *Consider* $P := \mathrm{LAYERED}_m(A, \{L_1, \ldots, L_m\})$ *and* P' *derived from* P *by replacing regular with trivial SV-semantics. Each value in* $\mathrm{dom}(A)$ *is mapped to a pair of integer level values.*

The elements of the L_i *are labeled with indexes:* $L_i := \{l_{i,1}, l_{i,2}, \ldots, l_{i,|L_i|}\}$. *Every element of* L_i *has to get a unique index value. Then, the level combination for each* $l_{i,j}$ *can be found with the following formula:*

$$l_{i,j} \to \left(\left| \bigcup_{x=1}^{i-1} L_x \right| - i + j, \left| \bigcup_{x=1}^{i} L_x \right| + 1 - (i+j) + |\{x \ : \ x \le i \land |L_x| = 1 \land |L_{x-1}| = 1\}| \right)$$

Proof. Consider three categorical values $l_{i,j}, l_{i,k}, l_{i+1,q} \in \mathrm{dom}(A)$ with $j < k$. A value $l_{x,y}$ is mapped to $(l_{x,y}[0], l_{x,y}[1])$. We have to prove that P' constructs the same order as P on elements of different layers and renders elements of the same layer indifferent. For readability, we will abbreviate $\left| \bigcup_{x=1}^{i-1} L_x \right|$ with s and $|\{x \ : \ x \le i \land |L_x| = 1 \land |L_{x-1}| = 1\}|$ with $t(i)$.

- $l_{i,j} \sim_{P'} l_{i,k}$:
  - $l_{i,j}[0] - l_{i,k}[0] = (s - i + j) - (s - i + k) = j - k \Rightarrow l_{i,j}[0] < l_{i,k}[0]$
  - $l_{i,j}[1] - l_{i,k}[1] = (s + |L_i| + 1 - (i+j) + t(i)) - (s + |L_i| + 1 - (i+k) + t(i)) = -j + k \Rightarrow l_{i,j}[0] > l_{i,k}[0]$

  With $l_{i,j}[0] < l_{i,k}[0] \wedge l_{i,j}[0] > l_{i,k}[0]$ it follows that $l_{i,j} \sim_{P'} l_{i,k}$.

- $l_{i+1,q} <_{P'} l_{i,j}$:
  - $l_{i,j}[0] \le l_{i+1,q}[0] \Leftrightarrow j \le |L_i| - 1 + q$
    This always holds as $j \le |L_i| \wedge (-1 + q) \ge 0$
    $\Rightarrow j = |L_i| - 1 - q \Leftrightarrow j = |L_i| \wedge q = 1$ and $j < |L_i| - 1 - q \Leftrightarrow j < |L_i| \vee q > 1$

- $l_{i,j}[1] \le l_{i+1,q}[1] \Leftrightarrow -j + t(i) \le |L_{i+1}| - 1 - q + t(i+1)$
  - case 1: $|L_i| = 1 \wedge |L_{i+1}| = 1 \Leftrightarrow t(i+1) = t(i) + 1 \Rightarrow j = 1 \wedge q = 1$
    $\Rightarrow -j + t(i) \le |L_{i+1}| - 1 - q + t(i) + 1$
  - case 2: $|L_i| > 1 \vee |L_{i+1}| > 1 \Leftrightarrow t(i+1) = t(i)$
    $\Rightarrow -j + t(i) \le |L_{i+1}| - 1 - q + t(i)$
    For $j = 1 \wedge q = |L_{i+1}|$, both sides are equal. As $(j \ge 1) \wedge (q - |L_{i+1}| \le 0)$, the inequation holds in all other cases, too.

To sum up the preceding points, we showed that $l_{i+1,q} <_{P'} l_{i,j}$ always holds:

$$
\begin{aligned}
j = 1 \quad \wedge \quad q = |L_{i+1}| &\Rightarrow l_{i,j}[0] < l_{i+1,q}[0] \wedge l_{i,j}[1] = l_{i+1,q}[1] \\
1 < j < |L_i| \wedge 1 < q < |L_{i+1}| &\Rightarrow l_{i,j}[0] < l_{i+1,q}[0] \wedge l_{i,j}[1] < l_{i+1,q}[1] \\
j = |L_i| \quad \wedge \quad q = 1 &\Rightarrow l_{i,j}[0] = l_{i+1,q}[0] \wedge l_{i,j}[1] < l_{i+1,q}[1]
\end{aligned}
$$

As we can see, all elements of the same layer are indifferent and better than all elements of (w.r.t. their indexes) higher layers. □

Example 6. Consider the color preference in Fig. 1. We derive a preference P' with the same sets but trivial instead of regular SV-semantics. Let dom(*color*) = {yellow, red, purple, blue, brown, black}. Figure 6 shows the BTG for P'. Nodes with invalid level combinations are white, nodes with other colors are labeled with the level combination the color is assigned to.

5   Summary and Outlook

In this paper we discussed the graphical representation of preference orders. In particular we considered WOPs, where the BTG forms a chain, and complex preferences like Pareto with its typical lattice structure. In addition we presented a method to embed base preferences with trivial substitutable values semantics into a lattice. For each kind of BTG we provided a set of measures for a detailed description of the graphical structure. The careful analysis of the underlying "better-than graph" allows the usage of typical lattice based algorithms. Furthermore, these measures can then be used to design new algorithms for preference evaluation. Nevertheless, embedding general strict partial orders and a combination of WOPs and general preferences is still an open problem. This is a challenging task and therefore remains for future work.

References

1. Stefanidis, K., Koutrika, G., Pitoura, E.: A survey on representation, composition and application of preferences in database systems. ACM Trans. Database Syst. **36**(4), 19:1–19:45 (2011)
2. Kießling, W.: Foundations of preferences in database systems. In: VLDB 2002: Proceedings of the 28th International Conference on Very Large Data Bases, pp. 311–322. VLDB Endowment, Hong Kong, China (2002)
3. Chomicki, J.: Preference formulas in relational queries. TODS 2003: ACM Trans. Database Syst. **28**, 427–466 (2003)
4. Arvanitis, A., Koutrika, G.: Towards preference-aware relational databases. In: ICDE 2012: Proceedings of the 28th International Conference on Data Engineering, Washington, DC, USA, April 2012
5. Davey, B.A., Priestley, H.A.: Introduction to Lattices and Order, 2nd edn. Cambridge University Press, Cambridge (2002)
6. Börzsönyi, S., Kossmann, D., Stocker, K.: The skyline operator. In: Proceedings of ICDE 2001, pp. 421–430. IEEE, Washington, DC, USA (2001)
7. Morse, M., Patel, J.M., Jagadish, H.V.: Efficient skyline computation over low-cardinality domains. In: Proceedings of VLDB 2007, pp. 267–278. VLDB (2007)
8. Preisinger, T., Kießling, W.: The hexagon algorithm for evaluating pareto preference queries. In: Proceedings of the 3rd Multidisciplinary Workshop on Advances in Preference Handling (2007)
9. Endres, M., Kießling, W.: High parallel skyline computation over low-cardinality domains. In: Manolopoulos, Y., Trajcevski, G., Kon-Popovska, M. (eds.) ADBIS 2014. LNCS, vol. 8716, pp. 97–111. Springer, Heidelberg (2014)
10. Endres, M., Kießling, W.: Semi-skyline optimization of constrained skyline queries. In: ADC 2011: Proceedings of the 22nd Australasian Database Conference, vol. 115, pp. 7–16. ACS (2011)
11. Lee, J., Hwang, S w.: BSkyTree: scalable skyline computation using a balanced pivot selection. In: Proceedings of the 13th International Conference on Extending Database Technology, EDBT 2010, pp. 195–206. ACM, New York, NY, USA (2010)
12. Han, H., Jung, H., Eom, H., Yeom, H.Y.: An efficient skyline framework for matchmaking applications. J. Netw. Comput. Appl. **34**(1), 102–115 (2011)
13. Endres, M., Roocks, P., Kießling, W.: Scalagon: an efficient skyline algorithm for all seasons. In: Renz, M., Shahabi, C., Zhou, X., Chemma, M.A. (eds.) DASFAA 2015. LNCS, vol. 9050, pp. 292–308. Springer, Heidelberg (2015)
14. Fishburn, P.C.: Intransitive indifference in preference theory: a survey. Oper. Res. **18**(2), 207–228 (1970)
15. Kießling, W., Endres, M., Wenzel, F.: The preference SQL system - an overview. Bull. Tech. Commitee Data Eng. IEEE Comput. Soc. **34**(2), 11–18 (2011)
16. Kießling, W.: Preference queries with SV-semantics. In: Proceedings of COMAD 2005, pp. 15–26. Computer Society of India, Goa, India (2005)
17. Endres, M., Roocks, P., Kießling, W.: Algebraic optimization of grouped preference queries. In: Proceedings of IDEAS 2014, pp. 247–256. ACM, New York, NY, USA (2014)
18. Preisinger, T., Kießling, W., Endres, M.: The BNL++ algorithm for evaluating pareto preference queries. In: Proceedings of the 2nd Multidisciplinary Workshop on Advances in Preference Handling, pp. 114–121 (2006)

User Requirements
and Database Evolution

Two Phase User Driven Schema Matching

Nick Bozovic[(⊠)] and Vasilis Vassalos

Department of Informatics, Athens University of Economics and Business,
Athens, Greece
{nbozovic, vassalos}@aueb.gr

Abstract. In recent years it has become apparent that schema matching is a labor intensive process that is very costly in resources; this has led to the development of various automated tools to substitute the human experts involved in it. To this end we propose two new ideas. The first is the separation of matching techniques into strong and weak ones, in what we call two phase schema matching. The second is using information a human expert can provide to the system during the process of schema matching, that is used to determine how to combine the various matching techniques. A system encompassing both our ideas is easily tunable and allows the human expert to become part of the matching process and help the system choose the best techniques to use. In extensive experiments we demonstrate that this approach is better than contemporary state of the art systems in relational databases. We also demonstrate that single purpose (or niche) matchers can be helpful in such a system where the system can opt to use them if appropriate.

Keywords: Data integration · Schema matching · Human interaction · Data mining · Artificial intelligence

1 Introduction

The problem of schema matching is a very complex and difficult one. Over the years many methodologies have been proposed by various teams. Surveys [1, 3, 13] show the focus is to use a computer to automate the process using composite systems; that is, more than one method is used to produce possible matches, and the results are then combined to produce the final matches. While these systems boost matching accuracy, an important question remains: Can matchers that are less reliable than others "unbalance" a system?

In this paper we propose two ways to address that. The first is what we dub "two phase schema matching", that aims to allow the simultaneous use of a wide array of matchers regardless of their performance in the specific schemas to be matched. We achieve this by separating the available matching techniques into "strong" and "weak" ones, and one using user feedback. The intuition is to use the right technique at the right time to maximize its usefulness while minimize the destabilizing effects it may have on the system.

What we call "strong" techniques are those that have consistently produced good matches and are less likely to be affected by differences in domains, implementation details and/or the size and quality of instance data. All other techniques are classified as "weak": they can be helpful if used, but their credibility is less certain.

© Springer International Publishing Switzerland 2015
T. Morzy et al. (Eds.): ADBIS 2015, LNCS 9282, pp. 49–62, 2015.
DOI: 10.1007/978-3-319-23135-8_4

The second is using a limited and easy-to-get amount of information from the human expert what will evaluate the results. This user interaction strategy is based on the fact that according to recent research [4] and our own experience, there is no safe way to perform schema matching tasks in a fully automated way, as there can be no guarantee of 100 % confidence and accuracy. While most recent work does recognize that machine based methods cannot solve the problem with 100 % precision there is little work published in the benefits of user-computer interaction and the impact it can have on solving the problem. Even if there was such a system, it would, in the end, need a human expert to evaluate the matches and authorize the mapping process. We therefore put forth the idea that since every outcome of an automated schema matching process must and will be evaluated by a human expert, it's to our advantage to bring some of that evaluation earlier in time. We use this evaluation to discover and use the knowledge hiding in that mid-step, to help determine the best way to match the remaining attributes.

Both approaches are aimed at large schema matching projects, for example the matching of two database schemata during a company merging or a company-wide system integration. The system is schema structure agnostic and thus able to find matches between structurally different schemata (for example between a relational database and a collection of flat files or an XML schema). In that sense where the term "schema" is used it should be understood not only as a relational schema but as data schema in general.

An overview of the two methods follows in Sects. 2 and 3, in Sect. 4 we perform experiments in both our systems and to currently available systems and comment on the outcome. Related work is presented in Sect. 5, and in Sect. 6 we present our conclusion and we ask some interesting questions for future consideration.

2   The Strong Vs. Weak Approach

In this approach, which we called ASID (Another Schema Integration Dashboard), matching is performed in two phases: one where only strong techniques are used and one where all matchers are used. An overview of this is presented in [Fig. 1]. The first-phase combiner uses their results to produce plausible matches. If there are unmatched attributes left or the score of a match is too low, these attributes are passed to the second-phase matching, where the "weak" techniques are also used. In the second phase, the results from all techniques are combined. This arrangement ensures that matches are not "unbalanced" by weak matching techniques, while the system can benefit from the extra matching ability they provide.

2.1   Matching Methods

In our system we use a mix both established and new matchers, that leverage information both on semantic and on data instance level. Multiple matchers, especially "niche" matchers that need specific information to provide good results, can be of great help but can also unbalance a system. In our setup with both the self-mute switch and

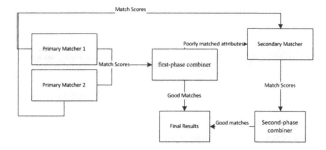

Fig. 1. Overview of the ASID system

the variable weights (described in the next section) we demonstrate that these methods do not hurt the results and can under some conditions improve them.

Name Matching. Name matching is a string matching technique between the names of the attributes, used extensively in schema matching. We use the Jaro metric for this task as recent research [14] has shown that it is one of the best algorithms for name matching. The Jaro metric is based on the number and order of the common characters between two strings.

Given strings $s = a_1 \ldots a_k$ and $t = b_1 \ldots b_L$, define a character a_i in s to be *common with t* if there is a $b_j = a_i$ in t such that $i - H \leq j \leq i + H$ where $H = \frac{\min(|s|,|t|)}{2}$.

Let $s' = a'_1 \ldots a'_{k'}$ be the characters in s which are common with t (in the same order they appear in s) and $t' = b'_1 \ldots b'_{L'}$ be defined the same way for t, Now define a transposition *for s', t'* to be a position i such that $a'_i \neq b'_i$. Let $T_{s',t'}$ be half the number of transpositions for s' and t'. The Jaro similarity metric for s and t is

$$Jaro(s,t) = \frac{1}{3}\left(\frac{|s'|}{|s|} + \frac{|t'|}{|t|} + \frac{|s'| - T_{s',t'}}{|s|}\right)$$

It is out of scope to further analyse the algorithm used. For more information, the reader could consult [14]. In our system, scores range in [0, 1].

Attribute Description Matching. This method tries to exploit simple forms of documentation that often exist in relational systems used in organizations. In most RDBMS deployments, each attribute has a description written in natural language that describes its contents, usually in a sentence or two.

Our matching method is a simple one: for each attribute of the source schema, we create a text corpus made of its description string and the description strings of all the attributes of the target schema that are possible matches (Fig. 2). Then we use TF/IDF weighting to compute vector similarity between the first member of the corpus and the remaining ones: the more similar the description, the higher the score. This method ensures that words appearing many times in the corpus, like "table", are not overly important to the computation of similarity, ensuring that the truly important words dominate the result.

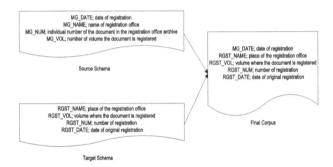

Fig. 2. Creating the corpus

Naïve Bayes Classifier. All available data from one schema are fed into a simple Naïve Bayes classifier; after the learning phase is complete, the sample data of the target attribute are classified. The total score for all attributes is normalized in [0,1] to be compliant in range to the previously computed score. The formula by which the normalization is achieved is a compromise between robustness and ease of implementation.

The transformation formula used is $s'_i = \frac{s_i - s_{\min}}{s_{\max} - s_{\min}}$ where s_i is the individual score of an attribute match and $s_{\min}$ and $s_{\max}$ respectively are the global minimum and maximum scores of said attribute. The intuition behind this is that the best score for an attribute will be transformed to 1 while the lowest will be 0. It is important to note that this procedure gives a boost to the best score that may lead to the production of false positives. When faced with the dilemma of producing some extra false positives or risk losing some good matches, we opted for the first choice, the rationale being that, in large-scale schema matching, a false positive is easier for a human to dismiss than it is for her to discover a missed true match.

TF-IDF Matching. This matcher is inspired by the WHIRL system, which "extends relational databases to reason about the similarity of text-valued fields using information-retrieval technology" [6]. We use only the similarity function used in WHIRL to create a matcher for ASID.

Specifically, data is inserted in the matcher and a collection of documents is created where each document consists of all the instances of any available target attribute. A "source" document is created from the data available for the source attribute. We use TF/IDF weighting to compute vector similarity scores between the source document in the collection and the target ones. The more similar the source document is to the corpus of the data, the higher the score. The last step is to normalize the results using the formula described above.

Datatype Matching. This matcher uses similarity between datatypes to determine similarity between attributes. The matcher deduces the datatypes from the data instances and not from the data definition. This is done because in most systems that are used for long periods of time the datatypes used differ from the datatypes defined in the documentation. The actual datatype of an attribute as determined by the system is the datatype that can hold every and all instances of the attribute in question. These datatypes are then used to compute similarity between attributes as described in the data similarity matrix presented in Table 1.

Table 1. Datatype similarity matrix

Datatype	String	Number	Integer	Currency	Date
String	1.0	0.6	0.6	0.6	0.4
Number	0.6	1.0	0.4	0.6	0
Integer	0.6	0.4	1.0	0.4	0
Currency	0.6	0.6	0.4	1.0	0
Date	0.4	0	0	0	1.0

Instance Pattern Matching. This matcher extracts patterns from the data instances. It aims to find patterns that can be hiding in the data of large schemas, patterns like passport numbers, license plates, VAT numbers etc. Similarity between these patterns is calculated with the edit distance method to penalize changes in stronger way than the more lenient Jaro metric. The patterns found are comprised of the characters found in all data instances and the places of the changing letters ("‡") and/or numbers ("№"). For example the IBAN Number of UK account would create a pattern of [GB№ №‡‡‡‡№№№№№№№№№№№№№№№№].

An alternate implementation of this would be to express the patterns of the source attributes in the form of a regular expression and award a score on those matches were the target attributes validate the regular expressions. For example from a source attribute containing IBAN numbers we can create the following regular expression:

$$[a - zA - Z]\{2\}[0 - 9]\{2\}[a - zA - Z0 - 9]\{4\}[0 - 9]\{7\}([a - zA - Z0 - 9]?)\{0, 16\}$$

with any target attribute containing IBAN numbers must adhere to. This method is more rigid than the first, since, contrary to the above regex that validates all known IBANs, a regex derived from UK IBANs will not validate German IBANS. In light of that for the implementation of our system we opted for the first approach.

3 The Human-in-the-Loop Approach

We can extend our two phase matcher system, which we will henceforth refer to it as ASID + , with a module that implements our proposed solution on how to increase the quality of the matching by using small pieces of information in the form of user feedback. We then proceeded and evaluated our approach and established that it can make a big difference in the performance of an automated schema matching tool. The design of the system is presented in Fig. 3.

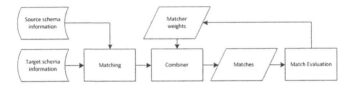

Fig. 3. 2-Phase matching with user feedback

3.1   The Human in the Loop Match Combiner

The human in the loop approach uses a simple voting match combiner to compute an overall score for each possible match. The sum of all matchers' scores is divided by the sum of their weights. If after the total evaluation a match is deemed probable (with a score greater than 0.5) the evaluator produces a "good match," presenting it to the user and removing the matched couple from the source and target schemata.

User Interaction and Weight Redistribution. As already suggested, combining the individual matchers' scores cannot involve training of the system since schema matching is more often than not a one-time task. Proposed solutions like a decision tree combiner fail to leverage the knowledge of the human expert, when apart from the schemata to be matched, this may be the only other available source information. In most systems interaction between the user and the system happens before (providing the data) or after (evaluating the results) the matching process. We propose an interactive matching method, based on our two phase idea, which aims to get the best out of both human and computer worlds.

After the system produces matches the user is asked to confirm or not the correctness of limited number of results. This feedback is then used to evaluate the performance of each technique and adjust the weights of each in our prediction combiner. The weights are initially set to values that are found to be good starting points but can be set to any number without affecting the final outcome of the system as they given enough iterations, will gradually shift to optimal values. That can even be as extreme as being zero to all but one technique, making the system work as a single matcher (Fig. 4).

The matches presented to the user are (pseudo)randomly selected from a pool of matches hovering just above or below a cutoff confidence point (in our experiments 0,5). The user is asked if a match is correct or not, as requiring them to manually search, construct and submit a correct match is tedious. When the system produces a true match the individual techniques that are correct increase their weight. If the match

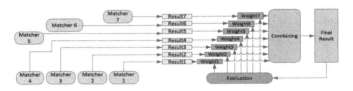

Fig. 4. User Interaction and weight redistribution

was wrong those that were wrong have their weight reduced. Both adjustment are proportional to their respective confidence score. The algorithm of the operation and an example of running it once is shown below (Fig. 5).

```
foreach (match in selectedmatches)
        {
        get user_responce
        If (user_responce) is true then multiplier=1
        else multiplier=-1
        foreach (matcher in matchersused)
                {
                if matcher.score >0.5
                then
                Matcher.weight= Matcher.weight+
                (Matcher.weight*matcher.score*multiplier/10)
                }
        }
```

Technique	1	2	3	4	5	6	7
Confidence	0,92	0,87	0,45	0,26	0,78	0,56	0,89
Initial Weight	50	50	50	50	50	50	50
New weight if correct	54,6	54,35	52,25	51,3	53,9	52,8	54,45
New weight if wrong	45,4	45,65	47,75	48,7	46,1	47,2	45,55

Fig. 5. Weight redistribution example

After the all random matches are evaluated by the user, the matching is repeated with the new values. It is of note that the weight of each technique has no meaning other than distinguishing the proportional strength of that one over the others.

3.2   Over-Fitting as a (Rare) Problem

During our experiments it became apparent that random sampling can in edge cases lead to over fitting. For example a source that was incrementally developed, had parts of it documented in English and parts of it in transliterated Greek. The documentation matcher did exceptionally well in the parts that were documented in English (which had corresponding documentation in the target schema) but underperform in the transliterated parts. When the matches presented to the user were all from the same type of documentation, the matcher was penalized or rewarded far beyond the desired point. For this every matcher has a self-mute switch that activates when it is no longer reliable (i.e. multiple documentation languages), and explicitly excludes itself from the process.

3.3 Confidence in the Human Expert

For the system to work the human expert must provide correct answers before the second matching step. We made certain choices and assumptions when we designed the system to interact with a single expert user. We designed and implemented the feedback phase aiming to minimize input errors by asking questions that can only be answered with a simple Yes/No/Ignore. We opted out of the more elaborate "Select and score the correct match for attribute x" as this is much more prone to errors as suggested in [25]. A central point of our design is that the user is just validating the results at a time earlier they usually have to, as a human expert would ultimately need to evaluate the results and make final decisions. With this in mind and the simple nature of the questions, his feedback is considered correct.

4   Experiments and Evaluation of the System

In this section we evaluate our two approaches against each other and against state of the art systems. We conduct experiments with various datasets and demonstrate the robustness of the two phase schema matching and the added value of human user interaction to the matching process. We also display that our novel niche matchers can be used to boost the systems performance.

4.1 Overview

Three distinct and very different schema collections were used in the experiments. The aim was to see the system perform matches for both small ($\sim$ 30 attributes), medium ($\sim$ 120 attributes) schemas and large schemas derived from open source projects.

For the small schema test we used part of the data used in [2] to evaluate the system.

For the medium schema experiments, we used real schemas from the Greek cadastre where different schemas, developed independently but address the same needs. With up to 120 attributes, they are representative of schemas used in medium applications. The large schemata were procured by two open source shopping cart solutions [15, 16] that are both large and complicated.

The systems we compare ASID + to are the Harmony matcher [20] as implemented in OpenII [21], and COMA ++ v3 community edition [22] (a derivative of [8–10, 17]). As ASID + outputs only the best match available we constrained Harmony to do the same and hence only evaluated the best, i.e. the one with the highest score, match produced by it. If we were to evaluate every match proposed by harmony above a reasonable threshold the number of false positives (and hence the number of pairs a human experts should have to evaluate and discard) would greatly increase. We also set Harmony's threshold to 0.5 to eliminate matches that would probably be false positives.

As our system is a self-tuning one we run all experiments using the default settings (For COMA ++ that meant using the $AllContextInstW dataflow with no other tuning).

In all experiments we ignored matches of attributes that exist only as non-real world unique identifiers as these depend on the design principles used and are not transferred between schemata.

For the rest of the paper we named our system using a priori knowledge the performance of each matcher ASID and it comes in two flavors, one with 4 matchers that excludes instance pattern matching and datatype matching and one that has all 6 matchers. ASID + is the name of our human in the loop implementation that also has the complete set of matchers. Also of note is that in these experiments ASID + was configured to only use a single feedback phase.

4.2 Small Schema Tests

The "small schema" dataset was kindly provided by AnHai Doan and is the relational dataset used in [2]. Table 2 has information on the schemas used).

Table 2. Small Schema information

Schema No.	Total Attributes	Data Instances
1	22	8,45 KB
2	20	4,03 KB
3	27	4,48 KB
4	28	6,02 KB

The correspondences between the attributes were sometimes, but not often, complex (rather than one-on-one). For creating the schemata in OpenII we used its own SQL DDL importer on the source schemata DDL scripts. Figure 6 summarizes the results of our evaluation and shows how our approach performs better in small schemas.

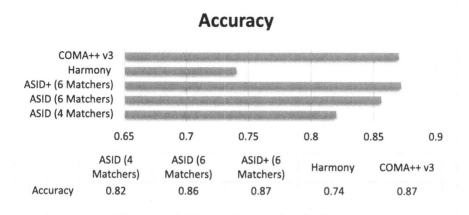

	ASID (4 Matchers)	ASID (6 Matchers)	ASID+ (6 Matchers)	Harmony	COMA++ v3
Accuracy	0.82	0.86	0.87	0.74	0.87

Fig. 6. Small schemas Accuracy (%)

Additionally to improving the overall number of correct matches there was a dramatic drop in false positives with almost all proposed matches being true positives. The experiments display the robustness of two phase method and that using feedback from a human expert can improve the system's performance.

4.3   Medium Schema Tests

The experiments were conducted using schemas created for the Greek cadastre. The system was supplied with the distinct values of more than 30.000 tuples in each case. All schemas and data were real (Table 3).

Table 3. Medium schema information

Schema No.	Total Attributes	Data Instances
1	108	2,39 MB
2	113	2,41 MB
3	92	1,29 MB

The results (Figs. 7 and 8) illustrate that, despite the increased complexity of real world schemas, accuracy remains high at ∼ 85 % The reduction in accuracy in this experiment for ASID + and ASID compared to their performance in the small schema experiments shows that user interaction can make the system more robust with regard to increased schema size.

For this comparison we decided against the use of the medium schemas because of them having documentation and attribute/table names in both the Greek (transliterated or not) and English language. Therefore the matching is heavily dependent on instance based matchers which the chosen open source schema matching systems lack, something that would skew these experiments would in favor of our system.

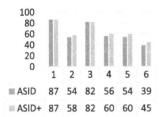

Fig. 7. Number of true positives on medium schemas (left)

Fig. 8. Medium schemas ASID accuracy (%) (right)

4.4   Large Schema Comparison

For the large schemas we used the 6-matcher, human in the loop ASID + while for Harmony we used the DDL importer to import the schemas but both DDL scripts of the e-shops had some non-standard datatypes substituted by their more common analogues.

The results in Fig. 9 suggest that ASID + is able to outperform Harmony by a margin of 20 per cent point. It is notable that it manages that by utilizing all of its matchers despite the fact that data instances available are of very low quality, confirming our assumption that a small user input can go a long way in increasing the robustness of our system. The same observation holds to a lesser extent for COMA + + where the difference is 7 per cent point.

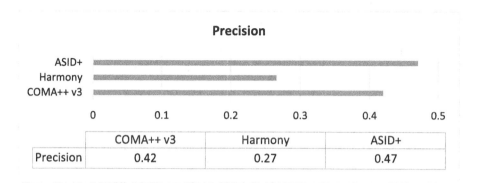

	COMA++ v3	Harmony	ASID+
Precision	0.42	0.27	0.47

Fig. 9. Large schemas systems comparison (Precision %)

4.5   Conclusions on Experiments

The experiments presented strongly suggest our approach can offer better results to automated schema matching. The experiments demonstrate both the robustness of two phase schema matching and the positive effect a user assisted prediction combining method can have to the schema matching process.

Moreover, the evaluation of the results of ASID + , irrespective of the status of the niche matchers, demonstrates that our method, by taking into account the user input to assign weights to the various matching methods, helps improve the results while at the same time allowing "niche" matching techniques to be used without results degradation.

5   Related Work

The systems proposed and/or build through the years can be divided for the purposes of this work in two major categories those that somehow enable a human in the matching process and those who do not. Of the later there are many works worth mentioning. Like iMap [11] that while able to identify complex matches (i.e. address = concat(city, state)), a feature rarely seen in schema matching tools, achieves 43-92 % success in matching attributes a range showing it may not be as stable as for industrial schema matching. Similarity Flooding [12] is based on the eponymous graph matching algorithm. The strength of the algorithm (and also its weakness) is the lack of knowledge it has for the attributes being matched. This had a side effect of producing out of context matches that need to be filtered out. The basic idea behind COMA ++ [9, 10] is to create a true composite system. This is a generic system, designed to be adaptable to many matching problems, achieving notable results not only in data integration and for that reason is one of the systems we tested ASID + against. The LSD system [5] is a demonstrator of a multi-strategy learning approach in schema matching. Results show it achieves 71–92 % accuracy but it demands training on target schemata, something that often is not possible. MKB [2] displays high accuracy, either autonomously or as an additional matcher but only when the existence of past mappings is available, as it requires a large amount of previously matched schemata, something not likely to be available in industrial RDBMS deployments. U-MAP [7] is again a technology demonstrator system for a novel idea: using information extracted from query logs to generate correspondences between the attributes of two different schemas and the mappings between them. YAM [23] is actually a schema matcher generator designed to produce a tailor-made matcher based on a knowledge base, not on human expert knowledge that is by definition more relevant to the domain at hand. Harmony [20] aims to speed up the task of finding correspondences across two data schemas. As implemented on the OpenII project [21] is uses a variety of rather simple matchers and incorporates but its one-to-many visualization solution tends to produce far too many correspondences, in real world schemas, making it counter intuitive.

Then there are the systems that enable human users to aid in the schema matching process. In [26] the authors focus on pay-as-you-go reconciliation in schema matching networks using a probabilistic matching network. In contrast to ours, the system in [26] is focused not on the costly and error prone task of one off schema matching but on improving the outcome of a lengthy, incremental process involving many schemata. As mentioned in [28] the authors build a system that aims to use user interaction to resolve matches; this happens at the start/end of the matching process and involves a large number of ordinary users. In [27] the authors propose the use of crowdsourcing techniques to reduce ambiguity in schema matching. In relation to the previous two works our system involves one human expert during the matching process. ASID does not use the knowledge gained to resolve matches or to retrain its matching techniques, but it uses it to decide which are (or are not) the best matchers in its disposal for a given problem - although we do plan to experiment with using the user feedback to also resolve matches in future work. There are also a number of other schema matching systems and techniques of note e.g., [7, 8, 19] designed to address different aspects of the schema matching problem.

6 Conclusions and Future Work

Our approach is based on the assumption that schema matching in business environments is under most scenarios a one-off task and hence learning-based approaches are of limited use. Based on this we proposed a two-phase method for building composite systems. We examined both the use of a priori knowledge in deciding what individual methods are most important and use of limited user input. To demonstrate the effectiveness of our approach we created a prototype system and used it in to match real small, medium and large schemata.

Conducting extensive experiments in a variety of schemata of variable size and domain, we demonstrate a significant improvement over existing open source systems, ranging from a 13 % to 19 % increase in accuracy in small schema matching a significant increase of precision in large schemas from 27 % to 47 %. The evaluation of the results also shows that taking user input into account to assign weights to the various matching methods helps improve the results accuracy by 5 %, while at the same time allowing "niche" matching techniques to be used without the risk of results degradation.

We plan on exploring additional strategies to initiate and use the human computer interaction that is the basis of our method. In particular, we plan to expand ASID + to not only use the feedback as a way to adjust the matcher's weights but also to treat the true positives attribute matches as match constraints that can possibly exclude some matches from even being generated. There is also the interesting path of extracting more information from the human user, instead of simply asking a random set of simple yes/no questions about matches hovering above or below our threshold.

References

1. Rahm, E., Bernstein, P.A.: A survey of approaches to automatic schema matching. VLDB J. **10**(4), 334–350 (2001)
2. Madhavan, J., Bernstein, P., Doan, A., Halevy, A.: Corpus-based schema matching. In: Proceedings of the 21st International Conference on Data Engineering, ICDE 2005, pp. 57–68. IEEE (2005)
3. Shvaiko, P., Euzenat, J.: A survey of schema-based matching approaches. In: Spaccapietra, S. (ed.) Journal on Data Semantics IV. LNCS, vol. 3730, pp. 146–171. Springer, Heidelberg (2005)
4. Peukert, E., Eberius, J., Rahm, E.: A self-configuring schema matching system. In: 2012 IEEE 28th International Conference on Data Engineering (ICDE), pp. 306–317. IEEE (2012)
5. Doan, A., Domingos, P., Halevy, A.: Learning to match the schemas of data sources: A multistrategy approach. Mach. Learn. **50**(3), 279–301 (2003)
6. Cohen, W.W., Hirsh, H.: Joins that generalize: text classification using WHIRL. In: Proceedings of the Fourth International Conference on Knowledge Discovery and Data Mining (KDD), pp. 169–173 (1998)
7. Elmeleegy, H., Lee, J., Rezig, E.K., Ouzzani, M., Elmagarmid, A.: U-MAP: a system for usage-based schema matching and mapping. In: Proceedings of the 2011 ACM SIGMOD International Conference on Management of data, pp. 1287–1290. ACM (2011)
8. Do, H.-H., Rahm, E.: COMA: a system for flexible combination of schema matching approaches. In: Proceedings of the 28th International Conference on Very Large Data Bases, pp. 610–621. VLDB Endowment (2002)
9. Aumueller, D., Do, H.-H., Massmann, S., Rahm, E.: Schema and ontology matching with COMA ++. In: Proceedings of the 2005 ACM SIGMOD International Conference on Management of Data, pp. 906–908. ACM (2005)

10. Massmann, S., Engmann, D., Rahm, E.: COMA ++: Results for the ontology alignment contest OAEI 2006. In: International Workshop on Ontology Matching, Collocated with the 5th ISWC-2006, p. 107. Athens, Georgia, USA (2006)
11. Dhamankar, R., Lee, Y., Doan, A., Halevy, A., Domingos, P.: iMAP: discovering complex semantic matches between database schemas. In: Proceedings of the 2004 ACM SIGMOD International Conference on Management of Data, pp. 383–394. ACM (2004)
12. Melnik, S., Garcia-Molina, H., Rahm, E.: Similarity flooding: a versatile graph matching algorithm and its application to schema matching. In: Proceedings of the 18th International Conference on Data Engineering, 2002, pp. 117–128. IEEE (2002)
13. Bernstein, P.A., Madhavan, J., Rahm, E.: Generic schema matching, ten years later. Proc. VLDB Endowment **4**(11), 695–701 (2011)
14. Bilenko, M., Mooney, R., Cohen, W., Ravikumar, P., Fienberg, S.: Adaptive name matching in information integration. IEEE Intell. Syst. **5**, 16–23 (2003)
15. osCommerce Online Merchant v2.3.3.4. http://www.oscommerce.com/Products
16. CubeCart free, v.5.2.8. http://www.cubecart.com/downloads/
17. Do, H.-H., Rahm, E.: Matching large schemas: approaches and evaluation. Inf. Syst. **32**(6), 857–885 (2007)
18. Mork, P., Rosenthal, A., Seligman, L., Korb, J., Samuel, K.: Integration Workbench: Integrating Schema Integration Tools, The MITRE Corporation, Case #06-0055, May 2006
19. Mork, P., Seligman, L., Rosenthal, A., Korb, J., Wolf, C.: The harmony integration workbench. In: Spaccapietra, S., Pan, J.Z., Thiran, P., Halpin, T., Staab, S., Svatek, V., Shvaiko, P., Roddick, J. (eds.) Journal on Data Semantics XI. LNCS, vol. 5383, pp. 65–93. Springer, Heidelberg (2008)
20. Seligman, L., Mork, P., Halevy, A., Smith, K., Carey, M.J., Chen, K., Wolf, C., Madhavan, J., Kannan, A., Burdick, D.: OpenII: an open source information integration toolkit. In: Proceedings of the 2010 ACM SIGMOD International Conference on Management of data, pp. 1057–1060. ACM (2010)
21. COMA Community Edition, Schema Matching Solution for Data Integration. http://sourceforge.net/projects/coma-ce/
22. Duchateau, F., Coletta, R., Bellahsene, Z., Miller, R.J.: (Not) yet another matcher. In: Proceedings of the 18th ACM Conference on Information and knowledge management, pp. 1537–1540. ACM (2009)
23. Sagi, T., Gal, A.: In schema matching, even experts are human: towards expert sourcing in schema matching. In: 2014 IEEE 30th International Conference on Data Engineering Workshops (ICDEW), pp. 45–49. IEEE (2014)
24. Nguyen, Q.V.H., Nguyen, T.T., Miklós, Z., Aberer, K., Gal, A., Weidlich, M.: Pay-as-you-go reconciliation in schema matching networks. In: 2014 IEEE 30th International Conference on Data Engineering (ICDE), pp. 220–231. IEEE (2014)
25. Zhang, C.J., Chen, L., Jagadish, H.V., Cao, C.C.: Reducing uncertainty of schema matching via crowdsourcing. Proc. VLDB Endowment **6**(9), 757–768 (2013)
26. McCann, R., Shen, W., Doan, A.: Matching schemas in online communities: a web 2.0 approach. In: IEEE 24th International Conference on Data Engineering, 2008 ICDE 2008, pp. 110–119. IEEE (2008)

CoDEL – A Relationally Complete Language for Database Evolution

Kai Herrmann[1]([✉]), Hannes Voigt[1], Andreas Behrend[2], and Wolfgang Lehner[1]

[1] Database Technology Group, Technische Universität Dresden, Dresden, Germany
{kai.herrmann,hannes.voigt,wolfgang.lehner}@tu-dresden.de
[2] Computer Science III, University of Bonn, Bonn, Germany
behrend@cs.uni-bonn.de

Abstract. Software developers adapt to the fast-moving nature of software systems with agile development techniques. However, database developers lack the tools and concepts to keep pace. Data, already existing in a running product, needs to be evolved accordingly, usually by manually written SQL scripts. A promising approach in database research is to use a declarative database evolution language, which couples both schema and data evolution into intuitive operations. Existing database evolution languages focus on usability but did not aim for completeness. However, this is an inevitable prerequisite for reasonable database evolution to avoid complex and error-prone workarounds. We argue that relational completeness is the feasible expressiveness for a database evolution language. Building upon an existing language, we introduce CoDEL. We define its semantic using relational algebra, propose a syntax, and show its relational completeness.

Keywords: Descriptive database evolution · Evolution language · Relational completeness

1 Introduction

Changes in modern software systems are no longer an exception but have become daily business. Following the mantra "Evolution instead of Revolution", agile software development centers the creativity and excellence of people to handle the unpredictably dynamic world of software development [3]. Agile methods are characterized by short development cycles, each with the goal of a shippable product. This provides constant feedback, which helps to establish a customer-oriented development process resulting in products that fit customer's true needs and yield high customer acceptance. It is in the very nature of agile development, that requirement specifications are in perpetual flux. Adjusting the software's design to updated requirements is as daily business as developing new features.

However, a major obstacle in this process are the database systems [2]. Whereas software development tools support developers in the process of designing changes with a comprehensive set of automatized refactoring features, the

© Springer International Publishing Switzerland 2015
T. Morzy et al. (Eds.): ADBIS 2015, LNCS 9282, pp. 63–76, 2015.
DOI: 10.1007/978-3-319-23135-8_5

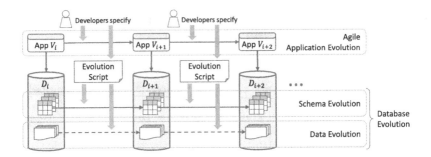

Fig. 1. Database evolution.

evolution of databases is usually realized by manually writing scripts of SQL-DDL and -DML operations. This manual database evolution is expensive and error-prone. Furthermore, many software projects show poor integration of the database developers. According to a survey [1], two third of the pooled software developers perform database-related changes without consulting the responsible database developers, which certainly increases the software developer's productivity but is not necessarily helping the quality of the resulting database.

To keep pace with agile software development, the database systems have to supply software-refactoring-like features. Such database evolution features need to evolve the database schema (schema evolution) and payload data (data evolution) in a single consistent step [15]. Such a database evolution processes as illustrated in Fig. 1. While evolving an application, the application developer specifies the corresponding database evolution with the help of *schema modification operations* (SMOs). In contrast to SQL-DDL and -DML statements, SMOs specify the evolution of the schema and the data in a descriptive, integrated way and ensure that the data is consistently evolved with the schema. SMOs are typically more compact than a script of DDL and DML operations resulting in the same evolution. On the user side, SMOs increase the developer's productivity while dealing with database evolution and reducing the chances of faulty evolution scripts and unintended data loss. On the database system side, SMOs open the opportunity to optimize and reduce the actual data movement involved in an evolution step or even invert evolution steps for database versioning. These benefits are enabled by the use of SMOs instead of DDL/DML.

A set of SMOs forms a *database evolution language* (*DEL*). Naturally, the design of a particular *DEL* determines its expressiveness. A powerful *DEL* lets the user easily specify all necessary evolution steps. In contrast, a weak *DEL* forces the user into more complicated evolution scripts or even to fall back on DDL/DML statements, which renders the *DEL* useless. In principal, a *DEL* should at least cover the power of DDL and DML of an ordinary database system. We argue, that a *DEL* for relational databases should at least be *relationally complete*: For any relational DDL/DML script, there exists a semantically equivalent sequence of SMOs. Relational DDL/DML scripts create, alter, and drop database objects, while conditions and the actual data are specified using expressions from a given DQL. The latter motivates the relational algebra [5] as the natural reference for determining the power of relational DDL and DML.

Given a relational database $D = \{R_1, \ldots, R_n\}$ with tables R_i, a DEL is relationally complete if it can transform D into any other relational database $D' = \{R'_1, \ldots, R'_m\}$ with each R'_i being computable from D with operators from the relational algebra. A minimal language providing relational completeness is $\mathcal{L}_{\min} = \{\text{ADD}\,(\cdot, \cdot), \text{DEL}\,(\cdot)\}$ with

$$\text{ADD}\,(R', \epsilon) \to D \cup \{R' = \epsilon\,(R_1, \ldots, R_n)\}$$
$$\text{DEL}\,(R) \to D \setminus \{R\}$$

The add operation adds a new table R' to the database D based on the given relational algebra expression ϵ. The delete operation removes the specified table R from D. Let $inst\,(\mathcal{L}_{\min})$ be the set of all operation instances of $\mathcal{L}_{\min}$ with valid parameters. Then obviously, a database D can be transformed into any other database D' with a sequence $s \in inst\,(\mathcal{L}_{\min})^+$. Hence, $\mathcal{L}_{\min}$ is relationally complete. From a practical standpoint however, $\mathcal{L}_{\min}$ is not very appealing, because it is rather unintuitive and not oriented on actual evolution steps. However, any other DEL which is as expressive as $\mathcal{L}_{\min}$ is relationally complete as well.

To the best of our knowledge, the most advanced DEL design is PRISM++ [6, 8]. PRISM++ provides SMOs to create, rename, and drop both tables and columns, to divide and combine tables both horizontally and vertically, and to copy tables. The PRISM++ authors claim practical completeness for their powerful DEL, by validating it against evolution histories of several open source projects. Although this evaluation suggests that PRISM++ is sufficient also for other software projects, it does not provide any reliable completeness guarantee. For instance, we do not see an intuitive way to remove all rows from a table A, which also occur in a table B using the PRISM++ DEL, since it does not offer any direct or indirect outer join functionality. Thus, we consider PRISM++ not to be relationally complete. Nevertheless, PRISM++ has an intuitive and field-proven design.

In this paper, we present a relationally *Complete DEL* (CoDEL), building on the set of PRISM++ SMOs to inherit its practical feasibility. However, CoDEL is relationally complete and equally expressive as $\mathcal{L}_{\min}$. Our contributions are:

1. We provide a formal definition of the semantics of all CoDEL operations and propose an SQL-like syntax. With that, CoDEL can serve as a reference language for the formal evaluation of other $DELs$.
2. We show the relational completeness of CoDEL. We show that all operations of the relational algebra – as presented in [5] plus selected extensions – can be expressed in CoDEL and whereby any $\mathcal{L}_{\min}$ expression, as well.
3. We lay the foundation for further research. CoDEL is a DEL, whose SMOs are compact with precisely defined semantics. Hence researchers can tackle their challenges on a per-SMO-level ("Divide and Conquer"). For instance, database versioning requires full invertibility of a database evolution. CoDEL allows to define invertibility locally for each operation, which greatly simplifies such research.

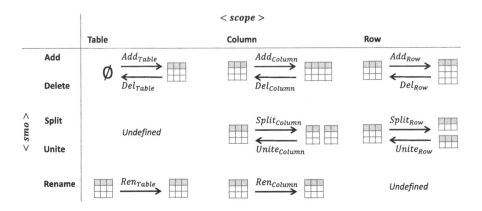

Fig. 2. Structuring of CoDEL.

We define CoDEL in Sect. 2, prove its relational completeness in Sect. 3, discuss related work in Sect. 4, and conclude the paper in Sect. 5.

2   CoDEL

Database evolution changes the schema of a database and/or the already existing data. A *DEL* contains operations to descriptively specify such changes as units, which clearly distinguishes it from SQL-DDL and -DML. PRISM++ limits itself to operations that modify individual tables – no PRISM++ operation accepts more than two tables. This keeps the PRISM++ *DEL* intuitive and easy to learn. CoDEL adopts this principle. However, CoDEL operations systematically cover all possible changes that can be applied to tables. Tables are the fundamental structuring element and the container for primary data in a relational database. Secondary database objects such as views, constraints, functions, stored procedures, indexes, etc. should be considered in database evolution as well. However, in this paper we focus on the evolution of primary data.

CoDEL defines SMOs of the pattern $\langle smo \rangle_{\langle scope \rangle}(\Theta)$, where $\langle smo \rangle$ is the type of operation, $\langle scope \rangle$ is the general database object the operation works on, and Θ is the set of parameters the SMO requires. Figure 2 gives a systematic overview of all SMOs in CoDEL. A relational database table is a two-dimensional structure consisting of columns and rows, hence, SMOs can operate on the level of columns, of rows, or of whole tables. On all three levels there are five basic operations: ADD, DEL, SPLIT, UNITE, and REN. We will now introduce the meaningful operations, as shown in Fig. 2. First, CoDEL has two basic operations to create (ADD_{table}) and drop (DEL_{table}) tables as a whole, similar to their counterparts in a standard DDL. Second, CoDEL has a set of operations to modify a table. Hence, CoDEL offers eight table modification SMOs $\langle smo \rangle_{\langle scope \rangle}$ with $\langle scope \rangle \in \{column, row\}$ and $\langle smo \rangle \in \{\text{ADD}, \text{DEL}, \text{SPLIT}, \text{UNITE}\}$. For instance, DEL_{column} removes a column from a given table and SPLIT_{row} partitions a table horizontally, while SPLIT_{column} partitions it vertically. CoDEL defines no SPLIT or UNITE of whole tables, since these operations are restricted to either column or row scope. Third, CoDEL includes two SMOs to rename a table (REN_{table}) and a column (REN_{column}). The renaming of rows is undefined.

Regarding relational completeness, REN_{column}, REN_{table}, DEL_{column}, and DEL_{row} are not necessary. However, they are very common [9] and included in CoDEL for usability's sake. To summarize, CoDEL is the *DEL* $\mathcal{L}_C$ with:

$$
\mathcal{L}_C = \begin{cases}
\text{ADD}_{table}, & \text{DEL}_{table}, \\
\text{ADD}_{column}, & \text{DEL}_{column}, \text{SPLIT}_{column}, \text{UNITE}_{column}, \\
\text{ADD}_{row}, & \text{DEL}_{row}, \quad \text{SPLIT}_{row}, \quad \text{UNITE}_{row}, \\
\text{REN}_{table}, & \text{REN}_{column}
\end{cases}
$$

All CoDEL SMOs require a set Θ of parameters. Let $inst\,(o, D)$ be the set of instances of the SMO o with a valid parameterization regarding the database D. For instance, the only parameter to remove a table with $\text{DEL}_{table}(\Theta)$ is the name of an existing table, so that $inst\,(\text{DEL}_{table}\,(\Theta)\,, D) = \{\text{DEL}_{table}(R) | R \in D\}$. Further, let $inst\,(\mathcal{L}, D) = \bigcup_{o \in \mathcal{L}} inst\,(o, D)$ be the set of all validly parameterized SMO instances of the *DEL* $\mathcal{L}$. Then, a CoDEL evolution script s for a database D is a sequence of instantiated SMOs with $s \in inst\,(\mathcal{L}_C, D_i)^+$, where D_i is the database after the application of the i-th SMO.

In the following, we specify the semantics of all CoDEL SMOs. Table 1 summarizes the definition of the semantics based on $\mathcal{L}_{min}$. The table also shows the SQL-like syntax we propose for the implementation of CoDEL. In the remainder, $R.C = \{c_1, \ldots, c_n\}$ denotes the set of columns of table R and R_i specifies the version i of the table R. Whenever an SMO does not change the table's name but its columns or rows, we increment this version counter i. CoDEL SMOs take tables as input and return tables. According to the SQL standard, tables are multisets. Our semantics definition with $\mathcal{L}_{min}$ is based on the relational algebra, though, where tables are sets. However, relational database systems internally manage row identifiers, which are at least unique per table. At the level of SMO implementation, we consider the row identifiers as part of the tables and hence, tables as sets. The corresponding multiset semantics of the SMOs can be achieved, by adding a multiset projection of the resulting tables that removes the row identifiers without eliminating duplicates.

ADD_{table} and DEL_{table}. The SMOs ADD_{table} and DEL_{table} are the simplified version of their $\mathcal{L}_{min}$ counterparts. $\text{ADD}_{table}(R, \{c_1, \ldots, c_n\})$ requires two parameters, a table name R and a set of column definitions c_i. It creates an empty table with the specified name and schema. $\text{DEL}_{table}(R)$ takes only a single parameter, the name of the table to be dropped.

ADD_{column} and DEL_{column}. ADD_{column} adds a new column to an existing table. As parameter $\text{ADD}_{column}(R_i, c, f(c_1, \ldots, c_n))$ takes the name R_i of the table, the column definition c of the new column, and a function f. The resulting table is R_{i+1}. ADD_{column} applies the function f to each row in R_i to calculate the row's value for the new column c. The function f receives all other column values

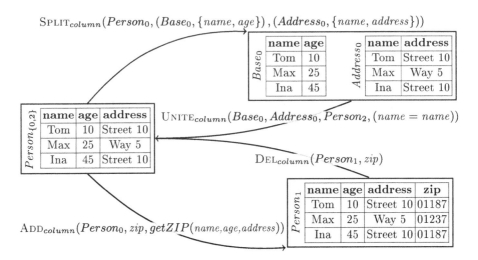

Fig. 3. Example for the operations on columns.

of the row as parameters. Figure 3 shows an example: $\text{ADD}_{column}(Person_0,\ zip,$ $\text{getZip}(name,\ age,\ address))$ adds a column zip to $Person_0$ by determining the zip code based on the currently available information.

DEL_{column} removes a column from a table. Specifically, $\text{DEL}_{column}(R_i, c)$ takes the name R_i of an existing table and the name $c \in R_i.C$ of the column that should be removed from R_i. The resulting table is R_{i+1}. Figure 3 shows an example, where we remove the column zip from table $Person_1$.

SPLIT$_{column}$ and UNITE$_{column}$. SPLIT_{column} partitions a table vertically and removes the original table. SPLIT_{column} has a generalized semantics, where the resulting partitioning is allowed to be incomplete and overlapping. $\text{SPLIT}_{column}(R, (S, \{s_1, \ldots, s_n\}), (T, \{t_1, \ldots, t_m\}))$ takes the name R of the original table, a pair of table name S and a set of column names s_i as specification of the first partition and optionally a second pair $(T, \{t_1, \ldots, t_m\})$ as specification of the second partition. The two sets of column definitions are independent. In case $S.C \cap T.C \neq \emptyset$, the columns $S.C \cap T.C$ are copied. In case $S.C \cup T.C \subset R.C$, the partitioning is incomplete. If the second partition is not specified, T is not created. Note that CoDEL prohibits empty sets of column definitions for S and T, since tables must have at least one column. Figure 3 shows an example with the SPLIT_{column} SMO. Table $Person_0$ is vertically partitioned to general information ($Base_0$) and address information ($Address_0$). The partitions overlap on the column $name$ to maintain the connection between addresses and person.

UNITE_{column} is the inverse operation of SPLIT_{column}. It joins two tables based on a given condition and removes the original tables. As parameters, $\text{UNITE}_{column}(R, S, T, cond, o)$ takes the names R and S of the original tables, the name T of the resulting table, a join condition $cond$ using SQL predicates without further nesting, and the optional request o for an outer join. In case

Table 1. Syntax and semantic of CoDEL operations.

SMO:	$\text{ADD}_{table}(R, \{c_1, \ldots, c_n\})$	$\text{DEL}_{table}(R)$
Semantic:	$\text{ADD}(R, \pi_{c_1, \ldots, c_n}(\emptyset));$	$\text{DEL}(R);$
Syntax:	`CREATE TABLE R (`c_1`, ..., `c_n`)`	`DROP TABLE R`

SMO:	$\text{ADD}_{column}(R_i, c, f(c_1, \ldots, c_n))$
Semantic:	$\text{ADD}\big(R_{i+1}, \pi_{R_i.C \cup \{c \leftarrow f(c_1, \ldots, c_n)\}}(R_i)\big); \; \text{DEL}(R_i);$
Syntax:	`ADD COLUMN c AS `$f(c_1, \ldots, c_n)$` INTO `R_i

SMO:	$\text{DEL}_{column}(R_i, c)$
Semantic:	$\text{ADD}\big(R_{i+1}, \pi_{R_i.C \setminus \{c\}}(R_i)\big); \; \text{DEL}(R_i);$
Syntax:	`DROP COLUMN c FROM `R_i

SMO:	$\text{SPLIT}_{column}(R, (S, \{s_1, \ldots, s_n\}), (T, \{t_1, \ldots, t_m\}))$
Semantic:	$\text{ADD}(S, \pi_{s_1, \ldots, s_n}(R)); \; [\text{ADD}(T, \pi_{t_1, \ldots, t_m}(R))];$
	$\text{DEL}(R);$
Syntax:	`DECOMPOSE TABLE R INTO S (`s_1`, ..., `s_n`) [, T (`t_1`, ..., `t_m`)]`

SMO:	$\text{UNITE}_{column}(R, S, T, cond, o)$
Semantic:	$o = \bot: \text{ADD}(T, R \bowtie_{cond} S); \qquad o = \top: \text{ADD}(T, R \,⟗_{cond}\, S);$
	$\text{DEL}(R); \; \text{DEL}(S);$
Syntax:	`[OUTER] JOIN TABLE R, S INTO T WHERE `$cond$

SMO:	$\text{ADD}_{row}(R_i, G, \{(a_1, f_1(G, V)), \ldots, (a_m, f_m(G, V))\}, S)$
Semantic:	S given: $\text{ADD}\big(S, \gamma_{G, \{f_j(G,V) \to a_j \mid 1 \leq j \leq m\}}(R_i)\big)$
	S not given: $\text{ADD}\big(R_{i+1}, R_i \cup \gamma_{G, \{f_j(G,V) \to a_j \mid 1 \leq j \leq m\}}(R_i)\big); \; \text{DEL}(R_i);$
Syntax:	`AGGREGATE TABLE `R_i` (`g_1`, ..., `g_n`) WITH `$a_1 = f_1(G, V)$`, ... [INTO S]`

SMO:	$\text{DEL}_{row}(R_i, cond)$
Semantic:	$\text{ADD}(R_{i+1}, \sigma_{\neg cond}(R_i)); \; \text{DEL}(R_i);$
Syntax:	`REMOVE FROM TABLE `R_i` WHERE `$cond$

SMO:	$\text{SPLIT}_{row}(R, (S, cond_S), (T, cond_T))$
Semantic:	$\text{ADD}(S, \sigma_{cond_S}(R)); \; [\text{ADD}(T, \sigma_{cond_T}(R))];$
	$\text{DEL}(R);$
Syntax:	`PARTITION TABLE R INTO S WITH `$cond_S$` [, T WITH `$cond_T$`]`

SMO:	$\text{UNITE}_{row}(R, S, T)$
Semantic:	$\text{ADD}\big(T, \pi_{R.C \cup \{\omega \to a_i \mid a_i \in S.C \setminus R.C\}}(R) \cup \pi_{S.C \cup \{\omega \to a_i \mid a_i \in R.C \setminus S.C\}}(S)\big);$
	$\text{DEL}(R); \; \text{DEL}(S);$
Syntax:	`MERGE TABLE R, S INTO T`

SMO:	$\text{REN}_{table}(R, R')$	$\text{REN}_{column}(R_i, c, c')$
Semantic:	$\text{ADD}(R', R); \; \text{DEL}(R);$	$\text{ADD}\big(R_{i+1}, \rho_{c'/c}(R_i)\big); \; \text{DEL}(R_i);$
Syntax:	`RENAME TABLE R INTO R'`	`RENAME COLUMN c IN `R_i` TO c'`

$o = \top$, UNITE_{column} performs an outer join, so that no rows from the original tables are lost. In case $o = \bot$ (or not specified) UNITE_{column} performs an inner join. With the inner join, UNITE_{column} loses all rows from R and S that do not find a join partner, since R and S are dropped after the join. Note that restricting the join to foreign key relations as other $DELs$ do, does not prevent this information loss. A foreign key does not guarantee that every row in the referenced table is actually referenced by at least one row in the referencing table. Figure 3 also shows an example of UNITE_{column}. The tables $Base_0$ and $Address_0$ are inner joined to the table $Person_2$ based on equal names. Since all persons have an address in this example, no rows are lost.

ADD$_{row}$ and **DEL$_{row}$.** ADD_{row} adds new rows to an existing table by aggregating the data in the current rows. As parameter $\text{ADD}_{row}(R_i, G, \{(a_j, f_j(G, V)) \mid 1 \le j \le m\}, S)$ requires the name R_i of the original table, the set of grouping columns $G = \{g_1, \ldots, g_n\} \subseteq R_i.C$, a set of pairs of column name a_j and aggregations function f_j, and optionally a new table name S. ADD_{row} produces new rows by grouping table R_i by all columns $g_k \in G$ and calculating the values for the columns a_j with the functions f_j. The functions f_j may contain constants, the values of the grouping columns G, and aggregate functions upon the remaining columns $V = R_i.C \setminus G$. If the new table name S is specified, ADD_{row} creates S with the newly produced rows and R_i remains available, which is particularly necessary, when the newly created rows have a different set of columns than $R_i.C$. Otherwise, ADD_{row} appends the new rows to R_i to form its new version R_{i+1}. In this case, we require the column definitions of the new rows to match the original table R_i, hence $\{g_1, \ldots, g_n\} \cup \{a_1, \ldots, a_m\} = R.C$. In general, the set of grouping columns is also allowed to be empty resulting in one group and hence, one new row.

DEL_{row} removes rows from a given table. $\text{DEL}_{row}(R_i, cond)$ takes the name of an existing table R_i and a condition $cond$. It removes all rows, which satisfy the condition and evolves the table to R_{i+1}.

SPLIT$_{row}$ and **UNITE$_{row}$.** SPLIT_{row} partitions a table horizontally. However, its semantics is more general than standard horizontal partitioning [4]. The SMO creates at most two partitions out of a given table – with the partitioning allowed to be incomplete and overlapping – and removes the original table. More precisely, $\text{SPLIT}_{row}(R, (S, cond_S), (T, cond_T))$ takes the name of the original table, a pair of table name S and condition $cond_S$ as specification of the first partition and optionally a second pair $(T, cond_T)$ as specification of the second partition. Both conditions $cond_S$ and $cond_T$ are independent. If the original tables contain rows that fulfill neither of the conditions, the resulting partitioning is incomplete. Rows that fulfill both conditions are copied resulting in overlapping partitions. In case both conditions hold for all rows, i.e., $cond_S = \top$ and $cond_T = \top$, T is a complete copy of S. Hence, SPLIT_{row} subsumes the functionality of a copy operations that can be found in other $DELs$. If $cond_T$ is not specified, SPLIT_{row} does not create table T.

UNITE_{row} is the inverse operation of SPLIT_{row}; it merges two given tables along the row dimension and removes the original tables. As parameters $\text{UNITE}_{row}(R, S, T)$ requires the names R and S of the original tables and the name T of the resulting table. The schema of R and S are not required to by equivalent. In case both schemas differ, T contains null values (ω) in the corresponding cells. UNITE_{row} eliminates duplicates in T. In case R and S contain equivalent rows, these rows will show up only once in T.

REN$_{table}$ and REN$_{column}$. The last two SMOs rename schema elements. REN$_{table}(R, R')$ renames the table with the name R into R'. REN$_{column}(R_i, c, c')$ renames the column c in table R_i into c', which results in table R_{i+1}.

We use the semantics definition, as summarized in Table 1, to show the relational completeness of CoDEL in the following section.

3 Relational Completeness

To show the relational completeness of CoDEL, we argue that it is at least as powerful as $\mathcal{L}_{\min}$ (Sect. 1), which is relationally complete by definition. There is always a semantically equivalent expression in CoDEL for any expression in $\mathcal{L}_{\min}$. The DEL (R) operation from $\mathcal{L}_{\min}$ is trivial, since it is equivalent to CoDEL's DEL$_{table}(R)$. On the contrary, ADD (R, ϵ) from $\mathcal{L}_{\min}$ is more complex, as ϵ covers the power of the relational algebra. Since both the relational algebra and CoDEL are closed languages, it is reasonable to address each operation of the relational algebra separately. We show that, for each operation from the relational algebra, there is a semantically equivalent sequence of SMOs in CoDEL.

We assume the basic relational algebra [5] and add common extensions like the extended projection, aggregation, and outer joins. However, we intentionally exclude other extensions like the transitive closure and sorting. CoDEL does not cover these extensions, since CoDEL is non-recursive and set-based. We maintain these characteristics, since they proved to be a reasonable trade-off between expressiveness and usability, however, they are open for further research. With respect to implementations based on current database management systems, the distinction between different types of null values [19] is not considered. For instance UNITE$_{row}$ adds null values in columns, which existed in only one input table, losing the information, whether a value was null before or did not exist at all. The following sections will consider all constructs from the relational algebra including the chosen extensions and show that CoDEL is capable to obtain the semantically equivalent results.

Relation: R The basic elements of the relational algebra are relations. They contain the data and are directly accessible by CoDEL as tables. Whenever one table is required multiple times within a relational algebra expression, CoDEL allows to copy them using SPLIT$_{row}(R, (S, \top), (T, \top))$.

Selection: $\sigma_{cond}(R)$ The selection returns the subset of rows from R, which satisfy the condition $cond$. CoDEL's SPLIT$_{row}(R, (S, cond))$ is semantically equivalent, which directly follows from the semantics definition in Table 1.

Rename: $\rho_{c'/c}(R_i)$ Renaming a column is subsumed by the extended projections, however, we include it here for completeness. CoDEL's obvious semantic equivalent according to Table 1 is REN$_{column}(R_i, c, c')$.

Extended Projection: $\pi_P(R)$ We will immediately consider the extended projection, as it subsumes the traditional projection. The extended projection defines a new set of columns, whose values are computed by functions depending on the existing columns. Assume the projection $P = \{f_k(R.C) \to a_k | 1 \le k \le m\}$ with $n = |R.C|$. The CoDEL sequence below, realizes such an extended projection. Without loss of generality, we use for-loops to iterate over the attribute sets. Since this is only schema depending and data independent, it does not extend the expressiveness of the DEL but is simply a short notation.

1: **for** $k = [1..m]$ **do**
2:     ADD$_{column}(R_{i+k-1}, a'_k, f_k(r_1, \dots, r_n))$;
3: **for** $r_j \in R.C$ **do**
4:     DEL$_{column}(R_{i+m+j-1}, r_j)$;
5: **for** $k = [1..m]$ **do**
6:     REN$_{column}(R_{i+m+n+k-1}, a'_k, a_k)$;
7: **for** $k = [i..(i+2m+n-1)]$ **do**
8:     DEL$_{table}(R_k)$;

$$R_{i+1} \overset{2}{=} \pi_{r_1,\dots,r_n,f_1(r_1,\dots,r_n)\to a'_1}(R_i) \tag{1}$$

$$R_{i+m} \overset{1,2}{=} \pi_{r_1,\dots,r_n,f_1(r_1,\dots,r_n)\to a'_1,\dots,f_m(r_1,\dots,r_n)\to a'_m}(R_i) \tag{2}$$

$$R_{i+m+1} \overset{4}{=} \pi_{r_2,\dots,r_n,a'_1,\dots,a'_m}(R_{i+m}) \tag{3}$$

$$R_{i+m+n} \overset{3,4}{=} \pi_{a'_1,\dots,a'_m}(R_{i+m}) = \pi_{f_1(r_1,\dots,r_n)\to a'_1,\dots,f_m(r_1,\dots,r_n)\to a'_m}(R_i) \tag{4}$$

$$R_{i+m+n+1} \overset{6}{=} \pi_{a'_1\to a_1, a'_2,\dots,a'_m}(R_{i+m+n}) \tag{5}$$

$$R_{i+m+n+m} \overset{5,6}{=} \pi_{a'_1\to a_1,\dots,a'_m\to a_m}(R_{i+m+n})$$
$$= \pi_{f_1(R_i.C)\to a_1,\dots,f_m(R_i.C)\to a_m}(R_i) \tag{6}$$

The first SMO adds a new column, with a masked name, for each column of the output table. This allows to compute the new values based on all existing ones. Afterwards, we drop the old columns, rename the new columns to their unmasked name, and remove all intermediate tables. Applying the semantics definitions of the CoDEL SMOs results in the desired extended projection, as shown above. The concrete line of the CoDEL sequence, which is applied in the semantics computation, is indicated by the numbers above the equal signs.

Outer Join: $R \bowtie_p S$ The outer join is another common extension to the traditional relational algebra. Beyond the rows according to an inner join, it also includes those rows in the result, which did not find a join partner. The missing values for columns of the other table are filled with null values ω respectively. Obviously, CoDEL's UNITE$_{column}(R, S, T, p, \top)$ is semantically equivalent, since we explicitly introduced the option to perform outer joins.

Cross Product: $R \times S$ The cross product produces a row in the output table for each pair of rows from the input tables. The following sequence of CoDEL SMOs is semantically equivalent as shown below.

1: $\text{ADD}_{column}(R_i, j, 1)$;
2: $\text{ADD}_{column}(S_k, j, 1)$;
3: $\text{UNITE}_{column}(R_{i+1}, S_{k+1}, T_0, R_{i+1}.j = S_{k+1}.j, \bot)$;
4: $\text{DEL}_{column}(T_0, j)$;

$$R_{i+1} \overset{1}{=} \pi_{r_1,\dots,r_n,1 \to j}(R_i) = \{(r_1,\dots,r_n,1) \mid (r_1,\dots,r_n) \in R_i\} \tag{7}$$

$$S_{k+1} \overset{2}{=} \{(s_1,\dots,s_m,1) \mid (s_1,\dots,s_m) \in S_k\} \tag{8}$$

$$T_0 \overset{3}{=} R_{i+1} \bowtie_{R_{i+1}.j=S_{k+1}.j} S_{k+1}$$
$$= \{(r_1,\dots,r_n,s_1,\dots,s_m,1) \mid (r_1,\dots,r_n) \in R_i, (s_1,\dots,s_m) \in S_k\} \tag{9}$$

$$T_1 \overset{4}{=} \{(r_1,\dots,r_n,s_1,\dots,s_m) \mid (r_1,\dots,r_n) \in R_i, (s_1,\dots,s_m) \in S_k\}$$
$$= R \times S \tag{10}$$

We add a new column j to both tables with $j \notin R_i.C$ and $j \notin S_k.C$ and the default value 1 to perform an inner join on j. Since its value is always 1, there will be one row in the output table for each pair of rows from the two input tables. We remove the additional column j and finally show the semantic equivalence between the relational cross product and the presented sequence of CoDEL SMOs.

Aggregate: $\gamma_{G,F}(R)$ The aggregation is another typical extension to the relational algebra. The rows are grouped by one set of columns $G = \{g_1,\dots,g_n\} \subseteq R.C$. Additional columns $A = \{a_i | 1 \le i \le p\}$ are computed by functions $F = \{f_i(G,V) \to a_i | a_i \in A\}$ with $V = \{v_1,\dots,v_m\} = R.C \setminus G$. These functions may contain values from grouping columns G, aggregate functions on the remaining columns in V, constants, and arithmetic functions. CoDEL contains a dedicated operation $\text{ADD}_{row}(R, G, F, S)$. It writes the result of the aggregation to the new table S. According to the semantics definition in Table 1, the semantics of ADD_{row} equals the discussed aggregation semantics from the relational algebra.

Union: $R \cup S$ The relational union, merges the rows from both input tables to the one output table including an elimination of duplicates. Using the SMO UNITE_{row}, CoDEL provides a semantic equivalent to the relational union operation.

1: $\text{UNITE}_{row}(R, S, T)$;

$$T \overset{1}{=} \pi_{R.C}(R) \cup \pi_{S.C}(S) = R \cup S \tag{11}$$

Please note, that the union in the relational algebra requires R and S to have identical sets of attributes ($R.C = S.C$), which justifies the simplification step.

Difference: $R \setminus S$ The relational difference returns all rows, which occur in the first, but not in the second table. Analogous to the union, it requires R and S to have identical sets of columns ($R.C = S.C$). The following CoDEL sequence is semantically equivalent to the relational difference.

1: $\text{ADD}_{column}(S_k, j, 1)$;
2: $\text{UNITE}_{column}(R_i, S_{k+1}, T_0, (R_i.c_1 = S_{k+1}.c_1 \wedge \ldots \wedge R_i.c_n = S_{k+1}.c_n), \top)$;
3: $\text{DEL}_{row}(T_0, j \neq \omega)$;
4: $\text{DEL}_{column}(T_1, j)$;

$$S_{k+1} \overset{1}{=} \pi_{s_1, \ldots, s_m, 1 \to j} (S_k) \tag{12}$$

$$\begin{aligned} T_0 &\overset{2}{=} R_i \bowtie S_{k+1} \\ &= \{(r_1, \ldots, r_n, 1) \mid (r_1, \ldots, r_n) \in R_i, (r_1, \ldots, r_n, 1) \in S_{k+1}\} \\ &\quad \cup \{(r_1, \ldots, r_n, 1) \mid (r_1, \ldots, r_n) \notin R_i, (r_1, \ldots, r_n, 1) \in S_{k+1}\} \\ &\quad \cup \{(r_1, \ldots, r_n, \omega) \mid (r_1, \ldots, r_n) \in R_i, (r_1, \ldots, r_n, 1) \notin S_{k+1}\} \end{aligned} \tag{13}$$

$$\begin{aligned} T_1 &\overset{3}{=} \sigma_{\neg(j \neq \omega)}(T_0) = \sigma_{(j = \omega)}(T_0) \\ &= \{(r_1, \ldots, r_n, \omega) \mid (r_1, \ldots, r_n) \in R_i, (r_1, \ldots, r_n, 1) \notin S_{k+1}\} \end{aligned} \tag{14}$$

$$\begin{aligned} T_2 &\overset{4}{=} \pi_{R.C}(T_1) \\ &= \{(r_1, \ldots, r_n) \mid (r_1, \ldots, r_n) \in R_i, (r_1, \ldots, r_n) \notin S_k\} = \underline{\underline{R_i \setminus S_k}} \end{aligned} \tag{15}$$

We add a new column j to S_k with $j \notin S_k.C$ and the default value 1. The outer join on all columns $c_i \in R_i.C = S_k.C$ is applicable, since the initial column sets are equal. Due to the nature of the outer join, the resulting table contains all rows which were in at least one of the two input tables. However, all rows, which occurred in S_k have the value 1 in the column j and are removed by the third SMO. All rows which occurred exclusively in R have a null value ω in the column j and remain as result. Applying the semantics definition of the SMOs finally leads to the relational difference operation. Please note, that $(r_1, \ldots, r_n) \notin S_k$ is equal to $(r_1, \ldots, r_n, 1) \notin S_{k+1}$ due to the first step.

Finally, we successfully showed that CoDEL provides a semantic equivalent for each relational algebra expression, which makes it equally expressive as $\mathcal{L}_{\min}$. Hence, it is relationally complete and a sound foundation for further research.

4   Related Work

Database evolution is a well recognized topic in the database research community [13, 18]. There are a number of approaches to increase comfort and efficiency in database evolution, for instance by defining a schema evolution aware query language [14]. Another approach is to define database evolution languages graph-based [12]. This allows modeling dependencies between different artifacts in the information system and applying changes globally. Furthermore, MeDEA [10] provides a general framework to describe database evolution in the context of evolving applications. MoDEF [17] basically introduces an IDE extension to automate the co-evolution of the evolving client schemas and the store.

Currently, PRISM [7] appears to provide the most advanced database evolution tool including an SMO-based *DEL*. PRISM was first introduced in 2008 and focused on the plain database evolution [8]. Later, the authors extended it to PRISM++, which includes the modification of constraints and update rewriting [6]. To benchmark database evolution languages and tools, researchers also analyzed the evolution histories of Wikimedia and other open source projects [9,16]. Finally, database versioning extends the ideas of database evolution to allow both forward and backward compatibility between the different versions of evolving schemas [15]. Another extension of PRISM takes a first step into this direction by answering queries on former schema versions according to the current data [11]. The presented *DEL* CoDEL inherits the principle style of SMOs from PRISM. However, PRISM is not relationally complete, while CoDEL is. This additional characteristic provided by CoDEL is highly valuable with respect to further research, particularly in the field of automated database versioning based on SMOs, where falling back on common DDL and DML evolution scripts is not an option.

5   Conclusion

Agile software development methods embrace the change. While software developers find support in refactoring methods to evolve their software, database developers still have to fiddle with DDL/DML scripts to evolve schema and data of a productive database consistently. Adding evolution support to a DBMS involves the design of a database evolution language (*DEL*). In this paper we considered the relational completeness of *DELs* for relational databases. Relational completeness is an important property of *DELs*. *DELs* that are incomplete in this respect, can force the user back to the manual evolution process based on DDL and DML limiting the utility of the evolution functionality. We presented the relationally complete *DEL* CoDEL. We detailed its formal definition and showed its relational completeness. CoDEL is to our best knowledge the first well-defined, relationally complete *DEL*. CoDEL can serve as a reference language for productive implementations of database evolution in DBMSs.

The solid formal base of CoDEL is also important for research and development beyond database evolution. For instance in database versioning, multiple clients access the same data in different schema versions. Database versioning requires invertible SMOs, so that the database system can translate data back and forth between schema versions. For the investigation of the invertibility of SMOs a solid formal definition of the SMOs is a prerequisite. Hence, CoDEL offers a good starting point towards database versioning. For the near future, however, we hope CoDEL helps to jump start more implementations of proper database evolution features in the DBMSs on the market, so that agile development methods final arrive at the database layer.

References

1. Ambler, S.W.: Whence data management? Dr. Dobb's J. **390**, 79 (2006)
2. Ambler, S.W., Sadalage, P.J.: Refactoring Databases: Evolutionary Database Design. Addison-Wesley Signature. Addison-Wesley, New York (2006). ISBN: 978-0321774514
3. Beck, K., Beedle, M., van Bennekum, A., Cockburn, A., Cunningham, W., Fowler, M., Grenning, J., Highsmith, J., Hunt, A., Jeffries, R., Kern, J., Marick, B., Martin, R.C., Mellor, S., Sutherland, J., Thomas, D., Schwaber, K.: Manifesto for Agile Software Development (2001)
4. Ceri, S., Negri, M., Pelagatti, G.: Horizontal data partitioning in database design. In: SIGMOD Conference, pp. 128–136 (1982)
5. Codd, E.F.: A relational model of data for large shared data banks. Commun. ACM **15**(3), 162–166 (1970)
6. Curino, C.A., Moon, H.J., Deutsch, A., Zaniolo, C.: Update rewriting and Integrity constraint maintenance in a schema evolution support system: PRISM++. VLDB Endow. **4**(2), 117–128 (2010)
7. Curino, C.A., Moon, H.J., Deutsch, A., Zaniolo, C.: Automating the database schema evolution process. VLDB J. **22**(1), 73–98 (2012)
8. Curino, C.A., Moon, H.J., Zaniolo, C.: Graceful database schema evolution: the PRISM workbench. VLDB Endow. **1**(1), 761–772 (2008)
9. Curino, C.A., Tanca, L., Moon, H.J., Zaniolo, C.: Schema evolution in wikipedia: toward a web information system benchmark. In: ICEIS, pp. 323–332 (2008)
10. Domínguez, E., Lloret, J., Rubio, Á.L., Zapata, M.A.: MeDEA: a database evolution architecture with traceability. Data Knowl. Eng. **65**(3), 419–441 (2008)
11. Moon, H.J., Curino, C.A., Ham, M., Zaniolo, C.: PRIMA - archiving and querying historical data with evolving schemas. In: SIGMOD Conference, pp. 1019–1022 (2009)
12. Papastefanatos, G., Vassiliadis, P., Simitsis, A., Aggistalis, K., Pechlivani, F., Vassiliou, Y.: Language extensions for the automation of database schema evolution. In: ICEIS, pp. 74–81 (2008)
13. Rahm, E., Bernstein, P.A.: An online bibliography on schema evolution. SIGMOD Rec. **35**(4), 30–31 (2006)
14. Roddick, J.F.: SQL/SE - a query language extension for databases supporting schema evolution. SIGMOD Rec. **21**(3), 10–16 (1992)
15. Roddick, J.F.: A survey of schema versioning issues for database systems. Inf. Softw. Technol. **37**(7), 383–393 (1995)
16. Skoulis, I., Vassiliadis, P., Zarras, A.: Open-source databases: within, outside, or beyond Lehman's laws of software evolution? In: Jarke, M., Mylopoulos, J., Quix, C., Rolland, C., Manolopoulos, Y., Mouratidis, H., Horkoff, J. (eds.) CAiSE 2014. LNCS, vol. 8484, pp. 379–393. Springer, Heidelberg (2014)
17. Terwilliger, J.F., Bernstein, P.A., Unnithan, A.: Worry-free database upgrades. In: SIGMOD Conference, p. 1191 (2010)
18. Terwilliger, J.F., Cleve, A., Curino, C.A.: How clean is your sandbox? In: Hu, Z., de Lara, J. (eds.) ICMT 2012. LNCS, vol. 7307, pp. 1–23. Springer, Heidelberg (2012)
19. Zaniolo, C.: Database relations with null values. J. Comput. Syst. Sci. **28**(1), 142–166 (1984)

Multidimensional Modeling and OLAP

Implementation of Multidimensional Databases in Column-Oriented NoSQL Systems

Max Chevalier, Mohammed El Malki[(✉)], Arlind Kopliku,
Olivier Teste, and Ronan Tournier

Université de Toulouse, IRIT UMR 5505, 118 Route de Narbonne,
31062 Toulouse, France
{Max.Chevalier,Mohammed.Malki,Arlind.Kopliku,
Olivier.Teste,Ronan.Tournier}@irit.fr

Abstract. NoSQL (Not Only SQL) systems are becoming popular due to known advantages such as horizontal scalability and elasticity. In this paper, we study the implementation of multidimensional data warehouses based on column-oriented NoSQL systems. To do this, we define a set of mapping rules that transform a conceptual multidimensional data model into a logical column-oriented model. We consider three ways (three sub-models) to structure conceptual multidimensional data model into a column-oriented model. Then, we show an implementation of the proposed rules. Finally, we focus, through experiment, on data loading, model-to-model conversion and OLAP cuboid computation.

Keywords: Data warehouse design · Multidimensional modelling · NoSQL databases · Model transformation rules · Column-Oriented NoSQL model

1 Introduction

NoSQL solutions have proven some clear advantages with respect to relational database management systems (RDBMS) [17]. Nowadays, research attention has turned towards using these systems for storing "big" data and analyzing it. This work joins substantial ongoing work on the area on the use of NoSQL solutions for data warehousing [4, 6, 18, 19].

In this paper, we study one category of NoSQL stores: column-oriented systems such as HBase [11] or Cassandra [13] and inspired by Bigtable [2]. Indeed, column-oriented systems are one of the most famous families of NoSQL systems. They allow more flexibility in schema design using a vertical data organization with column families and with no static non-mutable schema defined in advance, i.e. the data schema can evolve. However, although, column-oriented databases are declared schemaless (no schema needed), most use cases require some sort of data model.

When it comes to data warehouses, previous research has shown that it can be instantiated with different logical models [12]. Data warehousing relies mostly on

© Springer International Publishing Switzerland 2015
T. Morzy et al. (Eds.): ADBIS 2015, LNCS 9282, pp. 79–91, 2015.
DOI: 10.1007/978-3-319-23135-8_6

multidimensional data modelling which is a conceptual[1] model that uses facts to model an analysis subject and dimensions for analysis axes. Until now there is no work (only an initial attempt in [4]) that considers the direct mapping from the multidimensional conceptual model to NoSQL logical models. A possible way to do this is to map the multidimensional model to relational databases and then export the latter into a column oriented system. In this paper we study the direct transformation of the data warehouse conceptual model into a column-oriented system (see Fig. 1).

NoSQL models are more expressive than relational models i.e. we do not only have to describe data and relations; we also have a flexible data structure (e.g. nested elements). In this context, more than one approach is candidate as a mapping of the multidimensional model. Moreover, evolving requirements in terms of analyses or data query performance might demand switching from one logical model to another. Finally, analysis queries can be very time consuming and speeding their execution consists generally in precomputing these queries (called aggregates) and this pre-computation requires also a logical model.

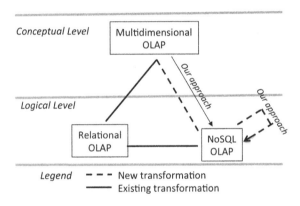

Fig. 1. From multidimensional conceptual model to logical models.

In this paper, we focus on data models for data warehousing. We study three logical column-oriented models. We provide a formalism for expressing each of these sub-models which enables us to generate mapping from the conceptual model. We show how we can instantiate data warehouses in column-oriented stores. Our study includes evaluation for these models on data loading, model-to-model conversions and the computation of pre-computed aggregates (also called OLAP cuboids grouped in an OLAP cube).

Our motivation is multiple. The implementation of OLAP systems with NoSQL systems is a new alternative [7, 8, 16]. These systems have several advantages such as increased flexibility and scalability. The increasing scientific research in this direction demands for formalization, common-agreement models and evaluations of different NoSQL systems.

---

[1] Conceptual level data models describe data in a generic way regardless the information technologies used, while logical level models use a specific technique for implementing the conceptual level.

We can summarize our contribution as follows:

- logical notations for NoSQL systems where structures and values are clearly separated;
- three column-oriented approaches to map conceptual multidimensional data warehouse schemas to a logical model;
- the conversions from one approach to another at the logical level through the definition of a set of transformation rules;
- the computation of the OLAP cube using NoSQL technologies.

2 State of the Art

Several research works have focused on translating data warehousing concepts into a relational (R-OLAP) logical level [3, 6] as multidimensional databases are mostly implemented using the relational technologies. Mapping rules are used to convert structures of the conceptual level (facts, dimensions and hierarchies) into a logical model based on relations. Moreover, many works have focused on implementing logical optimization methods based on pre-computed aggregates (also called materialized views) [1]. However, R-OLAP implementations suffer from scaling-up to large data volumes (i.e. "Big Data") and research is currently underway for new solutions such as using NoSQL systems [17]. Our approach aims at revisiting these processes for automatically implementing multidimensional conceptual models directly into NoSQL models.

Other studies investigate the process of transforming relational databases into a NoSQL logical model (see Fig. 1). In [14], the author proposed an approach for transforming a relational database into a column-oriented NoSQL database. In [18], the author studies "denormalizing" data into schema-free databases. However, these approaches never consider the conceptual model of data warehouses. They are limited to the logical level, i.e. transforming a relational model into a column-oriented model. More specifically, the duality fact/dimension requires guaranteeing a number of constraints usually handled by the relational integrity constraints and these constraints cannot be considered when using the logical level as starting point.

This study highlights that there are currently no approaches for automatically and directly transforming a data warehouse multidimensional conceptual model into a NoSQL logical model. It is possible to transform multidimensional conceptual models into a logical relational model, and then to transform this relational model into a logical NoSQL model. However, this transformation using the relational model as a pivot model has not been formalized as both transformations were studied independently of each other. Also, this indirect approach can be tedious.

We can also cite several recent works that are aimed at developing data warehouses in NoSQL systems whether columns-oriented [9], document-oriented [5], or key-value oriented [19]. However, the main goal of these papers is to propose benchmarks. These studies have not focused on the model transformation process and they only focus one NoSQL model. These models [5, 9, 19] require the relational model to be generated first before the abstraction step.

In our approach we consider the instatiation of datawarehouses on column-oriented stores. The conceptual model is mapped directly on three different logical models.

3 Multidimensional Conceptual Model and Cube

3.1 Conceptual Multidimensional Model

To ensure robust translation rules we present the multidimensional model used at the conceptual level [10, 16].

A **multidimensional schema**, namely E, is defined by $(F^E, D^E, Star^E)$ where:

- $F^E = \{F_1, \ldots, F_n\}$ is a finite set of facts,
- $D^E = \{D_1, \ldots, D_m\}$ is a finite set of dimensions,
- $Star^E: F^E \rightarrow 2^{D^E}$ is a function that associates facts of F^E to sets of dimensions along which it can be analyzed (2^{D^E} is the *power set* of D^E).

A **dimension**, denoted $D_i \in D^E$ (abusively noted as D), is defined by (N^D, A^D, H^D) where:

- N^D is the name of the dimension,
- $A^D = \{a_1^D, \ldots, a_u^D\} \cup \{id^D, All^D\}$ is a set of dimension attributes,
- $H^D = \{H_1^D, \ldots, H_v^D\}$ is a set hierarchies.

A **hierarchy** of the dimension D, denoted $H_i \in H^D$, is defined by $(N^{Hi}, Param^{Hi}, Weak^{Hi})$ where:

- N^{Hi} is the name of the hierarchy,
- $Param^{Hi} = <id^D, p_1^{H_i}, \ldots, p_{v_i}^{H_i}, All^D>$ is an ordered set of v_i+2 attributes which are called **parameters** of the relevant graduation scale of the hierarchy, $\forall k \in [1..v_i]$, $p_k^{H_i} \in A^D$.
- $Weak^{Hi}: Param^{Hi} \rightarrow 2^{A^D - Param^{Hi}}$ is a function associating with each parameter possibly one or more weak attributes.

A **fact**, $F \in F^E$, is defined by (N^F, M^F) where:

- N^F is the name of the fact,
- $M^F = \{f_1(m_1), \ldots, f_v(m_v)\}$ is a set of measures, each associated with an aggregation function f_i.

3.2 OLAP Cube

The **pre-computed aggregate lattice** or **OLAP cube** (also called sometimes the OLAP cuboid lattice) corresponds to a set of views or cuboids each being a subset of dimensions associated to a subset of measures of one fact. Technically, each view or cuboid corresponds to an analysis query. OLAP cuboids are pre-computed to speed up analysis query execution and thus facilitate analyzing data according to dimension combinations. Measure data is grouped according to the dimensions and aggregation functions are used to summarize the measure data according to these groups. Formally, an OLAP cuboid O is derived from E, $O = (F^O, D^O)$ such that:

- F^O is a fact derived from F ($F \in F^E$) with a subset of measures, $M^O \subseteq M^F$.
- $D^O \subseteq 2^{Star^E(F)} \subseteq D^E$ is a subset of dimensions of D^E. More precisely, D^O is one of the combinations of the dimensions associated to the fact F ($Star^E(F)$).

If we generate OLAP cuboids using all dimension combinations of one fact, we have an OLAP cuboid lattice [1, 3] (also called a pre-computed aggregate lattice or cube).

3.3 Case Study

We use an excerpt of the star schema benchmark [5]. It consists in a monitoring of a sales system. Orders are placed by customers and the lines of the orders are analyzed. A line consists in a part (a product) bought from a supplier and sold to a customer at a specific date. The conceptual schema of this case study is presented in Fig. 2.

The fact $F^{LineOrder}$ is defined by ($LineOrder$, {$SUM(Quantity)$, $SUM(Discount)$, $SUM(Revenue)$, $SUM(Tax)$}) and it is analyzed according to four dimensions, each consisting of several hierarchical levels (called detail levels or parameters):

- The Customer dimension ($D^{Customer}$) with parameters $Customer$ (along with the weak attribute $Name$), $City$, $Region$ and $Nation$,

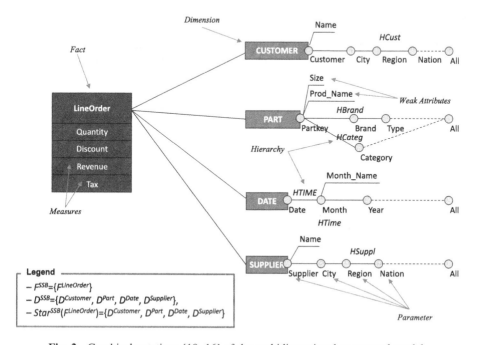

Fig. 2. Graphical notations [10, 16] of the multidimensional conceptual model.

- The Part dimension (D^{Part}) with parameters *Partkey* (with weak attributes *Size* and *Prod_Name*), *Category*, *Brand* and *Type*; organized using two hierarchies *HBrand* and *HCateg*,
- The Date dimension (D^{Date}) with parameters *Date*, *Month* (with a weak attribute, *MonthName*) and *Year*,
- The Supplier dimension ($D^{Supplier}$) with parameters *Supplier* (with weak attributes *Name*), *City*, *Region* and *Nation*.

From this schema, called E^{SSB}, we can define cuboids, for instance:

- ($F^{LineOrder}$, { $D^{Customer}$, D^{Date}, $D^{Supplier}$ }),
- ($F^{LineOrder}$, { $D^{Customer}$, D^{Date} }).

4   Modeling a Data Warehouse Using Column-Oriented Stores

4.1   Column-Oriented Data Model Formalism

Column-Oriented NoSQL models provide tables with a flexible schema (untyped columns) where the number of columns may vary between each record (called rows). Each row has a row key and a set of column families. Physical storage is organized according to these column families, hence a "vertical partitioning" of the data. A column family consists of a set of columns, each associated with a qualifier (name) and an atomic value. Every value can be "versioned" using a timestamp. The flexibility of a column-oriented NoSQL database enables managing the absence of some columns between the different table rows. However, in the context of multidimensional data storage, this rarely happens as data is usually highly structured. This implies that the structure of a column family (i.e. the set of columns of the column family) will be the same for all table rows.

The following notations are used for describing a NoSQL model with respect to the definition of conceptual models. In addition to attribute names and values that are also present in the conceptual model, we focus here on the structure of rows.

We define a row R^T as a combination of:

- T: the table where the row belongs
- F: the column families of the table
- K: all column names
- V: all atomic values of the column
- key: the row identifier
- P: all attributes mapped as a combination of row, column-family and column name. A attribute path $p \in P$ p $\in$ P. is described as $p = R^T.f.q:v$ where $f \in F$, $q \in K$ and $v \in V$.

The example displayed in Fig. 3 uses a tree-like representation and describes a row (r_i) identified by the key named Key (with a value v_0) in a table called *SSB*.

Table

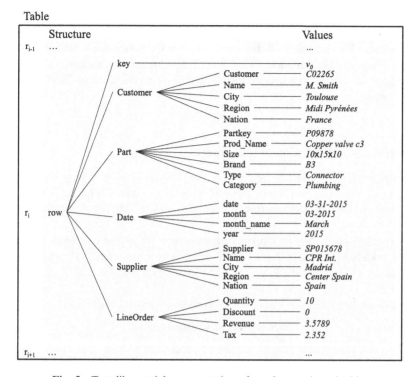

Fig. 3. Tree-like partial representation of a column-oriented table.

4.2 Column-Oriented Models for Data Warehousing

In column-oriented stores, the data model is determined not only by its attributes and values, but also by the column families that group attributes (i.e. columns). In relational database models, mapping from conceptual to logical structures is more straightforward. In column-oriented stores, there are several candidate approaches, which can differ on the tables and structures used. So far, no logical model has been proven better than another one and no mapping rules are widely accepted.

In this section, we present three logical column-oriented models. The first two models do not split data. Data contains redundancy as all the data about one fact and its related dimensions is stored in one table. The first model (*MLC0*) stores data grouped in a unique column family. In the second model (*MLC1*), we use one column family for each dimension and one dedicated for the fact. The third model (*MLC2*) splits data into multiple tables therefore reducing redundancy.

- **MLC0:** For each fact, all related dimensions attributes and all measures are combined in one table and one column family. We call this approach the "simple flat model".
- **MLC1** (inspired from [4]): For each fact, all attributes of one dimension are stored in one column family dedicated to the dimension. All fact attributes (measures) are stored in one column family dedicated to the fact attributes. Note that there are different ways to organize data in column families and this one of them.

- **MLC2:** For each fact and its dimensions, we store data in dedicated tables one per dimension and one for the fact table. We keep these tables simple: one column family only. The fact table will have references to the dimension tables. We call this model the "shattered model". This model has known advantages such as less storage space usage, but it can slow down querying as joins in NoSQL can be problematic.

4.3   Mappings with the Conceptual Model

The formalism that we have defined earlier enables us to define a mapping from the conceptual multidimensional model to each of our three logical models. Let $O = (F^O, D^O)$ be a cuboid for a multidimensional model E built from the fact F with dimensions in D^E.

Table 1 shows how we can map any measure m of F^O and any dimension D of D^O into all 3 models $MLC0$, $MLC1$ and $MLC2$. Let T be a generic table, T^D a table for the dimension D, T^F a table for a fact F and cf a generic column family.

Table 1. Transformation rules from the conceptual model to the logical models.

Conceptual: multidimensional model	Logical: Column-Oriented models		
	MLC0	MLC1	MLC2
$\forall D \in D^O, \forall d \in A^D$ (d is an attribute of D)	$d \rightarrow R^T.$ $cf{:}d$	$d \rightarrow R^T.$ $cf_D{:}d$	$d \rightarrow R^{T^D}.cf.d \wedge$ if $d = id^D$ then $\rightarrow$ $F^{T^F}.cf.d$
$\forall m \in F^O$	$m \rightarrow R^T.$ $cf.m$	$m \rightarrow R^T.$ $cf_F{:}m$	$m \rightarrow R^{T^F}.cf.m$

The above mappings are detailed in the following paragraphs.

Conceptual to MLC0.   To instantiate this model from the conceptual model, three rules are applied:

- Each cuboid O (F^O and its dimensions D^O) is translated into a table T with only one column family cf.
- Each measure $m \in F^O$ is translated into an attribute of cf, i.e. $R^T.cf{:}m$.
- For all dimensions $D \in D^O$, each attribute $d \in A^D$ of the dimension D is converted into an attribute (a column) of cf, i.e. $R^T.cf{:}d$.

Conceptual to MLC1.   To instantiate this model from the conceptual model, five rules are applied:

- Each cuboid O (F^O and their dimensions D^O) is translated into a table T.
- The table contains one column family (denoted cf_F) for the fact F.
- The table contains one column family (denoted cf_D) for every dimension $D \in D^O$.
- Each measure $m \in F^O$ is translated into an attribute (a column) in cf_F, i.e. $R^T.cf_F{:}m$.
- For all dimensions $D \in D^O$, each attribute $d \in A^D$ of the dimension D is converted into an attribute (a column) of cf_D, i.e. $R^T.cf_D{:}d$.

Conceptual to MLC2. To instantiate this model from the conceptual model, three rules are applied:

- Given a cuboid O, the fact F^O is translated into a table T^F with one column family cf and each dimension $D \in D^O$ is translated into a table T^D with one column family cf_D per table.
- Each measure $m \in F^O$ is translated into an attribute of the column family cf in the table T^F, i.e. $R^{T^F}.cf.m$.
- For all dimensions $D \in D^O$, each attribute $d \in A^D$ of the dimension D is converted into an attribute (a column) in the column family cf of the table T^D, i.e. $R^{T^D}.cf.d$. And if d is the root parameter (id^D), the attribute is also translated as an attribute in the column family cf of the table T^F, i.e. $R^{T^F}.cf.d$.

5 Experiments

Our goal is firstly to validate the instantiation of data warehouses with our three logical approaches. Secondly we consider model conversion from one model MLC_i to another MLC_j, with $j \neq i$. Thirdly we generate OLAP cuboids and we compare the computation effort required by each models. We use the Star Schema Benchmark, SSB [5, 15], that is popular for generating data for decision support systems. We use **HBase**, one of the most popular column-oriented system, as NosQL storage system.

5.1 Protocol

Data. Data is generated using an extended version of SSB to generate raw data specific to our models in normalized and denormalized formats. This is very convenient for our experimental purposes.

The benchmark models a simple product retail example and corresponds to a typical decision support star-schema. It contains one fact table "*LineOrder*" and 4 dimensions "*Customer*", "*Supplier*", "*Part*" and "*Date*" (see Fig. 2 for an excerpt). The dimensions are composed of hierarchies; *e.g.* Date is organized according to one hierarchy of attributes (d_date, d_month, d_year).

We use different scale factors (sf), namely sf=1, sf=10, sf=100 in our experiments. The scale factor sf=1 generates approximately 10^7 lines for the "*LineOrder*" fact, for sf=10 we have approximately 10^8 lines and so on. For example, using the split model we will have (sf x 10^7) lines for "*LineOrder*" and a lot less for the dimensions which is typical as facts contain a lot more information than dimensions.

Data Loading. Data is loaded into HBase using native instructions. These are supposed to load data faster when loading from files. The current version of HBase loads data with our logical model from CSV[2] files.

Lattice Computation. To compute the aggregate lattice, we use Hive on top of HBase to ease query writing as Hive queries are SQL-like. Four levels of aggregates are computed on top of the detailed facts (see Fig. 5). These aggregates are: all combinations of 3 dimensions, all combinations of 2 dimensions, all combinations of 1 dimension, and all data (detailed fact data). At each aggregation level, we apply aggregation functions: *max, min, sum* and *count* on all measures.

Hardware. The experiments are done on a cluster composed of 3 PCs (4 core-i5, 8 GB RAM, 2 × 2 TB disks, 1 Gb/s network), each being a *worker node* and one of them also acting as dispatcher (*name node*).

5.2　Experimental Results

In Table 2 we summarize data loading times by model and scale factor. We can observe at scale factor SF1, we have 10^7 lines on each line order table for a 997 MB disk memory usage for MLC2 (3.9 GB for both MLC0 and MLC1). At scale factor SF10 and SF100 we have respectively 10^8 lines and 10^9 lines and 9.97 GB (39 GB MLC0 and MLC1) and 97.7 GB (390 GB MLC0 and MLC1) for of disk memory usage. We observe that memory usage is lower in the MLC2 model. This is explained by the absence of redundancy in the dimensions. For all scale factors, the "dimension" tables "Customers", "Supplier", "Part" and "Date" have respectively 50000, 3333, 3333333 and 2556 records.

Table 2. Data loading time and storage space required for each model in HBase.

	MLC0	MLC1	MLC2
SF1 (sf=1, 10^7 lines)	380 s / 3.9 GB	402 s / 3.9 GB	264 s / 0.997 GB
SF10 (sf=10 10^8 lines)	3458 s / 39 GB	3562 s / 39 GB	2765 s / 9.97 GB
SF100 (sf=100, 10^9 lines)	39075 s / 390 GB	39716 s / 390 GB	33097 s / 99.7 GB

In Fig. 4, we show the time needed to convert one model to another model using SF1 data. When we convert data from MLC0 to MLC1 and vice-versa conversion times are comparable. To transform data from MLC0 to MLC1 we records are just split on the several columns families and during the reverse (MLC1 to MLC0), we fuse records. The conversion is more complicated when we consider MLC0 and MLC2. To convert MLC0 data into MLC2 we need to split data in multiple tables: we have to apply 5 projections on original data and select only distinct keys for dimensions. Although, we produce less data (in memory usage), more processing time is needed than when converting data to MLC1. Converting from MLC2 to MLC0 is the slowest process by far. This is due to the fact that most NoSQL systems (including HBase) do not natively support joins efficiently.

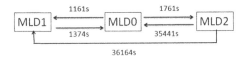

Fig. 4. Inter-model conversion times using SF1.

In Fig. 5, we sumarize experimental results concerning the computation of the OLAP cuboids at different levels of the OLAP lattice for SF1 using data from the model MLC0. We report the time needed to compute the cuboid and the number of records it contains.

We observe as expected that the number of records decreases from one level to the lower level. The same is true for computation time. We need between 550 and 642 s to compute the cuboids at the first level (using 3 dimensions). We need between 78 s and 480 s at the second layer (using 2 dimensions). And we only need between 2 and 23 s to compute the cuboids at the third and fourth level (using 1 and 0 dimensions).

OLAP cube computation using the model MLC1 provides similar results. The performance is significantly lower with the MLC2 model due to joins. These differences involve only the first layer of the OLAP lattice (the layer composed of cuboids constructed using 3 dimensions), as the other layers can be computed from the latter (aggregation functions used are all *commutative* [1]). Table 3 summarizes these differences in computation time. We also report the full results for computing all lattice aggregates using MLC0 in Fig. 5 where arrows show computation paths (e.g. the view

Table 3. Computation time of the first layer of OLAP lattice (3 dimension combinations).

	MLC1	MLC M2	MLC M0
CSD	556 s	4892 s	564 s
CSP	642 s	5487 s	664 s
CPD	573 s	4992 s	576 s
SPD	540 s	4471 s	561 s

Dimensions used: C = Customer,
S = Supplier, D = Date, P = Part
(i.e. Product)

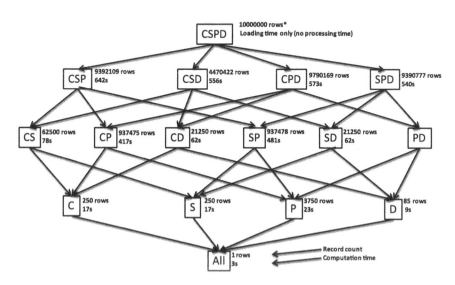

Fig. 5. Computation time and record count for each OLAP cuboid (letters are dimension names: C=Customer, S=Supplier, D=Date, P=Part/Product).

or cuboid CD can be computed from all cuboids that combine the C and D dimensions (*Customer* and *Date*): CSD and CPD).

Discussion. We observe that comparable times are required to load data in one model with the conversion times (except of MLC2 to MLC0). We also observe "reasonable[3]" times for computing OLAP cuboids. These observations are important. At one hand, we show that we can instantiate data warehouses in document-oriented data systems. On the other, we can think of cuboids of OLAP cube lattice that can be computed in parallel with a chosen data model.

6 Conclusion

In this paper, we studied instantiating multidimensional data warehouses using NoSQL column-oriented systems. We proposed three approaches to implement column-oriented logical model. Using a simple formalism that separate structures from values, we described mappings from the conceptual level (described using a multidimensional conceptual schema) to the logical level (described using NoSQL column-oriented logical schemas).

Our experimental work illustrates the instantiation of a data warehouse with each of our three approaches. Each model has its own weaknesses and strengths. The shattered model (MLC2) uses less disk space, but it is quite inefficient when it comes to answering queries (most requiring joins in this case). The simple models MLC0 and MLC1 do not show significant performance differences. Converting from one model to another is shown to be easy and comparable in time to "data loading from scratch". One conversion is significantly very time consuming and corresponds to merging data from multiple tables (MLC2) into one unique table. Interesting results were also obtained when computing the OLAP cuboid lattice using the column-oriented models and they are reasonable enough for a big data framework.

For future work, we will consider logical models in alternative NoSQL architectures, i.e. document-oriented models as well as graph-oriented models. Moreover, after exploring data warehouse instantiation across different NoSQL systems, we need to generalize across all these logical models. We need a simple formalism to express model differences and we need to compare models within each paradigm and across paradigms (e.g. document versus column). Finally we intend to study others query languages frameworks such as PIG or PHOENIX and compare them with Hive.

Acknowledgements. These studies are supported by the ANRT funding under CIFRE-Capgemini partnership.

References

1. Bosworth, A., Gray, J., Layman, A., Pirahesh, H.: Data cube: a relational aggregation operator generalizing group-by, cross-tab, and sub-totals. Technical Report MSR-TR-95-22, Microsoft Research February 1995

---

[3] "Reasonable time" for a Big Data environment running on commodity hardware (without an optical fiber network between nodes, i.e. the recommended 10,000 GB/s).

2. Chang, F., Dean, J., Ghemawat, S., Hsieh, W.C., Wallach, D.A., Burrows, M., Chandra, T., Fikes, A., Gruber, R.E.: Bigtable: a distributed storage system for structured data. ACM Trans. Comput. Syst. **26**(2), 4:1–4:26 (2008)
3. Chaudhuri, S., Dayal, U.: An overview of data warehousing and olap technology. SIGMOD Rec. **26**, 65–74 (1997)
4. Chevalier, M., El Malki, M., Kopliku, A., Teste, O., Tournier, R.: Implementing multidimensional data warehouses into NoSQL. In: 17th International Conference on Enterprise Information Systems, vol. DISI
5. Chevalier, M., El Malki, M., Kopliku, A., Teste, O., Tournier, R.: Benchmark for OLAP on NoSQL technologies, comparing NoSQL multidimensional data warehousing solutions. In: 9th International Conference on Research Challenges in Information Science (RCIS), IEEE
6. Colliat, G.: Olap, relational and multidimensional database systems. SIGMOD Rec. **25**(3), 64–69 (1996). http://doi.acm.org/10.1145/234889.234901
7. Cuzzocrea, A., Bellatreche, L., Song, I.Y.: Data warehousing and olap over bigdata: current challenges and future research directions. In: Proceedings of the Sixteenth International Workshop on Data Warehousing and OLAP, pp. 67–70. DOLAP 2013, ACM, New York, NY, USA (2013)
8. Cuzzocrea, A., Song, I.Y., Davis, K.C.: Analytics over large-scale multidimensionaldata: the big data revolution! In: Proceedings of the ACM 14th International Workshop on Data Warehousing and OLAP, pp. 101–104. DOLAP 2011, ACM, New York, NY, USA (2011)
9. Dehdouh, K., Boussaid, O., Bentayeb, F.: Columnar NoSQL star schema benchmark. In: Ait Ameur, Y., Bellatreche, L., Papadopoulos, G.A. (eds.) MEDI 2014. LNCS, vol. 8748, pp. 281–288. Springer, Heidelberg (2014)
10. Golfarelli, M., Maio, D., Rizzi, S.: The dimensional fact model: a conceptual modelfor data warehouses. Int. J. Coop. Inf. Syst. **7**, 215–247 (1998)
11. Harter, T., Borthakur, D., Dong, S., Aiyer, A.S., Tang, L., Arpaci-Dusseau, A.C., Arpaci-Dusseau, R.H.: Analysis of hdfs under hbase: a facebook messages casestudy. In: FAST, pp. 199–212 (2014)
12. Kimball, R., Ross, M.: The Data Warehouse Toolkit: The Definitive Guide to Dimensional Modeling. John Wiley & Sons, Inc. (2013)
13. Lakshman, A., Malik, P.: Cassandra: a decentralized structured storage system. SIGOPS Oper. Syst. Rev. **44**(2), 35–40 (2010)
14. Li, C.: Transforming relational database into hbase: a case study. In: International Conference on Software Engineering and Service Sciences (ICSESS), IEEE, pp. 683–687 (2010)
15. O'Neil, P., O'Neil, E., Chen, X., Revilak, S.: The star schema benchmark and augmented fact table indexing. In: Nambiar, R., Poess, M. (eds.) TPCTC 2009. LNCS, vol. 5895, pp. 237–252. Springer, Heidelberg (2009)
16. Ravat, F., Teste, O., Tournier, R., Zuruh, G.: Algebraic and graphic languages for OLAP manipulations. IJDWM **4**(1), 17–46 (2008)
17. Stonebraker, M.: New opportunities for new sql. Commun. ACM **55**(11), 10–11 (2012)
18. Vajk, T., Feher, P., Fekete, K., Charaf, H.: Denormalizing data into schema-free databases. In: 4th International Conference on Cognitive Infocommunications (CogInfoCom), IEEE, pp. 747–752 (2013)
19. Zhao, H., Ye, X.: A Practice of TPC-DS multidimensional implementation on NoSQL database systems. In: Nambiar, R., Poess, M. (eds.) TPCTC 2013. LNCS, vol. 8391, pp. 93–108. Springer, Heidelberg (2014)

A Framework for Building OLAP Cubes on Graphs

Amine Ghrab[1,2,3]($\boxtimes$), Oscar Romero[3], Sabri Skhiri[1],
Alejandro Vaisman[4], and Esteban Zimányi[2]

[1] EURA NOVA R&D, Mont-Saint-Guibert, Belgium
{amine.ghrab,sabri.skhiri}@euranova.eu
[2] Université Libre de Bruxelles, Brussels, Belgium
ezimanyi@ulb.ac.be
[3] Universitat Politècnica de Catalunya, Barcelona, Spain
oromero@essi.upc.edu
[4] Instituto Tecnológico de Buenos Aires, Buenos Aires, Argentina
avaisman@itba.edu.ar

Abstract. Graphs are widespread structures providing a powerful abstraction for modeling networked data. Large and complex graphs have emerged in various domains such as social networks, bioinformatics, and chemical data. However, current warehousing frameworks are not equipped to handle efficiently the multidimensional modeling and analysis of complex graph data. In this paper, we propose a novel framework for building OLAP cubes from graph data and analyzing the graph topological properties. The framework supports the extraction and design of the candidate multidimensional spaces in property graphs. Besides property graphs, a new database model tailored for multidimensional modeling and enabling the exploration of additional candidate multidimensional spaces is introduced. We present novel techniques for OLAP aggregation of the graph, and discuss the case of dimension hierarchies in graphs. Furthermore, the architecture and the implementation of our graph warehousing framework are presented and show the effectiveness of our approach.

1 Introduction

As the business and social environments become more interconnected and dynamic, graph-structured data become more prominent. Graphs have the benefit of revealing valuable insights from their topological properties. A new class of business facts and measures could be explored within the multidimensional space built from graphs. In addition, a multitude of emerging decision making problems can be represented using graph models and solved using graph algorithms. Common problems are fraud detection, trends prediction, real-time recommendation and Master Data Management just to name a few [1,2]. For example, by examining the eigenvector centrality in a social network, an analyst can detect influential people or communities. This information could then be

© Springer International Publishing Switzerland 2015
T. Morzy et al. (Eds.): ADBIS 2015, LNCS 9282, pp. 92–105, 2015.
DOI: 10.1007/978-3-319-23135-8_7

reused for recommendation and targeted advertising. In financial services, complex graph patterns could also be used to represent and detect complex rings which might lead to discover fraudulent transactions. Such scenarios rely mostly on the analysis of complex relationships between data entities, which is difficult to formulate and expensive to process using traditional relational systems [1]. We experience thus a growing need to integrate graph data within decision support systems. Such integration will help decision makers get an extended view and thus better understanding of their business environments and make more informed decisions.

Current decision support systems often rely on data stored in the organization's data warehouse. Data in the data warehouse is modeled following the multidimensional model, represented using the cube metaphor and interactively queried using the OLAP paradigm. However, traditional decision making systems, and particularly data warehousing solutions were initially developed to support relational data, and are not equipped for the efficient analysis and aggregation of graph properties. To extend current decision support systems with graph data and gain new insights over graphs, we need to design a novel OLAP technique aware of the specific properties of graphs.

Many approaches were proposed in the literature to extend current decision support systems with graphs [3–5]. They suggested the first foundations for building OLAP cubes on graphs. However, their techniques focused mostly on homogeneous graphs (i.e., graphs where all nodes are of the same type, and all edges are the same), and the OLAP analysis focus mainly on the graph topology as the measure of interest [3–5]. In such cases, all the attributes of the graph elements are considered as the dimensions and are used for aggregating the graph and performing its multi-perspective analysis. However, real-world graphs are complex and often heterogeneous. In this paper, we extend the state of-the-art to heterogeneous graphs (i.e., graphs where nodes and edges could be of different types to represent different real-world entities, and the different relationships between them). Therefore, not all attributes could be considered as dimensions through which the graph could be examined. We also examine a new class of measures to get additional insights from the graph topology. We extend the analysis capabilities on graphs by integrating GRAD, an analysis-oriented graph database model [6,7]. GRAD natively supports the representation of hierarchies and the analysis of the content of nodes. We use these characteristics to support dimension hierarchies and build additional OLAP cubes on graphs. We propose our novel technique for building OLAP cubes on graphs. Thereby equipping decision makers with the capability of performing effective multi-level/multi-perspectives analysis of their graph data and examining new business facts. Our main contributions in this paper are summarized as follows:

- We define the multidimensional concepts for graph data, and propose novel techniques for extracting the candidate multidimensional concepts and building graph cubes from property graphs.
- We present an extension of the property graph model, tailored for multidimensional analysis, and examine the additional candidate graph cubes brought by

this extension. We further extend our work to support dimension hierarchies within graphs.

– We suggest a graph data warehousing architecture, and provide an effective prototypical implementation of our techniques for building OLAP cubes.

The remaining of this paper is structured as follows: Sect. 2 presents our running example. In Sect. 3, we formally define the multidimensional structures on graphs. Section 4 presents our technique for extracting potential multidimensional spaces and building graph cubes on property graphs. In Sect. 5 we propose a technique for building OLAP cubes on novel graph database model, and extend our approach to support dimension hierarchies in graphs. Section 6 presents the architecture and implementation of our proof-of-concept graph warehousing framework. Section 7 discusses related work. Finally, Sect. 8 sketches future work and concludes the paper.

2   Running Example

We illustrate the analysis opportunities brought by graphs using a movie graph. The original dataset was published by the GroupLens research group.[1] The resulting graph contains movies with attributes, such as the year of release, titles, ratings and scores from different communities etc. Each movie is linked to its actors with an edge that contains the rank of the actor on the movie. We further enrich the dataset with information about actors' birth date and nationality, and movies country from the Movie Database website.[2] Figure 1(a) shows a subgraph of the movie graph. We start with a simple and flat multidimensional schema shown in Fig. 1(b). We introduce in Sect. 5 a more complete schema supporting hierarchies and enabling more advanced analysis.

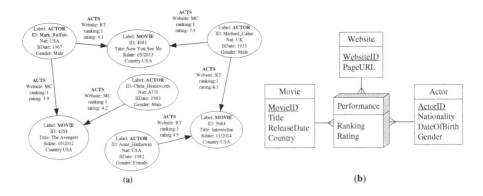

(a)   (b)

Fig. 1. A sample movie graph

---

[1] http://grouplens.org/datasets/movielens.

[2] https://www.themoviedb.org/.

3   Multidimensional Concepts on Graphs

In this section we formally define the multidimensional structures in the context of heterogeneous attributed graphs. We start with dimension levels.

Definition 1 [*Dimension Level*]. A level is a pair $L_i = \langle name, \mathcal{P}_i \rangle$, where *name* is the name of the level, and $\mathcal{P}$ is the aggregation pattern. $\mathcal{P} = (T, C)$ is a pair, where T is the pattern's topology and C are the constraints applied on its content (i.e. attributes). $\mathcal{P}$ is used to identify all graph elements that belong to the dimension's level and that should be merged after roll-up.   □

Dimensions provide the possible perspectives for the analysis of the graph topology and content. In graphs, we distinguish two types of dimensions: (1) Node-based dimensions, which are represented by the attributes of the nodes, and (2) Edge-based dimensions, which are represented by the attribute of the edges. We define a dimension as follows:

Definition 2 [*Dimension*]. A dimension is defined as $\mathcal{D} = \langle name, \mathcal{L}, \mathcal{R} \rangle$, where $\mathcal{L} = \{L_1, ..., L_n, All\}$ is the set of the dimension levels. $\mathcal{R}$ is a partial order on the elements of $\mathcal{L}$ and describes a directed acyclic graph defining the hierarchy and the aggregation direction between the dimension's levels L_i. The base level L_1 and highest level *All* are located at the ends of the partial order.   □

In the multidimensional model, a measure is the basic unit of data that is placed in the multidimensional space and examined through the dimensions.

Definition 3 [*Measures*]. A measure m is identified by the triple $\langle name, \mathcal{F}, \mathcal{A} \rangle$. It is computed over a graph $\mathcal{G} \in \mathbb{G}$ using a function $\mathcal{F}$ as follows: $\mathcal{F} : \mathbb{G} \longrightarrow Dom(m)$. In graphs, $\mathcal{F}$ could be a graph-specific function such the PageRank algorithm. $\mathcal{A}$ is the aggregation function (e.g., SUM, AVG etc.) used to compute an aggregated value of the measure.   □

Multiple classification for graph measures were proposed in the literature, such as the classification by the aggregation type (i.e., distributive, algebraic and holistic) [3]. Here we propose a new classification of graph measures, based on the type and the computation algorithm.

- **Content-Based Measures:** They are extracted from the attributes of graph elements. These measures are similar to the traditional measures and do not capture the graph topology. For example, the average rating of a movie and the average rank of an actor are content-based measures.
- **Graph-Specific Measures:** They capture the topological properties of graphs and are obtained by applying graph algorithms. They could be classified according to the type of the output as either (1) *numerical*, where the output is a numerical value such as the value of the page-rank, or (2) *topological*, where the measure is represented using graph structures such as the path between a pair of nodes. The second possible classification makes the distinction between (1) *local* measures, which are computed separately for graph

nodes or edges (e.g., the centrality of an actor), and (2) *global* measures which are computed for the whole graph (e.g., the diameter or number of cycles of the graph).

– **The Graph as a Measure:** As discussed by Chen et al. in [3], the graph itself could be considered as a measure examined from different perspectives and at different aggregation levels.

The cube metaphor is widely accepted as the underlying logical construct for conventional multidimensional models. Here we define the concept of cube using the notion of aggregate graphs defined as follows.

Definition 4 [*Aggregate Graph*]. An aggregate graph $\mathcal{G}'$ of an initial graph $\mathcal{G}$ is a graph obtained by condensing a subset of the nodes and edges of $\mathcal{G}$. Hence, each node corresponds to a set of nodes in $\mathcal{G}$, and each edge is the result of fusion of edges between pairs of aggregated nodes.   □

Definition 5 [*Graph Cube*]. A graph cube corresponds to a set of aggregate graphs obtained by restructuring the initial graph $\mathcal{G}$ in all possible aggregations. Each cuboid is therefore represented as an aggregate graph of $\mathcal{G}$. If an aggregation is performed from L_i to L_{i+1}, all graph elements that satisfy the aggregation pattern $\mathcal{P}_i$ are aggregated in the same node. The edges are constructed afterwards to link the pairs of nodes. Measures are then recomputed and placed on the aggregate graph.   □

In the next sections, we show how these formal definitions map to the specific graph structures of each model and illustrate them with examples applied on the movie graph. We discuss how to select a valid subset of attributes as the candidate dimensions or measures, and build the different graph cubes.

4   Building OLAP Cubes on Property Graphs

Many current graph databases represent graphs using the *property graphs* model [8]. We show in this section how we can use property graphs as a first foundation for building OLAP cubes. However, since property graphs describe basic graph structures (which are simple and oriented for storage and operational workloads), their analysis capabilities are limited. For advanced multidimensional modeling and analysis, richer graph structures are needed as we show later in Sect. 5.

Property graphs describe a directed, labeled and attributed multi-graph. Formally, we define a property graph as follows:

Definition 6 [*Property Graph*]. A property graph is represented as $\mathcal{G} = (\mathcal{V}, \mathcal{E}, \mathcal{L}_v, \mathcal{L}_e, \Lambda_v, \Lambda_e)$, where:

– $\mathcal{V}$ is the set of nodes.
– $\mathcal{E} \subseteq \mathcal{V} \times \mathcal{V}$ is the set of edges.
– $\mathcal{L}_v$ is the set of node labels and $\mathcal{L}_e$ is the set of edge labels.
– $\Lambda_v = \{a^1, a^2, ..., a^m\}$ is the set of node attributes represented as key/value pairs. Each node $v_i \in \mathcal{V}$ is associated with an attribute vector $\lambda_{v_i} = [a^1, a^2, ..., a^j]$.

– $\Lambda_e = \{b^1, b^2, ..., b^n\}$ is the set of edge attributes represented as key/value pairs. Each edge $e_i \in \mathcal{E}$ is associated with an attribute vector $\lambda_{e_i} = [b^1, b^2, ..., b^k]$. $\square$

A node $v_i \in \mathcal{V}$ is represented as $v_i = (l_i, \lambda_{v_i})$, where $l_i \in \mathcal{L}_v$ is the label and λ_{v_i} is the set of attributes. Similarly, an edge $e_j \in \mathcal{E}$ is represented as $e_j = (v_s, v_e, l_j, \lambda_{e_j})$, where v_s and v_e are the start and end nodes respectively, $l_j \in \mathcal{L}_e$ is its label and λ_{e_i} is its set of attributes. Each node (resp. edge) on the graph has exactly one label. We introduce the concept of **class** (denoted Σ_i) to describe a set of graph nodes that share the same label. For example, in the movie graph of Fig. 1(a), we have two classes which are MOVIE and ACTOR.

Given a property graph $\mathcal{G}$ and a pair of nodes from two connected but distinct classes of nodes, we explore the candidate dimensions, measures and cubes that could be built by exploring the graph of these two classes. We denote dimensions that span across two linked classes as inter-class dimensions, defined as follows.

Definition 7 [Inter-Class Dimensions]. Let $\mathcal{G}$ be a property graph, and let $v_s \in \Sigma_s$ and $v_e \in \Sigma_e$ be a pair of nodes from two distinct classes. Let $e_i = (v_s, v_e, l_i, \lambda_{e_i})$ be an edge that relates v_s and v_e. The node-based dimensions are the attributes of the two nodes v_s and v_e (i.e., $\lambda_{v_s} = [a^1, ..., a^k]$ and $\lambda_{v_e} = [a^1, ..., a^l]$). The candidate edge-based dimensions are a subset of the attributes of the edge e_i (i.e., $\lambda_{e_i} = [b^1, ..., b^k]$). $\square$

Example 1 (Analysis of Rating and Ranking of Actors per Website and Movie). Using the movie running example, the node-based dimensions are the attributes $\lambda_{v_{Movie}} = [ReleaseDate, Country]$ and $\lambda_{v_{Actor}} = [Nationality, DateOfBirth, Gender]$. For example, following the notation of Sect. 3, $\mathcal{D}_{Gender} = \langle Gender, \mathcal{L}, \mathcal{R} \rangle$, with the levels being the base level *Gender* and *ALL*. Therefore, all actor nodes could be at the base level where they all have the attribute *Gender*, and then could be grouped into two groups (i.e., a node for male actors, and a node for female actors), and finally grouped in one node regardless the gender. The edge-based dimension is represented by the $\lambda_{e_{ACTS}} = [Website]$ attribute of the ACTS edge relating actors and movies.

The graph lattice enumerates all possible OLAP aggregations of the graph, and is obtained by aggregating over all the inter-class dimensions. Figure 2 shows the graph lattice applied to the graph of Fig. 4, considering the dimensions of the previous example. Each node of the graph lattice represents an aggregate graph, that is, a cuboid of the graph cube. We distinguish three special kinds of aggregation on this graph (highlighted in Fig. 2), which are *Movie*-only aggregations (i.e., only movie nodes are kept not fully aggregated to the All level), *ACTS*-only aggregation and *Actor*-only aggregations.

Definition 8 [Inter-Class Measures]. Given a property graph $\mathcal{G}$ and a set of edges $E \subset \mathcal{E}$ relating nodes of the classes Σ_s and Σ_e, a content-based measure m_c is computed by applying an aggregation function on the attributes ($[b^1, ..., b^k]$) of the edges $e_i \in E$. The graph considered as a measure is obtained following the graph lattice, and the graph-specific measures are obtained by applying a graph algorithms on $\mathcal{G}$, or one of its aggregate graphs. $\square$

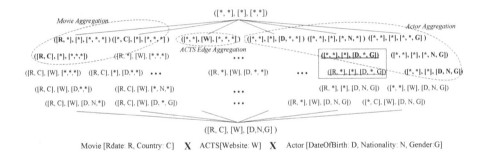

Fig. 2. The graph lattice of the movie graph

In order to analyze the properties of the relationships between the graph entities, we focus here on the potential measures existing within the edges. Clearly,
we cannot assume that all attributes of the edges are dimensions. As shown
by the multidimensional schema of Fig. 1(b), the attribute *Website* of the edge
labeled *ACTS* could indeed be a dimension. However, the attributes *ranking*
and *rating* are rather considered as measures in the current analysis scenario.
We should note that the distinction between attributes that are dimensions and
attributes that are measures is not straightforward, and thus requires a modeling
effort from the designer to distinguish them.

Now we apply these dimensions and measures on the property graph of
Fig. 1(a), and follow the graph lattice of Fig. 2 in order to study the graph
cube reflecting the ranking and rating of actors in the movie graph. Figure 3(a)
shows the aggregate graph (i.e., graph cuboid) where movies are grouped by

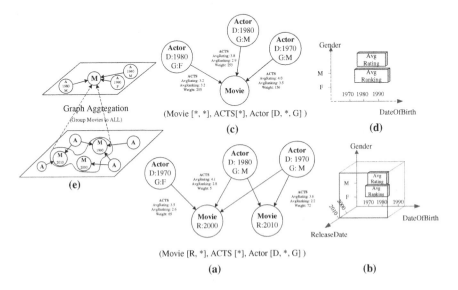

Fig. 3. OLAP aggregation of the movie graph and computation of the OLAP cubes

release date and actors are grouped by birth date and gender. A corresponding OLAP cube is shown in Fig. 3(b). The measures are *AverageRanking* and *AverageRating* of actors, which can be examined through the three dimensions left (i.e., *ReleaseDate*, *DateOfBirth* and *Gender*). We follow the graph aggregation as depicted by Fig. 3(e) to get the graph (Fig. 3(c)) and the cuboid (Fig. 3(d)) at the next aggregation level. On the lattice of Fig. 2, this aggregation corresponds to the two nodes underlined and put in rectangle. Note here that for graph-specific measures (e.g., closeness centrality of actors), the measures for the upper-level could not be computed directly from the cube at a lower level, as the computation function needs to traverse the aggregated graph itself to compute the new value of the graph-specific measure.

5   Building OLAP Cubes on GRAD

Many graph models were proposed in the literature to abstract different types of graphs and fit their particular analysis workloads [9]. In [6,7], we proposed GRAD, an analysis-oriented graph database model that extends property graphs with advanced graph structures, integrity constraints and a graph algebra. We use GRAD as the foundation for the OLAP cubes extraction techniques we present in this section.

As we discussed in the previous section, property graphs support OLAP analysis of inter-classes facts. However, they fall short from supporting OLAP analysis of the internal information stored within each node, or class of nodes. Therefore, we focus in this section on the additional cubes and analysis capabilities brought by GRAD. Note however that since GRAD extends property graphs, the candidate multidimensional spaces and cubes discussed in the previous section could similarly be built using GRAD.

5.1   OLAP Cubes on GRAD

Due to space limitations, we briefly introduce here the main components of the database model. In GRAD, we consider heterogeneous, attributed and labeled graphs. Complex attributes are supported on the nodes and rich semantics is explicitly expressed on the edges. The analysis process is centered around special analytical structures, namely *hypernodes* and *classes*. Hypernodes represent real world entities and are grouped within classes. Each analytics hypernode is an induced subgraph grouping an entity node, all its attribute and literal nodes, and all the edges between them. The core of a hypernode is the entity node which contains the label and the identifier attributes of the real world entity. Attribute nodes are attached to the entity node and denote the non-identifier, and potentially multi-valued attributes of each entity (e.g., budget, revenue). Literal nodes record the effective value of its corresponding attribute node. Rich semantics are embedded on the graph edges such as multiplicities, hierarchical and composition relationships.

Definition 9 [*GRAD Graph*]. A GRAD graph is denoted as $\mathcal{G} = (\mathcal{V}, \mathcal{E}, \mathcal{L}_v, \mathcal{L}_e, \Lambda_v, \Lambda_e)$, is formally defined as follows:

- $\mathcal{V} = (V_e \cup V_a \cup V_l)$ is the set of nodes, with V_e being the set of entity nodes, V_a the set of attribute nodes, and V_l the set of literal nodes.
- $\mathcal{L}_v = \{C_i, L_a\}$ is the set of labels on entity and attribute nodes respectively.
- $\Lambda_v = \{b^1, b^2, ..., b^n\}$ is the set of entity node attributes represented as key/value pairs. Each node is associated with a vector of j attributes $[b^1, b^2, ..., b^j]$.
- $\mathcal{E} = (E_e \cup E_a \cup E_l)$ is the set of edges, with E_e being the set of entity edges, E_a the set of attribute edges, and E_l the set of literal edges. All entity edges on the graph share the same label.
- $\Lambda_e = \{b^1, b^2, ..., b^m\}$ is the set of edge attributes represented as key/value pairs. Each edge is associated with a vector of k attributes $[b^1, b^2, ..., b^k]$. $\square$

Figure 4(a) illustrates a part of the movie graph modeled with GRAD. In this example, *Movie* is an entity node, while *Revenue* is an attribute node attached to *Movie*. The revenue has different values depending on a set of factors (location, time, language etc.), and each value is stored separately in a literal node.

In the previous section, we used property graphs to study the candidate multidimensional cubes between classes of nodes. Given a GRAD graph $\mathcal{G}$ and a class of entity nodes Σ_i, we explore in this section the candidate dimensions, measures and cubes that could be extracted from a single class Σ_i.

Definition 10 [*Intra-Class Dimension*]. Given a GRAD graph $\mathcal{G}$, a class of entity nodes Σ_i, and an entity node $u \in \Sigma_i$ with ID_u being the set of identifiers attributes of u. Then we can extract distinct sets of candidate dimensions. Each set of dimensions is the union between the attributes of the entity node and the attributes of literal edge of a given attribute node. For a given attribute node $v_i \in V_a$ linked to the entity node u, where $\lambda_i \subset \Lambda_e$ is the attributes of the literal edge $e \in E_l$ connected to v_i, $D_{v_i} = \{ID_u \cup \lambda_i\}$. $\square$

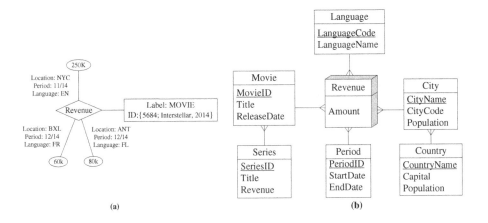

(a)   (b)

Fig. 4. Movies and actors' graph

Definition 11 [*Intra-Class Measures*]. They are defined by the triple $\langle name, \mathcal{F}, \mathcal{A} \rangle$ and are explored within each hypernode. The label of the attribute node is the name of the measure ($name \in L_a$). The actual values of these measures are embedded on the attributes of the literal nodes ($\mathcal{F}(v) \in [b^1, b^2, ..., b^k]$). □

Example 2 (Analysis of the Revenue of a Movie). Given the example of Fig. 4, suppose an analyst need to analyze the revenue of movies following the multi-dimensional schema of Fig. 4(b). Revenue is therefore considered as the name of the measure, which is the same as the label of the attribute node *Revenue*. The aggregation function is SUM. The values of the measures are stored within the literal nodes linked to the *Revenue* attribute node and the function computing the measure is the same as the one used to retrieve the value from the literal node. The dimensions for the revenue measure are named *Movie, Location, Period*, and *Language*. Given these dimensions, we can aggregate the graph to examine the value of revenue by navigating through the dimension hierarchy of the *Location* dimension from *City* to *Country* as shown in Fig. 5(a), or by rolling up to the level *ALL* of the language dimension as in Fig. 5(b). Concretely, at the graph level, this operation will incur merging the corresponding literal storing the measure values.

We distinguish here two types of graph aggregations: (1) Intra-hypernode aggregation, where literal nodes and edges of the same attribute node are merged, thus the dimensions is an attribute of the literal edges (e.g., revenue of a given movie by language), (2) Inter-hypernode aggregation, where entity nodes could be merged (e.g., revenue of all movies per given a city, period and language).

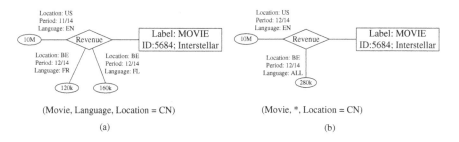

Fig. 5. Aggregation of revenue by language

5.2   Dimension Hierarchies on GRAD

In this subsection, we consider extending the OLAP analysis to support hierarchies within inter-class and intra-class dimensions.

– Dimension hierarchy for intra-class dimensions: Within each dimension (i.e., attribute location of revenue), we might have an inner hierarchy (e.g., City, Region, and Country). Therefore, we can extend the lattice with these new possible aggregations as shown in Fig. 5(a).

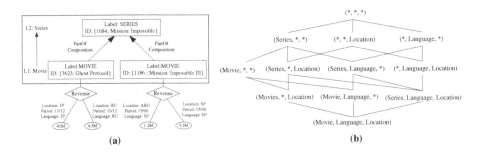

Fig. 6. Dimension hierarchy between classes

– Dimension hierarchy for inter-class dimensions: Explored between distinct classes of nodes. Within GRAD, specific types of edges such as composition and aggregation could be explicitly defined. Therefore, classes of nodes related by these specific relationships belong to the same dimension with the hierarchy following the child-parent direction of these relationships. Figure 6(a) shows the hierarchy of the movie dimension that is now composed by *Movie* and *Series* levels. The updated lattice is shown in Fig. 6(b).

6   Framework Architecture and Implementation

In this section, we present our prototypical implementation of the OLAP cubes extraction approach using Neo4j. The framework architecture is depicted in Fig. 7. The major components of our implementation are described as follows:

1. Graph ETL: The graph is extracted from external data sources that might have various formats (e.g., XML as for DBLP, or text files for MovieLens, etc.). For the running example, we have developed two modules for extracting and matching data from CSV files of MovieLens with data about actors from The Movie Database. The data is then formatted following GRAD and property graph structures before being loaded as the base graph on Neo4j.
2. Graph storage and materialization: The graph data is stored using multiple Neo4j graph database instances. We use two particular databases, one to store the graph at the base level and the other to keep the lattice. The other instances store the aggregate graphs. However, we needed a database-per-aggregate graph because Neo4j do not support materialized views on graph, and could not separate between subgraphs of the same database.
3. Graph lookup and update: This component acts as a middleware between the storage and processing layers. It loads the graph, at a given aggregation level, from a Neo4j database into HDFS to prepare it for distributed processing or aggregation. Once the processing is done, this layer stores the graph back into a new Neo4j instance if the graph was aggregated, or updates the original database if only some attributes were updated.

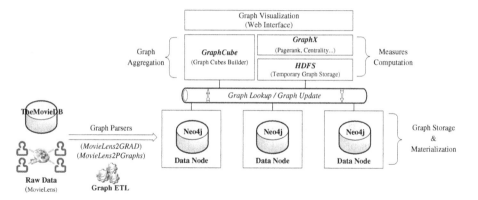

Fig. 7. Distributed OLAP cubes computation

4. Graph Aggregation and Measures Computation: Given a graph lattice, the *GraphCube* module performs the graph aggregation to generate potential graph cuboids as discussed through the paper. In order to efficiently compute of the graph-specific measures (e.g., PageRank or closeness centrality), we use the GraphX library. GraphX performs the iterative graph algorithms in-memory and thus outperforms the other distributed graph libraries on large scale graphs. Once the required graph measures are computed, the result is persisted in the corresponding Neo4j instance using the previous layer.

7 Related Work

Graph Data Warehousing: The challenge of designing graph data warehousing frameworks is part of the challenge of designing novel models and techniques for enabling multidimensional analysis of Big Data [10]. Big Data extracted from business and social environments is complex, scattered, dynamic, heterogeneous and unstructured. Most of it falls outside the decision maker's control. However, as motivated by Abelló et al. [11], incorporating such data into the decision process enables non-expert users to make well-informed decisions when required. Our work provides a foundation for extending decision support to graph data.

Graph Database Modeling: Graph database modeling and querying is the foundation for graph data warehousing. A survey of graph database models is provided by Angles et al. [9]. Multiple native graph models and query languages (e.g. GraphQL [12]) were developed to efficiently answer graph-oriented queries. In this paper, we leveraged and extended the database model we defined on [7] for graph data warehousing.

OLAP on Graphs: GraphOLAP is a conceptual framework for OLAP analysis of a collection of homogeneous graphs [3]. Attributes of the snapshots are considered as the dimensions. Aggregations of the graph are performed by over-laying a collection of graph snapshots. Dimensions are classified as topological and informational. Informational OLAP aggregations consist in edge-centric snapshot overlaying, thus only edges changes and no changes to the nodes are made. Topological OLAP aggregations consist of merging nodes and edges by navigating through the nodes hierarchy. Qu et al. introduced a more detailed framework for topological OLAP analysis of graphs [13]. GraphCube [4] is a framework for OLAP cubes computation and analysis through the different levels of aggregations of a graph. It targets single, homogeneous, node-attributed graphs. The framework introduced the cuboid and crossboid queries for building and analyzing the different graph cubes. Distributed Graph Cube is a distributed framework for graph cubes computation and aggregation implemented using Spark and Hadoop [14]. Pagrol is a Map-Reduce framework for distributed OLAP analysis of homogeneous attributed graphs [5]. Pagrol extended the model of GraphCube by considering the attributes of the edges as dimensions. These frameworks were designed to handle homogeneous graphs [3–5]. The attributes of the graph elements are considered as the dimensions, and the graph cubes are obtained by restructuring the initial graph in all possible aggregation. Yin et al. [15] introduced a data warehousing model for heterogeneous graphs focusing on edge-based dimensions. In this paper, we extended these frameworks to the general case of heterogeneous graphs, and we discussed various techniques for building graph cubes in different settings. In [16], authors introduced a framework for OLAP on RDF data. They proposed GOLAP, a graph model for OLAP on graphs, and FSPARQL an extension to SPARQL for OLAP querying of RDF data. GOLAP introduced a rule-based approach for defining new dimensions on the graph. The same technique could be integrated, as a pre-processing phase, within our work to provide more candidate dimensions and measures.

8   Conclusion

In this paper, we proposed our contribution to graph warehousing by designing novel techniques for building OLAP cubes on graphs. We applied our approach on both property graphs and a more advanced graph database model tailored for multidimensional modeling. We discussed techniques for OLAP aggregation of the graph and tackled the case of dimension hierarchies in graphs. In addition, we provided an overview of the architecture and implementation of our graph warehousing framework.

Graph data warehousing is an emerging research field that brings various challenges similar to traditional data warehousing (e.g. high dimensionality and cubes materialization). However, the structural properties and unstructured nature of graphs calls for the development of novel modeling and processing paradigms. Our immediate future work is to enable multidimensional concepts discovery on graphs within our framework. Yet, many remaining research directions are worth investigating to build industry-grade graph warehousing systems. Among these directions we cite OLAP analysis of dynamic graphs and the definition of a proper OLAP algebra and query language for graphs.

References

1. Robinson, I., Webber, J., Eifrem, E.: Graph Databases. O'Reilly Media Inc, Sebastopol (2013)
2. Petermann, A., Junghanns, M., Müller, R., Rahm, E.: Graph-based data integration and business intelligence with biiig. Proc. VLDB Endow. **7**(13), 1577–1580 (2014)
3. Chen, C., Yan, X., Zhu, F., Han, J., Yu, P.S.: Graph OLAP: a multi-dimensional framework for graph data analysis. Knowl. Inf. Syst. **21**(1), 41–63 (2009)
4. Zhao, P., Li, X., Xin, D., Han, J.: Graph cube: on warehousing and OLAP multidimensional networks. In: Proceedings of the 2011 ACM SIGMOD International Conference on Management of data, pp. 853–864. ACM (2011)
5. Wang, Z., Fan, Q., Wang, H., Tan, K.L., Agrawal, D., El Abbadi, A.: Pagrol: parallel graph olap over large-scale attributed graphs. In: 2014 IEEE 30th International Conference on Data Engineering (ICDE), pp. 496–507, March 2014
6. Ghrab, A., Skhiri, S., Jouili, S., Zimányi, E.: An analytics-aware conceptual model for evolving graphs. In: Bellatreche, L., Mohania, M.K. (eds.) DaWaK 2013. LNCS, vol. 8057, pp. 1–12. Springer, Heidelberg (2013)
7. Ghrab, A., Romero, O., Skhiri, S., Zimányi, E.: Analytics-Aware Graph Database Modeling, Technical report (2014) . http://research.euranova.eu/scientific-publications
8. Rodriguez, M.A., Neubauer, P.: Constructions from dots and lines. Bull. Am. Soc. Inf. Sci. Technol. **36**(6), 35–41 (2010)
9. Angles, R., Gutierrez, C.: Survey of graph database models. ACM Comput. Surv. **40**(1), 1:1–1:39 (2008)
10. Cuzzocrea, A., Bellatreche, L., Song, I.Y.: Data warehousing and OLAP over big data: current challenges and future research directions. In: DOLAP 2013 Proceedings of the Sixteenth International Workshop on Data Warehousing and OLAP, pp. 67–70. ACM, New York (2013)
11. Abelló, A., Darmont, J., Etcheverry, L., Golfarelli, M., Mazón, J.N., Naumann, F., Pedersen, T.B., Rizzi, S., Trujillo, J., Vassiliadis, P., Vossen, G.: Fusion cubes: towards self-service business intelligence. IJDWM **9**(2), 66–88 (2013)
12. He, H., Singh, A.: Query language and access methods for graph databases. In: Aggarwal, C.C., Wang, H. (eds.) Managing and Mining Graph Data. ADS, vol. 40, pp. 125–160. Springer, Heidelberg (2010)
13. Qu, Q., Zhu, F., Yan, X., Han, J., Yu, P.S., Li, H.: Efficient topological OLAP on information networks. In: Yu, J.X., Kim, M.H., Unland, R. (eds.) DASFAA 2011, Part I. LNCS, vol. 6587, pp. 389–403. Springer, Heidelberg (2011)
14. Denis, B., Ghrab, A., Skhiri, S.: A distributed approach for graph-oriented multidimensional analysis. In: IEEE International Conference on Big Data, pp. 9–16 (2013)
15. Yin, M., Wu, B., Zeng, Z.: HMGraph OLAP: a novel framework for multidimensional heterogeneous network analysis. In: Proceedings of the 15th International Workshop on Data Warehousing and OLAP, pp. 137–144. ACM (2012)
16. Beheshti, S.-M.-R., Benatallah, B., Motahari-Nezhad, H.R., Allahbakhsh, M.: A framework and a language for on-line analytical processing on graphs. In: Wang, X.S., Cruz, I., Delis, A., Huang, G. (eds.) WISE 2012. LNCS, vol. 7651, pp. 213–227. Springer, Heidelberg (2012)

A Generic Data Warehouse Architecture for Analyzing Workflow Logs

Christian Koncilia[1]([⊠]), Horst Pichler[1], and Robert Wrembel[2]

[1] Klagenfurt University, Institute of Informatics Systems, Klagenfurt, Austria
{christian.koncilia,horst.pichler}@aau.at
[2] Poznan University of Technology, Institute of Computing Science, Poznań, Poland
Robert.Wrembel@cs.put.poznan.pl

Abstract. This paper proposes an approach to represent and analyze the content of workflow logs in a data warehouse. When analyzing workflow logs one big problem arises: typically, an underlying workflow model consists of loops (frequently interleaving), often implemented by using goto-statements. These structures increase the number of possible execution paths significantly - in theory even indefinitely. In a naive Data Warehouse (DWH) implementation one would represent all possible execution paths by means of a dimension. However, this would lead to a huge or even infinite number of elements in the dimension. In this paper, we present a novel approach for analyzing workflow logs including loops and goto-statements.

1 Introduction

For over 20 years, data analysis has been performed by means of business intelligence (BI) architectures [10] that include a data warehouse (DW) and various on-line analytical processing (OLAP) applications (e.g., for trend analysis, trend prediction, data mining, and social network analysis). In a DW, data are usually organized as cubes, whose cells include values [38]. The cells are referenced by values of dimensions that set up the context of an analysis. Typically, cell values are computed by multiple aggregate and statistical functions.

Nowadays, information systems like workflow management systems, website clickstreams repositories, or public transportation infrastructures generate huge sets of data that are naturally ordered. This *order* implies the existence of various *patterns* that are formed by values of some attributes, which both frequently bear important information. For example, by discovering patterns in a workflow log one could check whether a given real workflow conforms to a defined model, or could enhance/improve/optimize an existing workflow [3]. By computing various statistical measures of a node in a workflow, one could discover bottlenecks in a system.

Data for which an order is important are commonly called *sequential data*. There are two main research directions of sequential data analytics. The first one focuses on developing data mining algorithms for discovering frequent or interesting patterns in sequential data, e.g., [14, 16, 25, 29–31]. The second one

© Springer International Publishing Switzerland 2015
T. Morzy et al. (Eds.): ADBIS 2015, LNCS 9282, pp. 106–119, 2015.
DOI: 10.1007/978-3-319-23135-8_8

focuses on analyzing sequential data in traditional SQL-like way and on searching for given (known) patterns, e.g., [1, 4, 23, 24, 32, 33, 36].

In this paper, we focus on the analysis of workflow logs by means of OLAP techniques. Workflow Management Systems (WfMS) are being widely used by many enterprises to improve business process definitions and automation. In WfMS, objects arrive to a given task at certain time, they are processed there for a certain period, and they may trigger follow-up tasks. A WfMS records in a *log* all steps of a business process execution. Thus, the workflow log contains processing information for all instances of activities of deployed workflow instances. For example, a workflow log records which task in which business process has been performed by which actor, and what was its processing time. This information is most valuable for business performance measurement. Therefore, being able to analyze workflow log data and explore orders between tasks processing is an important requirement. Unfortunately, most of the existing analytical tools either focus on process mining (process discovery), e.g., [39, 40] or offer only basic log analysis functionality based on some aggregate and statistical values [15]. For an OLAP-like analysis such functionalities are insufficient.

1.1 Motivation and Contribution

Standard data warehouses and OLAP techniques do not consider the sequential nature of data stemming from a workflow log. For example, they neglect the fact that a workflow execution consists of instances A, B, ..., and X and that there is a given order $A \rightarrow B \rightarrow ...X$. The only way to overcome this obstacle while computing some statistics about the workflow is to put all possible execution paths into a dimension in a DW cube. This however, is not feasible if an underlying workflow model consists of branches and/or loops, because this results in a significantly increased number of possible execution paths [15].

In order to have a sequential view on a process execution, execution paths need to be captured. Figure 1 shows a simple illustration of some process instances with variants. We can notice that some process instances are subjects to similar execution patterns/behavior. Through these variants, we can find out

CaseId	Event	TimeStamp	Type	UserId	Added data
1	New Ticket	01-02-2014 09:00	Start	101	
1	New Ticket	01-02-2014 09:04	Finish	101	
1	First Response	01-02-2014 10:13	Start	5	
2	New Ticket	01-02-2014 10:15	Start	123	
1	First Response	01-02-2014 10:22	Finish	7	Comments
1	Support Request	01-02-2014 10:23	Start	5	
2	First Response	01-02-2014 10:27	Finish	7	
2	Customer Clarification	01-02-2014 10:28	Start	123	
1	Support Request	01-02-2014 10:32	Finish	5	Solution
	...				

Fig. 1. Example of an event log

the most frequent patterns (i.e., the abstraction from a concrete form which keeps recurring in specific non-arbitrary contexts) as well as which of them perform well or badly. We can then define the best performing variants for a better and more consistent process performance.

Workflow users and process managers are interested in business process-oriented analysis. To produce the information of interest, huge amount of data needs to be: (1) transformed from a workflow log format into a format suitable for analysis, (2) loaded into a data warehouse, and (3) analyzed. **In this paper we contribute** a Sequence Warehouse (SeWA) architecture and OLAP tools to analyze data stemming from workflow logs, including **process variants**. We focus on analyzing workflow logs (sequences of activities) of known patterns, rather than discovering patterns in sequences. The Sequence Warehouse and the tools we provide allow to overcome the obstacles caused by branches and loops in a process. In details, our contribution includes:

1. a Data Manipulation Language (DML) and a Data Definition Language (DDL) for Sequential SQL (S-SQL);
2. an interface to the DDL based on the XES workflow standard [2,41];
3. a XES importer to import workflow logs into our Process Warehouse;
4. a data structure called *Sequence Cube* (SQ) that enables users, in combination with S-SQL, to analyze a log of any workflow (composed of loops and interleaving loops).

1.2 Related Work

As mentioned before, research on sequential data analytics has been focusing on: (1) discovering patterns and (2) OLAP-like analysis. Some representative research contributions on mining sequential patterns in a data warehouse include [20,25,27,31], whereas on mining data streams include [16,26,29,43]. An overview of these techniques can be found in [14,30]. The aforementioned techniques are able to discover patterns in sequences but they do not support typical OLAP-like analysis of sequential data, e.g., by means of SQL-like commands. Thus, they are complementary to the contribution of this paper.

The first SQL-like query language for sequential data analysis, called SEQUIN [35,36], allowed to aggregate sequences by means of the AGGREGATE operator. Another language, called Sorted Relational Query Language (SRQL), was proposed in [32]. This language supports the SEQUENCE BY operator applied to ordering events in a sequence and creating a Sequence Cuboid (a storage for sequences to be analyzed). The extension of SRQL is yet another language - Simple Query Language for Time Series (SQL-TS) [33,34]. SQL-TS extended SQL with new operators: CLUSTER BY - for clustering sequences and AS - for expressing a searched sequence pattern.

Sequential data analysis has been researched also with respect to data models and storage, e.g., [32,33,35,37]. [37] applies an object-relational model where sequences are modeled by an enhanced abstract data type. In [32] sequences are modeled as sorted relations. The query languages proposed in [33,35] allow typical OLTP selects on sequences and do not support OLAP analyzes. Further extension towards sequence storage and analysis has been made in [11] that proposed a general concept of a RFID warehouse. Unfortunately, no in-dept discussion on RFID data storage and analysis was provided.

[17,18] also focused on data generated by RFID devices. [17] addressed the problem of reducing the size of such data. They proposed techniques for constructing RFID cuboids and for computing higher level cuboids from lower level ones. Based on this foundation, [18] proposed a language for analyzing paths with aggregated measures. They focused on a relational implementation and applied three tables to storing RFID data and their sequential orders. The proposed approach lacks a formal data model and a query language for analyzing sequences.

In [24], the S-OLAP model was proposed. It is based on the set of operators for the purpose of analyzing patterns. The model was further extended in [12, 13] to analyze subway passenger flows. [12,13] contributed also an algorithm for supporting ranking pattern-based aggregate queries. Data in the S-OLAP model are organized in sequence cuboids, which represent answers to the so-called Pattern-Based Aggregate queries.

[7,8] proposed a formal model for time point-based sequential data with the definitions of a fact, measure, dimension, and a dimension hierarchy. Thus, the model allows to analyze sequential data in an OLAP-like manner. In [6] the Authors proposed an index supporting sub-sequence pattern queries. The functionality of pattern matching is still under development in [7,8].

In [39] the authors propose to store process instances in cells of a process cube. The process cube is constructed based on: (1) a *process cube structure* that defines the "schema" of the cube, (2) a *process cube view* that is analogous to a relational view, and (3) a *materialized process cube view* that is the process cube view filled in with data coming from an *event base*. The process cube is organized by 3 dimensions that are typical to a workflow, i.e., *class type, event class*, and *time window*. They redefined typical OLAP operators to work on the process cube, i.e., slice, dice, roll-up, and drill-down. The cells of the cube store process instances. Process discovery algorithms run on the cells. In order to analyze workflow logs coming from multiple WfMSs vendors, these logs have to be transformed into a common, standardized structure. To this end, [41] proposes an XML-based standard, called XES [2].

In the area of *Complex Event Processing* [9,42], stream cube [19] was developed to provide tools for OLAP analysis of stream data within a given time window. [22,23] presented a more advanced concept, called E-Cube. It allows to execute OLAP queries on data streams, also within a given time window. E-Cube includes: a query language (called SEQ) allowing to query events of a given pattern, a concept hierarchy allowing to compute coarser aggregates based on finer ones, a hierarchical storage with data sharing, and a query optimizer. SEQ enables grouping events by means of attributes and computing aggregate functions like COUNT. Nonetheless, this approach has been developed for the analysis of current data and is unable to perform OLAP-like analysis.

Among commercial systems, only Oracle [4] and Teradata Aster [1] support SQL-like pattern analysis in their OLAP engines. Teradata Aster was developed for the purpose of storing, processing, and analyzing big data based on MapReduce. The query language, called SQL-MapReduce (SQL-MR), includes the nPath clause that is applied to analyze sequences. Oracle together with *IBM* defined an ANSI SQL standard for finding patterns within sequences stored in tables [28]. To this end, the MATCH_RECOGNIZE clause was proposed. It allows to search for patterns, define patterns and pattern variables. Microsoft StreamInsight [5] is another system that allows to discover patterns in a data stream. However, the discovery has to be implemented in a procedural language.

To sum up, a unique feature that distinguishes our approach to analyzing sequential data from the discussed ones is that our approach allows to construct a sequence cube with a dimension representing patterns. The dimension is constructed dynamically on demand for a particular analysis being executed. The dimension represents patterns with loops, joins, and branches. Moreover, our approach, unlike most of the related ones, accepts workflow sequential data in the industry XES standard.

2   Running Example

2.1   Support Process

Let us assume that a software company implemented a support process management system for keeping track of support requests, claims, and bug reports for their product-lines. To keep track of support requests they implemented a process-based ticket system which is fed by customers or selected key users respectively. The process is depicted in Fig. 2. It contains interleaving loop-structures and consists of the following steps:

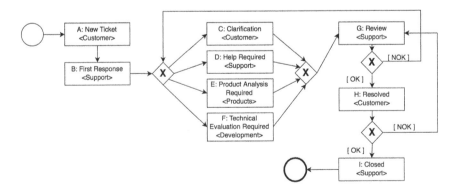

Fig. 2. The support process

- A - New Ticket (by role: Customers) - a new ticket has been reported, which opens a new case.
- B - First Response (by role: Support) - the ticket's content is quickly analyzed, corrected and remarked. Customers may be contacted by telephone for additional questions. Finally, the support user selects the next step.
- C - Customer Clarification (by role: Customers) - it is executed when the case is unclear or to fuzzy for support; additional information from the customer is required.
- D - Help Required (by role: Support) - a simple help-request that can be resolved by the first-level support.
- E - Product Analysis Required (by role: Products) - additional information about the product is required to resolve this case. A product manager adds information for a possible solution and returns the ticket to the support.
- F - Technical Evaluation Required (by role: Development) - deeper technical knowledge is required to assess the ticket (which is usually a bug or problem). An expert developer or component lead adds information for a possible solution and returns the ticket to the support.
- G - Review (by role: Support) - the supporter decides if the ticket can be resolved or whether further information is required.
- H - Resolved (by role: Customer) - the customer gets the information. If she/he is satisfied, then the case can be closed, otherwise another loop is entered.
- I - Resolved (by role: Customer) - the case is closed.

2.2   Event Log

Our event log is essentially an extraction and transformation of proprietary log data (cmp. [41]). This results in an XES event log of maturity level 4, which means that events have been recorded automatically, systematically, credibly, and completely. Furthermore, it is guaranteed that process instances (cases) and activity instances (steps, tasks) are supported in an explicit manner.

To save space, we do not show the XES-definition here, but only the concepts. On a case level (= XES-trace) we extract the following case relevant information entered by the reporter: ticket type, customer, priority, product affected, product version, and component affected. On a task level (= XES-event) we store the information presented in Fig. 1, which shows an example sequence of events belonging to different cases. Each record represents a life cycle event (here only either start or finish) of one of the steps in our process. Additionally, each event type may hold additional data, entered by users throughout the process.

2.3   Example Queries

We aim at providing a tool for answering specific strategic questions which encompass knowledge, or at least a strong suspicion, about the influence of (common) executions patterns on critical performance factors (like customer satisfaction or the qualification of support employees). To this end, the tool that we provide enables users to answer different OLAP-Query scenarios. For instance, regarding the running example described in Sect. 2 our approach would be able to answer the queries like:

- What are the cases where more than two loops over step 'Review' where required?
- What are the cases where a deeper inquiry (second or third level) and possibly several loops where necessary before they could be resolved as a simple 'Support Case' in the step 'Help Required'?
- What are the delayed cases with a long run time, with several ping-pongs in between and a customer-declination at the end?
- What are the cases with a customer-declination with a prior 'Customer Clarification'?

Naturally, our approach allows a vice-versa view, which means that high-level instance types can also be a result of any query, which can then be further investigated by applying common OLAP-operations (e.g., drilling down into an instance type to get more specific data).

3   Architecture

The SeWA architecture consists of three building blocks as depicted in Fig. 3: (a) the user interface, (b) the three parts of the SeWA Engine, and (c) the SeWA Warehouse. In this section, we will discuss these parts.

3.1   User Interface

The user interface enables the user to query the SeWA Warehouse. In our prototypical implementation, the user interface is a web application that allows the user to submit Sequential SQL (S-SQL) queries and depicts the result of a given query. The S-SQL query language will be discussed later in Sect. 3.3.

3.2   DWH Model

The main part of our data warehouse model is the *Sequence Cube* (SQ). A SQ consists of three parts: (a) a set of facts, (b) a set of dimensions, and (c) a set of sequences. Facts and dimensions are defined as in any traditional data warehouse. Facts are numerical measures that define what a user may analyze. Dimensions define how facts may be analyzed. Each dimension has a hierarchical structure composed of levels. For example, dimension $Time$ may consist of levels Day, $Month$, and $Year$, such that Day rolls up to $Month$ that rolls up to $Year$. Dimension Members (DM) are instances of a dimension hierarchy.

Each cube may consist of several sequences. These sequences are represented in a set of sequences. The notion of a set of sequences requires more details about sequences and patterns and will be discussed later.

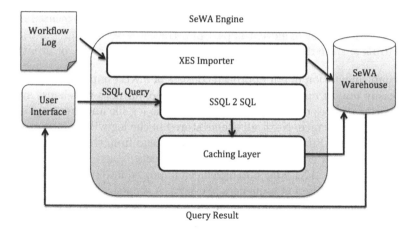

Fig. 3. The SeWA architecture

Example: for our running example this would lead to a sequence cube with *Time*, and *Employees* modeled as traditional data warehouse dimensions. Furthermore, we add a set of sequences called *Instances* to our sequence cube representing the instances of the workflow.

3.3   SeWA Engine

The SeWA Engine is the core of our architecture. It consists of three building blocks applied to: (a) importing workflow logs into the data warehouse, (b) parsing and executing queries stated against the warehouse, and (c) caching queries in order to improve performance. The next three sections will describe the SeWA Engine in details.

XES Importer Each meaningful implementation of a workflow system provides detailed information about all workflow instances by means of a log file. Such a log may be stored in various formats, e.g., as a traditional *.log* file, in a database, or in any other format. One widely known and implemented log format is MXML [40]. However, the MXML format has several limitations, as discussed in [41]. Therefore, [41] proposes to use XES (eXtensible Event Stream) as standard format for workflow logs.

[2] defines XES as "an XML-based standard for event logs". The primary purpose of XES was to serve as a standard format for process mining. However, XES may be used to store different kinds of event based logs. XES consists of four different building blocks:

1. **Logs**: the log object contains information associated with a process. Such a process may be browsing a web site or a defined workflow.
2. **Traces**: a trace object describes one specific instance, e.g., it may contain data about executing a workflow instance.

3. **Events**: an event represents a single activity which has been executed during the process. At least one event is part of a trace.
4. **Attributes**: store data assigned to a give element (log, trace, or event).

The implementation of our approach consists of a XES importer enabling the user to easily import an XES formatted log file into our data warehouse system. This importer iterates over all traces defined in the log file and creates a unique $instance_id$ for each trace. Next, all events stored in the log will be imported into an event repository. Finally, all attributes will be read from the log and assigned to the corresponding entities, i.e. a trace or an event.

After one or several XES log files have been imported into the data warehouse, the user may define different sequences of events. A sequence basically consists of three different elements: a unique name $SCname$, a set of grouping attrributes $\mathbb{A}_{\mathbb{G}} = \{A_1, A_2, \ldots, A_n\}$, and a set of ordering attributes $\mathbb{A}_{\mathbb{O}} = \{A_1, A_2, \ldots, A_n\}$. Grouping attributes are used to cluster events and ordering attributes are used to order the events in a cluster. Each element of $\mathbb{A}_{\mathbb{G}}$ and $\mathbb{A}_{\mathbb{O}}$ is an element of the attributes defined the XES file. In order to create a sequence we provide the following DDL statement:

$Create\ Sequence(SNname, \mathbb{A}_{\mathbb{G}}, \mathbb{A}_{\mathbb{O}}, XESname)$;

where $SNname$, $\mathbb{A}_{\mathbb{G}}$, and $\mathbb{A}_{\mathbb{O}}$ correspond to the elements described in the last paragraph. $XESname$ represents the name of an imported XES log file.

Example: For our running example, we would create a sequence named $sequ1$, using the grouping attribute $CaseId$ and the ordering attribute $TimeStamp$. Thus, if the XES log file has been named $ticketing$, we would create a sequence with the following statement:

$createsequence('sequ1', \{CaseId\}, \{TimeStamp\}, 'ticketing')$;

S-SQL to SQL. We developed a query language, called Sequential Structured Query Language (S-SQL), as an extension of standard SQL. S-SQL enables the user to easily query a data warehouse that stores data about traces and events, as described in Sect. 3.3. S-SQL includes both, DDL and DML commands.

The DDL part has been extended with the $Create\ Sequence()$ statement (cf. Sect. 3.3). The DML part has been extended with several keywords enabling users to formulate queries in order to analyze sequences using different functions, like for instance $HEAD()$, $TAIL()$, or $PATTERN()$.

We defined the S-SQL syntax using a parsing expression grammar (PEG). Without giving details on how PEG for standard SQL looks like, S-SQL extends SQL by means of $SSQLClause$ that is defined as follows:

```
SSQLClause          ::= (PBSClause ("AND" | "OR"))*
                        PBSClause
PBSClause           ::= Patternclause Bindclauses
                        Sequenceclause
Patternclause       ::= "PATTERN" Patternstring
Bindclauses         ::= (Bindclause ',' Bindclauses) |
                        Bindclause
```

```
Sequenceclause     ::= "ON SEQUENCE" String
Bindclause         ::= "BIND" "(" CommaSepStrings ")"
                       "TO" DimensionLevel
Patternstring      ::= "'" WildcardCSString "'"
DimensionLevel     ::= String "." String
WildcardCSString   ::= ((String | Wildcard)',')*
                       (String | Wildcard)
String             ::= [a-z A-Z 0-9 _]+
CommaSepStrings    ::= (String ',')* String
Wildcard           ::= "*" | "?" | "+"
```

In the context of analyzing workflow data, the *PATTERN()* keyword is of special interest. A pattern is a powerful way of querying sequential data in general and workflow logs in particular. It is a string which describes a pattern that traces have to fulfill in order to show up in the query result. A pattern is defined as a comma separated string of placeholders. Each placeholder may be either a string composed of alphanumerical characters or a wildcard. Valid wildcards are ?, *, and +, where ? means any single event matches, * means any number of events matches, and + means any number of events greater zero matches. Furthermore, each placeholder has to be bound to a dimension. Multiple patterns may be defined and combined using boolean operators *AND* and *OR*.

As S-SQL is an extension of SQL, standard SQL clauses may be used to aggregate, filter, group, and sort data.

Example: let us consider an example S-SQL statement that select all traces from our log file that fulfill the pattern '*a*,*,*b*', as shown in Listing 1.1. All elements of the pattern, i.e., *a* and *b*, represent a single event - the execution of a single activity. Therefore, this S-SQL statement would select all traces that are of any length and where the duration of the first event took no longer than 10 time units and where the duration of the last event took longer than 20 time units.

Listing 1.1. Sample S-SQL query

SELECT * FROM t1
WITH PATTERN 'a,*,b'
     BIND (a,b) TO instances ON SEQUENCE sequ1
WHERE a.duration < 10 AND b.duration > 20;

Example: for our running example, we could also answer query "find cases where more than two loops over step 'Review' where required", as discussed in Sect. 2. It simplementation in S-SQL is given in Listing 1.2.

Listing 1.2. S-SQL Query1

SELECT * FROM t1
WITH PATTERN 'a,b,?,g,h,gh,*'
     BIND (a,b,g,h) TO instances ON SEQUENCE sequ1;

Of course we could use a $GROUPBY$ statement and aggregation functions like AVG or SUM to add dimensions and aggregate the resulting facts along different hierarchies.

Set of Sequences. As described before, each cube may consist of several sequences which are stored in a set of sequences. The basic idea behind the set of sequences is, that - when analyzing workflow logs - workflow instances will not be spanned along a traditional dimension but will be considered as sequence dimension.

Hierarchies of this sequence dimension could be built automatically on the fly for each query stated. Currently, the prototype implemented does this in a naive way, e.g. for a query stated that uses a pattern (as discussed in Sect. 3.3) it will simply iterate through all elements (except the first and last) of the pattern and group them pairwise. Each group will be replaced with a wildcard '$*$'. Thus, a query that uses a pattern A, B, C, D, E will create the dimension members $A, B, *, E$ and $A, *, E$. The generated dimension members will then form a dimension "$A, *, E$" $\rightarrow$ "$A, B, *, E$" $\rightarrow$ "A, B, C, D, E" where $X \rightarrow Y$ means that Y rolls-up to X.

Caching Layer. We implemented a simple caching layer that is triggered each time a query that consists of a pattern clause is stated. Due to space limitations, this layer can only be briefly described here.

Basically it works as follows. First, a S-SQL query is decomposed into a pattern clause and other elements. Second, the caching layer checks if the pattern or its super-pattern is already in the cache. If one of the two conditions is true, we consider this as a cache hit. For instance, the pattern $S, *, E$ is a super-pattern of the patterns $S, A, *, E$ or $S, A, B, *, E$ or S, A, B, C, E.

In case of a cache hit, we can simply use the result of this query as an intermediate query result. This intermediate result will then be used as an input for the query stated in order to apply all other clauses, i.e. $WHERE, GROUPBY$, etc.

In case of a cache miss, the query will be stated against the data warehouse without any filtering clauses, e.g. without applying the $WHERE$ clause. The intermediate result of this query will than be stored in the cache with an assigned timestamp. Furthermore, this intermediate result serves as input for the query stated in order to apply all other clauses. The currently implemented cache management strategy is the Least Recently Used (LRU).

3.4   Implementation

All parts of the SeWA architecture (XES Importer, User Interface, as well as the S-SQL DDL and DML extensions) have been implemented using PHP, the *PHP PEG* package[1] (a package used for defining PEGs - parsing expression grammar - and parsing strings into PHP objects), and PostgreSQL.

---

[1] PHP PEG has been developed by Hamish Friedlander. Available at: https://github.com/hafriedlander/php-peg.

The S-SQL DDL extensions consist of two PHP scripts: (a) a script that parses and imports a XES file into several tables in the PostgreSQL database and (b) a script that generates sequences of events. The latter one is triggered with each call to the *Create Sequence*() DDL extension.

The S-SQL DML extensions work in such a way, that each S-SQL statement is parsed and translated into PHP objects. After successfully parsing the statement, it is translated into an SQL query. For instance, the example query from Sect. 3.3 is translated into a SQL query with over 40 lines of code. Other S-SQL queries that we tested resulted in SQL queries with up to 160 lines of code.

We would like to emphasize that although we implemented a simple caching layer (cf. Sect. 3.3), this implementation does not focus on query performance. Instead, it serves as implementation proof of concept only.

4   Summary

In this paper, we presented a novel approach that uses a query language called S-SQL (Sequential SQL) to analyze workflow logs. S-SQL supports patterns with wildcards, thus enabling the user to analyze workflow logs that consists of loops and/or arbitrary goto-transitions, without the need to predefine relevant queries.

Our approach works with any blocked or non-blocked workflow model. In fact, the approach presented in this paper does not require any information about the underlying workflow model. As we do not store or rely on any workflow model, our approach is immune to concept shifts and model evolution. In this paper, we discussed the analysis of finished executions only. However, our approach is able to analyze both, finished and non-finished executions.

Future work comprises of several tasks: support for reqular expressions in pattern queries, the application of interval OLAP [21] to extend the power of analyzing workflow logs or performance issues when it comes to analyze huge workflow logs. The last task mentioned will consist of the integration of indexing strategies for sequential data and the integration of indexing strategies for information stemming from workflow logs especially.

References

1. http://www.teradata.com/Teradata-Aster/overview/. Accessed 04 December 2014
2. http://www.xes-standard.org/. Accessed 04 December 2014
3. Process mining manifesto. IEEE CIS Task Force on Process Mining. http://www.win.tue.nl/ieeetfpm/doku.php?id=shared:process_mining_manifesto. Accessed 04 December 2014
4. Sql for pattern matching. https://docs.oracle.com/database/121/DWHSG/pattern.htm#DWHSG8956. Accessed 04 December 2014
5. Streaminsight. http://msdn.microsoft.com/en-us/library/ee391416. Accessed 04 December 2014
6. Andrzejewski, W., Bębel, B.: FOCUS: An Index FOr ContinuoUS subsequence pattern queries. In: Morzy, T., Härder, T., Wrembel, R. (eds.) ADBIS 2012. LNCS, vol. 7503, pp. 29–42. Springer, Heidelberg (2012)

7. Bębel, B., Morzy, M., Morzy, T., Królikowski, Z., Wrembel, R.: OLAP-like analysis of time point-based sequential data. In: Castano, S., Vassiliadis, P., Lakshmanan, L.V.S., Lee, M.L. (eds.) ER 2012 Workshops 2012. LNCS, vol. 7518, pp. 153–161. Springer, Heidelberg (2012)

8. Bebel, B., Morzy, T., Królikowski, Z., Wrembel, R.: Formal model of time point-based sequential data for OLAP-like analysis. Bull. Pol. Acad. Sci. Tech. Sci. **62**(2), 331–340 (2014)

9. Buchmann, A.P., Koldehofe, B.: Complex event processing. Inf.Technol. **51**(5), 241–242 (2009)

10. Chaudhuri, S., Dayal, U., Narasayya, V.: An overview of business intelligence technology. Commun. ACM **54**(8), 88–98 (2011)

11. Chawathe, S.S., Krishnamurthy, V., Ramachandran, S., Sarma,S.: Managing RFID data. In: Proceedings of the International Conference on Very Large Data Bases (VLDB) (2004)

12. Chui, C.K., Kao, B. Lo, E.Cheung, D.: S-OLAP: an olap system for analyzing sequence data. In: Proceedings of ACM SIGMOD International Conference on Management of Data (2010)

13. Chui, C.K. Lo, E., Kao, B., Ho, W.-S.: Supporting ranking pattern-based aggregate queries in sequence data cubes. In: Proceedings of ACM Conference on Information and Knowledge Management (CIKM) (2009)

14. Dong, G., Pei, J.: Sequence Data Mining, vol. 33. Springer, New York (2007)

15. Eder, J., Olivotto, G.E., Gruber, W.: A data warehouse for workflow logs. In: Han, Y., Tai, S., Wikarski, D. (eds.) EDCIS 2002. LNCS, vol. 2480, pp. 1–15. Springer, Heidelberg (2002)

16. Ezeife, C., Monwar, M.: Ssm : A frequent sequential data stream patterns miner. In: Proceedings of IEEE Symposium on Computational Intelligence and Data Mining (2007)

17. Gonzalez, H., Han, J., Li, X.: FlowCube: constructing RFID flowcubes for multi-dimensional analysis of commodity flows. In: Proceedings of the International Conference on Very Large Data Bases (VLDB) (2006)

18. Gonzalez, H., Han, J., Li, X., Klabjan, D.: Warehousing and analyzing massive RFID data sets. In: Proceedings of the International Conference on Data Engineering (ICDE), pp. 83-93 (2006)

19. Han, J., Chen, Y., Dong, G., Pei, J., Wah, B.W., Wang, J., Cai, Y.D.: Stream cube: an architecture for multi-dimensional analysis of data streams. Distributed and Parallel Databases **18**(2), 173–197 (2005)

20. Han, J.-W., Pei, J., Yan, X.-F.: From sequential pattern mining to structured pattern mining: a pattern-growth approach. J. Comput. Sci. Technol. **19**(3), 257–279 (2004)

21. Koncilia, C., Morzy, T., Wrembel, R., Eder, J.: Interval OLAP: analyzing interval data. In: Bellatreche, L., Mohania, M.K. (eds.) DaWaK 2014. LNCS, vol. 8646, pp. 233–244. Springer, Heidelberg (2014)

22. Liu, M. Rundensteiner, E., Greenfield, K., Gupta, C., Wang, S., Ari, I., Mehta, A.: E-Cube: multi-dimensional event sequence analysis using hierarchical pattern query sharing. In: Proceedings of ACM SIGMOD International Conference on Management of Data (2011)

23. Liu, M., Rundensteiner, E.A.: Event sequence processing: new models and optimization techniques. In: Proceedings of SIGMOD Ph.D. Workshop on Innovative Database Research (IDAR) (2010)

24. Lo, E., Kao, B., Ho, W.-S., Lee, S.D., Chui, C.K., Cheung, D.W.: OLAP on sequence data. In: Proceedings of ACM SIGMOD International Conference on Management of Data (2008)
25. Mabroukeh, N.R., Ezeife, C.I.: A taxonomy of sequential pattern mining algorithms. ACM Comput. Surv. **43**(1), 1–41 (2010)
26. Marascu, A., Masseglia, F.: Mining sequential patterns from data streams: a centroid approach. J. Intell. Inf. Syst. **27**(3), 291–307 (2006)
27. Masseglia, F., Teisseire, M., Poncelet, P.: Sequential pattern mining. In: Wang, J. (ed.) Encyclopedia of Data Warehousing and Mining. IGI Global, Hershey (2009)
28. Melton, J. (ed.).: Working draft database language sql - part 15: Row pattern recognition (sql/rpr). ANSI INCITS DM32.2-2011-00005 (2011)
29. Mendes, L.F., Ding, B., Han, J.: Stream sequential pattern mining with precise error bounds. In: Proceedings of the IEEE International Conference on Data Mining (ICDM) (2008)
30. Mooney, C.H., Roddick, J.F.: Sequential pattern mining - approaches and algorithms. ACM Comput.Surv. **45**(2), 19 (2013)
31. Pei, J., Han, J., Mortazavi-asl, B., Pinto, H., Chen, Q., Dayal, U., Hsu, M.-C.: Prefixspan: Mining sequential patterns efficiently by prefix-projected pattern growth. In: Proceedings of Internatiional Conference on Data Engineering (ICDE) (2001)
32. Ramakrishnan, R., Donjerkovic, D., Ranganathan, A., Beyer, K.S., Krishnaprasad, M.: SRQL: Sorted relational query language. In: Proceedings of Internatonal Conference on Scientific and Statistical Database Management (SSDBM) (1998)
33. Sadri, R., Zaniolo, C., Zarkesh, A., Adibi, J.: Optimization of sequence queries in database systems. In: Procedings of ACM SIGMOD-SIGACT-SIGART Symposium on Principles of Database System (PODS) (2001)
34. Sadri, R., Zaniolo, C., Zarkesh, A.M., Adibi, J.: A sequential pattern query language for supporting instant data mining for e-services. In: Proceedings of International Conference on Very Large Data Bases (VLDB) (2001)
35. Seshadri, P., Livny, M., Ramakrishnan, R.: Sequence query processing. SIGMOD Record **23**(2), 430–441 (1994)
36. Seshadri, P., Livny, M., Ramakrishnan, R.: SEQ: A model for sequence databases. In: Proceedings of International Conference on Data Engineering (ICDE) (1995)
37. Seshadri, P., Livny, M., Ramakrishnan, R.: The design and implementation of a sequence database system. In: Proceedings of Interntional Conference on Very Large Data Bases (VLDB) (1996)
38. Vaisman, A., Zimányi, E.: Data Warehouse Systems. Springer, Heidelberg (2014). ISBN 978-3-642-54655-6
39. van der Aalst, W.M.P.: Process cubes: slicing, dicing, rolling up and drilling down event data for process mining. In: Song, M., Wynn, M.T., Liu, J. (eds.) AP-BPM 2013. LNBIP, vol. 159, pp. 1–22. Springer, Heidelberg (2013)
40. van Dongen, B., van der Aalst, W.M.P.: A meta model for process mining data. In: Proceedings of of CAiSE Workshops (2005)
41. Verbeek, H.M.W., Buijs, J.C.A.M., van Dongen, B.F., van der Aalst, W.M.P.: XES, XESame, and ProM 6. In: Soffer, P., Proper, E. (eds.) CAiSE Forum 2010. LNBIP, vol. 72, pp. 60–75. Springer, Heidelberg (2011)
42. Wu, E., Diao, Y., Rizvi, S.: High-performance complex event processing over streams. In: Proceedings of ACM SIGMOD International Conference on Management of Data (2006)
43. Zheng, Q., Xu, K., Ma, S.: When to update the sequential patterns of stream data? In: Whang, K.-Y., Jeon, J., Shim, K., Srivastava, J. (eds.) PAKDD 2003. LNCS (LNAI), vol. 2637, pp. 545–550. Springer, Heidelberg (2003)

ETL

HBelt: Integrating an Incremental ETL Pipeline with a Big Data Store for Real-Time Analytics

Weiping Qu[(⊠)], Sahana Shankar, Sandy Ganza, and Stefan Dessloch

Heterogeneous Information Systems Group,
University of Kaiserslautern, Kaiserslautern, Germany
{qu,s_shankar12,s_ganza,dessloch}@informatik.uni-kl.de

Abstract. This paper demonstrates a system called HBelt which tightly integrates a distributed, key-value data store HBase with an extended ETL engine Kettle. The objective is to provide HBase tables with real-time data freshness in an efficient manner. A distributed ETL engine is extended and integrated as an overlay of HBase. Meanwhile, we extend this ETL engine with the capability of processing incremental ETL flows in a pipelined fashion. Delta batches are defined by the MVCC component in HBase to flush the incremental ETL pipeline for multiple concurrent read requests.Experimental results show that high query throughput can be achieved in HBelt for real-time analytics.

1 Introduction

Nowadays, many scalable, distributed data stores have been developed to deliver large scale data analytics over high volume of structured/unstructured data for valuable results. Data is first extracted, transformed and loaded (ETL) from heterogeneous sources into a centralized data store using ETL tools.

In order to meanwhile keep track of updates happening at the sources side, incremental ETL [9,10] has been widely used to propagate only deltas to the analytical systems instead of re-loading source data from scratch. Incremental ETL normally runs the maintenance flows periodically, i.e. hourly, or in micro-batches (minutes). However, for time-critical decision making, it is desirable to have real-time databases which provide queries with up-to-date state of touched tables. This forces ETL engines to propagate deltas to the target system in a very fast pace even with high update ratio in the external sources.

Background and Motivation. In our previous work [1], we introduced a demand-driven bulk loading scheme to allow early uptime for analytical systems by first offloading large amounts of *cold* data into a distributed, scalable, big data store HBase [2]. Data resides in HBase initially and becomes incrementally available in the target system according to the access priorities. Meanwhile, there are more and more updates collected from a variety of external sources. To achieve data freshness for time-critical decision making, an efficient maintenance mechanism is needed to refresh the data that are still buffered in HBase.

© Springer International Publishing Switzerland 2015
T. Morzy et al. (Eds.): ADBIS 2015, LNCS 9282, pp. 123–137, 2015.
DOI: 10.1007/978-3-319-23135-8_9

In this work, we propose our HBelt system which tightly integrates HBase with a pipelined data integration engine extended by an open-source ETL tool (Pentaho Data Integration (Kettle) [3], shortly Kettle) for real-time analytics. HBelt enables HBase tables to keep track of concurrent data changes in external data sources and provides each analytical query with a consistent view of both the base data and the latest deltas preceding the submission of the query. Data changes are propagated to HBase in a query-driven manner. The contributions of this paper are as follows:

- We deploy a Kettle environment directly in the same cluster shared by HBase. A copy of an ETL flow instance runs on each HBase working node. Besides, a HBase-specific partitioner is implemented in Kettle to distribute captured deltas to the correct HBase working nodes.
- We define our consistency model in HBelt and embed the Multi-Version Consistency Control (MVCC) component of HBase into Kettle. The MVCC component is used to define delta batches that need to be propagated to the target HBase tables for answering specific query requests.
- We propose a pipelined Kettle engine to process different delta batches in parallel. Kettle is geared towards data pipelining for high throughput of an ETL flow.

The remainder of this paper is as follows. We relate our work to several recent attractive work in different domains in Sect. 2. We give a brief introduction of key components in HBase and Kettle in Sect. 3. In Sect. 4, we introduce our HBelt system which integrates HBase with Kettle in terms of consistency and performance. Experiments are conducted and discussed in Sect. 5.

2   Related Work

PigLatin [7] is a script language developed in the Pig project. Pig scripts can be used to perform batch ETL jobs that run as MapReduce [8] jobs and thereby can be seen as a distributed ETL engine. Map/Reduce tasks are executed remotely directly over data stored in cluster nodes, thus delivering high scalability and parallelism. Furthermore, Pig also supports loading data into HBase through its pre-defined HBaseStorage class. Regarding function shipping, HBelt is similar to Pig which executes ETL flows directly on remote data nodes. However, HBelt allows each query/request to access up-to-date state of data by integrating MVCC component into Kettle. Meanwhile, we implemented pipelined version of ETL flows to enable HBase to efficiently react to trickle-feeding updates instead of batch processing.

Real-time databases result from the trend of merging OLTP & OLAP workloads, also known as *one-size-fits-all* databases. Hyper [13] is a typical example of these databases and is designed as an in-memory database. In Hyper, updates in OLTP workloads are performed in sequence in a single thread while each OLAP query session will see a snapshot of the current table state in a child thread forked from the parent update thread. Another example related to our work is

R-Store [6] which stores both real-time data and historical cubes in HBase. Historical cubes are used for OLAP queries and get incrementally maintained with the updates captured from real-time OLTP data by a streaming MapReduce called HStreaming. One difference between HBelt and R-Store is the location of OLTP data. Real-time data resides in R-Store while HBelt assumes a more general situation that real-time deltas are captured from external OLTP sources using the extract component in ETL.

Golab et al. proposed temporal consistency and scheduling algorithms in their real-time stream warehouse [11, 12]. Each real-time query always accesses the latest value preceding the submission time of the query. In their stream warehouse, data is divided into multiple partitions based on consecutive time windows. Each partition represents data in a certain time window and there are three consistency levels defined for queries, i.e. open, closed and complete. A partition is marked as open if data currently exists in or is expected to exist in the partition. From the query perspective, a closed partition implies that the scope of pending data has been fixed, whereas data is expected to arrive in a limited time window. This means that the query can be executed over base data that might be incomplete. The complete level is the strongest query consistency and all the data has arrived in the partition. We reuse this notion of temporal consistency in our work for consistency control by extending the MVCC component in HBase.

3   Background

In this section, we give a brief introduction of HBase and Kettle as background and describe only the components which are relevant to our work.

3.1   HBase

HBase [2] is a scalable, distributed key-value store that is widely used to deliver real-time access to big data. It follows a master/slave architecture. In HBase, a table is horizontally partitioned into a set of regions with non-overlapping key ranges. Each region contains a set of in-memory key-value lists called *memStore* and multiple on-disk *storeFiles*. Once a memStore fills up, it is flushed onto disk as a new storeFile. All data (regions) reside only in slave nodes called HRegion-Servers while the master node has only meta-data information which specifies how the regions with different key ranges are partitioned across HRegionServers.

As a data store, it provides only primitive operations (i.e. put, get and scan) based on a given row key. Based on the meta-data information (row key-HRegionServer mappings), a master node delegates all the put/get operations to corresponding HRegionServers where the actual operations take place. For large scale data analytics over HBase, there have already been efforts that implements an extra SQL layer over HBase which accesses tables stored in HBase through these primitive operations [4, 5].

In HBase, only two transaction isolation levels are guaranteed, i.e. read uncommitted and read committed. In order to achieve consistency between concurrent reads and writes, a component called Multi-Version Consistency Control (MVCC) is used. Each region contains a MVCC instance which maintains an internal *write queue*. A write queue is a list of Write Entry (*we*) elements which is used to assign a unique write number to an individual write or a batch of writes. Writes are not allowed to commit until their preceding writes have committed in this write queue. In this way, sequential writes are guaranteed in HBase. When a get/scan operation is issued with read committed as the transaction isolation level, the MVCC component returns the latest committed write number to this thread as read point *readPt* for fetching key-values whose write numbers are lower than or equal to this value in this region.

3.2 Kettle

Kettle [3] (or PDI) is an open-source ETL tool that has been widely used in the research community and provides a full-fledged set of transformation operations (called *step* in Kettle). A stream or batch of files are taken as input and processed row by row in each step. During flow execution, each step is running as an individual thread. Step threads are connected with each other through an in-memory queue called *RowSet*. The results of a preceding step are put in its output rowset which in turn is the input rowset of its subsequent step where rows get fetched. Step threads kill themselves once they are finished with their batch of files.

Kettle also enables a cluster execution mode in which multiple copies of the same flow instance can run in parallel over distributed nodes for better performance. The cluster environment follows a master/slave architecture. The input files of the flows running on the slave nodes are constructed by partitioning and distributing rows in master node according to a user-defined partition schema.

4 HBelt System

In this Section, we introduce our HBelt system, which integrates a distributed, HBase big data store with an extended, pipelined data integration engine based on Kettle for real-time analytics. Analytical queries are issued to a relational database layer over HBase in which actual target tables reside. In order to keep track of concurrent data changes at the source side, the internal consistency in HBase is maintained by multiple Kettle pipeline instances before each query is executed. A single query sees a consistent view which consists of the base data and the latest deltas preceding the submission time of this query. Furthermore, we try to reduce the synchronization delay by introducing two kinds of parallel computing techniques: data partitioning and data pipelining. Therefore, the objective of HBelt is to ensure both consistency and performance. The architecture is illustrated in Fig. 1.

4.1   Architecture Overview

As described in Sect. 3, a table stored in HBase are horizontally partitioned to a set of regions with non-overlapping key ranges and distributed over multiple HRegionServers. Current Kettle implementation (since Version 5.1) has provided a so-called HBase Output step to maintain a HBase table by using a single flow instance. All calculated deltas have to go through this step to arrive in target HRegionServers. However, since both HBase and Kettle follow the master/slave architecture, it is desirable to utilize the essence of distributed processing from both systems in terms of integration. In HBelt, the same number of the flow instance copies are constructed as the number of HRegionServers and further executed directly on each single HRegionServer node.

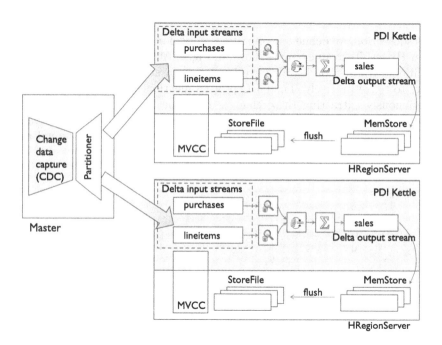

Fig. 1. HBelt architecture

Take a logical ETL flow as an example, which processes data changes captured from external *purchases* and *lineitems* sources to maintain the target *sales* table in HBase. In the master node (at the left side of Fig. 1), a change data capture (CDC) step uses methods like log-sniffing [14] or timestamps to capture the source deltas. In order to forward source deltas to the right HRegionServers for further flow execution, both the keys in the deltas and the key ranges of regions stored in HBase tables need to be considered. This is done by a component called *Partitioner*. In this example, *purchase* rows have $purc\_id$ as key and both *lineitems* rows and the *sales* table have compound keys ($purc\_id$, $item\_id$). The partitioner component fetches cached meta-data of the *sales* table from HBase in the same master node and forms a user-defined partition schema in Kettle. This meta-data shows the mapping from row keys to HRegionServers, based on which the *lineitems* deltas can be distributed to server nodes correctly. For a *purchases* row whose *purc_id* might span across regions in multiple HRegionServers, copies of this *purchases* row are sent to HRegionServers along with *lineitems*. In this way, we guarantee that calculated deltas for the target *sales* table should reside on the correct HRegionServer.

So far, we have introduced a sub-flow which consists of two steps: CDC and Partitioner. This sub-flow runs independently of query requests on HBase tables and feeds source deltas continuously to the delta input streams in HRegion-Servers to reflect the concurrent updates on the source side.

4.2   Consistency Model

In this subsection, we define our consistency model in HBelt for real-time analytics over HBase. Take an example shown in Fig. 2. At the upper left side, there is a traditional transaction log file recording five transactions ($T_1 \sim T_5$) committed from t_1 to t_5, respectively. The CDC process mentioned in previous subsection is continuously extracting these changes from the log file and sending corresponding deltas to the delta input streams of both of the Kettle flow instances (in this case only two flow instance copies are running on two individual HRegionServers).

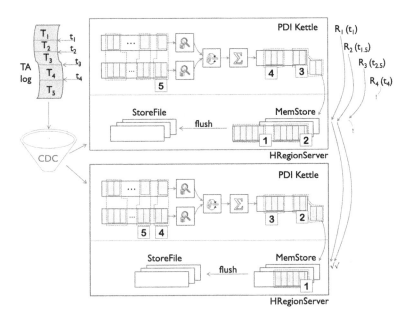

Fig. 2. Consistency model

Meanwhile, four distinct requests have been issued to HBase to perform scan operations over regions stored in these two HRegionServers. The first scan request R_1 occurs at timestamp t_1 which forces HBelt to refresh existing HBase table using changes (e.g. insertions, updates and deletions) derived from the first transaction T_1 which is committed at t_1. Once these changes have been successfully propagated and stored into the memStores in these two HRegionServers, R_1 is triggered to started immediately. Although the second scan request R_2 is issued at later time $t_{1.5}$, it still precedes the committing time of the second transaction T_2 (at t_2). Hence, it shares the same state of the HBase table as R_1. The third scan request R_3 has its occurring time $t_{2.5}$ which succeeds the committing time of T_2 and precedes the committing time of T_3. Since the deltas from T_2 are only available in the first HRegionServer, R_3 first completes the scan operation over regions in the first HRegionServer and waits for the regions in the second HRegionServer to be refreshed by T_2's committed changes. To answer the fourth request R_4, relevant regions stored in both HRegionServers need to be upgraded by the deltas from T_1 to T_4. Since neither of Kettle flows has finished propagated these deltas to HBase, R_4 is suspended until the HBase table is refreshed with correct deltas.

4.3 MVCC Integration for Delta Batches

In this subsection, we show how maintenance flows and query requests are scheduled in each HRegionServer to achieve the consistency we defined in previous subsection. Recall that in HBase the consistency in each region is maintained by a Multi-Version Consistency Control instance (see Sect. 3) where a local *write queue* is used to ensure sequential writes. A write queue maintains a list of **open** Write Entries *we* for assigning unique write numbers to batches of writes during insertions. Writes are only visible after they are committed and corresponding *we*s are marked as **complete**. However, in order to make each query request see a consistent view of base data and deltas, the current MVCC implementation in HBase has to be extended to meet our needs.

At any time, there is always one and only one **open** write entry *we* in the write queue. While source deltas continuously arrive in each HRegionServer, instead of triggering the maintenance flow to start immediately, deltas are first buffered in input streams and all of them are assigned the write number of this open *we*. We define that all the deltas sharing the same write number belong to a delta batch with a batch id. Once a read request is issued by an analytical query, this *we* is first marked as **closed** instead of **complete** (Here we embedded the temporal consistency described in Sect. 2 in our work). The **closed** state indicates that the maintenance flow now gets started to digest this delta batch with *we*'s write number as batch id and the final calculated deltas with this batch id have not yet completely arrived in HBase. Therefore, the read request awaits the completion of its maintenance flow and gets pushed into a waiting list *read queue*. Meanwhile, a new write entry *we'* is created and inserted into write queue to *paint* newly incoming deltas with *we''*'s write number.

At the time the last row with (*we*'s) batch id gets successfully inserted into HBase memStore by the final maintenance step, *we* is finally marked as `complete` and gets removed. All waiting reads in the *read queue* are notified of this event and check whether the complete batch id matches their local ones. The read request which waits for exactly this event gets started to continue with either a get or scan operation. Even though during the scan operation more new delta batches are inserted into the same regions, this read request would not be interfered with since it has an older batch id which restricts the access of rows with newer batch ids. In this way, we guarantee that each read request always sees the latest value of a consistent view of base data and deltas preceding its submission time.

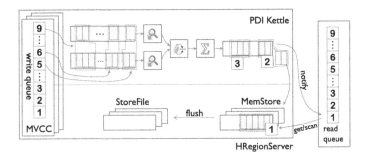

Fig. 3. MVCC integration for delta batches on HRegionServer

Figure 3 illustrates a snapshot taken at the time nine read requests have been issued by analytical queries. The arrival of these requests forces MVCC to group the corresponding deltas into nine batches which were once buffered in the streams before each occurrence of request. These read requests are waiting in the read queue until their delta batches get finished through the maintenance flow. Meanwhile, nine pending batches are denoted by the write entries stored in the write queue of MVCC component. They are all marked as `closed` except the first one since the first delta batch has been successfully moved to HBase memStore and can be made accessible to the first request. Thus, the first request is reactivated by the final maintenance step and continues with the get/scan operation. The second and third batches have already been put into the output streams and their requests are about to start. Note that, due to high request rate, delta batches 5–9 are still buffered in the input streams since the maintenance flow is still processing previous batches.

4.4   Pipelining Delta Batches in Kettle

As we can see from the previous subsection, the maintenance flow could be busy with processing different batches issued by multiple requests, especially with a high request rate. Hence, there is a need to speed up the performance of the maintenance flow. For each read request, in order to keep track of concurrent updates at the source side, the synchronization latency incurred by the maintenance flow is fixed. However, another potential optimization opportunity is to increase the throughput of the system. To address this, a pipelined flow engine based on Kettle is proposed.

As described in Sect. 3, the original Kettle implementation simply takes a stream/batch of data as input with no comprehension of different consecutive batches. It is important to distinguish different batches for specific transformation operations e.g. sort, aggregation, etc. in our work. Otherwise a maintenance flow could generate incorrect deltas for each read request, leading to inconsistent analytical results. For example, if a sort operation would receive rows from two delta batches and process them at the same time, the results coming out of this operation would be totally different from the results of sorting two batches separately. This also holds for aggregation operations like sum() or avg().

Algorithm 1. Step Implementation in Pipelined Kettle

   Input: rq // read queue which bufferes waiting read requests.
         in // intput rowsets
         out // output rowsets
   Init: $readPt$ // local read point
          $index$ // index used to iterate read queue.

1 **while** $true$ **do**
2 | **if** rq is empty $\parallel$ in is empty **then**
3 | | wait();
4 | $readPt \leftarrow rq[index{+}{+}]$;
5 | init(); // clear local caches, counters, etc.
6 | **while** in.getRow().$batch$ID $==$ $readPt$ **do**
7 | | r $\leftarrow$ processRow();
8 | | out.add(r);
9 | | out.notify();
10 | depose();

In this work, we extended Kettle to a pipeline flow engine which is able to react to different mini-batch jobs at the same time while still guaranteeing consistency. The extension of a single step thread is given above (see Algorithm 1). All steps in the maintenance flow share the same *read queue* which holds a list of pending read requests mentioned in previous subsection. Furthermore, each step maintains a local index which points at certain read request in the queue as a local read point *readPt*. This *readPt* is actually the batch id of the delta batch that needs to be processed. Once a step successfully fetches a batch id that matches the id of the rows in its input rowset, this step first initializes itself by clearing local caches and counters. After a row is processed, in addition to putting the result into the output rowset, it notifies its subsequent step of the existence of the output. When the batch is finished, instead of killing itself

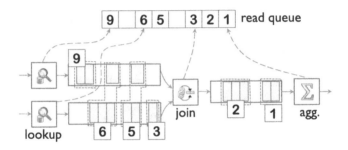

Fig. 4. Pipelined Kettle

as in the original implementation, it deposes itself (e.g. release used database connections) and tries to fetch the next read request in the queue.

As shown in Fig. 4, a pipelined Kettle flow is being flushed by nine delta batches. Due to diverse operational costs, the lookup step in the upper branch of the join step has already started to work on the ninth batch while another lookup step in the lower branch is still working on the sixth one. However, the join step would not continue with processing the rows in subsequent (e.g. fifth or sixth) batches until it makes sure that there is no more row of batch id 3 existing in neither of its input rowsets. Even though the fifth and sixth batches are already available, they are still invisible to the join step since the current *readPt* is still three. Data pipelining is introduced here to increase the throughput of the maintenance flow. However, the synchronization latency for each request is not improved or sometimes even increased, for example, the fifth batch cannot start until the join step finished with all the deltas in the third batch. We will examine it in the experiments.

5   Experimental Results

The objective of HBelt is to provide get/scan operations in HBase with real-time data access to the latest version of HBase's tables by tightly integrating an ETL engine, i.e. Kettle, with HBase. Though current Kettle (since Version 5.1) has implemented "HBase Output" step towards Big Data Integration, in our scenario, sequential execution of a single Kettle flow at once to maintain target HBase tables for time-critical analytics could lead to long data maintenance delay at high request rate. In this section, we show the advantages of our HBelt system by comparing its performance in terms of maintenance latency and request throughput with the sequential execution mode. We mainly examine the performance improvements by using data partitioning and data pipelining techniques in HBelt.

In the experiments, our HBelt ran on a 6-node cluster where a node (2 Quad-Core Intel Xeon Processor E5335, 4×2.00 GHz, 8 GB RAM, 1TB SATA-II disk) served as the master and the rest five nodes (2 Quad-Core Intel Xeon Processor X3440, 4×2.53 GHz, 4 GB RAM, 1TB SATA-II disk, Gigabit Ethernet) were the slave nodes running HRegionServer and Kettle threads (see Subsect. 4.1). Meanwhile, the same cluster was used to accommodate an original version (0.94.4) of HBase connected with a Kettle engine (Version 5.1) running on a client node (Intel Core i7–4600U Processor, 2×2.10 GHz, 12 GB RAM, 500GB SATA-II disk) to simulate the sequential execution mode.

We used TPC-DS benchmark [15] in our test. A *store_sales* table (with SF 10) resided in HBase and was maintained by a Kettle flow with the update files *purchases* (♯: 10K) and *lineitems* (♯: 100K) generated by TPC-DS *dsdgen*. The maintenance flow is depicted in Fig. 5. Purchases and lineitems are the delta files and are joined together in an incremental fashion after applying several surrogate key lookup steps. The intermediate join results are further aggregated as the final delta rows for the target store sales table. In sequential execution mode, the source delta files (purchases & lineitems) resided in the client node and were used as input for the Kettle flow to populate the store sales table in the 6-node HBase cluster using HBase Output. However, in HBelt mode, these source delta files were initially stored in the master node and later continuously distributed and fed to the five slave nodes where two input rowsets were used to buffer delta rows as delta input streams (instead of CSV Input steps). Furthermore, in contrast to sequential execution mode, each region was the target output instead of "HBase Output" step.

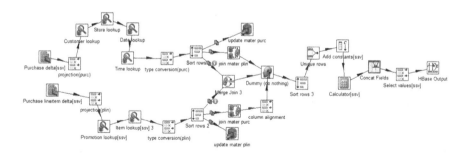

Fig. 5. Test maintenance flow in kettle

Data Pipelining: We first examined the performance improvement associated with the data pipelining technique implemented in the Pipelined Kettle component of our HBelt. The *store sales* table was not split in HBase and had only one region. Thus, only one HRegionServer was activated to serve issued request load and only one pipelined Kettle instance was dedicated to refresh the target table with *purchases* and *lineitems* delta files. Moreover, the delta files were split evenly to 210 chunks to emulate the input deltas to maintain the target table for 210 read requests occurring consecutively in a small time window.

The maintenance latency for each request is shown in Fig. 6. In sequential execution mode (SEQ), the same Kettle flow ran 210 times at the client side one flow at once to refresh target HBase table with 210 delta chunks. The latency difference between two adjacent requests is the duration of one flow execution.

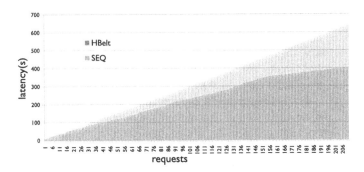

Fig. 6. Maintenance latencies of 210 consecutive read requests on single node

Since each flow execution took the same size of delta chunk as input, the maintenance latency grows linearly. The last request has to wait for the completions of preceding 210 flow executions (~10.5 min). Using HBelt, the flow pipeline shown in Fig. 5 was flushed by 210 delta batches at the same time. The latency difference between two adjacent requests depends on the slowest step in the pipeline rather than one complete flow execution. In summary, HBelt outperforms SEQ in terms of maintenance latency even though only one region existed in the HBase cluster, i.e. no data partitioning parallelism. Each request started earlier than in SEQ. The synchronization delay for the last request is 400 s, thus increasing the performance by ~30 %. This proves that HBelt is able to deliver high throughput at a high request rate or in case of "hotspot" issue in HBase, i.e. a single HRegionServer has a higher load than others.

Data Partitioning: We show another advantage of HBelt here: running one pipelined Kettle instance directly on each individual HRegionServer. Firstly, the *store sales* tables were evenly pre-split to 10 regions with non-overlapping row key ranges over 5 HRegionServers, thus each HRegionServer was active and managed 2 regions. Secondly, the request load consisted of a thousand scan operations in which each individual Region[1→10] was scanned by 50 scan operations, subsequent 100 operations scanned Regions (1~3), 100 operations scanned Regions (4~6), 100 operations scanned Regions (6~8), 100 operations scanned Regions (8~10) and the rest 100 operations scanned the entire table. Hence, each request required in average only 2/7 portion of the table to become up-to-date before it was executed. Finally, we generated a set of delta files *purchases* and *lineitems* of ten sizes {♯: (10 K & 120 K), (20 K & 240 K), ..., (100 K & 1200 K)} each of which were further split to 1000 chunks to simulate the delta inputs for the 1000 scan requests. In each chunk only 2/7 portion in average is needed to refresh the necessary regions for one request.

The request throughputs with different delta size settings are shown in Fig. 7. As the baseline, the request throughput in SEQ decreases steadily from 2.78 (♯requests/s) to 0.46 (♯requests/s) with increasing delta sizes, which indicates growing maintenance overhead. The throughput in SEQ mode is much lower than that in HBelt since two scan operations have to be executed sequentially

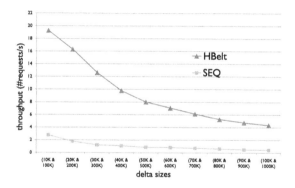

Fig. 7. Request throughput after issuing 1000 requests using diverse delta sizes

no matter how many deltas are really needed to answer certain request. HBelt provides much higher throughput (19.28 to 4.35 ♯requests/s). The efficiency is two fold. Due to data partitioning, HBelt is able to propagate deltas for concurrent requests with non-overlapping key ranges at the same time. For example, a scan operation which accesses Region(1~3) has no conflict with another scan operation which touches Region(4~6). Separate ETL pipeline can refresh independent regions at the same time. Meanwhile, since deltas were split and distributed over multiple ETL pipeline instances, the size of input deltas dropped drastically and the latency became less as well. In addition to data partitioning, pipelined Kettle still provides data pipelining parallelism for multiple concurrent requests arriving at the same HRegionServer.

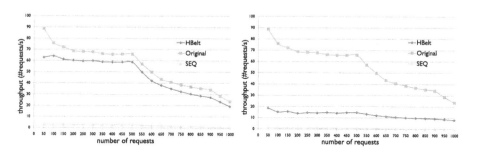

Fig. 8. Request throughput with small deltas (10 K purchases & 100 K lineitems)

Fig. 9. Request throughput with large deltas (50 K purchases & 500 K lineitems)

Figures 8 and 9 compare the throughput with increasing requests among three settings: HBelt, sequential execution mode and an original HBase setting which does not have maintenance overhead incurred by our ETL pipelines. With small delta sizes (10 K purchases & 100 K lineitems), HBelt achieves performance much similar to original HBase which does not guarantee data freshnees. However, as the size of delta grows, the request throughput of HBelt dropped significantly while it still outperforms the sequential execution mode.

6 Conclusion

In this work, we introduced our HBelt system which integrates an ETL engine Kettle with a big data store HBase to achieve real-time analytics over tables stored in HBase. The integration utilized the architectural essence of both systems, i.e. master/slave architecture. A copy of the Kettle flow instance runs directly on each HBase data node. File inputs are partitioned using our HBase-specific partitioner and further distributed over these data nodes, thus allowing multiple Kettle flow instances to work synchronously for concurrent non-conflicting requests. In this way, we provide data partitioning parallelism in HBelt. Furthermore, we defined the notion of our consistency model to enable each request to see the latest version of tables preceding the request submission time. The consistency component in HBase is embedded in Kettle to identify correct delta batches for answering specific HBase requests. Moreover, we extended Kettle to a pipelined version which is able to work on multiple distinct delta batches at the same time. A pipelined Kettle flow can be flushed by a large number of delta batches, thus increasing request throughput. Finally, the experimental results show that HBelt is able to reduce maintenance overhead and raise request throughput for real-time analytics in HBase.

References

1. Qu, W., Dessloch, S.: A demand-driven bulk loading scheme for large-scale social graphs. In: Manolopoulos, Y., Trajcevski, G., Kon-Popovska, M. (eds.) ADBIS 2014. LNCS, vol. 8716, pp. 139–152. Springer, Heidelberg (2014)
2. http://hbase.apache.org
3. Casters, M., Bouman, R., Van Dongen, J.: Pentaho Kettle Solutions: Building Open Source ETL Solutions with Pentaho Data Integration. John Wiley & Sons, Indianapolis (2010)
4. https://wiki.trafodion.org/
5. http://phoenix.apache.org/
6. Li, F., Ozsu, M.T., Chen, G., Ooi, B.C.: R-Store: a scalable distributed system for supporting real-time analytics. In: IEEE 30th International Conference on Data Engineering, ICDE 2014, pp. 40–51. IEEE, March 2014
7. Olston, C., Reed, B., Srivastava, U., Kumar, R., Tomkins, A.: Pig latin: a not-so-foreign language for data processing. In: Proceedings of the 2008 ACM SIGMOD International Conference on Management of Data, pp. 1099–1110. ACM, June 2008
8. Dean, J., Ghemawat, S.: MapReduce: simplified data processing on large clusters. Commun. ACM **51**(1), 107–113 (2008)
9. Vassiliadis, P., Simitsis, A.: Near real time ETL. In: Kozielski, S., Wrembel, R. (eds.) New Trends in Data Warehousing and Data Analysis. AIS, pp. 1–31. Springer, Cambridge (2009)

10. Jörg, T., Dessloch, S.: Near real-time data warehousing using state-of-the-art ETL tools. In: Castellanos, M., Dayal, U., Miller, R.J. (eds.) BIRTE 2009. LNBIP, vol. 41, pp. 100–117. Springer, Heidelberg (2010)
11. Golab, L., Johnson, T., Shkapenyuk, V.: Scheduling updates in a real-time stream warehouse. In: IEEE 25th International Conference on Data Engineering, ICDE 2009, pp. 1207–1210. IEEE, March 2009
12. Golab, L., Johnson, T.: Consistency in a stream warehouse. In: CIDR, Vol. 11, pp. 114–122 (2011)
13. Kemper, A., Neumann, T.: HyPer: a hybrid OLTP&OLAP main memory database system based on virtual memory snapshots. In: IEEE 27th International Conference on Data Engineering, ICDE 2011, pp. 195–206. IEEE, April 2011
14. Kimball, R., Caserta, J.: The Data Warehouse ETL Toolkit. John Wiley & Sons, Indianapolis (2004)
15. http://www.tpc.org/tpcds/

Two-ETL Phases for Data Warehouse Creation: Design and Implementation

Ahlem Nabli[1], Senda Bouaziz[1], Rania Yangui[2(✉)], and Faiez Gargouri[2]

[1] MIRACL Laboratory, Faculty of Sciences, Sfax University, 1171 Sfax, Tunisia
`ahlem.nabli@fsegs.rnu.tn, bouaziz.senda@hotmail.fr`
[2] MIRACL Laboratory, Institute of Computer Science and Multimedia,
Sfax University, 1030 Sfax, Tunisia
`yangui.rania@gmail.com, faiez.gargouri@isimsf.rnu.tn`

Abstract. Building the ETL process is potentially one of the biggest tasks of building a warehouse. In fact, it is complex, time consuming, and consumes most of data warehouse projects implementation efforts, costs, and resources. Nevertheless, the difference on data structures imposes new requirements on the ETL process implementation and maintenance. What makes these tasks even more challenging is the fact that data continue to grow rapidly and business requirements change over time. In this paper, we propose a method that contains Two-ETL phases, one treats the pre-treatment phase and another deals with the actual ETL. Our method consists on determining the correspondence table, modeling new operations using the Business Process Modeling Notation (BPMN) and implementing these operations with Talend Open Source (TOS). In addition, our method allows the design of ETL process in an earlier stage, which enormously facilitates the implementation of this process. Another advantage of our proposal is the use of the BPMN which allows to cover a deficit of communication that often occurs between the design and implementation of business processes.

Keywords: Extract transform and load · Business process modeling notation · Data warehouse design · Transformation operations · Correspondence table

1 Introduction

Business Intelligence (BI) solutions are very important as they require the implementation and the design of complex ETL process. This process is a software which allows the alimentation of a DW and its periodic refreshment from different sources. It is often used to get back various information to feed regularly the DW. New applications, such as, real-time data warehousing, require agile and flexible tools that allow BI users to take suitable decisions based on extremely up-to-date data. This is the case of the BWEC[1] (Business for Women of Emerging Country)

---

[1] Towards a new Manner to use Affordable Technologies and Social Networks to Improve Business for Women in Emerging Countries.

© Springer International Publishing Switzerland 2015
T. Morzy et al. (Eds.): ADBIS 2015, LNCS 9282, pp. 138–150, 2015.
DOI: 10.1007/978-3-319-23135-8_10

project that aims at improving the social economic situation of Handicraft women in Algerian and Tunisian countries involved in this research works.

To feed the DW, data must be identified and extracted from their original locations. Consequently, the data must be transformed and verified before being loaded into the DW. The large amount of data from multiple sources causes a high probability of errors and anomalies. This increases the need of a new ETL tool which are able to be adapted with the constant changes, to produce and to modify executable code quickly.

Recall that in the literature [1,2], the main stages for the DW design methodologies can be summarized as follows: requirement analysis, conceptual design, logical design, ETL process design and physical design. In fact, it was recognised that ETL process is a very time-consuming step, it takes about 80 % of the total time of the decision-making implementation due to its difficulty and complexity [3]. The design and the implementation of an ETL process usually involve the development of very complex tasks imposing high levels of interaction with a vast majority of the components of a DW system architecture. The implementation and the maintenance of such processes face various design drawbacks, such as the change of business requirements, which consequently leads to adapt existing data structures and reuse existing parts of ETL system.

Several works [5,9,11,12] have dealt with ETL process modeling and they don't focus on incorporating pre-processing phase of ETL process since the conceptual modeling phase of the DW. Furthermore, it has been noticed that while trying to design the ETL process, people tend to overlook the work done in the conceptual phases and which contain a useful knowledge for the ETL process. In this paper, we propose a method called two-ETL phases for DW creation where the first phase is carried out since the conceptual design of DW. This phase handles the determination of the correspondence table and the modelling of the transformation operations. The second phase deals with the implementation of the ETL process.

The remainder of this paper is organized as follows. Section 2 reviews some related works concerning the ETL modeling process. Section 3 describes our proposed method to create a DW. Section 4 details the first ETL phase. Section 5 presents the second ETL phase. Finally, Sect. 6 gives a conclusion and some future research directions.

2   Related Works

Various approaches for designing and optimizing ETL process have been proposed in the last few years [5,9,11,12]. This approaches can be classified into three main groups. The first group uses UML (Unified Modeling Language) to model the ETL process. The second one uses MDA (Model Driven Architecture). The third group uses BPMN (Business Process Modeling Notation).

UML Based ETL Process Modeling: Trujillo and Luján-Mora [4] have proposed an extension of the UML language to model the ETL process. Also, Mallek et al. [5] proposed the use of the UML activity diagram for the modeling of the

ETL process named ETL-WEB. More recently, El-Sappagh et al. [6] proposed an entity-mapping diagram (EMD) framework, consisting in a new notation and a new set of constructs for ETL conceptual modelling.

MDA Based ETL Process Modeling: Munoz et al. [7,8] presented the modeling of the ETL process of DW with MDA by formally defining a set of transformation rules QVT (Query, View, and Transformation). The PIM is modeled using the UML activity diagram. Atigui et al. [9] have proposed an approach where the designer built his unified conceptual model PIM which describes the multidimensional structures and related ETL process.

BPMN Based ETL Process Modeling: El Akkaoui et al. [10] provide an independent platform for the conceptual modeling of an ETL process based on the BPMN. Using the same BPMN objects presented by [10,11] proposes a correspondence between the ETL process and the needs of decision-makers to easily identify which data are necessary and how include them in the DW. Oliveira et al. [12] proposed to extend a previous work [10] by defining specific conceptual models that takes into account of the capture of evolutionary data, the change of dimensions, the treatment of the substitutions keys and the data quality. Wilkinson et al. [13], for instance, presented a method to guide BPMN specifications in the definition of conceptual models of ETL systems. Table 1 highlights a summary of the literature review which is based on five criteria:

- C1-Modeling of ETL process. This criterion is relative to the level of abstraction adopted in the ETL modeling approach: **"C"** (Conceptuel modeling of ETL process) and **"L"** (Logical modeling of ETL process).
- C2-Modeling language: the language used for the modeling of the ETL process.
- C3-ETL operations such as:
  - **"D"** Operations on the predefined data: all types of transformations realized with the data such as aggregation, filter, join, concatenation, etc.
  - **"C"** Operations expressing the constraints: all types of transformations and declarations errors or constraints such as Incorrect, Log, etc.
  - **"U"** Operations defined by the user: where the designer can define new operations.
  - **"S"** Operations for structuring: to unify the structure of the inputs.
- C4-ETL Level: indicates the level of the ETL process versus the DW design methodologies.
- C5-Design approaches: take into account the needs in the ETL process.

Notice that, the majority of works proposed the conceptual or/and logical design for the ETL process for data driven approaches. Only the work of Jovanovic et al. [14] and El-Akkaoui et al. [11] take into consideration the business requirement in the ETL process.

For the modeling language of ETL process, BPMN notation seem to be a good choice since it can cover a deficit of communication that often occurs between the design and implementation of business process. For that, we adopt this modeling language in our method.

As general rules, the ETL process starts after the logical modeling of the DW schema like the works of [10–13]. Nevertheless, Jovanovic et al. [14] propose to start the ETL process at the logical level.

To facilitate and minimize the complexity of ETL process, we propose to start the ETL process from the conceptual design phase of the DW and to take into account the business requirements in the ETL process. In fact, starting an ETL modeling at an earlier stage allows to benefit from the knowledge generated during the conceptual modeling of data warehouses by saving the traceability of the data in a correspondence table and modeling the transformation operations with BPMN.

Table 1. Summary of the literature review.

Approaches	C1	C2	C3				C4	C5
			D	C	U	S		
Munoz et al., 2008	C/L	UML (Activity Diagram)	Yes	No	No	No	Fourth step	No
Wilkinson et al., 2010	C/L	BPMN	Yes	No	No	No	Fourth step	No
El-Sappagh et al., 2011	C	UML	Yes	No	Yes	Yes	Fourth step	No
Atigui et al., 2012	C/L	UML	Yes	Yes	No	No	Fourth step	No
El Akkaoui et al., 2012	C	BPMN	Yes	Yes	No	No	Fourth step	Yes
Oliveira and Belo, 2012	C	BPMN	Yes	Yes	No	No	Fourth step	No
Jovanovic et al., 2012	C	/	Yes	Yes	No	No	Third step	Yes
Mallek et al., 2014	C	UML (Activity Diagram)	Yes	Yes	No	Yes	Fourth step	No
Our approach	C	BPMN	Yes	Yes	Yes	Yes	Second step	Yes

3   Two-ETL Phases Method

In this paper, we propose a method called Two-ETL phases for DW creation from heterogeneous sources. This method is composed of two phases to accomplish the ETL process and overcome its complexity. The first phase is done in a parallel way with the conceptual DW design and the second phase is realized to ensure the implementation of the specified ETL. In this method we propose to advance the ETL that is on the fourth level of the DW design methodologies into the second level (*cf.* Fig. 1).

As input to our method, we have heterogeneous data sources with different schemes and business requirements. Our method contains three main steps (*cf.* Fig. 2.): (i) DW design process, (ii) first ETL phase and (iii) second ETL phase. Step (i) and the step (ii) are operating in parallel to allow the design of DW conceptual and logical shemes, the correspondence table and the transformation operations. Finally, the third step (iii) consists on the implementation of new operations or the use of predefined ones in order to create the DW.

The Two-ETL phases as defined greatly facilitates the ETL process by minimizing the complexity and the time allocated to the implementation based on the explicit knowledge stored in the CT and modeling step. In the following, we will detail the Two-ETL phases.

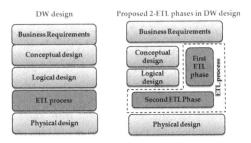

Fig. 1. Steps of DW design methodologie

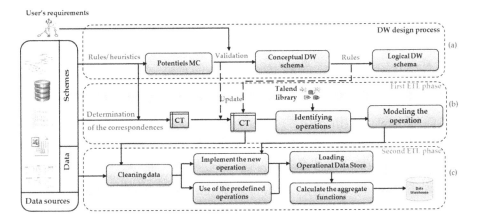

Fig. 2. Generic view of our proposed method

4   First ETL Phase

In the DW (mixed or data driven) design approaches, a set of rules/heuristics is used to identify potentials Multidimensional Concepts (MC) from the available data sources in order to obtain the DW conceptual schema. When starting this step, it is very important to save the data traceability of the used rules. For that, we propose to save this traceability on a table called Correspondence Table (CT).

Since the CT is well identified, we carry on the identification of the transformation operations.The last step of first ETL phase is the conceptual modeling of identified operations. As output of the first ETL phase we have the correspondence table with full documentation of all transformation operations. The explained process is modeled in BPMN language (cf. Fig. 3).

This flow uses pools of BPMN which provide a high expressivity for modeling. This pool encloses three lanes. This lanes focuses on the identification of the DW design shema, the determination of the correspondence, definition and modeling of the transformation operations, which allow to generate the correspondence table.

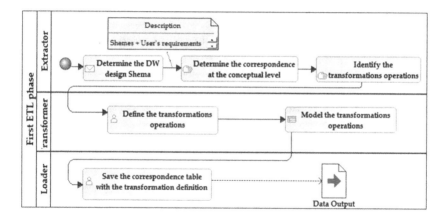

Fig. 3. The first ETL phase

In the following, we will detail the preparation of the correspondence table, the identification of the transformation operations and the modeling of these operations.

4.1 Excerpt of the Correspondence Table

The identification of the Correspondence Table (CT) at the earliest phases of DW design is very important for the ETL process. This table stores the DS attributes names and the corresponding potential MC according to the used rule. At this stage, CT contains the first two columns of Table 2. The CT is then updated when the validation of potentials MC with business requirement is done. At this level, we obtain the valid DW conceptual schema and the valid CT. Finally, a set of rules is used to derive a logical schema. We recover the result of applied rules on our CT. Table 2 presents an excerpt of the correspondence table. The name of data sources (ONAT, Postal codes of Tunisia, ontology) will be presented in the Sect. 5.

4.2 Identification of Transformation Operations

After the determination of the CT, we propose to clarify the various types of transformation operations. In this context, we make a distinction between two types of operations: the **defined** transformation operations, which are supported by ETL tools (i.e. mapping, filtering, etc.) and the **undefined** transformation operations, which are unsupported by the ETL tools. In fact, the defined operations can not cover all the possible transformations because they depend on the data sources and the conceptual model of DW. Therefore, we need to apply some other transformation operations which depend on the context of work. These transformation operations are carried out through the addition of a new

Table 2. Excerpt of the correspondence table

Data sources		Target data	operations	
Name of the DS	Field	(data warehouse)	Operations names	Operations types
ONAT	Full_Name	Name First Name	Decomposition operation	Undefined
ONAT	BirthDate	Age_Group	Discrimination operation	Undefined
ONAT	Date	Day Month Year Week Quarter Semester	Explosion operation	Undefined
ONAT	Sex	Sex	Mapping operation	Defined
ONAT and postal codes of Tunisia	Postal_Code	Postal_Code Office_Desig Governorate Country	Join operation	Defined
ONAT	Address	Street_Number Street_Desig	Decomposition operation	Undefined
ONAT and ontology	Activity	Activity Activity_Desig	Mapping operation	Defined
ONAT and ontology	Activity_Group	Activity_Gr Activity_Gr_Desig Raw_Mat_Desig Raw_Mat_Price	Join operation	Defined

expressions that contain the composition of two or more predefined functions or calling a routine which contains a program according to the transformation operation.

Based on the operator library, we alter the CT by a new two columns (*cf.* Table 2): operation name and operation type. In the first one we indicate the name of the operation (i.e. join, split, etc.) and the second one if the operation is supported by ETL tool or not (defined or undefined).

4.3   Modeling of the Transformation Operations

In the ETL tools (Talend[2], Pentaho Data Integration[3], etc.) many transformation operations are available to transform data such as Mapping operation, Filtering operation, etc. But in the real case study we can needs others operations unsupported by those ETL tools for that we should add new operations.

This section is dedicated to present same proposed operations identified in the correspondence table such as: decomposition operation, explosion operation and discrimination operation.

Decomposition Operation: According to the correspondence table, the decomposition operation occurs when we need to decompose a field of the source into

---

[2] http://www.talend.com.
[3] http://www.pentaho.fr/explore/pentaho-data-integration/.

multiple target attributes in the DW. This operation is modeled in the workflow as a succession of tasks which are: select the field to decompose, define the criteria by which we will do the decomposition, execute the operation decomposition and save the resulted attribute into a temporary table.

Example 1. As mentioned in Table 2 the attribute **Ful_ Name** will be decomposed based on a regular expression to generate **First_Name** and **Name** as suggested in the DW (*cf.* Fig. 4).

Example 2. This type of operation is applied to the field **Address** to determine the parameter **Street_Number** and weak attribute **Street_Desig** of the dimension **Artisan** (*cf.* Fig. 5).

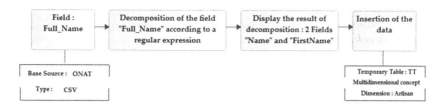

Fig. 4. BPMN modeling of decomposition operation for Full_Name attribute.

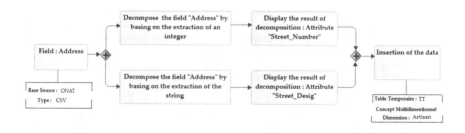

Fig. 5. BPMN modeling of decomposition operation for address attribute.

Discrimination Operation: This kind of operations occurs when we have to assign a categorical attribute from a set of numeric values. In our case of study, we have to define the parameter **Age_Group** from the **Birth_date** attribute. This operation must be preceded by a conversion operation. The conversion operation calculates the **Age** from the **Birth_date**. Once the age is determined, we move to the discrimination operation. The workflow of this operation is as follows: select the attribute, define the conversion operation, execute this operation, define the operation of discrimination, execute this operation and save the result in the temporary table. An example of modeling is presented in Fig. 6.

Explosion Operation: The explosion operation aims at defining a multiple attributes from a single field. A concrete example of this operation is manifested

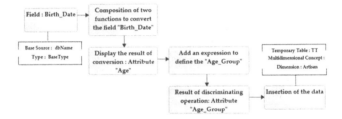

Fig. 6. BPMN modeling of discrimination operation for Birth_Date attribute.

by the generation of several attributes from the field **Date**. The workflow of this operation is: select the date, define the applied operation, execute the operation and save the attributes in the temporary table. Figure 7 illustrates the steps in the explosion operation of the field **Date** in **Day**, **Month**, **Half_year**, **Quarter**, **Week**, and **Year**.

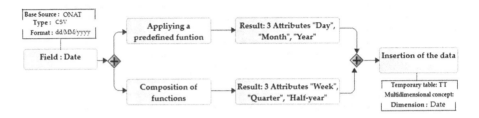

Fig. 7. BPMN Modeling of explosion operation for Date attribute.

5   Second ETL Phase

This phase consists of the implementation of the ETL process based on the correspondence table and the modeled operations. So, we start by loading data from data sources into a temporal database based on CT then the cleaning (For example: treat the missing data and the null values) step is realized. After that, we use the supported operations and implement the unsupported ones. Finally, a set of aggregation functions used and then loaded into a DW shema. In Fig. 8, we present the sequence flow of this phase.

Our method is experimented in the real case study BWEC project. Many information are collected for this project about handicraft women such as those about profiles, productions and the ability to use new technologies. These information are represented through: the ONAT[4] data source which contains a list of artisans and their information, an ontology which contains the list of the raw materials of production and the data source of Postal Codes of Tunisia. Figure 9 presents the DW schema to be loaded (b), the excerpt of ontology (c) and the excerpt of ONAT (a).

---

[4] http://www.onat.nat.tn/accueil/.

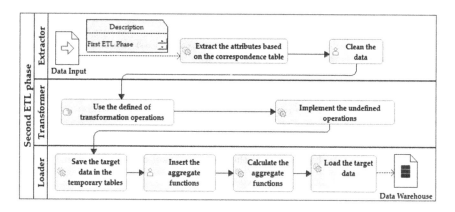

Fig. 8. The second ETL phase

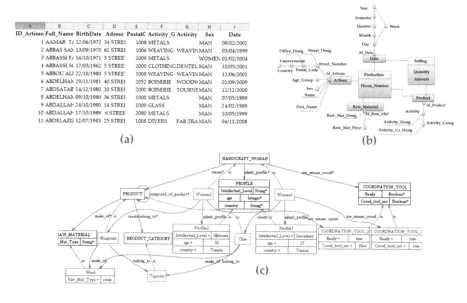

Fig. 9. Excerpt of the DS (a), (c) and the DW schema (b)

We have choisen Talend Open Studio (TOS) for the creation of our DW. TOS is based on the creation of a "job" to maintain the execution of the data process. The user can apply the various components of the palette to build the work on the design side and view the generated code.

Available components in Talend realize some operations, but we notice the absence of some operations detected in the source analysis phase. TOS allows adding new operations. In the folowing, we detaied the realization of the new operations.

Fig. 10. Expression of the discrimination operation

Fig. 11. Expression which determines the parameter quarter

Realization of the Discrimination Operation: The realization of the discrimination operation requires two steps: the first step is the conversion of *Birth_date* in *Age* and the second step is the discrimination of *Age_Group* from the calculated *Age*. Figure 10 describes the insertion of two expressions where we define the different age groups.

Realization of the Explosion Operation: The realization of the explosion operation requires the addition of expressions that we have to determine the parameters *Day*, *Month*, *Half_year*, *Quarter*, *Week*, and *Year*. To do this, we take the internal base ONAT input and output the Date table in Microsoft SQL Server. Figure 11 shows the insertion of the phase to generate a *Quarter* of each date of the source database.

Figure 12 present the execution of the ETL process (defined and implemented), the statistics appear on the graphical interface elements. These statistics indicate the success of our method.

6   Conclusion

In this paper, we have proposed two-ETL phases as part of an integrated and global approach for DW creating. Our method allows to paralyze the design of ETL process with the conceptual DW design, which facilitates the implementation of this process. In this method, we have used the BPMN that allows to cover

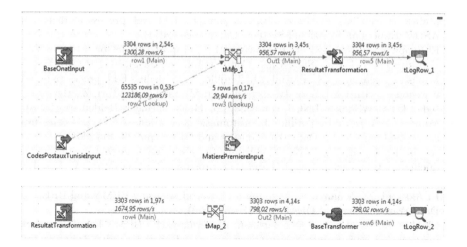

Fig. 12. General process of all operations

a deficit of communication that often occurs between the design and implementation of business processes. The first ETL phase allows the determination of correspondences and the identification of transformation operations. This step is performed with the DW design process. The result of this work provides the correspondence table that saves the traceability of data. From the CT, we have performed modeling of the operation to facilitate and minimize the complexity of the second ETL phase. Finally we implemented these operations and loaded them in the operational data store. Future works include developing a validation procedure for the produced models using this framework. This will allow to produce a rigorous comparison between the outcome of this methodology, and other ones, not only in terms of workflow structure, but also in terms of flexibility, adaptability to change, usability, and performance. Changes can occur during the lifecycle of the warehouse, not only in sources, but also within the warehouse.

References

1. Golfarelli, M.: From user requirements to conceptual design in data warehouse design-a survey. In: Data Warehousing Design and Advanced Engineering Applications: Methods for Complex Construction, pp. 6–11 (2010)
2. Nabli, A.: Approche d'aide à la conception automatisée d'entrepôt de données: Guide de modèlisation. Presses Acadmiques Francophones (2013)
3. Favre, C., Bentayeb, F., Boussaid, O., Darmont, J., Gavin, G., Harbi, N., Kabachi, N., Loudcher, S.: Les entrepôts de données pour les nuls. ou pas!. In: 2éme Atelier aide à la Décision à tous les Etages (EGC/AIDE), Janvier 2013
4. Trujillo, J., Luján-Mora, S.: A uml based approach for modeling ETL processes in data warehouses. In: Song, I.-Y., Liddle, S.W., Ling, T.-W., Scheuermann, P. (eds.) ER 2003. LNCS, vol. 2813, pp. 307–320. Springer, Heidelberg (2003)

5. Mallek, H., Walha, A., Ghozzi, F., Gargouri, F.: ETL-web process modeling. In: ASD Advances on Decisional Systems Conference (2014)
6. El-Sappagh, A., Hendawi, A., Bastawissy, H.: A proposed model for data warehouse ETL processes. J. King Saud Univ. Comput. Inf. Sci. **23**(2), 91–104 (2011)
7. Muñoz, L., Mazón, J.-N., Pardillo, J., Trujillo, J.: Modelling ETL processes of data warehouses with UML activity diagrams. In: Meersman, R., Tari, Z., Herrero, P. (eds.) OTM-WS 2008. LNCS, vol. 5333, pp. 44–53. Springer, Heidelberg (2008)
8. Munoz, L., Mazon, J., Trujillo, J.: Automatic generation of ETL processes from conceptual models. In: Data Warehousing and OLAP, pp. 33–40 (2009)
9. Atigui, F., Ravat, F., Teste, O., Zurfluh, G.: Using OCL for automatically producing multidimensional models and ETL processes. In: Cuzzocrea, A., Dayal, U. (eds.) DaWaK 2012. LNCS, vol. 7448, pp. 42–53. Springer, Heidelberg (2012)
10. El Akkaoui, Z., Zimanyi, E.: Defining ETL worfklows using BPMN and BPEL. In: Data Warehousing and OLAP, pp. 41–48 (2009)
11. El Akkaoui, Z., Mazón, J.-N., Vaisman, A., Zimányi, E.: BPMN-based conceptual modeling of ETL processes. In: Cuzzocrea, A., Dayal, U. (eds.) DaWaK 2012. LNCS, vol. 7448, pp. 1–14. Springer, Heidelberg (2012)
12. Oliveira, B., Belo, O.: BPMN patterns for ETL conceptual modelling and validation. In: Chen, L., Felfernig, A., Liu, J., Raś, Z.W. (eds.) ISMIS 2012. LNCS, vol. 7661, pp. 445–454. Springer, Heidelberg (2012)
13. Wilkinson, K., Simitsis, A., Castellanos, M., Dayal, U.: Leveraging business process models for ETL design. In: Parsons, J., Saeki, M., Shoval, P., Woo, C., Wand, Y. (eds.) ER 2010. LNCS, vol. 6412, pp. 15–30. Springer, Heidelberg (2010)
14. Jovanovic, P., Romero, O., Simitsis, A., Abelló, A.: Requirement-driven creation and deployment of multidimensional and ETL designs. In: Castano, S., Vassiliadis, P., Lakshmanan, L.V.S., Lee, M.L. (eds.) ER 2012 Workshops 2012. LNCS, vol. 7518, pp. 391–395. Springer, Heidelberg (2012)

Direct Transformation Techniques for Compressed Data: General Approach and Application Scenarios

Patrick Damme[✉], Dirk Habich, and Wolfgang Lehner

Database Systems Group, Technische Universität Dresden, 01062 Dresden, Germany
{Patrick.Damme,Dirk.Habich,Wolfgang.Lehner}@tu-dresden.de

Abstract. Lightweight data compression techniques like dictionary or run-length compression play an important role in main memory database systems. Having decided for a compression scheme for a dataset, the transformation to another scheme is very inefficient today. The common approach works as follows: First, the compressed data is decompressed using the source decompression algorithm resulting in the materialization of the raw data in main memory. Second, the compression algorithm of the destination scheme is applied. This indirect way relies on existing algorithms, but is very inefficient, since the whole uncompressed data has to be materialized as an intermediate step. To overcome these drawbacks, we propose a novel approach called *direct transformation*, which avoids the materialization of the whole uncompressed data. Our techniques are cache optimized to reduce necessary data accesses. Moreover, we present application scenarios, where such direct transformations can be efficiently applied.

Keywords: Lightweight data compression · Main memory database systems · Efficient algorithms

1 Introduction

As a consequence, e.g., of the developments in the main memory domain, modern database systems are very often in the position to store their entire data in main memory. Aside from increased main memory capacities, a further driver for in-memory database systems was the shift to a column-oriented storage format in combination with compression techniques. Using both mentioned software concepts, large datasets can be held in main memory with a low memory footprint. That means, modern in-memory database systems have to manage and process large compressed datasets. For compression, lightweight compression techniques have been established in this domain [1,3,5,6,9]. These lightweight techniques provide a good compression rate and they are less CPU intensive than heavyweight approaches like Huffman [4]. Examples of lightweight compression techniques are: dictionary compression [1,9], run-length encoding [1,6] and null

© Springer International Publishing Switzerland 2015
T. Morzy et al. (Eds.): ADBIS 2015, LNCS 9282, pp. 151–165, 2015.
DOI: 10.1007/978-3-319-23135-8_11

suppression [1,6]. Moreover, recent research in the field of lightweight compression techniques increases the performance by the utilization of parallelization concepts like SIMD capabilities of modern CPUs [5,7,8].

Based on the availability of various different lightweight compression schemes, the complexity of the physical database design increases. That means, for each column an appropriate compression scheme has to be identified. Abadi et al. [1] have proposed a decision tree to heuristically decide which compression scheme to use for a column. As they have shown [1], the optimal lightweight compression scheme depends on various influencing factors like the number of distinct values, data locality or access pattern. However, these influencing factors usually change over time. To react in an appropriate way on the physical database layer, efficient techniques to transform compressed data from one compression scheme to another are required. To the best of our knowledge, this aspect has not been considered before for lightweightly compressed data. Therefore, this paper primarily focuses on this aspect.

A naïve transformation approach would be the indirect way from a source to a destination compression scheme. First, the compressed data is decompressed using the source decompression algorithm resulting in the materialization of the raw data in main memory. Second, the compression algorithm of the destination scheme is applied. This indirect way relies on existing algorithms and can be realized for arbitrary pairs of source and destination compression schemes. However, the naïve approach is very inefficient and the whole uncompressed data has to be materialized as an intermediate step. To overcome these drawbacks, we contribute a novel *direct transformation* approach in this paper:

- Our novel *direct transformation* techniques convert compressed data in scheme X to another compression scheme Y in a direct and interleaved way.
- We avoid the materialization of the whole uncompressed data as in the naïve approach. Furthermore, our direct techniques are cache optimized to reduce necessary memory accesses.
- We introduce different direct transformation algorithms in detail.
- In our evaluation, we show that our direct transformation techniques outperform the indirect, classical way to convert the compression scheme.
- Furthermore, we present different application scenarios for our direct transformation techniques.

The remainder of the paper is organized as follows: The next section briefly reviews related work in the context of lightweight compression techniques. In Sect. 3, we describe transformation approaches in general. Then, we present different examples of direct transformation techniques in Sect. 4. Section 5 shows the results of our empirical evaluation. Before we conclude the paper in Sect. 7, we highlight different applications requiring efficient transformation techniques.

2   Related Work

The field of lightweight compression has been studied for decades. The main archetypes of lightweight compression techniques are dictionary compression

(DICT) [1,9], delta coding (DELTA) [5,6], frame-of-reference (FOR) [3,9], null suppression (NS) [1,6], and run-length encoding (RLE) [1,6]. DICT replaces each value by its unique key. DELTA and FOR represent each value as the difference to its predecessor respectively a certain reference value. These three well-known techniques try to represent the original data as a sequence of small integers, which is then suited for actual compression using a scheme from the family of NS. NS is the most well-studied kind of lightweight compression. Its basic idea is the omission of leading zeros in small integers. Finally, RLE tackles uninterrupted sequences of occurrences of the same value, so-called runs. In its compressed format, each run is represented by its value and length, i.e., by two uncompressed integers. Therefore, the compressed data is a sequence of such pairs (see Fig. 1).

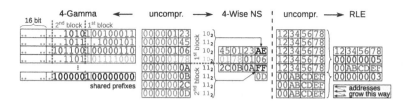

Fig. 1. Examples of some uncompressed data and its representations in the described formats of NS (left) and RLE (right). The data of 4-Gamma is given in binary, whereby gray dots mean leading zero bits. All other data is presented in hexadecimal notation.

In recent years, research in the field of lightweight compression has mainly focussed on the efficient implementation of these schemes on modern hardware. For instance, Zukowski et al. [9] introduced the paradigm of patched coding, which especially aims at the exploitation of pipelining in modern CPUs. Another promising direction is the vectorization of compression techniques by using SIMD instruction set extensions such as SSE and AVX. Numerous vectorized techniques have been proposed, e.g., in [5,7,8]. The techniques 4-Wise Null Suppression and 4-Gamma Coding introduced by Schlegel et al. in [7] are especially important to understand this paper.

4-Wise NS eliminates leading zeros at *byte* level and processes blocks of four values at a time. During compression, the number of leading zero bytes of each of the four values is determined. This yields four 2-bit descriptors, which are combined to an 8-bit compression mask. The compression of the values is done by a SIMD byte permutation bringing the required lower bytes of the values together. This requires a permutation mask being looked up in an offline-created table using the compression mask as a key. After the permutation, the code words have a horizontal layout, i.e., code words of subsequent values are stored in subsequent memory locations. Thus, the compressed data is a sequence of compressed blocks (see Fig. 1). The decompression simply reads the compression mask, looks up the appropriate permutation mask which reinserts the leading zeros bytes and applies the permutation.

4-Gamma eliminates leading zeros at *bit* level and processes blocks of four values at a time. The compression algorithm first determines the minimum number of bits required to represent the highest of the four values. This number is the shared prefix of the block. All values are represented by that many bits and stored using a vertical layout, i.e., each of the four code words is stored to a separate memory word. This requires shift and logical operations, which are done using vectorized instructions. Finally, the unary representation of the shared prefix is stored to a separate memory location. Again, the compressed data is a sequence of compressed blocks (see Fig. 1). The decompression determines the length of the shared prefix and applies appropriate logical and shift operations to the compressed block in order to extract the original values.

3   Transformation Algorithms in General

The aim of transformation algorithms is to change the compressed format some data is represented in. Therefore, the transformation takes data represented in its *source format* as input and outputs the representation of the data in its *destination format*. Note that this is a lossless process, i.e., after the transformation, the original uncompressed data can still be obtained by applying the decompression algorithm of the destination format. We differentiate between two different types of transformations: *indirect* and *direct*, which are described next. For our novel *direct transformation*, we introduce two variants in Sect. 3.2.

3.1   Indirect Vs. Direct Transformations

For the implementation of a transformation, two different approaches exist: (1) *indirect transformations* and (2) *direct transformations*.

Indirect transformations constitute a naïve approach. First, the compressed input data is decompressed using the decompression algorithm belonging to the source format. In this case, the *entire* uncompressed data is materialized in main memory. Finally, the compression algorithm of the destination format is applied to the uncompressed data in order to obtain the representation of the data in the destination format.

Since indirect transformations rely solely on existing compression and decompression algorithms, they can easily be implemented for arbitrary pairs of source and destination formats. However, they suffer from a major inefficiency: The materialization of the uncompressed data as an intermediate step. This requires a lot of expensive load and store operations. Furthermore, it results in a suboptimal cache utilization: when the uncompressed data is read by the recompression, it is not in the caches anymore.

In order to perform transformations efficiently, we propose to employ *direct transformations*. The decisive criterion for direct transformations is that no uncompressed data is written to main memory. Ideally, all intermediate data of a transformation can reside in CPU registers or at worst in the L1 cache. This allows for high-speed access to these intermediate data. We expect considerable speed ups of direct transformations compared to indirect ones. Our experimental results presented in Sect. 5.1 prove this expectation correct. Direct transformations can, for instance, be accomplished by a tightly interleaved execution of parts of the decompression algorithm of the source format and parts of the compression algorithm of the destination format within the body of a loop iterating over the input data. Thereby, intermediate stores and loads to and from memory can be omitted.

We propose to investigate such direct transformations as a new class of algorithms, which is closely related to compression algorithms. Figure 2 again contrasts the data flows of indirect and direct transformations.

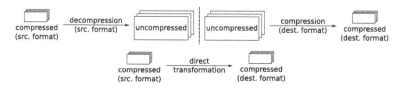

Fig. 2. A comparison of the data flows of indirect (top) and direct (bottom) transformations.

3.2 Precise Vs. Imprecise Transformations

In the literature there exists a multitude of compressed formats and associated compression algorithms. Whereas a compressed format specifies the structure of the compressed data, the respective compression algorithm tells how to make use of this structure in order to obtain the best compression rate possible in the given format. Hence, the output of the compression of the destination format could be considered to be a reference for the transformation.

If a transformation algorithm produces a result which equals the result of the compression of its destination format bit by bit, we call it a *precise* transformation. Indirect transformations are always precise, since they actually use the compression algorithm of their destination format. On the other side, direct transformations do not necessarily need to be precise.

For certain combinations of source and destination formats this *bitwise* equality might require a disproportionately high effort. At the same time, data represented in a certain compressed format does not need to use the size reduction potential of the format to its maximum extent. As an example, consider run-length encoding (RLE) [1,6], which replaces each run in the uncompressed data by its value and its length. The uncompressed sequence $[7, 7, 7, 7, 7, 4, 4]$ would be represented as $[(7, 5), (4, 2)]$ in a precise way. It could, however, also be represented as $[(7, 2), (7, 3), (4, 2)]$ and still be decompressable. Making use of this observation, we introduce a relaxed definition of transformations:

We call a (direct) transformation *imprecise*, if its output O satisfies:

1. O is a valid instance of the destination format.
2. There is some valid input data, such that O is not bitwise equal to the output of the destination format's compression.
3. An application of the decompression of the destination format to O yields uncompressed data that is bitwise identical to the uncompressed data that can be obtained from the output of the respective precise transformation.

The third criterion is especially crucial, since it guarantees that imprecise transformations are in fact *lossless* and do not require any changes to the decompression algorithm. Usually, the result of an imprecise transformation has a bigger size than that of a precise transformation for the same source and destination formats with the same input data. We expect that imprecise transformations might perform better than precise ones for certain pairs of source and destination formats.

In the following section, we present some of our direct transformation algorithms including some imprecise variants.

4   Example Techniques

Figure 3 provides an overview of all transformation techniques, we have investigated so far.[1] However, in this paper we present only a selection of these techniques, namely Rle2FourNs, FourNs2Rle, and FourNs2FourGamma covering all aspects which have to be considered. Currently, we focus on unsigned 32-bit integers as the data type of the uncompressed data. Our algorithms use vectorization through SIMD instructions, since they employ fragments of the vectorized (de)compression algorithms of the involved formats.

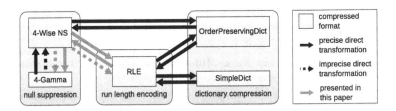

Fig. 3. An overview of the transformation techniques investigated by us so far.

---

[1] Our source code is downloadable at https://wwwdb.inf.tu-dresden.de/team/staff/patrick-damme-msc/.

4.1 Rle2FourNs

The foundation of the direct transformation from the format of RLE to that of 4-Wise NS is the observation that runs of equal *single values* in the uncompressed data yield (shorter) runs of byte-wise equal homogeneous[2] *compressed blocks* in the compressed format of 4-Wise NS.

The transformation algorithm iterates over its RLE-compressed input data and performs the following steps for each pair of run value and run length:

1. The run value and run length are loaded from the compressed input data.
2. The number of compressed blocks of 4-Wise NS necessary to represent the run is calculated by dividing the run length by four.
3. One block consisting of four copies of the run value is compressed the same way 4-Wise NS would do it. Note that this is done *only once per run*. After this step, the compressed block resides in a vector register in the CPU, i.e., no data is stored to main memory.
4. The compressed block is appended to the output data as often as necessary. This is done by storing the content of the vector register of the previous step to memory multiple times, which does not require any load instructions.

In practice, this procedure gets more complicated, since the run length cannot be assumed to be a multiple of four. In the vicinity of the border between two adjacent runs as well as at the end of the input buffer, it can be necessary to process heterogeneous blocks.

For small run values, this approach can be further accelerated. Storing the compressed block to memory is done using a vectorized store instruction, which writes 16 bytes of vector register content to memory at once. If the run value has exactly one effective byte[3], then the compressed block including the compression mask spans only five bytes. That is, it fits three times into a 16-byte vector register. Hence it is possible to store out three compressed blocks at once. A similar improvement can be made for run values having exactly two effective bytes. In that case, three copies fit into two vector registers. We implemented these optimizations by modifying the permutation masks used by 4-Wise NS to not only permute, but also copy the data within the vector register.

4.2 FourNs2Rle

The direct transformation in the inverse direction, i.e., from the format of 4-Wise NS to the format of RLE, makes use of the fact that runs of byte-wise equal homogeneous *compressed blocks* in the compressed input data mean (longer) runs of equal *single values* in the uncompressed format.

The transformation iterates over all compressed blocks of 4-Wise NS in its input, performing the following steps for each block:

---

[2] We call a block *homogeneous*, if it contains just one distinct value. Otherwise we call it *heterogeneous*.

[3] Following Schlegel et al. [7], we use the term *effective bits* to denote all but the leading zero bits of a value. The analogous holds for the term *effective bytes*. By definition, the value zero also has one effective bit respectively one effective byte.

1. The compressed block is checked for homogeneity. First, the compression mask is examined. Only if it indicates that all four values have the same length, the actual values are compared *in the compressed form*. If the block is homogeneous, the algorithm continues with step 2, otherwise with step 4.
2. The number of subsequent occurrences of the compressed block is determined *in the compressed input data*, i.e., without decompression. This is done by a simple loop starting at the first byte of the compressed block in the input data. In every iteration, it compares one byte to the corresponding byte in the next block, whose position can be calculated as the block size is known from the compression mask.
3. The one value is extracted from the compressed block and appended to the output as a run value *once*. The run length is obtained by multiplying the number of subsequent occurrences of the compressed block from the previous step by four and appended to the output as well. The algorithm proceeds to the next compressed block and returns to step 1.
4. Since the current compressed block contains more than one distinct value, it is not of interest if it is repeated. Instead, the single block is decompressed to a temporary buffer residing in the L1 cache and immediately recompressed using RLE. The algorithm continues with the next compressed block and returns to step 1.

Hitherto, this yields only an imprecise transformation, which is given the name FourNs2RleImprecise. The reason why the produced output might differ from the output of a direct compression with RLE, is the coarse-grained view on the data. Runs are only determined at block-level, but in fact, 4-Wise NS might partition a run in the uncompressed data into up to three parts: The run might start in a heterogeneous block, run through arbitrarily many homogeneous blocks and finally end in a heterogeneous block again. What FourNs2RleImprecise lacks, is to stitch these parts together. Doing so, however, causes additional overhead. Avoiding this, is the justification for the imprecise technique.

4.3   FourNs2FourGamma

The main idea of the transformation from the format of 4-Wise NS to that of 4-Gamma is a temporary decompression of one compressed block of 4-Wise NS immediately followed by the recompression with 4-Gamma.

The main loop of the algorithm processes each compressed block of 4-Wise NS in the input data as follows:

1. The 8-bit compression mask is loaded and the respective permutation mask for decompression as well as the size of the compressed block are looked up in the tables for the decompression of 4-Wise NS.
2. The decompressing permutation is executed. Note that after this step, the uncompressed block resides in a vector register and does not need to be stored to main memory.

3. The shared prefix of 4-Gamma is determined like in the compression of 4-Gamma, i.e., by computing the maximum number of effective bits via the number of leading zero-bits of the bitwise OR of the four values. This is done based on the register contents from the previous step, i.e., without accessing memory. The calculation requires four extractions of a 32-bit element from a vector register, three scalar bitwise OR operations, and one invocation of a scalar count-leading-zeros operation.
4. The four values are shifted to right and stored to the values section of the output data, while the shared prefix is stored to the prefix section.

The precise calculation of the maximum number of effective bits of the four uncompressed values in step 3 requires many instructions and therefore costs a lot of time. In order to reduce these costs, the imprecise transformation FourNs2FourGammaImprecise relaxes the strict interpretation of the shared prefix. It approximates the maximum number of effective *bits* of the four values by the maximum number of effective *bytes* increased by eight times. The crucial point is that the latter number can directly be obtained from the 8-bit compression mask of 4-Wise NS by looking it up in a table indexed with the compression mask. This table has a total size of 256 bytes and is created offline. Note that the output data produced this way is perfectly decompressable by 4-Gamma without any changes done to its decompression algorithm. Figure 4 contrasts the result of the precise and the imprecise transformation for an example block.

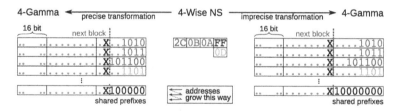

Fig. 4. A comparison of the outputs of the precise and the imprecise variant of the direct transformation FourNs2FourGamma.

5   Experimental Evaluation

We implemented our direct transformation algorithms as well as the corresponding (de)compression algorithms in C++ and compiled them with g++ 4.8 using the optimization flag -O3. Our experiments were conducted on a machine equipped with an Intel Core i7-4710MQ at 2.5 GHz and 16 GB of RAM. The L1 data, L2 and L3 caches have a capacity of 32 KB, 256 KB and 6 MB, respectively.

In all experiments, the underlying uncompressed data consisted of 100 M unsigned 32-bit integers. We use synthetic test data in order to be able to freely specify the properties of the data, especially the distribution of values and the lengths of runs within the data. We report speeds in terms of million integers per second (mis), whereby *integer* refers to an underlying uncompressed value.

5.1   Indirect Vs. Direct Transformations

To find out if direct transformations reach higher speeds than indirect ones, we ran both on the same data and measured the required run times. The (de)compression algorithms employed in the indirect transformations are vectorized and hand-tuned to allow a fair comparison. The results are presented in Fig. 5. The top row of diagrams shows the speeds side by side, while the bottom row explicitly provides the speed ups.

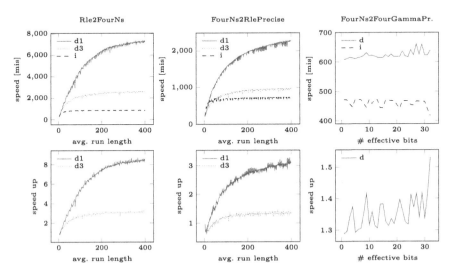

Fig. 5. Comparison of the presented direct transformations (d) to the indirect ones (i).

The first two columns correspond to the transformations involving RLE. Here the data was generated such that it contains runs. The values given at the horizontal axis are the average run lengths. The length of each individual run was chosen uniformly from the interval $avg \pm 2$. We show the results of the direct transformation when all original values have one (d1) or three (d3) effective bytes each, and of the indirect transformation (i), for which the influence of the number of effective bytes would not be visible at the scale of the diagrams.

Both directions of the transformation exhibit the same general trends: (1) the speed increases as the average run length increases, and (2) the more effective bytes the values have, the lower the maximum speed. Except for very small average run lengths, the direct transformations outperform the indirect ones, whereby FourNs2RlePrecise requires run lengths that are a little longer than Rle2FourNs in order to overtake the indirect transformation. The speed ups observed reach up to 8.6 and 3.2 for Rle2FourNs and up to 3.2 and 1.4 for FourNs2RlePrecise, if all values have one respectively three effective bytes.

The third column of Fig. 5 provides the results for FourNs2FourGamma-Precise. In this case, the data was generated such that all values have the same number of effective bits, which is given at the horizontal axis. It can clearly be seen that the direct transformation is faster than the indirect one for all numbers of effective bits. The speed up achieved is between 1.3 and 1.5, which is still considerable.

Our experimental results show that direct transformations are much faster than indirect ones and should thus be employed instead of the latter. We conducted similar experiments for all other direct transformations shown in Fig. 3 and obtained similar results.

5.2 Precise Vs. Imprecise Transformations

In addition to precise transformations, we suggested that it could be faster to perform imprecise transformations in certain cases. We experimentally compared both variants. The results are given in Fig. 6. The top and bottom row of diagrams are concerned with the precise (pr) and imprecise (im) variants of FourNs2Rle and FourNs2FourGamma, respectively. The columns report speeds, speed ups, and compression rates, from left to right.

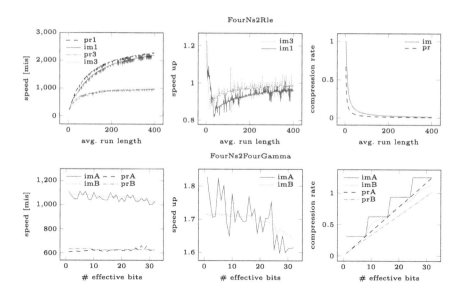

Fig. 6. Comparison of the precise transformations (pr) to the imprecise ones (im).

The results show that the imprecise variant of FourNs2Rle is faster than the precise one only for low average run lengths. A look at the compression rates of the output of the precise and imprecise transformations reveals the reason. As expected, the imprecise variant yields a worse compression rate than the precise one and thus has to store more data. For average run lengths between about 20 and 100, this difference is most significant. For this reason, the precise transformation is clearly faster here. However, this difference in compression rates becomes negligible for long runs. As a consequence, the speed up of the imprecise variant converges on 1.0 again. That is, the imprecise variant yields at least only a slight slow down.

For FourNs2FourGamma we used two different distributions of the original data: (A) all uncompressed values have the same number x of effective bits, and (B) the number of effective bits is chosen uniformly from the interval $[1, x]$, whereby x is the number given at the horizontal axis. In this case, the facts are much clearer. The imprecise variant significantly outperforms its precise counterpart for both distributions and all possible xs yielding speed ups between 1.6 and 1.8, although it leads to a worse compression rate of the output.

To sum it up, the experiments revealed that imprecise direct transformations can indeed be faster then precise ones. However, this is not generally the case as not all combinations of source and destination formats as well as data characteristics seem to be suited. Still, imprecise transformations remain to be an interesting concept and will be promising for other transformation techniques.

6   Application Scenarios

In this section, we stress the usefulness of our direct transformation techniques by presenting two interesting applications requiring efficient transformations.

6.1   Indirect Compression

One possible application of direct transformation techniques is the acceleration of the actual compression. Assume we want to represent some uncompressed data in the compressed format Y. In the classical case, i.e., without considering transformations, there is only one way to achieve this: a *direct compression* by applying the compression algorithm of Y to the uncompressed data. However, also taking transformations into account, there are far more possibilities. We can, for instance, first apply the compression algorithm of some intermediate format X and then perform a transformation from the format X to the format Y. One can trivially see that such an *indirect compression* can only lead to a speed up if it employs a direct transformation. This is due to the fact that an indirect transformation would undo the intermediate compression to the format X as its first step and thus render it to be pure overhead.

The results presented in Fig. 7 prove that such indirect compressions can indeed outperform direct ones. In the example, the compressed format of 4-Wise NS was obtained from uncompressed data by either using the compression algorithm of 4-Wise NS directly (dc) or by the indirection via the format of RLE (ic). Again, the experiment was run for uncompressed values having one or three effective bytes each. The difference of the speed of the direct compression that is subject to the number of effective bytes is negligible at the scale of the diagrams. While the speed of the direct compression is not affected by the average run length, the indirect compression gets faster as the run length increases until it overtakes the direct one at run lengths of about 50 or 150, reaching speed ups of up to 1.8 and 1.3 for one or three effective bytes per uncompressed value.

Unsurprisingly, this does not work for all possible indirect compressions. We conducted the same experiment for the compression to the format of RLE and to the format of 4-Gamma via the format of 4-Wise NS. Both resulted in slow downs compared to the direct compression. Nevertheless, indirect compressions still remain an interesting approach.

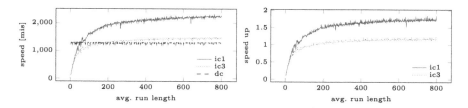

Fig. 7. Comparison of the direct compression (dc) to the format of 4-Wise NS to the indirect one (ic) via the format of RLE.

6.2   Transformations During Query Processing

Another, even more promising application for direct transformations, is the change of the compressed format during query processing. Currently, a shift towards in-memory database systems as the prevailing technology for analytical data processing is taking place. These systems keep *all* their data in main memory, so accessing intermediate results is as expensive as accessing the base data. Thus, intermediate results must be treated efficiently. One way to do so, is to compress not only the base data, but also intermediate results.

Compressed data offers advantages, such as reduced transfer times, better cache utilization, and a higher TLB hit rate. Moreover, many plan operators can directly process compressed data without decompression. On the other side, compression has two major disadvantages. Firstly, it introduces a certain computational overhead, which makes efficient implementations crucial. Recent research [5,7–9] has proven that this is manageable. Secondly, a compressed format has to be chosen. Approaches exist to make this decision wisely, as, e.g. [1], but do not consider the necessity to change the format later.

The optimal format depends on the properties of the data. While the properties of the base data might change only incrementally over time caused by DML operations, the properties of intermediate results usually change dramatically during the processing of a single query. Consequently, operators should be able to output data in another format than their input. For example, a selection might get dictionary-compressed data as input and let only small values pass, such that afterwards a null suppression scheme would be more appropriate. Not adapting the format of the operator's output implies a waste of performance potential. At this point, transformations can be a solution. They could be applied to the output of an operator or even inside an operator. This idea is especially well combinable with the *transient decompression* strategy proposed by Chen et al. in [2]. The authors suggest that operators which are unable to process compressed data directly should temporarily decompress their input, but use the compressed form for the output, again. If a recompression has to be done anyway, it could, of course, provide another format. Note that transformations during query processing must be as efficient as possible, since they are applied online. Due to that, our novel direct transformation techniques are inevitable.

7   Conclusions

In-memory database management systems are of increasing importance in both, business and science. They regularly combine a column-oriented storage format with lightweight compression techniques. The efficient implementation of lightweight compression algorithms as well as the decision for an optimal compressed format have been studied in the literature. However, if the characteristics of the compressed data change over time or during query processing, efficient transformations to other compressed formats can be beneficial, but have not been investigated before. In order to fill this gap, we proposed to use direct transformations that avoid to materialize any uncompressed data in main memory and are cache optimized. Furthermore, we presented precise and imprecise transformations as two variants of lossless direct transformations. Besides a conceptual introduction of such techniques, we also described three concrete algorithms. We conducted an experimental evaluation proving that our new techniques outperform the naïve approach of a complete decompression and recompression. To highlight the usefulness of our direct transformations, we described two possible application scenarios: indirect compression and transformations during query processing.

Acknowledgments. This work was funded by the German Research Foundation (DFG) in the context of the project "Lightweight Compression Techniques for the Optimization of Complex Database Queries" (LE-1416/26-1).

References

1. Abadi, D., Madden, S., Ferreira, M.: Integrating compression and execution in column-oriented database systems. In: SIGMOD, pp. 671–682 (2006)
2. Chen, Z., Gehrke, J., Korn, F.: Query optimization in compressed database systems. SIGMOD Rec. **30**(2), 271–282 (2001)
3. Goldstein, J., Ramakrishnan, R., Shaft, U.: Compressing relations and indexes. In: ICDE, pp. 370–379 (1998)
4. Huffman, D.: A method for the construction of minimum-redundancy codes. Proc. IRE **40**(9), 1098–1101 (1952)
5. Lemire, D., Boytsov, L.: Decoding billions of integers per second through vectorization. In: CoRR abs/1209.2137 (2012)

6. Roth, M.A., Van Horn, S.J.: Database compression. SIGMOD Rec. **22**(3), 31–39 (1993)
7. Schlegel, B., Gemulla, R., Lehner, W.: Fast integer compression using simd instructions. In: DaMoN Workshop, pp. 34–40 (2010)
8. Stepanov, A.A., Gangolli, A.R., Rose, D.E., Ernst, R.J., Oberoi, P.S.: SIMD-based decoding of posting lists. In: CIKM, pp. 317–326 (2011)
9. Zukowski, M., Heman, S., Nes, N., Boncz, P.: Super-scalar RAM-CPU cache compression. In: ICDE, pp. 59–70 (2006)

Transformation, Extraction
and Archiving

Analysis of the Blocking Behaviour of Schema Transformations in Relational Database Systems

Lesley Wevers$^{(\boxtimes)}$, Matthijs Hofstra, Menno Tammens,
Marieke Huisman, and Maurice van Keulen

University of Twente, Enschede, The Netherlands
{l.wevers,m.huisman,m.vankeulen}@utwente.nl
{m.hofstra,m.j.tammens}@student.utwente.nl

Abstract. In earlier work we have extended the TPC-C benchmark with basic and complex schema transformations. This paper uses this benchmark to investigate the blocking behaviour of online schema transformations in PostgreSQL, MySQL and Oracle 11g. First we discuss experiments using the data definition language of the DBMSs, which show that all complex operations are blocking, while we have mixed results for basic transformations. Second, we look at a technique for online schema transformations by Ronström, based on triggers. Our experiments show that pt-online-schema-change for MySQL and *DBMS_REDEFINITION* for Oracle can perform basic transformations without blocking, however, support for complex transformations is missing. To conclude, we provide a solution outline for complex non-blocking transformations.

1 Introduction

Software is in constant need of maintenance, adaptation and extension. For applications storing and maintaining data in a database, a software change often involves restructuring of data, i.e., a schema change with an accompanying conversion of the data. To ensure that no concurrency conflicts occur, many relational database systems block access to the data during a schema change. The effect is that concurrent transactions completely halt until the execution of the schema change has finished, which could take many hours to days for large databases. This is a real problem for systems that need 24/7 availability, such as telecommunication systems, payment systems and control systems [5,7].

Goals. We experimentally investigate the blocking behaviour of online schema transformations in current DBMSs. We look at the capabilities provided by the standard SQL data definition language (DDL) as implemented by the DBMSs, and we investigate a method developed by Ronström [6], which can perform non-blocking schema changes on any DBMS that supports triggers. We investigate basic transformations provided by the SQL DDL such as adding columns and indexes, and we look at complex transformations that require multiple DDL operations, such as changing the cardinality of a relationship, or changing the primary key of a table. While the basic transformations are the most common, these complex transformations are often needed in realistic transformations.

© Springer International Publishing Switzerland 2015
T. Morzy et al. (Eds.): ADBIS 2015, LNCS 9282, pp. 169–183, 2015.
DOI: 10.1007/978-3-319-23135-8_12

Contenders. We investigate PostgreSQL, MySQL and Oracle 11g, which represent a large fraction of the DBMSs used in industry. We now provide a brief overview of their capabilities for online schema transformations. First, PostgreSQL does not provide non-blocking DDL, but it is interesting as it can perform many DDL operations instantaneously. Next, MySQL has recently added support for online DDL[1]. In addition, a number of tools have been developed in industry to perform online schema changes on MySQL using Ronström's method, including pt-online-schema-change[2], oak-online-alter-table[3], and the online-schema-change tool developed at Facebook[4]. As these tools have similar capabilities, we investigate pt-online-schema-change in our experiments as a representative. Finally, Oracle 11 g does not provide online DDL, but it can perform non-blocking schema changes using the *DBMS_REDEFINITION* package[5].

Approach. For our experiment we have developed a benchmark [8] that extends the standard TPC-C benchmark[6] with basic and complex schema transformations. We run the standard TPC-C workload, while concurrently executing a schema transformation, and measure the impact on the TPC-C throughput. An important aspect of our benchmark is that schema transformations should be correct, i.e., they should satisfy the ACID properties, they should be composable to allow the execution of complex transformations, and ideally, transformations should be specified declaratively. We briefly discuss our requirements and the benchmark in Sect. 2, and we discuss our experimental setup in Sect. 3.

Results. In Sect. 4 we discuss our experimental results for online transformations using the DDL provided by the DBMSs. We see mixed results for basic transformations, while all complex transformations block the TPC-C workload. In Sect. 5 we discuss the experimental results for Ronström's approach using pt-online-schema-change for MySQL and Oracle's *DBMS_REDEFINITION* package. We see that pt-online-schema-change can perform all basic DDL operations without blocking, but it can not perform complex transformations. Oracle can perform some complex transformations, but is limited to operations on a single table. We summarize our results in Sect. 6, and in Sect. 7 we discuss a solution outline to support complex non-blocking schema transformations.

Contributions. The contributions of this paper are:

- An experimental investigation of the blocking behaviour of basic and complex schema transformations using the DDL in PostgreSQL, MySQL and Oracle 11g, and using Ronström's method as implemented by pt-online-schema-change and Oracle's *DBMS_REDEFINITION*.
- A solution outline for complex non-blocking schema transformations.

---

[1] http://dev.mysql.com/doc/refman/5.6/en/innodb-online-ddl.html.

[2] http://www.percona.com/doc/percona-toolkit/2.1/pt-online-schema-change.html.

[3] http://openarkkit.googlecode.com/svn/trunk/openarkkit/doc/html/oak-online-alter-table.html.

[4] https://www.facebook.com/notes/mysql-at-facebook/online-schema-change-for-mysql/430801045932.

[5] http://docs.oracle.com/cd/B19306_01/appdev.102/b14258/d_redefi.htm.

[6] http://www.tpc.org/tpcc/spec/tpcc_current.pdf.

2   Benchmark

In an earlier paper we have defined requirements for non-blocking schema transformations, based on which we have extended the standard TPC-C benchmark to measure the impact of various types of schema transformations on the TPC-C workload. In this section we briefly discuss the requirements and the benchmark. More details can be found in our earlier paper [8].

Requirements. We have defined requirements on the functionality of schema transformations, and on their performance characteristics.

In terms of functionality, we assert that a schema transformation should satisfy the ACID properties like any other transaction that updates the database. Moreover, ideally, schema transformations should be specified declaratively. Similar to queries, a user should not have to be concerned with *how* a transformation is executed, but only *what* the result of a transformation should be. For instance, an implementation of the DDL satisfies this requirement if it provides ACID guarantees for transactionally composed DDL operations. Moreover, the system should provide a mechanism to update applications as part of the schema transformation, e.g., by replacing stored procedures transactionally.

In terms of performance, a schema transformation should have minimal impact on the performance of concurrent transactions. In particular, regular transactions should not be blocked, should not experience excessive slowdown, and should be able to complete without aborting. Moreover, the schema transformation itself should be able to commit while concurrent transactions are running, and the time to commit from the start of the transformation should be minimal. In our benchmark we measure the impact of schema transformations on the OLTP throughput, and the time-to-commit of the transformation.

Transformations. Our benchmark contains basic transformations as provided by the SQL data definition language. Additionally, we also investigate bulk data updates without changing the schema, which is required in many complex transformations. Furthermore, our benchmark also contains a number of complex transformations, which generally consist of multiple DDL statements. In particular, we look at creating a column derived from another column, changing the cardinality of a relationship, and changing a primary key. Most transformations involve the largest table in the TPC-C schema, and update the stored procedures to allow the TPC-C workload to keep running on the transformed schema. A detailed description of the benchmark cases can be found in our earlier paper [8].

Benchmark Process. The execution of a benchmark case is done in four phases. First, during the *setup* phase, we create a TPC-C database. Some benchmark cases require a modification to the TPC-C schema, which we also perform in this phase. Next, during the *intro* phase, we start the TPC-C benchmark load. We wait for 10 min before starting the transformation, while measuring the baseline TPC-C performance. Next, we start the *transformation* phase, where we execute the benchmark transformation. We wait for it to complete, while logging the begin and end time of the transformation. Finally, we wait for another 10 min while measuring the TPC-C throughput in the *outro* phase.

Benchmark Results. As seen in Fig. 1, we present the result of a benchmark as a line graph that plots the TPC-C transaction execution rate over time. We mark the start and commit time of the transformation with vertical lines, and we show the time-to-commit under the x-axis. Moreover, we plot aborted and failed transactions in red. The y-axis starts at zero transactions per second, which corresponds to blocking behaviour. We do not show the absolute TPC-C throughput as we are only interested in blocking behaviour and the relative performance of TPC-C during and after a schema transformation compared to the intro phase.

3   Experimental Setup

An implementation of our benchmark, and all experimental results can be found on our website[7]. We use the TPC-C implementation HammerDB[8] to create the TPC-C database and to provide stored procedures. We use HammerDB to generate one database for each DBMS, which we backup once, and then restore in the setup phase of every experiment. Before starting the introduction phase of the experiment, we let the TPC-C benchmark run for ten seconds, as to give the DBMS some time to warm up. To generate load on the system, and to measure the TPC-C performance, HammerDB provides a driver script. However, as this script does not perform logging of transactions, we have ported the script to Java and we have added logging facilities. For all experiments, we generate a database of 30 warehouses, and we use 64 threads of load on the database. We do not spawn new threads to start other transactions while a thread is blocked. For the experiments we have used a quad-core Intel i7 machine with 16 GB of RAM and a solid-state drive. For the software we used Ubuntu Linux kernel 3.20.0, PostgreSQL version 9.1.14, MySQL version 5.6.20, pt-online-schema-change version 2.2.11, oracle 11 g release 11.2.0.3.0, and HammerDB version 2.14.

Stored Procedures. Many of our benchmark cases update the TPC-C stored procedures so that the workload can keep running after the transformation. As such, we need support from the DBMS to change stored procedures as part of a schema transformation. PostgreSQL provides transactional DDL which also supports transactional upgrades of stored procedures. In contrast, MySQL does not have transactional DDL, and does not provide a mechanism to upgrade stored procedures safely. This means that stored procedure upgrades in our MySQL experiments are not atomic. Oracle provides *editions*, which allow switching between different versions of stored procedures safely. However, we found it difficult to automate our tests using editions, and chose to use non-atomic updates of stored procedures. This does not affect the results of our experiments.

---

[7] http://wwwhome.ewi.utwente.nl/~weversl2/?page=ost.

[8] http://hammerora.sourceforge.net/.

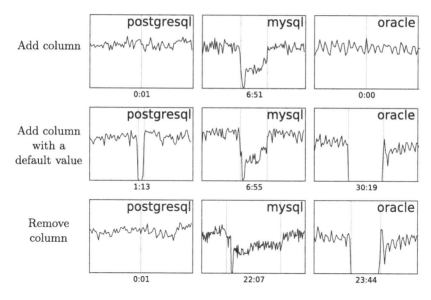

Fig. 1. Adding and removing columns.

4 Experimental Results: Data Definition Language

This section shows our experimental results for online schema transformations using the data definition language in PostgreSQL, MySQL and Oracle 11 g. First, we look at basic operations, including column operations, index operations and bulk data updates. To conclude, we investigate composition of DDL statements to perform complex transformations.

4.1 Basic Transformations

Adding and Removing Columns. Figure 1 shows the impact of basic column operations on the TPC-C workload. Both PostgreSQL and Oracle can add a column instantaneously, without noticeably interrupting the TPC-C workload. MySQL can not add a column instantaneously, but uses its online schema change functionality. Despite this, MySQL still shows a short period of blocking at the start of the operation, and we see a significant reduction in throughput. When adding a column with a default value, PostgreSQL and Oracle now materialize the column being created, which results in a period of blocking. For MySQL we see the same behaviour as the previous case. When removing a column, PostgreSQL can perform this operation instantaneously, and MySQL can use its online schema change feature. Interestingly, DROP COLUMN causes Oracle to block. Oracle also allows a column to be marked as unused, which effectively removes the column without reclaiming disk space. Disk space can be reclaimed using DROP UNUSED COLUMNS, however, this is still a blocking operation.

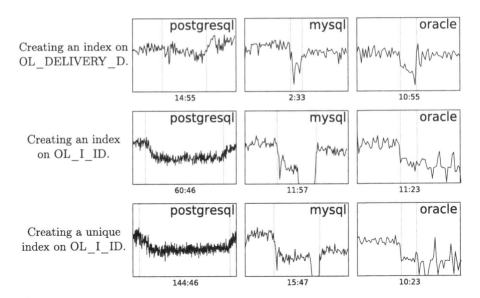

Fig. 2. Creating normal and unique indexes.

Creating Indexes. Figure 2 shows the impact of creating indexes on the TPC-C workload. We have created indexes on two columns with different workload: the OL_DELIVERY_D column which is nullable and is not written on insertion, while the OL_I_ID is being written to on insertion. All tested DBMSs allow online creation of indexes. PostgreSQL shows a small impact on TPC-C throughput, but behaves well. Oracle commits more quickly than PostgreSQL, but shows periods of significant blocking after the commit, suggesting that Oracle is creating the index in the background. We have run the experiment for three hours after the commit, and have seen that this behaviour persists during this period. Despite supporting online index creation, MySQL blocks for a significant amount of time on when indexing the OL_I_ID column. We see that creating a unique index has similar characteristics to creating a regular index, but the time to commit for PostgreSQL and MySQL is longer. Removing indexes is an instantaneous operation in all three DBMSs, so we don't show their results.

Bulk Data Transformations. For some transformations it is essential that we can update data in bulk. An update statement differs from an ALTER TABLE statement in that the schema is not changed. However, semantically it is a schema transformation. Updating prices in a database to use a different currency is an example of such a transformation. Moreover, bulk data operations are important in many complex transformations to transform data or to move data between tables. Where stored procedures may simply fail on a schema that it does not expect, for bulk data updates this is not the case. As such, it is important that bulk data transformations satisfy the ACID properties.

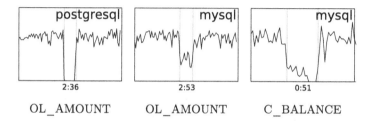

Fig. 3. Bulk data transformations in PostgreSQL and MySQL.

Figure 3 shows the impact of a bulk data update on the column in the caption using PostgreSQL and MySQL. We do not show results for Oracle, because it could not execute the bulk update due to concurrency conflicts. In both cases, we use the serializable transaction level to guarantee correctness. We see that PostgreSQL takes a table lock to guarantee serializability, and blocks the TPC-C workload. Interestingly, MySQL does not block the workload when updating the OL_AMOUNT column, because it only locks the OL_AMOUNT column, which is not being updated by the TPC-C workload. We ran the experiment on the C_BALANCE column, which is being updated, and see that MySQL now blocks toward the end of the operation. During the transformation, transactions can still execute, as MySQL doesn't take a complete table lock.

4.2   Complex Transformations

Transactional Composition. A natural way to construct a complex transformation from DDL operations is to wrap them into a transaction. If every DDL operation is non-blocking, commits instantaneously, and does not block other transformations from starting after committing, then the composed transformation can also be non-blocking and instantaneous. However, all complex transformations that we have considered involve bulk data updates, which, as we have seen in the previous section, is blocking in current DBMSs. When composing an instantaneous transformation with a bulk data update, the instantaneous operation can take a table lock, which is held during the bulk data update.

We see this behaviour in PostgreSQL, as shown in Fig. 4 (top row). In the leftmost experiment we have added a column OL_TAX whose value is derived from an existing column. First, we add the new column, which is non-blocking and instantaneous, and then we fill the column using UPDATE, which results in a table lock. We see the same behaviour in all complex cases that we have tested.

Non-transactional Composition. MySQL and Oracle 11 g do not provide transactional DDL: they auto-commit after each DDL operation. However, MySQL does support online DDL. Can we use this to perform complex transformations correctly? As many operations require bulk data operations that can not be performed without blocking, this is not possible in general.

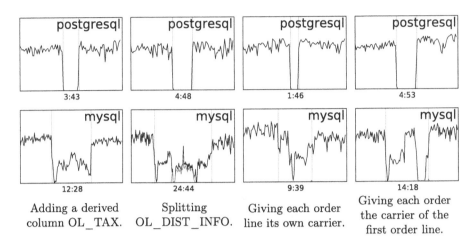

Fig. 4. Complex transformations in PostgreSQL and MySQL.

When composing non-blocking transformations non-transactionally, intermediate schemas are visible to concurrent transactions. If we keep using the original stored procedures on these intermediate states, they can fail to execute, perform erroneous operations, or encounter lost updates, which could damage the integrity of the database. We could update the stored procedures directly after the commit of each transformation step to handle intermediate states. However, this does not work for bulk data transformation, as the original stored procedures will keep executing while the bulk data is in progress, which results in concurrency conflicts. For instance, if we want to add a derived column, we can first create a new column, and then fill it using a bulk update statement. However, while the update statement is in progress, the original transactions can continue executing on the source column, which updates are not reflected in the derived column, thus resulting in lost updates. To solve this, we could attempt to update the stored procedures *before* the transformation starts, but this does not solve the problem, as the new transactions can be blocked from writing to the derived column while the bulk update is in progress.

Figure 4 (bottom row) shows results when performing complex transformations using the online DDL in MySQL, where we only update the stored procedures after the transformation. While the transformations are mostly non-blocking, their results are incorrect in all cases because the TPC-C transactions keep executing on intermediate transformation states, which results in lost updates. In the second transformation, we also see many erroneous transactions because we do not update the stored procedures after every transformation step.

4.3   Conclusions

Our experiments with MySQL, PostgreSQL and Oracle show that support for non-blocking transformations using the DDL is rather weak. Most problematic are adding columns with default values and performing bulk data updates. As complex transformations regularly require bulk data updates, non-blocking complex transformations are currently not possible at all. If non-blocking bulk updates where possible, many complex transformations could in principle be performed by adapting the stored procedures to intermediate states. However, this would also be very costly to implement in terms of development effort.

5   Experimental Results: Ronström's Method

Ronström proposed a method that allows changing of columns, adding indexes, and horizontally and vertical splitting and merging of tables by using database triggers [6]. The method works as follows. First, an *interim* table that matches the desired schema is created. Next, triggers are created on the original table that propagate any changes on the original table to the interim table. Next, data is copied to the new tables in small batches, while performing the desired schema transformation on the data. Finally, after copying the data, the original table is replaced by the interim table. A benefit of Ronström's method is that it can be implemented on top of existing database systems that support triggers.

In this section, we investigate pt-online-schema-change, which implements Ronström's method for MySQL, and we look at the *DBMS_REDEFINITION* package provided by Oracle 11g. While tools similar to pt-online-schema-change could be implemented for PostgreSQL, to our knowledge, at the time of writing no such tools are available.

5.1   Pt-online-schema-change for MySQL

The pt-online-schema-change tool from the Percona Toolkit implements Ronström's method for MySQL. The tool accepts a single ALTER TABLE statement, which it executes transactionally. Multiple transformations can be performed using a single ALTER TABLE statement, but multi-table transformations and data transformations are not supported. It creates a new table with the new schema, and copies the rows from the source table to this new table. Copying is done in chunks of a certain size, which can be configured using two strategies. First, a fixed chunk size can be specified, and second, a fixed time per chunk can be specified. While a chunk is being copied, the copied rows are locked for writing. A larger chunk size impacts concurrent transactions more due to locking, while a shorter chunk size slows down the schema transformation.

Effect of Chunk Size and Load. The chunk size and the TPC-C benchmark load have a large effect on the performance of pt-online-schema-change. This is because pt-online-schema-change executes transactions in *LOW PRIORITY* mode, to minimize slowdown for concurrent transactions. Figure 5a shows the behaviour of MySQL when pt-online-schema-change is used to add a column with a TPC-C load of 64 threads and chunk size 1,000. The time to commit is very long, about 102 min, much longer than the 6:51 used by MySQL to perform the same operation. If we lower the TPC-C load from 64 threads to only 4 threads, and keep the chunk size at 1,000, pt-online-schema-change commits in only 14:44, as shown in Fig. 5b. If we increase the chunk size to 10,000, pt-online-schema-change completes in 4:17, as shown in Fig. 5c, however, we also see a reduction in TPC-C performance.

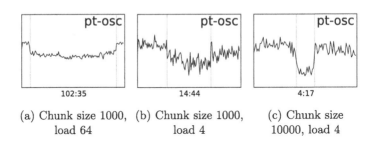

(a) Chunk size 1000,   (b) Chunk size 1000,   (c) Chunk size
    load 64   load 4   10000, load 4

Fig. 5. Adding a column using pt-online-schema-change.

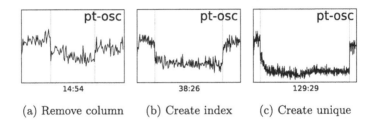

(a) Remove column   (b) Create index   (c) Create unique

Fig. 6. Experimental results for pt-online-schema-change.

Results. Figure 6 shows experimental results for pt-online-schema-change on several basic transformations. All basic DDL operations could be performed using pt-online-schema-change, and impact on performance is generally acceptable. Interestingly, pt-online-schema-change does not suffer from the initial period of blocking that we have seen in experiments using MySQL's online DDL. However, pt-online-schema-change does not support bulk data operations, and can not be used to perform transformations consisting of multiple DDL statements.

5.2 DBMS_REDEFINITION for Oracle

Since version 9i, Oracle provides the *DBMS_REDEFINITION* package, which allows schema transformations to be performed using Ronström's method. To use *DBMS_REDEFINITION*, the following steps have to be followed. First, an interim table has to be created with the desired schema. Next, the transformation is started by defining a mapping from fields in the original table to fields in the interim table, and by specifying a key that must be present in both the original and interim table, which is used to propagate updates on the original table to the interim table. Next, after the transformation is complete, objects such as indexes, constraints and stored procedures can be added to the table. The package provides a method to copy all existing objects from the original table to the interim table. Finally, the transformation can be finished to replace the original table with the interim table. This is a blocking operation, and takes longer if the interim table is not synchronized with the original table.

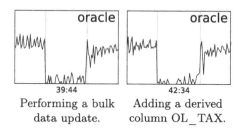

| Performing a bulk | Adding a derived |
| data update. | column OL_TAX. |

Fig. 7. Experimental results for Oracle's DBMS_REDEFINITION.

Figure 7 shows experimental results where we use *DBMS_REDEFINITION* to perform a bulk data update, and to add a column whose value is derived from another column. In general, we see that transactions can continue executing during the transformation, but performance is poor, and there are periods during the transformation where the throughput drops to zero. Despite this, *DBMS_REDEFINITION* allows any single-table transformation to be performed. Both of these cases can not be handled by pt-online-schema-change. However, transformations that involve multiple tables can not be performed, as the source of the transformation must be a single table. Compared to using the data definition language, this approach is more verbose, as the interim table must be defined, and all objects on the table must be copied.

6   Analysis Results

Our experiments with basic DDL operations in PostgreSQL, MySQL and Oracle 11 g show mixed results. PostgreSQL can add and remove columns instantaneously and it can create indexes online, but blocks when adding a column with a default value, and when performing bulk updates. MySQL provides online DDL for adding and removing columns, but blocks for a significant period of time at the start of the transformation. MySQL also supports online creation of indexes, but our experiments show long periods of blocking at the end of the transformation. Similar to PostgreSQL, Oracle 11 g can add columns instantaneously, however, adding columns with default values and removing columns takes very long, and blocks concurrent transactions. Bulk data updates are a problem in all tested DBMSs. PostgreSQL and MySQL simply block, while Oracle 11 g can not execute the operation due to concurrency conflicts.

Using the DDL for complex non-blocking transformations is not possible in any of the DBMSs. Using transactional DDL, PostgreSQL can generally perform all operations correctly, but blocks access to all affected tables during the transformation. MySQL and Oracle do not support transactional DDL. Composing non-blocking DDL operations non-transactionally is possible in general by updating stored procedures after each transformation step, however, MySQL and Oracle can not perform non-blocking data updates, which prevents us from performing most complex transformations. Moreover, such an approach is non-declarative, and can be costly to implement.

As an alternative to the DDL provided by the DBMSs, we have investigated Ronström's method. This method is interesting, as it can perform non-blocking schema transformations based on blocking transformations in any DBMS that implements triggers. The pt-online-schema-change tool shows that Ronström's method is a promising approach for basic online transformations: it can perform all basic schema transformations without blocking, with the exception of bulk data updates. However, complex transformation cases can not be handled by pt-online-schema-change, as it only supports a single ALTER TABLE statement at a time, and there is no support for UPDATE statements. Oracle's *DBMS_REDEFINITION* shows that Ronström's method can also be used for more complex single-table operations, but its implementation shows a significant amount of blocking.

7   Solution Outline

Native Support. With the existence of Ronström's method, it could be argued that DBMSs do not need to provide native support for online schema transformations, but only have to provide support for triggers and atomic updates of schema meta-data. This is the approach that Oracle has taken with edition-based redefinition[9] and the *DBMS-PARAL-LEL_EXECUTE* package[10]. Edition-based redefinition allows atomic updates of schema meta data and provides cross-edition triggers that can transform data between versions of the schema. The *DBMS_PARALLEL_EXECUTE* package can be used to avoid full table locks while transforming data between versions.

A drawback of Ronström's approach is that transformations can take a long time to execute. Native implementations of Ronström's method in DBMSs can potentially be more efficient than external tooling. For instance, Løland and Hvasshovd present Log Redo as an alternative implementation for Ronström's approach that avoids the use of triggers, and has minimal impact on performance [3]. However, while more efficient implementations of Ronström's method could reduce execution time, this does not scale to very large databases.

On-the-fly Transformations. An interesting alternative to Ronström's method is to perform transformations lazily, or on-the-fly. The basic idea is to commit

---

[9] http://docs.oracle.com/cd/E11882_01/appdev.112/e41502/adfns_editions.htm.
[10] http://docs.oracle.com/cd/E11882_01/appdev.112/e40758/d_parallel_ex.htm.

a transformation before transforming the data, and transform the data before it is accessed. From the viewpoint of the user, this allows a transformation to be executed instantaneously. Moreover, data can be transformed in the background during idle time. Lazy transformations have already been investigated in the context of Object-Oriented database systems [1]. Additionally, Neamtiu has shown that many relational schema changes can be performed on-the-fly [4].

Depending on the implementation, on-the-fly transformations have two advantages over Ronström's method. First, they compose naturally: two on-the-fly operations executed in sequence form an on-the-fly operation. Second, on-the-fly transformations can be implemented to execute in-place or incrementally: reusing storage space or garbage collecting parts of the original table that are already transformed. This avoids the problem of additional memory consumption for intermediate tables as seen in Ronström's method.

A drawback of on-the-fly transformations is that they must be implemented in the DBMS. Moreover, they provide overhead on data access, which increases latency. Most importantly, instantaneous transformations and on-the-fly transformations are limited to operations that can produce results on-the-fly. For instance, it is not possible to create an index or check a constraint instantaneously, and lookups in very large tables that do not have an index can not be done instantaneously. Consequently, composing a blocking transformations with an on-the-fly transformation leads to a blocking transformation. This shows that on-the-fly transformation by themselves are not a full solution to the problem.

Complex Operations using Ronström's Method. In his original paper, Ronström has proposed the use of SAGAs to compose basic transformations into more complex transformations [6]. The idea of SAGAs is to execute the individual operations of a transaction as a sequence of transactions, where for each operation an undo operation is provided that can be used to rollback the complete sequence of operations [2]. While SAGAs provide failure atomicity for composed operations, they expose intermediate states of the transformation to concurrent transactions. This requires applications that use the database to handle these states, which is non-declarative and requires additional development effort.

However, we think that almost any relational transformation can be performed atomically using Ronström's method without the use of SAGAs. The following is a sketch of the solution. First, to compose transformations, we can chain interim tables, i.e., triggers on the original table propagate updates to the first interim table, while triggers on the first interim table propagate updates to a second interim table, and so on. Using multiple interim tables can require a lot of memory. However, sequential transformations could potentially be combined to use a single interim table. Second, we can define triggers on multiple tables to propagate updates to one or more tables. This allows multi-table transformations. Finally, update propagation is inefficient for operations that require lookups on tables that are not indexed. This can be solved by dividing a transformation into two steps, where indexes are constructed in the first step, and where the transformation is performed in the second step using these indexes.

From a practical viewpoint, manually implementing transformations using this approach is quite complex, and optimizing such transformations even more so. One has to reason about updates on all involved tables, and how these should be propagated to interim tables. Data could be lost if certain triggers are missing or wrongly implemented. To solve this, tooling could be developed to transform declarative transformation specification into optimized execution plans.

Solution Outline. Rönstrom's method is essentially an optimistic concurrency control method: it performs operations on a snapshot of the state, and repairs any conflicts that arise from concurrent operations. As such, Ronström's method never blocks access to the state, but it requires additional memory to maintain multiple versions of the state. Moreover, it can only commit after the transformation has been completely executed. On the other hand, an on-the-fly method is essentially a pessimistic concurrency control method: it avoids conflicts by transforming data before access, i.e., it blocks access to parts of the database until the transformation for that part has been executed. However, on-the-fly methods can commit immediately, and require less memory compared to Ronström's method as they can perform transformations in-place or incrementally.

A solution to minimize time to commit would combine both approaches by first using Ronström's method to check constraints and prepare indexes, and then performing the remainder of the transformation using on-the-fly methods. However, if time to commit is not crucial, Ronström's method could be preferable in situations where predictable low-latency access to data is crucial.

Similar to declarative query support, we envision that DBMSs allow us to perform arbitrary schema transformations declaratively. As such, a DBMS should provide a schema transformation optimizer that can construct a non-blocking execution plan from a declarative specification of a schema transformation with the goal of minimizing throughput reduction, access latency, memory consumption and time to commit.

References

1. Ferrandina, F., Meyer, T., Zicari, R.: Implementing lazy database updates for an object database system. In: VLDB 1994, pp. 261–272 (1994)
2. Garcia-Molina, H., Salem, K.: Sagas. In: SIGMOD 1987. pp. 249–259. ACM (1987)
3. Løland, J., Hvasshovd, S.-O.: Online, non-blocking relational schema changes. In: Ioannidis, Y., Scholl, M.H., Schmidt, J.W., Matthes, F., Hatzopoulos, M., Böhm, K., Kemper, A., Grust, T., Böhm, C. (eds.) EDBT 2006. LNCS, vol. 3896, pp. 405–422. Springer, Heidelberg (2006)
4. Neamtiu, I., Bardin, J., Uddin, M.R., Lin, D.Y., Bhattacharya, P.: Improving cloud availability with on-the-fly schema updates. In: COMAD 2013, pp. 24–34. Computer Society of India (2013)
5. Neamtiu, I., Dumitras, T.: Cloud software upgrades: challenges and opportunities. In: MESOCA 2011, pp. 1–10. IEEE (2011)
6. Ronström, M.: On-line schema update for a telecom database. In: ICDE 2000, pp. 329–338. IEEE (2000)

7. Sockut, G.H., Iyer, B.R.: Online reorganization of databases. ACM Comput. Surv. **41**(3), 14:1–14:136 (2009)
8. Wevers, L., Hofstra, M., Tammens, M., Huisman, M., van Keulen, M.: A benchmark for online non-blocking schema transformations. In: DATA 2015 (2015)

A Benchmark for Relation Extraction Kernels

João L.M. Pereira[(✉)], Helena Galhardas, and Bruno Martins

INESC-ID and Instituto Superior Técnico, Universidade de Lisboa,
Lisbon, Portugal
{joaoplmpereira,helena.galhardas,bruno.g.martins}@tecnico.ulisboa.pt

Abstract. Relation extraction from textual documents is an important task in the context of information extraction. This task aims at identifying relations between pairs of named entities and assigning them a type. Relation extraction is often approached as a supervised classification problem, involving pre-processing steps such as text segmentation, entity recognition, and morphological and syntactic annotations. In previous studies, the way data is pre-processed differs among them, thus making the comparison of classification techniques for relation extraction unfair and inconclusive. Some of these classification techniques for relation extraction involve the use of kernels, which enable the comparison of complex structures. We propose a benchmark for the comparison of different kernels for relation extraction. Specifically, we propose the application of a common pre-processing stage, together with the use of an online learning algorithm to train Support Vector Machines with kernels designed for the classification of candidate pairs of related entities. We also report the results of the systematic experimental validation we have performed, using well known datasets in the area.

Keywords: Relation extraction · Benchmark · SVMs · Kernels · Online learning

1 Introduction

Textual corpora available in digital format are growing fast. These documents contain valuable information that, if properly identified and structured, can be used by several applications (e.g., in news aggregators). Information Extraction consists in automatically obtaining structured data from textual documents. This activity is typically composed of several tasks, namely segmentation, entity extraction, normalization, co-reference resolution and relation extraction [11].

This paper focuses on the Relation Extraction (RE) task. For example, given the text: *"The Taliban group Tehreek-e-Taliban Pakistan claimed the attack on the Karachi airport in southern Pakistan"*, a RE system should be able to identify the relation between the terrorist entity *"Tehreek-e-Taliban Pakistan"* and the location entity *"Karachi"*, and to classify this relation as being of the type outrage. The result will be the tuple ⟨ *"Tehreek-e-Taliban Pakistan"*, *"Karachi"*⟩ of a relation with schema outrage(terrorist, location).

© Springer International Publishing Switzerland 2015
T. Morzy et al. (Eds.): ADBIS 2015, LNCS 9282, pp. 184–197, 2015.
DOI: 10.1007/978-3-319-23135-8_13

There are different types of techniques for RE. These techniques are usually divided in two main groups: (i) rule-based, that specify logical inferences manually designed by specialists; and (ii) Machine Learning (ML)-based, that extract relations through the automatic analysis of patterns and/or correlations in data [11]. Several techniques, specially the ML-based, depend on linguistic and/or lexico-syntactic data annotations (e.g., part-of-speech tags for words). The tasks for obtaining these annotations constitute a pre-processing step.

Currently, many of the techniques for RE are based on supervised ML which identifies patterns in the previously annotated data that constitute a training dataset. A RE task can be seen as a supervised ML problem by considering the representation of a pair of entities as a data instance, and the relation type (or a non-relation) as the class that a data instance can belong to. To provide the dataset in the format usually accepted by a supervised ML technique, we have to transform each data instance representing a pair of entities into a vector that contains the relevant features (e.g., words or grammatical dependencies) of the pair of entities. Representing such information in a vector is a very demanding task in terms of execution time and memory. Moreover, it results in large vectors because the space of all possible relevant features may be huge. Thus, supervised ML techniques, and the models that they generate, can be inefficient since their execution time highly depends on the length of the vectors. One way to circumvent this situation is to use complex data structures, which organize relevant features in a more efficient way, and give them directly as input to the supervised ML techniques and the generated models. Then, we use kernel functions that enable specific ML techniques to compare complex data structures. The use of kernel functions is a common practice in the ML community, in particular in RE systems. For example, Support Vector Machines (SVMs) constitute a supervised ML technique that can be based on kernels.

Over the years, some competitions (e.g., MUC [5], ACE [6] and SemEval [8]), aiming at comparing RE systems, took place. Several RE systems based on kernels were evaluated with datasets from these competitions. However, these competitions did not enable a fair comparison of RE kernels. In fact, these competitions used different pre-processing stages, distinct SVM training approaches, and datasets that belong to the same domain, thus making the comparison of kernels unfair. Inspired by the work of Marrero et al. [10], who proposed a benchmark for entity extraction systems, we propose, in this paper, a benchmark for kernels for the RE task. This benchmark was developed using a RE framework named REEL [1][1]. The benchmark is composed of: (i) two sets of documents from distinct domains: AImed[2] and SemEval[3]; (ii) a common preprocessing step, composed of segmentation, tokenization, normalization, capitalization, part-of-speech tagging and dependency analysis; (iii) a set of SVMs based on state-of-the-art RE kernels; (iv) an online learning algorithm that trains

---

[1] http://reel.cs.columbia.edu/
[2] ftp://ftp.cs.utexas.edu/pub/mooney/bio-data/interactions.tar.gz
[3] http://semeval2.fbk.eu/semeval2.php?location=data

the SVMs involving various kernels; and (v) a validation process that measures convergence, quality and execution time.

The rest of this paper is organized as follows. Section 2 presents the fundamental concepts. Section 3 describes the proposed benchmark. Section 4 describes the experiments we performed and discusses the obtained results. Finally, in Sect. 5, we conclude with final remarks and directions for future work.

2   Fundamental Concepts

In this section, we present the fundamental concepts required to understand the rest of the paper. In Sect. 2.1, we introduce SVMs and online learning. In Sect. 2.2, we describe the main RE kernels described in the literature.

2.1   SVMs and Online Learning

SVMs [9] constitute one of the most popular supervised ML techniques. A SVM is a binary classification model that uses an hyperplane to separate the data that belongs to two distinct classes. The biggest challenge when generating SVMs is to find an optimal hyperplane that properly separates the data in two classes. An hyperplane is represented by its normal vector $\mathbf{w}$ and the goal is to find the optimal vector, $\mathbf{w}^*$. For that, we define an objective function $f(S; \mathbf{w})$ that represents the proximity between two vectors $\mathbf{w}$ and $\mathbf{w}^*$ for a given training dataset S. Then, we apply optimization techniques [12] (e.g., gradient descent) to this objective function, to obtain a vector $\mathbf{w}$ close to the optimal vector $\mathbf{w}^*$.

The training dataset is composed of data instances. These data instances are pairs of entities. A data instance $(\mathbf{x}, y)$ from the training dataset S is constituted by an input vector $\mathbf{x}$ and a scalar y that represents the binary class the input vector belongs to. A position of the vector $\mathbf{x}$ corresponds to a relevant feature of the data instance. A SVM uses the vector $\mathbf{w}$ to predict the class of a data instance $\mathbf{x}$ by computing the inner product $\langle \mathbf{w}, \mathbf{x} \rangle$. The objective function is then given by $f(S; \mathbf{w}) = \frac{\lambda}{2}||\mathbf{w}||^2 + \sum_{(\mathbf{x},y) \in S} loss(\mathbf{x}, y; \mathbf{w})$ where $||\mathbf{w}||$ is a regularization method applied to $\mathbf{w}$ that penalizes the model complexity degree (i.e., rewards fewer patterns in a model) to prevent overfitting, and $loss(\mathbf{x}, y; \mathbf{w})$ is a loss function given by $\max(0, 1 - y\langle \mathbf{w}, \mathbf{x} \rangle)$ that returns a penalty value if the model predicts the wrong class of a data training instance $\mathbf{x}$. In order to evaluate the model's predictive performance against the training dataset S, the objective function applies the loss function to each training data instance $(\mathbf{x}, y) \in S$ and sums the returned penalty values. The λ parameter is introduced by the user in order to adjust the two model components: the regularization and the loss function computed for the training dataset. In practice, it balances the model complexity degree with the model predictive performance for the training dataset.

In the context of RE, representing a data instance (i.e., a pair of entities and its features) through a vector is not a straightforward task. In order to obtain a vector, we need to transform a complex data structure that represents the data instance into a vector. In this process, we have to generalize and omit several features that, comparatively to the original structure, result in weak representations of the data. A more satisfactory alternative than to relay on a single vector is to represent the data instances by more complex and informative data structures (i.e., graphs or sequences). A kernel [11] can then be used to compare two data instances represented by complex data structures. Kernels return a similarity value between two complex structures and they behave as inner products in non-linear feature spaces. Moreover, optimization techniques can compute a vector $\mathbf{w}$ close to the optimal vector $\mathbf{w}^*$ by calculating inner products between vectors without accessing directly to the vector positions. Since the optimization techniques do not directly access vector positions, we can replace the inner products by kernels and generate a kernel based model.

Online learning techniques are ML techniques that use online optimization algorithms to train a ML model. These algorithms find the best set of parameters for an objective function by processing the data instances in an online fashion (i.e., process a data instance at a time), thus enabling us to analyze the ML model at a specific point of the training process. Online optimization techniques are generally faster than other types of optimization techniques [12].

So far, we have presented binary SVM classifiers. However, most classification problems involve more than two classes. These problems are called multi-class classification problems and they are not solved directly using one SVM classifier. We can use heuristics that combine several SVM classifiers, for example *One-VS-One* or *One-VS-All* [9]. In Sect. 3.3, we provide more details about the *One-VS-One* heuristic that we used in this work.

2.2   Relation Extraction Kernels

This section describes three state-of-the-art kernels. Each data instance (i.e., pair of entities) is initially described by a sentence composed of words and the location of the words that compose each entity. This is an initial representation given as input to a pre-processing stage that outputs complex structures to be used by each kernel. In this paper, we considered the following kernels:

Subsequences Kernel (SSK): Bunescu and Mooney [4] developed a kernel based on subsequences. SSK is composed by the sum of three sub-kernels that compare sparse word subsequences (i.e., word sequences not necessarily contiguous) in different locations of a sentence: before, after and between the pair of entities. This kernel is defined by a function that computes the similarity between words based on the number of features that the words have in common (e.g., word stem or word grammatical category).

Shortest Path Kernel (SPK): Bunescu and Mooney [3] also proposed a kernel based on the comparison of dependency graphs. A dependency graph is a directed graph that represents the grammatical dependencies between words in a sentence. Each node represents a word and each edge represents a dependency between the two nodes. SPK has the particularity of using, for comparison, only the sub-graph that contains the shortest path between the entities. Only graphs with the same number of nodes and similar edges in the shortest path are compared (i.e., for the others, the kernel returns zero). The final value returned by the kernel is calculated by multiplying the values returned by the comparison of the nodes. The nodes that are in the same position of the sub-graphs are compared using the same SSK function that computes the similarity between words. This function counts the number of features that the words have in common.

Bag-Of-N-Grams Kernel (BNK): Giuliano et al. [7] proposed a kernel that combines two simpler kernels, namely: (i) a global context kernel that considers textual information related to the whole sentence and (ii) a local context kernel that considers only the words around the entities. Similarly to SSK, the global context kernel is separated in three sub-kernels that evaluate the similarity of three independent sequences of words located before, between and after the pair of entities. This kernel is based on n-grams instead of subsequences. N-grams are small sets of contiguous words of size n (usually $n = 3$). Each sub-kernel compares the number of common n-grams between two sequences of words. The local context kernel compares the sequence of words centered at each entity and constituted by 6 words (i.e., 3 words before the entity and 3 words after the entity). The similarity between two words is obtained by the number of common features, analogously to SSK and SPK.

Table 1. Statistical characterization of the datasets.

	Sentences	Candidate entities pairs	Entities	Relations	Relation types
AImed	1159	5471	3754	996	1
SemEval	10717	10717	21434	8853	9

3   Benchmark

This section presents the development process of a benchmark for RE kernels. This benchmark enables an impartial evaluation process to assess the suitability of each kernel for the RE task. This evaluation process is composed of the following tasks: (i) pre-processing of datasets described in Sect. 3.1 using the techniques from the tools presented in Sect. 3.2; (ii) training and execution of SVMs based on kernels using the learning techniques described in Sect. 3.3; and (iii) evaluation of the kernels using the measures described in Sect. 3.4.

3.1   Datasets

This section describes the two datasets considered in the benchmark: AImed and SemEval. We chose these datasets because they belong to two distinct domains and they enclose different classification problems. Both datasets are composed of English documents. Table 1 presents a statistical summary for these datasets.

AImed: is composed of 255 Medline article abstracts, 200 of which referring to interactions between proteins, and with a remaining 25 which do not refer to any interaction. In total, AImed contains 5471 proteins pairs, of which 996 correspond to interactions and 4475 correspond to pairs without interaction.

SemEval (Semantic Evaluation) [8]: is an ongoing series of evaluations for computational semantic analysis systems. It proposed, in 2010, a classification challenge for relations between entity pairs. The dataset associated to this challenge is composed of 8000 training sentences and 2717 test sentences. Each sentence is clearly identified and contains a single annotated candidate pair. This challenge includes the following nine asymmetric relation types: *Member-Collection, Cause-Effect, Component-Whole, Instrument-Agency, Entity-Destination, Product-Producer, Message-Topic, Entity-Origin,* and *Content-Container*.

3.2 Linguistic Pre-processing Techniques

RE techniques generally use various types of linguistic and/or lexico-syntactic annotations. Therefore, it is necessary to pre-process the text in order to obtain these annotations. For pre-processing the datasets, we use the following techniques and tools:

1. **Sentence Segmentation:** Separates the text into sentences. We processed the text with the *Apache OpenNLP*[4] library, which is a tool for natural language processing based on ML.
2. **Tokenization:** Identifies the words of every sentence, keeping their order in the text. We processed the sentences with the *Apache OpenNLP* library.
3. **Token Normalization:** Labels the words with their stem, in such a way that words of the same family have the same label (e.g., *claimed* and *claims* have the same stem, *claim*). We used the *Porter Stemming* algorithm from the Snowball[5] framework, which finds the stem for each word.
4. **Capitalization:** Labels every word with two different representations, transforming them according to two patterns. The first pattern replaces each character by one of four specific symbols, depending on whether it is a capital letter, a lowercase character, a number or another character. In the second pattern, a sequence of characters of the same type is replaced by the symbol associated with the character type followed by the character +. For example, the normalization of the word *"Karachi"* results in *Cccccc* and *Cc+*. We used our own implementation of the two patterns[6].
5. **Part-of-Speech Tagging:** Produces Part-of-Speech (POS) tags and Generic Part-of-Speech (GPOS) tags for each word. A POS tag represents the grammatical class (e.g., *claimed* labeled with verb past tense) that this word has in the text, while a GPOS tag (e.g., verb) represents a high-level grammatical class. We directly map the POS tags into GPOS tag and we used the *Apache OpenNLP* library to obtain the POS tags.

---

[4] http://opennlp.apache.org/
[5] http://snowball.tartarus.org/
[6] http://web.tecnico.ulisboa.pt/joaoplmpereira/Capitalization.html

6. **Dependency Analysis:** Produces a dependency tree or graph for each sentence. We used the *Stanford CoreNLP*[7] library, which contains a dependency parser for identifying syntactical dependencies between words.

The *Apache OpenNLP* library uses a Maximum Entropy Model (MEM) [2] previously trained with data in English for each pre-processing task: a MEM for sentence segmentation[8], a MEM for tokenization[9], and a MEM for POS tagging[10]. We use sentence segmentation, tokenization, and dependency analysis to produce data structures that can be compared by kernels. Token normalization, capitalization and POS tagging are optional pre-processing steps introduced in the form of word features that improve the accuracy of the kernels.

3.3 Learning Techniques

We used an efficient online learning technique called Pegasos [12] to train the SVMs. This involves an iterative process that performs multiple passes over the training data.

In RE problems where the goal is to search for a single relation type (e.g., protein interaction extraction from AImed), it is sufficient to use a binary classifier. However, if the problem involves searching for various types of relations (e.g., the RE task in SemEval), it is necessary to use multi-class classifiers. To produce a multi-class classifier using binary classifiers, we used an *One-VS-One* heuristic. This technique makes use of as many binary classifiers as pairs of classes. We train each binary classifier with the subset of the training data that is annotated with the corresponding pair of classes. In the test phase, a data instance is evaluated by all the classifiers. Each classifier votes in the class associated to the binary class that it assigns to the data instance. Then, we choose the most voted class as the label.

3.4 Metrics

In this section, we present the metrics we use to evaluate the different kernels:

Convergence: Measures the proximity of a SVM to the optimal SVM. This proximity is given by applying the objective function for SVMs with a training dataset and the **w** of a SVM as arguments, as described in Sect. 2.1. This metric enables us to understand the impact of changing the parameters of the learning phase (i.e., λ and the number of iterations that indicate the number of passes over the data) on the generated SVMs.

Quality of Binary Classification: The measures commonly used to evaluate classification techniques are: precision, recall, and the F_1-measure. Defining r as

---

[7] http://nlp.stanford.edu/software/corenlp.shtml

[8] http://opennlp.sourceforge.net/models-1.5/en-sent.bin

[9] http://opennlp.sourceforge.net/models-1.5/en-token.bin

[10] http://opennlp.sourceforge.net/models-1.5/en-pos-maxent.bin

a relation type, *precision* is the fraction of correctly extracted relations of type r over the total number of extracted relations of type r. *Recall* is the fraction of correctly extracted relations of type r over the total number of relations of type r present in the dataset. These measures are usually contradictory: by increasing the recall value, we reduce the precision value and vice versa. To globally evaluate a classification technique, it is necessary to combine these two measures. A solution to combine these two scores is to use the F_1-*measure* that corresponds to the harmonic mean of precision and recall.

Quality of Multi-class Classification: In multi-class classification, we use new measures obtained by averaging the values of precision and recall over the multiple classes, namely macro-averages and micro-averages. *Macro-averages* calculate directly the system precision or recall averages for the various classes. To calculate the *Micro-averages*, first, we count, for all relations of type r, the values used in the formulas of precision and recall: the number of correctly extracted relations, the total number of extracted relations, and the total number of relations present in the dataset. We then compute the Micro-averages by using these summed values into the precision or the recall formulas.

Execution Times: We measure two execution times: the time to train a SVM and the time required to predict the class type for a pair of entities. Execution times are extracted in nano seconds of the CPU time. Then, we convert them into other units for easier comparison (i.e., seconds, minutes or hours).

4 Benchmark Results

In this section, we present the configurations and the results of the experiments we performed using the benchmark presented in Sect. 3. In Sect. 4.1, we describe the settings used in the experiments. In Sect. 4.2, we present the analysis of the convergence values. In Sect. 4.3, we analyze the quality of the results obtained for the various models. In Sect. 4.4, we analyze the execution times for training a SVM and for classifying a data instance.

4.1 Setup

The experimental setup that we have used is as follows:

Datasets: We used the datasets described in Sect. 3.1, namely AImed and SemEval. Both are pre-processed through the techniques and tools described in Sect. 3.2. The AImed dataset was split into 10 folds over which we applied a cross-validation process. The AImed dataset contains a single type of relation, which means that extracting relations in this dataset is a binary classification problem. The SemEval competition provided distinct training and testing splits. Thus, for SemEval, it was not necessary to use cross-validation. Extracting relations from SemEval is a multi-class problem. The data is annotated with 9 types of asymmetric relations, which gives a total of 19 distinct classes (two for each type plus another for a non-relation).

Learning Techniques: We used the learning techniques described in Sect. 3.3. In particular, we implemented the Pegasos technique extended to work directly with kernels[11]. We used the extended version of Pegasos to train the SVMs: one for the AImed dataset; and one for each of the 172 binary classifiers used by the *One-VS-One* heuristic for the SemEval dataset.

Kernels: We used the *SSK*, *SPK* and *BNK* kernels described in Sect. 2.2, since they use different data structures to represent the data instances.

Parameters: The parameter λ, which controls the importance of the regularization versus the loss function in the SVM training, took values in $\{10^{-4}, 10^{-5}, 10^{-6}, 10^{-7}, 10^{-8}\}$. The number of iterations T, which is introduced in the online learning technique to indicate how many passages should be made over the training instances, took values in $\{50, 100, 150, 200\}$.

Software and Hardware: We developed the benchmark using the framework for RE named REEL [1], which provides an implementation of SSK, SPK and BNK. We performed the experiments on a machine with an Intel Core i5 CPU M 460, 2.53 GHz and 4 GB of memory RAM.

4.2   Convergence

In this section, we evaluate the impact of the λ and T parameters over the training of SVMs with RE kernels. To do this, we analyze the variation of the objective function values with these parameters (see Fig. 1). The experiments show that there is no substantial difference in the results obtained for different kernels. Therefore, we only present the results obtained for BNK.

In general, the number of iterations required to stabilize the objective function values increases when we decrease the λ parameter values. In fact, for small values of λ, we do not reach stable values for the objective function. Regarding the number of iterations T, the objective function values start to slowly decrease after iteration number 100, and stabilize after iteration 150. For the AImed dataset (see Fig. 1(a)), the models that obtain the lowest objective function values are the models trained with a λ value of 10^{-6}. For the SemEval dataset, the models that obtain the lowest objective function values are the models trained with a λ value of 10^{-5}. This situation occurs because, for datasets in technical language domains such as the AImed dataset, models tend to be more complex (i.e., enclose more patterns), since it is difficult to find a suitable generalization (i.e., few patterns that can explain the data).

From the results reported in this section, we conclude that: (i) models trained with very low values for the λ parameter obtain worse results in terms of convergence when compared to higher values, and (ii) after 150 iterations, there was no significant variation in the objective function values.

---

[11] http://web.tecnico.ulisboa.pt/joaoplmpereira/OnlineLearning.html

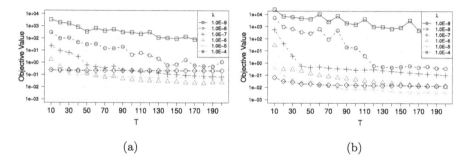

Fig. 1. Variation of the objective function values for BNK models with parameters λ and T. (a) AImed; (b) Binary classifier for the pair of entities *Content-Container(e2,e1)* vs *Message-Topic(e1,e2)* in the SemEval dataset

4.3 Quality of the Obtained Extractions

In this section, we evaluate the quality of the results obtained by the SVMs using the kernels BNK, SPK and SSK. We begin by evaluating the impact of the variation of the λ and T parameters in the quality of SVMs using the AImed dataset (see Fig. 2), and then using the SemEval dataset (see Fig. 3).

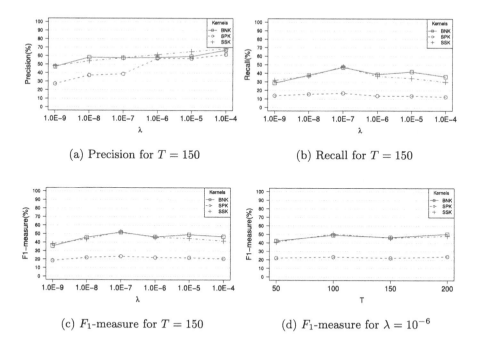

Fig. 2. Variation of the quality of the results obtained through SVMs with kernels BNK, SPK and SSK, for the AImed dataset.

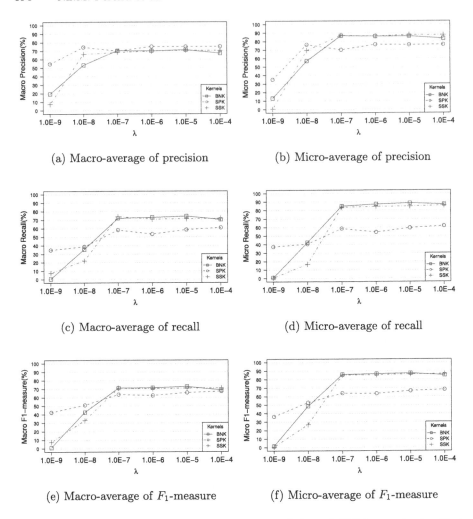

(a) Macro-average of precision   (b) Micro-average of precision

(c) Macro-average of recall   (d) Micro-average of recall

(e) Macro-average of F_1-measure   (f) Micro-average of F_1-measure

Fig. 3. Variation of the quality of the results obtained through SVMs over the SemEval test dataset for kernels BNK, SPK and SSK in $T = 150$.

For both datasets, BNK and SSK produce similar values for the F_1-measure (less than 5 % difference), as shown in Figs. 2(c) and (d), 3(e) and (f). The precision values of SSK are slightly better than the ones of BNK (from 1 % to 5 %) as observed in Figs. 2(a), 3(a) and (b), but the opposite occurs for the recall values (see Figs. 2(b), 3(c) and (d)). Generally, BNK presents more balanced precision and recall values than SSK. In order to explain the differences observed we need to consider the data structures that were used. Both kernels analyze the sentences, splitting them in a similar way. However, in SSK, the features are structured into subsequences which assign a fixed order to the words. BNK is more flexible because it uses data structures with sets of n-grams, which are composed of n-grams independently of their position in the sentence.

In general, in both datasets, SPK gets worse results than the remaining kernels (e.g., between 5 % and 20 % less). For the AImed dataset, this difference is more significant than in the SemEval dataset. Since the precision values obtained by SPK are comparable to those of the other two kernels (see Figs. 2(a), 3(a) and (b)), the differences in terms of quality derive from its low recall. SPK obtains the best macro-average of precision for SemEval. SPK obtains good results in terms of precision because it compares shortest paths and it assigns zero to paths with different size. So, this kernel becomes more inflexible to compare data samples. For AImed, SPK obtains low quality results than SSK and BNK (less than 30 %), because this dataset is composed of technical medical documents, and the dependency parser used by SPK was trained with news articles. Therefore, the dependencies found are inappropriate for this dataset and consequently lead to the low quality performance of SPK. Due to the inadequate use of the dependency parser, we consider unfair to compare SPK with the other kernels in the AImed dataset.

In this section, we conclude that: (i) BNK and SSK have similar behaviors; (ii) BNK is better in recall and returns more balanced precision and recall values; (iii) SPK is the kernel that obtains, in general, worse results, but nevertheless, stands out for the SemEval dataset in terms of precision.

4.4   Execution Time

In this section, we compare the kernels in terms of training and testing execution times. The pre-processing execution times were not considered because they heavily depend on the pre-processing techniques and tools used. For calculating the training execution times, we consider an estimate of the CPU time. We calculate this estimate by multiplying the number of calls to the kernel with the average execution time of comparing two data instances. We did not take the CPU time of the online optimization technique into consideration, because it is significantly lower than the time needed to calculate the kernels (i.e., less than 1 % of the total execution time). For the AImed dataset the execution times correspond to the classifier training time. For the SemEval dataset, the execution time is the sum of the training execution times of all classifiers used in the multi-class problem.

Figures 4 and 5 report the training execution times for AImed and SemEval, respectively. For both datasets, SPK is the fastest kernel. However, the pre-processing step of SPK is usually much more expensive than the pre-processing step of the other kernels, because it needs a dependency analysis process. SPK only takes into account the shortest path between the entities in the dependency graph therefore: (i) it compares less sentence features than the other two kernels; (ii) it ignores a large quantity of comparisons since it only compares sentences whose number of nodes in the shortest path between the two entities is equal.

For the AImed dataset, SSK is faster than BNK. However, for SemEval, the inverse situation occurs. This difference is due to the size of the sentences in each

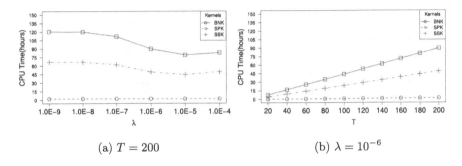

(a) $T = 200$ (b) $\lambda = 10^{-6}$

Fig. 4. Average execution times for training over the 10 folds of AImed dataset.

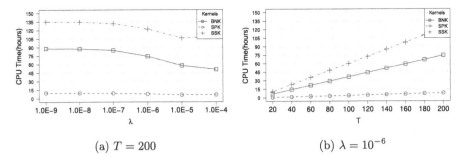

(a) $T = 200$ (b) $\lambda = 10^{-6}$

Fig. 5. Execution times for training over the SemEval dataset.

of the datasets i.e., the average size of a sentence in AImed is significantly higher than in SemEval. SSK compares more subsets of tokens than BNK, but it uses a dynamic programming algorithm that performs faster comparisons, even when the sentence size increases. BNK is very penalized when the sentence length and the diversity of words increase, especially in the AImed dataset.

Regarding the variation of the λ parameter, we conclude that there is a relationship between the λ values and the training execution times. Models trained with lower λ values have lower execution times (see Figs. 4(a) and 5(a)). In Sect. 4.2, the λ parameter was associated with the stability of the objective function and convergence of the model. In fact, we observed that the objective function values stabilize for low values of λ. When we make the λ parameter constant and vary the number of iterations, we observe that the execution times grow linearly with the number of iterations (see Figs. 4(b) and 5(b)).

The testing execution times are generally constant for both datasets. With the AImed dataset, the testing execution time is 168 ms for BNK, 8 ms for SPK, and 290 ms for SSK. For the SemEval dataset, it takes 11 ms for BNK, 0.68 ms for SPK, and 6 ms for SSK to classify each data instance. Not surprisingly, the results obtained are similar to the training phase results.

In this section, we conclude that: (i) SPK is the fastest in training and testing, (ii) BNK and SSK have execution times with the same orders of magnitude. However, (iii) SSK is faster than BNK in datasets of technical domains, such as

AImed, and (iv) BNK is faster than SSK for datasets in the news domains such as the SemEval dataset.

5   Discussion and Future Work

In this paper, we proposed a benchmark to compare kernels for RE tasks, in which we train and analyze SVMs leveraging the kernels to be compared. We conducted an extensive experimental analysis using our benchmark over three state-of-the-art kernels, which enabled us to take the following main conclusions: (i) SVMs stabilize after $T = 100$ iterations for models trained with lower values of λ; (ii) BNK and SSK have a similar performance (i.e., quality and execution time) for the AImed and SemEval datasets; (iii) SPK obtains high quality results only when the dependency parser is trained with data of the same domain as the dataset; (iv) SPK is the kernel that has lowest execution times.

As future work, we plan to develop techniques that are able to effectively combine the various kernels for RE. This method should be domain independent and should be used with different sets of kernels.

Acknowledgements. We would like to thank Gonçalo Simões for the fruitful discussions, and for advice on preliminary versions of this paper.

This work was supported by *Fundação para a Ciência e a Tecnologia*, under Project UID/CEC/50021/2013, and under Project DataStorm (ref. EXCL/EEI-ESS/0257/2012).

References

1. Barrio, P., Simões, G., Galhardas, H., Gravano, L.: REEL: a relation extraction learning framework. In: JCDL (2014)
2. Berger, A.L., Pietra, V.J.D., Pietra, S.A.D.: A maximum entropy approach to natural language processing. Comput. Linguist. **22**, 39–71 (1996)
3. Bunescu, R., Mooney, R.J.: A shortest path dependency kernel for relation extraction. In: HLT-EMNLP (2005)
4. Bunescu, R., Mooney, R.J.: Subsequence kernels for relation extraction. In: CoNLL (2006)
5. Chinchor, N.A.: Named entity task definition. In: MUC-7 (1998)
6. Doddington, G.R., et al.: The automatic content extraction (ACE) program - tasks, data, and evaluation. In: LREC (2004)
7. Giuliano, C., Lavelli, A., Romano, L.: Exploiting shallow linguistic information for relation extraction from biomedical literature. In: EACL (2006)
8. Hendrickx, I., et al.: SemEval-2010 task 8: multi-way classification of semantic relations between pairs of nominals. In: SemEval (2010)
9. Hsu, C.W., Lin, C.J.: A comparison of methods for multiclass support vector machines. IEEE Trans. Neural Netw. **13**, 415–425 (2002)
10. Marrero, M., Sanchez-Cuadrado, S., Lara, J.M., Andreadakis, G.: Evaluation of named entity extraction systems. Res. Comput. Sci. **41**, 47–58 (2009)
11. Sarawagi, S.: Information extraction. Found. Trends Databases **1**, 261–377 (2008)
12. Shalev-Shwartz, S., Singer, Y., Srebro, N.: PEGASOS: primal estimated sub-GrAdient SOlver for SVM. In: ICML (2007)

Web Content Management Systems Archivability

Vangelis Banos[(✉)] and Yannis Manolopoulos

Department of Informatics, Aristotle University, 54124 Thessaloniki, Greece
vbanos@gmail.com

Abstract. Web archiving is the process of collecting and preserving web content in an archive for current and future generations. One of the key issues in web archiving is that not all websites can be archived correctly due to various issues that arise from the use of different technologies, standards and implementation practices. Nevertheless, one of the common denominators of current websites is that they are implemented using a Web Content Management System (WCMS). We evaluate the Website Archivability (WA) of the most prevalent WCMSs. We investigate the extent to which each WCMS meets the conditions for a safe transfer of their content to a web archive for preservation purposes, and thus identify their strengths and weaknesses. More importantly, we deduce specific recommendations to improve the WA of each WCMS, aiming to advance the general practice of web data extraction and archiving.

1 Introduction

The web has moved from small informal websites to large and complex systems, which require strong software systems to be managed effectively [4]. The increasing needs of organisations and individuals in this area have led to the rise of a new type of software, i.e. Web Content Management Systems (WCMSs) [15]. WCMSs are created in various different programming languages, using many new web technologies [9]. There are millions of websites using WCMSs; for instance, Wordpress is used by 74.6 million websites[1], whereas Drupal is used by more than 1 million websites[2]. WCMSs have a large contribution in the development of the web. However, we must not overlook the fact that the web is an ephemeral communication medium. The average lifetime of a web page is below 100 days [14], making it necessary to archive the web to preserve information for current and future generations. Web archiving is the process of retrieving material from the web and preserving it in an archive for perpetual availability for access and research [16].

One of the open issues in web archiving is that not all websites are amenable to being archived with correctness and accuracy. To define and measure this website behavior we have previously introduced the metric of *Website Achivability*

---

[1] https://managewp.com/14-surprising-statistics-about-wordpress-usage.
[2] https://www.drupal.org.

© Springer International Publishing Switzerland 2015
T. Morzy et al. (Eds.): ADBIS 2015, LNCS 9282, pp. 198–212, 2015.
DOI: 10.1007/978-3-319-23135-8_14

(WA). WA is defined as the extent to which a website meets the conditions for the safe transfer of its content to a web archive for preservation purposes [1,2]. The open and continuously evolving nature of the web makes it difficult to predict the WA of a website. There are a very large number of different combinations of technologies, standards and development approaches used in web development which affect WA.

We believe that the wide adoption of WCMSs has benefits for web archiving and needs to be taken into consideration. WCMSs constitute a *common technical framework* which may facilitate or hinder web archiving for a large number of websites. If a web archive is compatible with a certain WCMS, it is highly probable that it will be able to archive all websites built with this WCMS.

In this work, we evaluate the WA of 12 prominent WCMSs to identify their strengths and weaknesses and propose improvements to improve web content extraction and archiving. We conduct an experimental evaluation using a non-trivial dataset of websites based on these WCMSs and make observations regarding their WA characteristics. We also come up with specific suggestions for each WCMS based on our experimental data.

Our aim is to improve the web archiving practice by indicating potential issues to the WCMS development community. If our findings result in advances in WCMS source code upstream, all web archiving initiatives will benefit as the websites based on these WCMSs will become more archivable. The main contributions of this work are:

- Specific observations regarding the WA Facets of 12 prominent WCMSs,
- Recommendations to improve each WCMS source code upstream to improve their WA.

This paper is organised as follows: Sect. 2 presents work related to WCMS archiving. Section 3 presents the CLEAR+ method to evaluate WA. Section 4 presents the ArchiveReady WA evaluation system, our experimental method and the results which are discussed further in Sect. 5.

2   Related Work

WCMSs have been already studied in the context of web archiving due to their wide scale usage. According to W3Techs, 38 % of the top 1 million websites is created using a WCMS. Gomez et al., the creators of the Portuguese web archive, report that there are millions of websites which are supported by a small number of publishing platforms. During the development of their web crawling process, they have come up with specific rules to harvest specific WCMSs, such as Joomla, because they did not allow crawlers to harvest all their files [11]. Faheem et al. have also presented an approach to create Application Aware Helpers (AAH) that fit into the archiving crawl process chain to perform intelligent and adaptive crawling of web applications. In their work they are focusing on specific WCMSs [8]. Pinsent et al. have dedicated a chapter in the Preservation of Web Resources Handbook 2008 regarding Content Management Systems Archiving.

They mention that some WCMSs may present problems to a web crawler. The content gathering may be incomplete or the web crawler may get stuck in a 'loop' as it constantly requests pages. This behaviour depends on the specific implementation and the WCMS used [18]. Rumianek proposes a procedure to overcome the problems faced by archivists of database-driven websites such as WCMSs [19]. Although interesting, this approach is not practical as it requires the implementation of new systems on top of existing web archiving platforms.

There has also been some work to try archiving content from blogs, which are a special kind of WCMSs. Pennock et al. have created ArchivePress, a specialised Wordpress CMS archiving tool, which creates plugins for WordPress to make it operate as an archiving tool [17]. Kelly et al. have also investigated multiple alternatives on archiving blogs [13]. The BlogForever project has created a new approach to harvest, preserve, manage and reuse blog content [12]. Blanvillain et al. have presented their BlogForever crawler which concentrated on techniques to automatically extract content such as articles and comments from blog posts using a simple and robust algorithm to generate extraction rules based on string matching using the blog's web feed in conjunction with blog hypertext [3].

There is interest in finding methods to archive WCMS-based websites but, according to our knowledge, there has been no attempt to evaluate the feasibility of archiving different WCMSs and highlight their strengths and weaknesses regarding web archiving. In our work, we try to conduct such an analysis using a substantial experimental dataset and a novel method.

3   CLEAR+: A Credible Live Evaluation of Archive Readiness Plus

The Credible Live Evaluation of Archive Readiness Plus method (CLEAR+) is an approach to produce on-the-fly measurement of WA, which is defined as the extent to which a website meets the conditions for the safe transfer of its content to a web archive for preservation purposes [1,2]. We use the latest iteration of the CLEAR+ method as of 02/2015.

In short, the basic operating principles of our method is that we communicate through standard HTTP requests and responses with the target website in a similar manner as regular web archiving systems and retrieve information that we evaluate using recognized practices in digital curation (e.g. using adopted standards, validating formats, and assigning metadata) to generate a credible score representing the target WA. We measure WA from several different perspectives, which we call *WA Facets*, by conducting specific *Evaluations* on *Website Attributes* (Fig. 1). For instance: "What is the percentage of valid versus invalid hyperlinks? "Do the CSS files used in a website comply with W3C standards? "What is the percentage of corrupt image files, if any?

The score for each WA Facet is computed as the weighted average of the scores of the evaluations associated with this Facet. The significance of each evaluation defines its weight. The WA Facets can be summarised as follows:

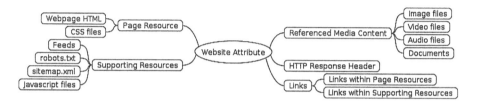

Fig. 1. Website attributes used for WA evaluation [1]

- F_A: *Accessibility* indicates the facilitation of web archiving systems to access and retrieve website content via standard web communication methods.
- F_C: *Cohesion* is the robustness of a website against the failure of different web services. This Facet is concerning websites which are dispersed across different services (e.g. different servers for images, javascript widgets, and other resources) in different domains.
- F_M: *Metadata Usage* indicates the adequate provision of metadata [6]).
- F_S: *Standards Compliance* indicates the encoding of digital resources using known and transparent standards.

The outcomes of the WA evaluation are different scores in the range of [0–100] for F_A, F_C, F_M and F_S. The final WA score is the average of the WA Facets scores.

4   Evaluation

We present the ArchiveReady system to evaluate WA and the method we follow to define and evaluate a significant corpus of websites. Finally, we present detailed results and we identify specific characteristics of different WCMSs.

4.1   ArchiveReady WA Evaluation System

ArchiveReady [1,2] is a real-time WA evaluation system, which is the reference implementation of the CLEAR+ method. It is available at http://archiveready. com. ArchiveReady is based on standard open source software and uses specialised tools to evaluate websites such as JHOVE for media file validation [7], W3C HTML[3], CSS[4] and RSS[5] validation services, as well as the PhantomJS headless WebKit browser to access and process websites[6]. The WA evaluation process can be summarised as follows:

---

[3] http://validator.w3.org/.
[4] http://jigsaw.w3.org/css-validator/.
[5] http://validator.w3.org/feed/.
[6] http://phantomjs.org/.

1. ArchiveReady receives a target URL and performs an HTTP request to retrieve the webpage hypertext.
2. After analysing it, multiple HTTP connections are initiated in parallel to retrieve all web resources referenced in the target webpage, imitating a web spider.
3. In stage 3, Website Attributes (Fig. 1) are evaluated. In more detail: (a) HTML and CSS analysis and validation, (b) HTTP response headers analysis and validation, (c) Media files (images, other objects) retrieval, analysis, and validation. (d) Sitemap.xml and Robots.txt retrieval, analysis and validation, (e) RSS feeds detection, retrieval, analysis and validation, (f) Network transfer performance evaluation.
4. The metrics for the WA Facets are calculated according to the CLEAR+ method and the final WA rating is produced.

ArchiveReady provides a simple REST API to enable WA evaluation from third party applications.

4.2 Website Corpus Evaluation Method

We use 5.821 random WCMS samples from the Alexa top 1 million websites[7] as our experimental dataset. We use this dataset because it contains high quality websites from multiple domains and disciplines. This dataset is also used in other related research [11,20]. We select our corpus with the following process:

1. We implement a simple python script to visit each homepage and look for the <meta name="generator" content="software name" /> tag.
2. For each website having the required meta tag, we evaluate if it belongs to one of the WCMSs listed in wikipedia[8]. If yes, we record it in our database.
3. We continue this process until we have a significant number of instances for 12 WCMSs (Blogger, DataLife Engine, DotNetNuke, Drupal, Joomla, Mediawiki, MovableType, Plone, PrestaShop, Typo3, vBulletin, Wordpress).
4. We evaluate each website using the ArchiveReady REST API and record the outcomes in our database.
5. We analyse the results using SQL to calculate various metrics.

The generator meta tag is not used universally on the web due to a variety of reasons, such as security. Thus, we have skipped a large number of websites, which did not indicate the system they use. Also, we do not take into consideration the version number of each WCMS as it would be impractical. There would be too many different variables in our experiment to conduct useful research. Also, it is highly improbable that the top internet websites would use legacy versions of their WCMS. The Git repository for this paper[9] contains all the captured data and the necessary scripts to reproduce all the evaluation experiments.

---

[7] http://s3.amazonaws.com/alexa-static/top-1m.csv.zip.

[8] http://en.wikipedia.org/wiki/List_of_content_management_systems.

[9] https://github.com/vbanos/wcms-archivability-paper-data.

4.3   Evaluation Results and Observations

For each WCMS, we present the average and standard deviation for each WA Facet, as well as their cumulative WA (Fig. 2). First of all, our results are consistent. While the WA Facet range is 0–100 %, the standard deviation of all WA Facet values for each WCMS ranges from 4.2 % (Blogger, F_A) to 13.2 % (Mediawiki, F_S). There are considerable differences between different WCMSs regarding their WA. The top WCMS is DataLife Engine with a WA score of 83.52 %, with Plone and Drupal scoring also very high (83.06 % and 82.08 %). The rest of the WCMSs score between 80.3 % and 77.2 %, whereas the lowest score belongs to Blogger (65.91 %). In many cases, even though two or more WCMSs may have similar WA score, their WA Facet scores are significantly different and each WCMS has different strengths and weaknesses. Thus, it is beneficial to look into each WA Facet differences.

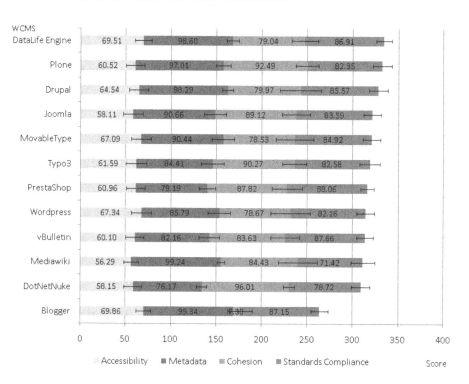

Fig. 2. WA Facets average values and standard deviation for each WCMS

F_A: Top value is around 69.85 % for Blogger and 69.51 % for DataLife Engine, whereas the minimum value is below 60, at 56.29 % for Mediawiki and 58.15 % for DotNetNuke.

F_M: Top value is 99.24 % for Mediawiki, whereas the minimum value is 76.17 % for DotNetNuke. The difference between the minimum and the maximum value is around 23 points, which almost twice the difference between F_A range (13).

F_C: Appears to have the greatest differentiation among the WCMSs. The minimum value is only 7.38 % for Blogger, whereas the maximum value is 96.01 % for DotNetNuke. At first sight, there seems to be an issue in the way Blogger is using multiple online services to host its web resources. Other WCMSs also vary from 78.5 % (MovableType) to 92 % (Plone), which is a considerable variation.

F_S: Range is between 71.42 % for Mediawiki and 88.06 % for PrestaShop. Again these differences should be considered significant.

F_A has the smallest differentiation and F_C has the greatest one among all WA Facets.

We continue our research with more detailed observations regarding specific evaluations. Due to the large number of WA evaluations and the space restrictions, we choose to discuss only highly significant rules. Similar research is easy to be conducted by anyone interested using the full dataset and source code available on github. We present our observations grouped by the four different WA Facets.

Table 1. A_1 The percentage of valid URLs. Higher is better.

WCMS	Valid URLs	Invalid URLs	Correct (%)
Blogger	45425	1148	97 %
Mediawiki	39178	1763	96 %
Drupal	52501	2185	96 %
MovableType	22442	1009	96 %
vBulletin	104492	5841	95 %
PrestaShop	57238	3287	94 %
DataLife Engine	31981	2342	93 %
Plone	25719	1856	93 %
Wordpress	47717	3515	93 %
DotNetNuke	38144	2791	93 %
Typo3	30945	3747	89 %
Joomla	37956	4886	88 %

$\boldsymbol{F_A}$: **Accessibility** refers to the web archiving systems' ability to traverse all website content via standard HTTP protocol requests [10].

A_1: The percentage of valid versus invalid hyperlink and CSS URLs (Table 1). These are critical for web archives to retrieve all WCMS published content. Hyperlinks are created not only by users but also by WCMS subsystems. In any case, some WCMSs check if they are valid, whereas others don't. In addition, some WCMSs may be incurred with invalid hyperlinks due to bugs. The results show that not all WCMSs have the same frequency of invalid hyperlinks. Joomla and Typo3 have a high percentage (88 % and 89 %), whereas Blogger, Mediawiki, Drupal and MovableType have the highest percentage of invalid hyperlinks (97 % and 96 %).

A_2: The number of inline JavaScript scripts per WCMS instance (Table 2). The excessive use of inline scripts in modern web development results in web archiving problems. Plone, MovableType and Typo3 have the lowest number of inline scripts per instance (4.82, 6.82 and 6.89). The highest usage by far comes from Blogger (27.11), while Drupal (15.09) and vBulletin (12.38) follow.

Table 2. A_2 The number of inline scripts per WCMS instance. Lower is better.

WCMS	Instances	Inline scripts	scripts/instance
Plone	431	2076	4.82
MovableType	295	2011	6.82
Typo3	624	4298	6.89
Mediawiki	408	3753	9.20
DataLife Engine	321	3159	9.84
Wordpress	863	8646	10.02
DotNetNuke	598	6028	10.08
Joomla	501	5163	10.31
PrestaShop	466	5130	11.01
vBulletin	462	5721	12.38
Drupal	528	7969	15.09
Bogger	324	8783	27.11

The sitemap.xml[10] protocol is meant to create files that include references to all the webpages of the website. Sitemap.xml files are generated automatically by WCMSs when their content is updated. The results of the A_3 evaluation (Table 3) indicate that most WCMS lack proper support for this feature. Only DataLife Engine has a very high score (86 %). Also, Wordpress and Drupal score over 60 %. All other WCMSs perform very poorly, which is surprising.

F_C: **Cohesion** is relevant to the level of dispersion of files comprising a single website to multiple servers in different domains. The lower the dispersion of a website's files, the lower the susceptibility to errors because of a failed third-party system. We evaluate the performance for two F_C related evaluations.

C_1: The percentage of local versus remote images is presented in Table 4. Blogger is suffering from the highest dispersion of images. On the contrary, Plone, DotNetNuke, PrestaShop, Typo3 and Joomla have the higher F_C, over 90 %.

---

[10] http://www.sitemap.org/.

Table 3. A_3 Sitemap.xml is present. Higher is better.

WCMS	Instances	Issues	Correct
DataLife Engine	321	46	86 %
Wordpress	863	272	68 %
Drupal	528	189	64 %
PrestaShop	466	237	49 %
MovableType	295	152	48 %
Typo3	624	322	48 %
Plone	431	249	42 %
vBulletin	462	329	29 %
Joomla	501	359	28 %
Blogger	324	240	26 %
DotNetNuke	598	461	23 %
Mediawiki	408	335	18 %

Table 4. C_1 The percentage of local versus remote image. Higher is better.

WCMS	Local images	Remote images	Percent
Plone	7833	290	96 %
DotNetNuke	13136	680	95 %
PrestaShop	19910	1187	94 %
Typo3	15434	897	94 %
Joomla	14684	1251	92 %
MovableType	8147	1388	86 %
Drupal	16636	3169	84 %
vBulletin	11319	2314	83 %
Wordpress	20350	4236	83 %
Mediawiki	4935	1127	81 %
DataLife Engine	9638	2356	80 %
Blogger	1498	8121	16 %

C_2: The percentage of local versus remote CSS (Table 5). Again, Blogger has a very low score (2 %), whereas all other WCMSs perform very well.

F_S: **Standards Compliance** is a necessary precondition in digital curation practices [5]. We evaluate S_1: Validate if the HTML source code complies with the W3C standards using the W3C HTML validator and present the results in Table 6.

Table 5. C_1 The percentage of local versus remote CSS. Higher is better.

WCMS	Local CSS	Remote CSS	Percent
DotNetNuke	5243	101	98%
Typo3	3365	154	96%
Plone	1475	72	95%
Joomla	4539	222	95%
DataLife Engine	919	56	94%
PrestaShop	5221	400	93%
MovableType	578	42	93%
vBulletin	1459	104	93%
Mediawiki	1120	84	93%
Drupal	2320	354	87%
Wordpress	5658	1019	85%
Blogger	18	954	2%

Table 6. S_1 HTML errors per instance. Lower is better.

WCMS	Instances	Errors	Errors/Instance
Plone	431	12205	28.32
Mediawiki	408	14032	34.39
Typo3	624	23965	38.41
Wordpress	863	35805	41.49
Joomla	501	26609	53.11
PrestaShop	466	30066	64.52
DotNetNuke	598	43009	71.92
Drupal	528	47131	89.26
vBulletin	462	46466	100.58
MovableType	295	29994	101.67
DataLife Engine	321	34768	108.31
Blogger	324	71283	220.01

Plone has the lower number of errors (28.32), followed by Mediawiki (34.39) and Typo3 (34.41). On the contrary, Blogger has the most errors per instance (220.01), followed by far by DataLife Engine (108.31) and MovableType (101.67).

S_3: The usage of Quicktime and Flash formats is considered problematic for web archiving because web crawlers cannot process their contents to extract information, including web resource references. Results show that their use is very low in all WCMS (Table 7).

Table 7. S_2 The lack of use of proprietary files (Flash, QuickTime). Higher is better.

WCMS	Instances	No proprietary files	Success
PrestaShop	466	460	99 %
Mediawiki	408	398	98 %
Blogger	324	310	96 %
Plone	431	412	96 %
Wordpress	863	821	95 %
Typo3	624	592	95 %
vBulletin	462	434	94 %
Drupal	528	494	94 %
DotNetNuke	598	548	92 %
DataLife Engine	321	294	92 %
MovableType	295	263	89 %
Joomla	501	439	88 %

Table 8. A_5: valid feeds. Higher is better.

WCMS	valid feeds	invalid feeds	Correct
Blogger	872	83	91 %
DataLife Engine	240	57	81 %
Wordpress	1283	317	80 %
Joomla	556	141	80 %
vBulletin	299	96	76 %
MovableType	271	120	69 %
Drupal	133	74	64 %
PrestaShop	82	112	42 %
Typo3	124	191	39 %
Plone	116	184	39 %
DotNetNuke	2	14	13 %
Mediawiki	10	521	2 %

S_4: Check if the RSS feed format complies with W3C standards. The results (Table 8) indicate that Blogger has mostly correct feeds (91 %), whereas every other WCMS has various levels of correctness. The lowest scores belong to Mediawiki (2 %) and DotNetNuke (13 %). In general, the results show that there is a problem with RSS feed standard compliance.

F_M: **Metadata Usage**: The lack of metadata impairs the archive's ability to manage content effectively. Web sites include a lot of metadata, which need to be communicated in a correct manner to be utilised by web archives [6].

Table 9. M_1: HTTP content-type header. Higher is better.

WCMS	Instances	Exists	Success
Blogger	324	324	100 %
Drupal	528	527	100 %
MovableType	295	294	100 %
vBulletin	462	458	99 %
Plone	431	427	99 %
Typo3	624	618	99 %
Joomla	501	494	99 %
DotNetNuke	598	589	98 %
Mediawiki	408	401	98 %
DataLife Engine	321	315	98 %
PrestaShop	466	456	98 %
Wordpress	863	841	97 %

Table 10. M_2: HTTP caching headers. Higher is better.

WCMS	Instances	Issues	Percentage
Blogger	324	3	99 %
Mediawiki	408	12	97 %
Drupal	528	23	96 %
DataLife Engine	321	16	95 %
Plone	431	49	89 %
MovableType	295	106	64 %
Joomla	501	186	63 %
Wordpress	863	466	46 %
Typo3	624	364	42 %
vBulletin	462	326	29 %
PrestaShop	466	388	17 %
DotNetNuke	598	569	5 %

M_1: Check if the HTTP Content-type header exists (Table 9). There is virtually no issue with HTTP Content-Type in all WCMSs. Their performance is excellent.

M_2: Check if any HTTP Caching headers (Expires, Last-modified or ETag) are set. HTTP Caching is highly relevant to accessibility and performance. Blogger, Mediawiki, Drupal, DataLife Engine and Plone have very good support of HTTP Caching headers (Table 10).

5 Discussion and Conclusions

We evaluated the WA and presented specific statistics regarding 12 prominent WCMSs. We concluded that not all WCMSs are considered equally archivable. Each one has its own strengths and weaknesses, which we highlight in the following:

1. *Blogger* has by far the worst overall WA score (65.91 %, Fig. 2), mainly due to the very low F_C. Blogger files are dispersed in multiple different web services, which is increasing the possibility of errors in case one of them fails. In addition, Blogger scores very low in many metrics such as the number of inline scripts per instance (Table 2) and HTML errors per instance (Table 6). On the contrary, Blogger scores very high regarding F_M and F_S.
2. *DataLife Engine* has the highest WA score (83.52 %). One aspect that they should look into is HTML errors per instance (Table 6), where it has the second worst score.
3. *DotNetNuke* has the second worst WA score in our evaluation (77.2 %). F_C is their strong point (96.01 %) but they have issues in every other area. We suggest that they look into their RSS feeds (13 % Correct) (Table 8), and lacking HTTP caching support (5 %) (Table 10).
4. *Drupal* has the third highest WA score (82.08 %). It has good overall performance; the only issue is the existence of too many inline scripts per instance (15.09) (Table 2).
5. *Joomla* WA score is average (80.37 %). It has a large number of invalid URLs per instance (12 %) (Table 1) and it has also the highest usage of proprietary files (12 %) (Table 7) which is not good for accessibility and preservation.
6. *Mediawiki* WA score is low (77.81 %). This can be attributed to mostly invalid feeds (only 2 % are correct according to standards) and very low sitemap.xml support (18 %), Table 3.
7. *MovableType* WA score is average (80.02 %). It does not stand out in any evaluation either in a positive or a negative way. General improvement in all areas would be welcome.
8. *Plone* has the second highest WA score (83.06 %). It must be commented for having the lowest number of HTML errors per instance (28.32) (Table 6) and very high F_C scores (96 % for images, Table 4 and 95 % for CSS, Table 5).
9. *PrestaShop* WA score is average (79 %). It has average scores in all evaluations; however, it should be commented for not using any proprietary files (top score: 99 % at Table 7).
10. *Typo3* WA score is average (79 %). It has the largest number of invalid URLs per instance (12 %) (Table 1).
11. *vBulletin* WA score is consistenly low (78.37 %). General improvement in all areas would be welcome.
12. *Wordpress* WA score is average (78.47 %). We cannot highlight a specific area, where it should be improved. As this is currently the most popular WCMS, Wordpress developers should look into all WA Facets and try to improve.

We recommend that the WCMS development communities investigate the presented issues and resolve them as many are easy to be fixed without causing any issues with existing users and installations. If the situation regarding the highlighted issues is improved in the next releases of the investigated WCMS, the impact would be significant. A large number of websites which could not be archived correctly would no longer have these issues once they update their software and newly created websites based on these WCMS would be more archivable. Web archiving operations around the world would see great improvement, resulting in general advancements in the state of web archiving.

References

1. Banos, V., Kim, Y., Ross, S., Manolopoulos, Y.: CLEAR: a credible method to evaluate website archivability. In: Proceedings 10th International Conference on Preservation of Digital Objects (iPRES) (2013)
2. Banos, V., Manolopoulos, Y.: A quantitative approach to evaluate website archivability using the clear+ method. Int. J. Digital Libr. (2015)
3. Blanvillain, O., Kasioumis, N., Banos, V.: Blogforever crawler: techniques and algorithms to harvest modern weblogs. In: Proceedings 4th International Conference on Web Intelligence, Mining & Semantics (WIMS) (2014)
4. Boiko, B.: Understanding content management. Bull. Am. Soc. Inf. Sci. Technol. **28**(1), 8–13 (2001)
5. Coalition, D.P.: Institutional strategies - standards and best practice guidelines (2012). http://www.dpconline.org/advice/preservationhandbook/institutional-strategies/standards-and-best-practice-guidelines. Accessed 10 November 2014
6. Day, M.: Metadata, curation reference manual (2005). http://www.dcc.ac.uk/resources/curation-reference-manual/completed-chapters/metadata. Accessed 10 November 2014
7. Donnelly, M.: JSTOR/Harvard Object Validation Environment (JHOVE). Digital Curation Centre Case Studies and Interviews (2006)
8. Faheem, M., Senellart, P.: Intelligent and adaptive crawling of web applications for web archiving. In: Daniel, F., Dolog, P., Li, Q. (eds.) ICWE 2013. LNCS, vol. 7977, pp. 306–322. Springer, Heidelberg (2013)
9. Fernández-Garcia, N., Sánchez-Fernandez, L., Villamor-Lugo, J.: Next generation web technologies in content management. In: Proceedings (companion) 13th International Conference on World Wide Web (WWW), pp. 260–261 (2004)
10. Fielding, R., Gettys, J., Mogul, J., Frystyk, H., Masinter, L., Leach, P., Berners-Lee, T.: Hypertext transfer protocol-http/1.1 (1999). http://tools.ietf.org/html/rfc2616. Accessed 10 November 2014
11. Gomes, D., Costa, M., Cruz, D., Miranda, J., Fontes, S.: Creating a billion-scale searchable web archive. In: Proceedings (companion) 22nd International Conference on World Wide Web (WWW), pp. 1059–1066 (2013)
12. Kasioumis, N., Banos, V., Kalb, H.: Towards building a blog preservation platform. World Wide Web **17**(4), 799–825 (2014)
13. Kelly, B., Guy, M.: Approaches to archiving professional blogs hosted in the cloud. In: Proceedings 7th International Conference on Preservation of Digital Objects (iPRES) (2010)

14. Lawrence, S., Pennock, D.M., Flake, G.W., Krovetz, R., Coetzee, F.M., Glover, E., Nielsen, F.Å., Kruger, A., Giles, C.L.: Persistence of web references in scientific research. IEEE Comput. **34**(2), 26–31 (2001)
15. McKeever, S.: Understanding web content management systems: evolution, lifecycle and market. Ind. Manage. Data Syst. **103**(9), 686–692 (2003)
16. Niu, J.: An overview of web archiving. D-Lib Magazine, 18(3/4) (2012)
17. Pennock, M., Davis, R.: Archivepress: a really simple solution to archiving blog content. In: Proceedings 6th International Conference on Preservation of Digital Objects (iPRES) (2009)
18. Pinsent, E., Davis, R., Ashley, K., Kelly, B., Guy, M., Hatcher, J.: PoWR: the preservation of web resources handbook (2010)
19. Rumianek, M.: Archiving and recovering database-driven websites. D-Lib Magazine 19(1/2) (2013)
20. W3Techs. Usage of content management systems for websites (2014). http://w3techs.com/technologies/overview/content_management/all. Accessed 10 November 2014

Modeling and Ontologies

Evidence-Based Languages for Conceptual Data Modelling Profiles

Pablo Rubén Fillottrani[1,2] and C. Maria Keet[3]([⊠])

[1] Departamento de Ciencias e Ingeniería de la Computación,
Universidad Nacional del Sur, Bahía Blanca, Argentina
prf@cs.uns.edu.ar

[2] Comisión de Investigaciones Científicas, La Plata, Provincia de Buenos
Aires, Argentina

[3] Department of Computer Science, University of Cape Town,
Cape Town, South Africa
mkeet@cs.uct.ac.za

Abstract. To improve database system quality as well as runtime use of conceptual models, many logic-based reconstructions of conceptual data modelling languages have been proposed in a myriad of logics. They each cover their features to a greater or lesser extent and are typically motivated from a logic viewpoint. This raises questions such as what would be an evidence-based common core and what is the optimal language profile for a conceptual modelling language family. Based on a common metamodel of UML Class Diagrams (v2.4.1), ER/EER, and ORM/2's static elements, a set of 101 conceptual models, and availing of computational complexity insights from Description Logics, we specify these profiles. There is no known DL language that matches exactly the features of those profiles and the common core is small (in the tractable $\mathcal{ALNI}$). Although hardly any inconsistencies can be derived with the profiles, it is promising for scalable runtime use of conceptual data models.

1 Introduction

Database and information system development and use can be aided by conceptual data models that have a logic-based underpinning, both at the analysis stage and during runtime. Automated reasoning over isolated conceptual data models, such as EER and UML Class Diagrams, to improve their quality and avoid bugs aims to tackle this problem by various means. Notably, Description Logics (DLs) is used (among many: [1,5,10]), but also other techniques, such as constraint programming [8], OCL [27], CLIF [26], and Alloy [7]. There are also scenarios for using the models at runtime, such as for scalable test data generation [28] and for designing [6] and executing [13] queries with the conceptual model's vocabulary rather than quirky database table names and columns. Logic-based approximations of conceptual models are used also for querying databases during the stage of query compilation [32].

All these efforts face the same issue: how to formalise the diagrams in which logic? Even just zooming in on DLs shows that at times it is claimed that any

© Springer International Publishing Switzerland 2015
T. Morzy et al. (Eds.): ADBIS 2015, LNCS 9282, pp. 215–229, 2015.
DOI: 10.1007/978-3-319-23135-8_15

one of the languages in the $\mathcal{DLR}$ family is good for representing and unifying the conceptual data modelling languages [12], the much leaner DL-Lite family of languages [1], or using $\mathcal{SROIQ}$ (OWL 2 DL) instead [33]. While one could choose one's pet language, from a scientific viewpoint, it would be good to know *which DL (or other logic) is most appropriate, and why?* Here, 'most appropriate' is cast in the light of the needs from the viewpoint of the modelling languages, and what features those conceptual modellers bother to use in their models. This raises the following questions:

1. What is the profile of a common core of language features among the main conceptual data modelling languages (CDMLs)?
2. Is there an optimal language profile to capture each of UML, ER and EER, and ORM and ORM2, based on a set of publicly available diagrams?
3. Are any language features missing from the many extant DL languages, given a set of actual conceptual models, or too much in any case?

To establish a common core, harmonisation of terminology across CDMLs is needed. This has been done with a unifying metamodel of the static, structural entities (including constraints) of UML v2.1.4 (Class Diagrams), ER, EER, ORM, and ORM2 [16,20]. The feature overlap that can be determined from the metamodel is augmented by a classification of the entities in the models of a dataset of 101 conceptual data models of the three language families. These models were collected from projects, scientific papers, textbooks, and online diagrams; the dataset and analysis are available online [15,21].Together with the known computational complexity of various DL languages and formalisation trade-offs, a common core and profiles for each of the UML, ER/EER, and ORM/ORM2 families have been specified. This ranges from $\mathcal{ALNI}$ of the core to the "$\mathcal{DLR}_{ifd}$ without disjointness and completeness" for ORM2, with a good approximation with $\mathcal{CFDI}_{nc}^{\forall-}$. Remarkably, these conceptual model profiles/DL fragments are all tractable, and therewith are very suitable for scalable runtime usage of conceptual models. The only possible complication are the (sparsely occurring) advanced datatype constraints, which DLs do not support well, and promising computational complexity results are yet to be obtained. The remainder of the paper is structured as follows. We first introduce preliminaries about the metamodel, the dataset, and fundamental formalisation choices (Sect. 2). The core and CDML profiles are described and motivated in Sects. 3 and 4, respectively. Related works are analysed in each profile section. We discuss in Sect. 5 and conclude in Sect. 6.

2   Preliminaries

To put the profiles in context, we first describe the input we used, being the unifying metamodel, the dataset, motivation for the logic chosen, and some insights from philosophy that clarifies CDML formalisations. We assume the reader is familiar with the basic DL notation; see [4] for details.

2.1 Unifying Metamodel and Dataset

As the three CDML families under consideration—UML v2.1.4 Class Diagrams, ER and EER (henceforth abbreviated as (E)ER), and ORM and ORM2 (henceforth abbreviated as ORM/2)—originate from different sub-fields in database and information systems development, they each have their own vocabulary with syntactic and semantic differences. This has been investigated and a terminology comparison table and a unified metamodel are presented in [20], which therewith facilitates cross-language comparisons as well as categorisation of entities of models in those languages into the harmonised terminology. Further, it neatly demonstrates the intersection of entities across the languages, which has been extended in [16] also with constraints. Its top-type is Entity, which has four direct subclasses: Relationship with 11 subclasses, Role, Entity type with 9 subclasses, and Constraint with 49 subclasses. All entities also have constraints specified among them on how they may be used, e.g., each relationship must have at least two roles and a disjoint object type constraint is only declared on class subsumptions.

We have used this metamodel to classify the entities of the models in a set of 101 UML, (E)ER, and ORM/2 models. Their average 'model size' (vocabulary+subsumption) is about 50 entities/model, with at total of 8036 entities of which 5191 (i.e., 64 %) are entities that were classified in an entity (language feature) that appears in all three language families and 1108 (13.8 %) in ones that are unique to a language family (e.g., UML's aggregation) [21]. While one would prefer industry models, they are not publicly available. Only one paper presents quantitative results on industry models, being a set of 168 ORM diagrams that were made by a single engineer in the proprietary modelling tool from LogicBlox [28]. Our model data for ORM is similar to theirs [21].

2.2 General Logic-Based Reconstruction Design Choices

There are two important considerations: which logic family to use, and what to do with the relationships.

Concerning the language(s) to create a logic-based reconstruction of the three main CDML families under consideration, and to compare them, one could go for some 'arbitrary' very expressive logic, such as FOL, or one of its serialisations (e.g., Common Logic's CLIF), or a priori a decidable one (DLs) with CDM features ($\mathcal{DLR}$ family of DLs) or in line with the Semantic Web (OWL species). There is no best fit with respect to various requirements, as the comparison in Table 1 demonstrates, other than that DLs give us a view on decidability and computational complexity of concept satisfiablity, which is therefore chosen.

A formalisation decision that applies to each CDML family concerns the relationships, which is due to two distinct ontological commitments as to what they are, being the so-called *standard view* and *positionalism*. The standard view uses directionality—or: a natural language 'reading' direction—of the relationship where the participating objects have a fixed order, as formalised with the n-ary predicate ($n \geq 2$), conflating the verbalisation with the name of the relationship. In the positionalist commitment, relationships have (are composed of)

Table 1. Selection of languages, requirements, and their evaluation for formalising UML, (E)ER, and ORM/2; "–": negative evaluation; "+": positive. (OntoIOP is in the process of standardisation with OMG, which aims to link logical theories represented in the same or different languages.)

$\mathcal{DLR}_{ifd}$	OWL 2 DL	FOL
– no implementation	+ several reasoners, relatively scalable	– few reasoners, not really scalable
– no interoperability	+ linking with ontologies doable	– no interoperability with existing infrastructures
– no integration	+ 'integration' with OntoIOP	+ 'integration' with OntoIOP
+ formalisation exist	– formalisation to complete	± formalisation exist
+ little feature mismatch	– what to do with OWL 2 DL features not in the CDM languages and vv.	+ little feature mismatch
– modularity infrastructure	+ modularity infrastructure	– modularity infrastructure
± EXPTIME-complete	± N2EXPTIME-complete	– undecidable

argument places that are entities of themselves, which are filled by the participating objects, and those positions have no order in the relationship; refer to [18, 22] for theoretical details. The three selected CDML families are positionalist [20]—UML associations have association ends, ORM/2 fact types have roles, (E)ER has components of a relationship—, but most DLs are standard view, except for the $\mathcal{DLR}$ family. The $\mathcal{DLR}$ family has only one proof-of-concept implementation [9], however, whereas the former do in so far as they are OWL 2 DL or proper fragments thereof. Therefore, we need to assess how to convert positionalist relationships into standard view ones. There are several options, each with its trade-offs that may affect the complexity of the language. We use the diagrams in Fig. 1 as illustration to discuss them.

The UML standard v2.4.1 [25] and earlier versions require named association ends (DL role components), like the teacher and taughtBy in Fig. 1, but not a name of the association (DL role). Options to formalise it:

(1) make each association end a DL role, teacher and taughtBy, then choose:
    (a) declare them inverse of each other with teacher ≡ taughtBy⁻,
    (b) do not declare them inverses.
(2) choose to 'bump up' either teacher or taughtBy from association end to DL role, and use the other through a direct inverse (ObjectInverseOf() in OWL 2) and omit the extension of the vocabulary with the other (e.g., teacher and teacher⁻ cf. adding also taughtBy).

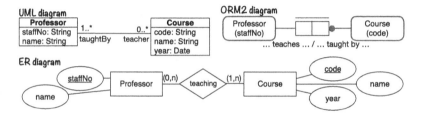

Fig. 1. Sample UML, EER and ORM2 diagrams, representing that a course is taught by at least one professor, and a professor may teach zero or more courses (for space limitations, some value types in the ORM diagram are suppressed).

The explicit inverses (Item 1a) is essentially a workaround for having made two relationships where only one existed, trying to keep the two somehow related so as to make up for the 'splitting'. Arguably, declaring them inverses is not strictly necessary, and omitting it could be considered comparable to omitting the identification constraint across the roles of a reification of an n-ary into n binaries in OWL, which is generally tolerated. Either way, one can deterministically and automatically generate a formalisation of the UML Class Diagram.

Item 2's need for a choice among association ends can be done economical in the formalisation by taking the one that requires a cardinality constraint; in the example, the preference is for taughtBy, not teacher (the latter has only a 0..*), generating a domain and a range axiom for taughtBy, and a Course $\sqsubseteq$ $\exists$taughtBy.Professor. This can be automated for cases like the example, but not if Professor were to have also a 1..* (or more) multiplicity, which then would make it an arbitrary choice again, and therewith, still not a single, unique formalisation.

In favour of the latter main option, is that it has been shown that using Item 2-inverses compared to Item 1a-inverses results in better automated reasoner performance, reducing time by more than a third [19]. Adding inverses to a language may change its computational complexity, however, and a few popular ones do not have inverses; e.g., $\mathcal{ALCQ}$ and $\mathcal{ALCQI}$ are both PSpace-complete [29], and OWL 2 EL [24] does not have inverse object properties.

ER is also positionalist, but it has a different practical issue cf. UML. It is customary to give the relationship a name that is 'non-directional', like the teaching in Fig. 1 or its infinitive, rather than naming the relationship components. Morphing it into the standard view then requires either:

(i) a renaming of the relationship to prevent an ambiguous DL role name in the formalisation, or
(ii) an arbitrary domain and range assignment.

This user-mediated step favours using an ObjectInverseOf() rather than adding a second new name if more than one cardinality constraint is not (0, n), but this means also here it cannot be guaranteed it will result in exactly one formalisation of the diagram. (Some UML models have association names, not named association ends, but the same problem does not exist, for an association name has a filled arrow-tip for the reading direction.)

ORM's fact type readings can be useful candidates for naming DL roles, but only one is required in a diagram, not n for the n participating entities. The software assigns auto-generated identifiers to the ORM roles and to the fact types (relationships) by default, but a modeller also can name them, which is then shown in the diagram. Due to this freedom in modelling, one single rule is not possible, but a sequence of possible cases—and choices—is needed. Thus, it cannot be guaranteed that there will be only one formalisation.

In sum, no matter which formalisation option is chosen regarding relationships, the CDML families each require their own transformation algorithm, and due to the options, it is possible to construct different profiles based on the formalisation choices. We will return to this in Sects. 4 and 5.

3   Core Profile

The Core Profile is composed by the elements of the metamodel that belong to the three main families of languages: UML Class Diagrams, (E)ER and ORM/2, and that are extensively used in the analysed models. Interoperability of model semantics between models expressed in these different modelling languages can be assured by restricting models to this set of entities. An important criterion here was to find a 'simple' a language as possible whilst covering the main common entities used in conceptual data models.

- Object Type C. This is represented by concept C in DL.
- Binary Relationship R between object types C and D. This is represented by a DL role R together with the inclusion assertion $\top \sqsubseteq \forall R.C \sqcap \forall R^-.D$ to type the relationship. This formalisation reflects the standard view of relationships. We restrict it to binary relationships only, because general n-ary relationships are rarely used in the whole set of analyzed models. (The (E)ER and ORM/2 models exhibit a somewhat higher incidence of n-aries, so they are included in the respective profiles; see below.)
- Attribute a of datatype T within an object type C, including the transformation of ORM's Value type following the transformation rule given in [17]. This is represented in a DL by a role a between concepts C and T, together with the inclusion axiom $C \sqsubseteq \forall a.T$. Formalisation of CDM datatypes in DL as concrete domains or datatypes [3,23] generally translate a datatype into a DL concept, and a datatype value as a DL nominal or instance, which lead to high undecidability results. Although datatypes and concepts share some properties (both can participate in inheritance and conjunction, both can be attached with cardinality constraints), there are also important differences between them: a datatype cannot participate in relationships, cannot be defined by quantifiers or negation over other datatypes, while concepts cannot be composed (which is not the same as union) and cannot be filtered with facets. Identity of a nominal is inherently different as identity of a datatype value, and this is reflected in counting quantifiers.

- Subsumption between two object types C and D. This is represented in DL by the inclusion axiom $C \sqsubseteq D$.
- Object Type cardinality $m..n$ in relationship R with respect to object type. This is represented by the inclusion axiom $C \sqsubseteq \geq n\,R'.\top \sqcap \leq m\,R'.\top$ where R' is either R or R^- depending on C being the first or the second object type in R. R is a unique name in the conceptual model (otherwise qualified cardinality is needed).
- Mandatory constraint. This is a special case of the previous one, with $n = 1$. It is interpreted as $C \sqsubseteq \exists R'.\top$, with R' as before.
- Single identification (in object types with respect to an attribute, and 1:1 mandatory). Let C be an object type identified by attribute a. Then this is interpreted in a DL by the inclusion axioms $C \sqsubseteq \exists a.\top \sqcap \leq 1\,a.\top$ and $\top \sqsubseteq \leq 1\,a^-.C$

In total, all the entities in the core profile sum up to 87.57 % of the entities in all the analysed models, covering The following entities, despite that they belong to all three CDMLs, are not part of the core profile because of their very low participation in the dataset: Role and Relationship Subsumption, Completeness constraints, and Disjointness constraints. Note that this means that it is not possible to express union of concepts in this Core Profile.

Reasoning over this Core Profile is quite simple. Since completeness and disjointness constraints are not present, negation cannot be directly expressed. It is possible to code negation only with cardinality constraints [4, chapter 2], but then we need to reify each negated concept as a new idempotent role. Another form of getting contradiction in this context is by setting several cardinality constraints on the same relationship participation, which is unusual in modelling languages. In any case, the main reasoning problems on the conceptual model only are class subsumption and class equivalence. The description logic $\mathcal{ALNI}$ (which is called $\mathcal{PL}_1$ in [14], and has polynomial subsumption) is expressive enough to represent this profile, since we only need $\top$, $\sqcap$, inverse roles and cardinalities constraints. Its data complexity is unknown.

The core profile shows that a relatively small set of entities concentrates most usage on conceptual models, and that these entities are consistent by assuming just single pairs of maximum and minimum cardinality constraints.

4   Specific CDML Profiles

We describe first the extension of the core so as to cover UML Class Diagrams v2.4.1, and subsequently (E)ER and ORM/2.

4.1   UML Class Diagram Profile

The UML Class Diagram Profile is composed by the Core Profile plus the following entities:

- Shared, Composite Aggregation. No axiomatisation is added for these relationships since the UML 2.4.1 standard [25] does not include additional static constraints, so they are coded as simple binary relationships.
- Subsumption between two UML associations R and S. Since we only have binary relationships, this can be represented in DL as role inclusion $R \sqsubseteq S$.
- Attributive Property Cardinality and Attribute Value Constraint. Cardinalities on attributes can be represented as cardinalities on relationships, but in order to represent value constraints it is necessary to include in the formalisation some datatype facilities to define new datatypes. The attribute is assigned a new datatype which is derived from the original one plus the constraining facets (in terms of XML Schema) on its values.

In total, 99.44 % of all the elements in the analysed UML models are covered by this profile. To formalise this profile in DLs we need to add role hierarchies and datatypes (concrete domains) to the $\mathcal{ALNI}$ logic for the Core Profile. This yields the logic $\mathcal{ALNHI}(\mathcal{D})$ that, as far as we know, has not been studied yet. If we assume unique names and some reasonable (at least from the conceptual modelling point of view) restrictions on the interaction between role inclusions and cardinality constraints, we can represent this profile in $DL\text{-}Lite_{core}^{\mathcal{HN}}$, which is NLOGSPACE for subsumption and AC^0 for data complexity [2].

Typical UML elements like qualified relationship, completeness constraints, and disjointness constraints do not belong to this profile. On the one hand, it is possible to say that the extra expressiveness that is not being used by modellers limit the formal meaning of their models. But since two of these rarely used features are necessary for proving the EXPTIME-hardness of reasoning on UML class diagrams [5], then reasoning over such limited diagrams becomes much more efficient.

4.2 (E)ER Profile

The (E)ER Profile is composed by the Core Profile plus the following entities:

- Composite and Multivalued attribute. Multivalued attributes can be represented with attribute cardinality constraints, and composite attributes with the inclusion of the union datatype derivation operator.
- Weak Object Type, Weak Identification. Each object type (entity type) in (E)ER is assumed by default to have at least one identification constraint. In order to represent external identification, we can use functionality constraints on roles as in $\mathcal{DLR}_{ifd}$ [11], or in $\mathcal{CFD}$ [31].
- Ternary relationships. This is described below and in Algorithm 1.
- Associative Object type. This is formalised by the reification of the association as a new DL concept with two binary relationships.
- Multiattribute identification. This can be formalised as a new composite attribute with single identification.

99.06 % of all the elements in the set of (E)ER models belong to this profile.

The only DL language family with arbitrary n-aries and the advanced identification constraints is $\mathcal{DLR}_{ifd}$, which happens to be positionalist. However, the DL role components are not strictly needed for (E)ER, and one may wish to pursue an n-ary DL without DL role components but with identification constraints, like in the $\mathcal{CFD}$ family of languages. Therefore, we provide here Algorithm 1, which summarises the procedure to go from (E)ER straight to the standard view. The main steps involve binaries vs. higher arities, and recursive ones that generally do have their named relationship components vs 'plain' binaries that have only the relationship named.

Algorithm 1. *(E)ER to standard view and common core*

D_R: domain of R; R_R range of R; n set of R-components
if R is binary **and** $D_R \neq R_R$ **then**
    Rename R to two 'directional' readings, Re_1 and Re_2
    Make Re_1 and Re_2 a DL role each
    Type role with $\top \sqsubseteq \forall Re_1.D_R \sqcap \forall Re_1^-.R_R$
    Declare inverses with $Re_1 \equiv Re_2^-$
else
    if R is binary **and** $D_R = R_R$ **then**
        for all i, with $i \in n$ **do**
            if i is named **then**
                $Re_i \leftarrow i$
            else
                $Re_i \leftarrow$ user-added label or auto generated label
            end if
            Make Re_i a DL role
        end for
        Type one Re_i, i.e., $\top \sqsubseteq \forall Re_i.D_R \sqcap \forall Re_i^-.R_R$
        Declare inverses among all Re_i
    end if
else
    Reify R into $R' \sqsubseteq \top$
    for all i, $3 \geq i \geq n$ **do**
        $Re_i \leftarrow$ user-added label or auto generated label
        Make Re_i a DL role,
        Type Re_i as $\top \sqsubseteq \forall Re_i.R' \sqcap \forall Re_i^-.R_R$, where R_R is the player ((E)ER entity type) in n
        Add $R' \sqsubseteq \exists Re_i.\top$ and $R' \sqsubseteq\, \leq 1\, Re_i.\top$
    end for
    Add external identifier $\top \sqsubseteq\, \leq 1\, (\sqcup_i Re_i)^-.R'$
end if

Using this translation, and since we do not have covering constraints in the profile, we can represent the (E)ER Profile in the description logic $DL\text{-}Lite_{core}^{\mathcal{N}}$ [2] which has complexity NLOGSPACE for the satisfiability problem. This low complexity is in no small part thanks to its unique name assumption, whereas most logics operate under no unique name assumption. A similar result is found in [1] for ER_{ref}, but it excludes composite attributes and weak object types.

4.3   ORM/2 Profile

For ORM, there is no good way to avoid the ORM roles (DL role components), as they are used for several constraints that have to be included. They can be transformed away (discussed below) such that an ORM/2 Profile is obtained by joining the features of the Core Profile. The following entities from the unifying metamodel are added, noting that the starred ones include the formalisation after the transformation from positionalist to standard view:

- Unary role, which is formalised as a boolean attribute. $\star$
- Subsumption between roles; formalised by using DL role hierarchies. $\star$
- n-ary relationships ($n \geq 2$). This is formalised similarly as for (E)ER (see Algorithm 1).
- Subsumption between relationships. This is formalised with an inclusion assertion between the reified concepts.
- Disjoint constraints between ORM roles R and S. This is formalised as two inclusion assertion for roles: $R \sqsubseteq \neg S$ and $S \sqsubseteq \neg R$. $\star$
- Nested object type. The nested object type is identified with the reified concept of the relationship.
- Value constraints. We need to define a new datatype with the constraints, as done in UML profile.
- Disjunctive mandatory constraint for object type C in roles R_i. This is formalised as the inclusion axiom $C \sqsubseteq \sqcup_i \exists R_i$. $\star$
- Internal Uniqueness constraint for roles $R_i, 1 \leq i \leq n$ over relationship objectified with object type R as described below. We need an identification axiom (**id** $C\ 1\ R_1 \ldots 1\ R_n$) as in $\mathcal{DLR}_{ifd}$.
- External Uniqueness constraint between roles $R_i, 1 < i \leq n$ not belonging to the same relationship. Let C be the connected object type between all the R_i, if it exists, or otherwise a new object type representing the reification of a new n-ary relationship between the participating roles. Then we can formalise the constraint with the identification axiom (**id** $C\ 1\ R_1 \ldots 1\ R_n$).
- External identification. This is the same as the previous one, with the exception that we are now sure such C exists (i.e., the mandatoryness is added cf. simple uniqueness).

This profile contains 98.69 % of all the elements in the analysed ORM/2 models. This is still a high coverage considering the assortment of entities available in the language. We decided not to include any ring constraint in this profile. Although the irreflexivity constraint counts for almost half of all appearances of ring constraints, its participation is still too low to be relevant.

In order to formalise this ORM/2 profile we need both n-aries and identification constraints, as in the (E)ER Profile. It differs from the (E)ER profile, in that ORM needs the argument positions for some constraints. We map this positionalist commitment into a standard view. This is motivated by the observation that typically fact type readings are provided, not user-named ORM role names, and only 9.5 % of all ORM roles in the 33 ORM diagrams in our dataset had a user-defined name, with a median of 0. We process the fact type (relationship R) readings and ignore the role names as follows. $\mathcal{DLR}$'s relationship is typed, w.l.o.g. as binary and in $\mathcal{DLR}$-notation, as $R \sqsubseteq [r_c]C \sqcap [r_d]D$, with r_c and r_d variables for the ORM role names and C and D the participating object types. Let $read_1$ and $read_2$ be the fact type readings, like the **teaches** and **taughtby** in Fig. 1, then use $read_1$ to name DL role Re_1 and $read_2$ to name DL role Re_2, and type R as $\top \sqsubseteq \forall Re_1.C \sqcap \forall Re_2.D$. This turns, e.g., a disjoint constraints between ORM roles r_c of relationship R and s_c of S into $Re_1 \sqsubseteq \neg Se_1$ and $Se_1 \sqsubseteq \neg Re_1$.

Concerning complexity of the ORM/2 Profile, this is not clear either. The EXPTIME-complete $\mathcal{DLR}_{ifd}$ is the easiest fit, but contains more than is strictly needed: neither concept disjointness and union are needed (but only among roles), nor its **fd** for complex functional dependencies. The PTIME $\mathcal{CFDI}_{nc}^{\forall-}$ [30] may be a better candidate if we admit a similar translation as the one given in Algorithm 1, but giving up arbitrary number restrictions and disjunctive mandatory on ORM roles.

5 Discussion

As mentioned in Sect. 2.2, other design choices could have led to another 'core profile'. This concerns two choices in particular: i) we used inverses and therewith could avoid qualified cardinality restrictions (thanks to typing of the relationship), and ii) transforming the positionalist into a standard view representation. The advantages are that there are clear indications that the current core profile is computationally better behaved and it can be used more easily with most implemented languages. A disadvantage is that for the ORM/2 profile, a positionalist DL language is needed and for (E)ER, it would make it easier for it fits more nicely with a known language ($\mathcal{DLR}_{ifd}$). Transformations are very well doable, as shown in Algorithm 1, but it adds an extra step in any implementation. The alternative is to create a 'positionalist core', but this is likely to be computationally less well-behaved, and does not enjoy wide software support when it comes to formal characterisations of the CDMLs.

5.1 On Missing and 'Useless' Features

As will be clear, there is no 'ideal' DL language for the CDMLs, not one that captures exactly and only the needed features, that is positionalist to avoid forcing the artificial standard view encoding, and has a usable implementation. The major mismatches regarding implementations have to do with n-aries, DL role components, advanced identifiers, and attributes with their data types, dimensions and value constraints. Data types are being investigated for DLs (e.g., [3,23]), and the results obtained here may serve as a motivational use case. Further, dimensional value types are yet to be addressed; e.g., a ternary 'attribute', say, height, consisting of the class it is measured for, the data type, and its measurement unit. Others could be a 'nice to have', notably arbitrary n-aries and, for ease of transformation algorithms from CDMLs to a logic, more implementations of positionalism (a DL with DL role components).

Viewed from the perspective as to what can safely be omitted from a logic for CDMLs, then, notably, nominals—computationally costly—are certainly not needed (recall also Sect. 3), and disjointness and completeness are used remarkably few times. Whether the latter is due to a real perceived irrelevance for conceptual data modelling or merely due to unfamiliarity by modellers is a separate line of investigation. The few disjointness and completeness constraints encountered, however, were predominantly in models taken from courses, textbooks, and from the UML standard. Also, there are multiple relationships in the models where properties, such as transitivity and reflexivity, certainly could apply, and if one can declare them (as in ORM) it is done, but it is unclear why it has been done so few times (23 in total in the 33 ORM/2 models). A conjecture is that this is due to their limited implementation support.

5.2   Answering the Research Questions

The results obtained in the previous sections provide the answers to the three questions posed in the introduction. Concerning question 1, the profile of a common core of language features among the main CDMLs has been specified in Sect. 3, covering UML Class diagrams v2.4.1, ER and EER, and ORM and ORM2. Although an important criterion was to keep the logic as 'simple' as possible, it is, perhaps, remarkable that a language with such low expressiveness as $\mathcal{ALNI}$ sufficed when taking into account the intersection of the languages and the usage of the CDML features in actual conceptual data models. $\mathcal{ALNI}$ in in PTIME, and possibly even better computationally well-behaved with the unique name assumption, as no unique name assumption together with number restrictions increases complexity, as shown with the DL-Lite family in [2]. Either way, this makes it certainly promising for scalable implementations for interoperability or conceptual model-mediated analysis and management of large scale data systems, including Ontology-Based Data Access, and to augment query compilation with the 'background knowledge' of the conceptual model, meeting requirements such as aimed for in [6,13,17,28,32].

Regarding question 2 on CDML profiles: based on the profiles defined that took input from the set of 101 conceptual models, there is no optimal language with known complexity that matches exactly a CDML profile to capture each of the UML, (E)ER, and ORM/2 languages (recall Sect. 4), where each profile had its own version of a mismatch. Of the three profiles, the one for UML is closest to the core profile, mainly thanks to the removal of relational properties of the aggregation associations from the UML standard (transitivity and asymmetry were asserted in earlier versions of the standard), and that qualified associations were hardly used.

This brings us to the answer to Question 3 on missing features and too many. A shortcoming of the available DLs is the limited support for constraints on datatypes. Conversely, nominals, negation and union, and most relational properties do not seem to be needed. Using nominals to encode values is suboptimal (see Sect. 3), whose in-depth argument is omitted due to space limitations.

6   Conclusions

Conceptual data model language-specific profiles for their logic-based reconstruction have been defined, as well as a common core. No CDM profile matches fully with an existing DL language. The common core capturing most entities occurring in the dataset of models amounts to $\mathcal{ALNI}$ which is PTIME. This means that efficient translations between models in these languages can be done preserving most of their elements and meaning. Even for the most expressive CDM language ORM2, the vast majority of entities can be captured with a $\mathcal{DLR}_{ifd}$ without disjointness and union or $\mathcal{CFDI}_{nc}^{\forall-}$ with arbitrary number restrictions. No features are really missing from any DL, other than advanced datatype constraints, but rather tend to have too many constructs. Given the absence of negation, there is little TBox reasoning of interest, other than cardinality constraints. The lean common core and profiles pave the way for a modelling-informed single language for model interoperability, and for their runtime usage in scalable databases and information systems.

Acknowledgments. This work is based upon research supported by the National Research Foundation of South Africa (Project UID90041) and the Argentinean Ministry of Science and Technology.

References

1. Artale, A., Calvanese, D., Kontchakov, R., Ryzhikov, V., Zakharyaschev, M.: Reasoning over extended er models. In: Parent, C., Schewe, K.-D., Storey, V.C., Thalheim, B. (eds.) ER 2007. LNCS, vol. 4801, pp. 277–292. Springer, Heidelberg (2007)
2. Artale, A., Calvanese, D., Kontchakov, R., Zakharyaschev, M.: The DL-Lite family and relations. J. Artif. Intell. Res. **36**, 1–69 (2009)
3. Artale, A., Ryzhikov, V., Kontchakov, R.: DL-Lite with attributes and datatypes. In: Proceedings of ECAI 2012, pp. 61-66. IOS Press (2012)
4. Baader, F., Calvanese, D., McGuinness, D.L., Nardi, D., Patel-Schneider, P.F. (eds.): The Description Logics Handbook - Theory and Applications, 2nd edn. Cambridge University Press, Cambridge (2008)
5. Berardi, D., Calvanese, D., De Giacomo, G.: Reasoning on UML class diagrams. Artif. Intell. **168**(1–2), 70–118 (2005)
6. Bloesch, A.C., Halpin, T.A.: Conceptual queries using ConQuer-II. In: Embley, D.W. (ed.) ER 1997. LNCS, vol. 1331, pp. 113–126. Springer, Heidelberg (1997)
7. Braga, B.F.B., Almeida, J.P.A., Guizzardi, G., Benevides, A.B.: Transforming OntoUML into Alloy: towards conceptual model validation using a lightweight formal methods. Innov. Sys. Softw. Eng. **6**(1–2), 55–63 (2010)

8. Cadoli, M., Calvanese, D., De Giacomo, G., Mancini, T.: Finite model reasoning on uml class diagrams via constraint programming. In: Basili, R., Pazienza, M.T. (eds.) AI*IA 2007. LNCS (LNAI), vol. 4733, pp. 36–47. Springer, Heidelberg (2007)
9. Calvanese, D., Carbotta, D., Ortiz, M.: A practical automata-based technique for reasoning in expressive description logics. In: Proceedings of IJCAI 2011, pp. 798–804. AAAI Press (2011)
10. Calvanese, D., De Giacomo, G., Lenzerini, M.: On the decidability of query containment under constraints. In: Proceedings of PODS 1998, pp. 149–158 (1998)
11. Calvanese, D., De Giacomo, G., L., M.: Identification constraints and functional dependencies in description logics. In: Proceedings of IJCAI 2001, pp. 155–160. Morgan Kaufmann (2001), seattle, Washington, USA, August 4–10 (2001)
12. Calvanese, D., Lenzerini, M., Nardi, D.: Unifying class-based representation formalisms. J. Artif. Intell. Res. **11**, 199–240 (1999)
13. Calvanese, D., Keet, C.M., Nutt, W., Rodríguez-Muro, M., Stefanoni, G.: Web-based graphical querying of databases through an ontology: the WONDER system. In: Proceedings of ACM SAC 2010, pp. 1389–1396. ACM (2010)
14. Donini, F., Lenzerini, M., Nardi, D., Nutt, W.: Tractable concept languages. In: Proceedings of IJCAI 1991, vol. 91, pp. 458–463 (1991)
15. Fillottrani, P.R., Keet, C.M.: Ontology-driven unification of conceptual data modelling languages (2012–2015). http://www.meteck.org/SAAR.html
16. Fillottrani, P.R., Keet, C.M.: KF metamodel formalisation. Technical report 1412.6545v1 December 2014, arxiv.org
17. Fillottrani, P.R., Keet, C.M.: Conceptual model interoperability: a metamodel-driven approach. In: Bikakis, A., Fodor, P., Roman, D. (eds.) RuleML 2014. LNCS, vol. 8620, pp. 52–66. Springer, Heidelberg (2014)
18. Fine, K.: Neutral relations. Philos. Rev. **109**(1), 1–33 (2000)
19. Keet, C.M., d'Amato, C., Khan, Z.C., Lawrynowicz, A.: Exploring Reasoning with the DMOP Ontology. In: Proceedings of ORE 2014. CEUR-WS, vol. 1207, pp. 64–70 (2014)
20. Keet, C.M., Fillottrani, P.R.: Toward an ontology-driven unifying metamodel for UML class diagrams, EER, and ORM2. In: Ng, W., Storey, V.C., Trujillo, J.C. (eds.) ER 2013. LNCS, vol. 8217, pp. 313–326. Springer, Heidelberg (2013)
21. Keet, C.M., Fillottrani, P.: An analysis and characterisation of publicly available conceptual models. In: Proceedings of ER 2015. LNCS, Springer (2015). (in print)
22. Leo, J.: Modeling relations. J. Phil. Logic **37**, 353–385 (2008)
23. Lutz, C., Areces, C., Horrocks, I., Sattler, U.: Keys, nominals, and concrete domains. J. Artif. Intell. Res. **23**, 667–726 (2005)
24. Motik, B., Grau, B.C., Horrocks, I., Wu, Z., Fokoue, A., Lutz, C.: OWL 2 Web Ontology Language Profiles. W3C recommendation, W3C 27 October 2009. http://www.w3.org/TR/owl2-profiles/
25. Object management group: superstructure specification. standard 2.4.1, object management group (2012). http://www.omg.org/spec/UML/2.4.1/
26. Pan, W., Liu, D.: Mapping object role modeling into common logic interchange format. In: Proceedings of ICACTE 2010, vol. 2, pp. 104–109. IEEE Computer Society (2010)
27. Queralt, A., Artale, A., Calvanese, D., Teniente, E.: OCL-Lite: finite reasoning on UML/OCL conceptual schemas. Data Knowl. Eng. **73**, 1–22 (2012)
28. Smaragdakis, Y., Csallner, C., Subramanian, R.: Scalable satisfiability checking and test data generation from modeling diagrams. Autom. Softw.Eng. **16**, 73–99 (2009)

29. Tobies, S.: Complexity results and practical algorithms for logics in knowledge representation. Ph.D. thesis, RWTH-Aachen, Germany (2001)
30. Toman, D., Weddell, G.: On adding inverse features to the description logic $CFD_{nc}^{\forall}$. In: Pham, D.-N., Park, S.-B. (eds.) PRICAI 2014. LNCS, vol. 8862, pp. 587–599. Springer, Heidelberg (2014)
31. Toman, D., Weddell, G.E.: Applications and extensions of ptime description logics with functional constraints. In: IJCAI. pp. 948–954 (2009)
32. Toman, D., Weddell, G.E.: Fundamentals of Physical Design and Query Compilation. Synthesis Lectures on Data Management. Morgan & Claypool, San Rafael (2011)
33. Wagih, H.M., Zanfaly, D.S.E., Kouta, M.M.: Mapping object role modeling 2 schemes into $\mathcal{SROIQ(D)}$ description logic. Int. J. of Comp. Theory Eng. **5**(2), 232–237 (2013)

Ontological Commitments, DL-Lite Logics and Reasoning Tractability

Mauricio Minuto Espil[1]($\boxtimes$), Maria Gabriela Ojea[2,3], and Maria Alejandra Ojea[3]

[1] ARBA, Buenos Aires, Argentina
mauriciominutoespil@yahoo.com.ar
[2] Colegio del Salvador, Buenos Aires, Argentina
[3] GCABA, Buenos Aires, Argentina

Abstract. We propose a particularly suitable family of description logics for ontology definition, STOC-DL-Lite, which results a variant to the well-known DL-Lite family. Our logics: *a*- augment DL-Lite logics with *ontological commitments*, sentences that assert that there is an a-priori agreement that certain concepts may exist as universal entities, and *b*- modify the usual Tarskian notion of satisfiability, enforcing any individual in the domain of a model to conform at least one *licit type*, i.e. a set of concepts literals that characterizes an universal entity. We show that, regarding time complexity, the satisfiability problem for the *boolean* fragment of the presented family is tractable, in marked contrast to DL-Lite$_{bool}^{\mathcal{N}}$, and its NP-complete satisfiability problem.

1 Introduction

Bringing entities an accurate description is essential for ontology building, particularly when providing semantics for managing Web data [8], or when designing a Semantic Web application [13], or a data warehouse [12]. The publishing of ontologies through conceptual theories in the Web shows a continuous development, with special languages like OWL-DL being conceived for this purpose. Nonetheless, a description language should be able to distinguish universal entities from mere concepts or attributes, in order to result appropriate in describing real domains. Unfortunately, description logic languages with fixed Tarskian-like semantics interpret universal variables (concepts or unary predicates) as subsets of a non-specific set (the universe). As a consequence, domain specific universal entities are not recognized per se in the semantics of those languages. The languages of description logics in the family of DL-Lite [2] do not escape the problem: being pure logic languages, they are all entity-unaware. Let us present some examples of the problem:

Example 1: Two philosophers P1 and P2 are reunited, discussing problems concerning mutual interests. P1 asks P2 whether he agree on the existence of women and fishes, to what P2 answers affirmatively. Next, P1 asks P2 whether he agree that both women and fishes are animals, to what, again, P2 answers positively. Finally, P1 poses P2 the following question:

© Springer International Publishing Switzerland 2015
T. Morzy et al. (Eds.): ADBIS 2015, LNCS 9282, pp. 230–244, 2015.
DOI: 10.1007/978-3-319-23135-8_16

"According to what both of us agree, and supposing that you accept my saying that there are a woman and a fish in the kitchen, what is the minimum number of animals you consider are present in the kitchen?"

P2 answers that the minimum number is *one*, and justifies the answer translating the agreements into a set of DL-Lite axioms: $woman \sqsubseteq animal$, $fish \sqsubseteq animal$, $woman(x)$, $fish(y)$, and applying DL-Lite semantics to solve the query. For P2, since DL-Lite interpretations admits considering an individual o in the domain as a member of both the interpretations of concepts *woman* and *fish*, the legitimate answer in one. In opposition, P1 supports that *two* would be the correct answer, because they never agreed on the possible existence of a woman being also a fish, or a fish being also a woman, i.e. the possible existence of *mermaids*. The position of P1 consists in considering concept $woman \sqcap fish$ in itself as referring to a distinct universal entity, and accepts the intersection of the sets interpreting concepts *woman* and *fish* as the interpretation of concept $woman \sqcap fish$ (as DL-Lite semantics prescribe), only in the case that there is an agreement in the discourse that an individual may conform $woman \sqcap fish$. Since concept $woman \sqcap fish$ does not occur by itself in any of the axioms, nor the axioms $woman \sqsubseteq fish$ nor $fish \sqsubseteq woman$ were part of the agreement (thus *woman* implying *fish*, or *fish* implying *woman*), the discourse does not entail the possibility that $woman \sqcap fish$ may refer to anything other than the empty set. Hence, P1 considers that two is the legitimate answer.

Example 2: We can certainly assert that *lions* are *vertebrate*, and *ants* are not. However, using the semantic notion of ordinary people, how can we handle concept *vertebrate* and its negation applied to *mountains*? Clearly, it is not the case that a *mountain* is a *vertebrate*, but, for ordinary people, *invertebrate* (or *not vertebrate*) said of a *mountain* does not make any sense; *vertebrate* and $\neg vertebrate$ are admissible when said of animals only, not of an individual whatever in the universe. Now consider the following set of axioms: $vertebrate \sqsubseteq animal$; $\neg vertebrate \sqsubseteq animal$; $mountain(Everest)$. Under DL-Lite semantics, any model of the axioms would entail $animal(Everest)$, since *animal* amounts to the whole universe. Under the position of philosopher P1, however, concepts $animal \sqcap mountain$ and $animal \sqcap \neg mountain$ are not mentioned in the axioms, and thus are void of meaning, i.e. they refer to nothing. As a consequence, $animal(Everest)$ does not hold in any model of the axioms. Moreover, with the same argument, P1 would consider $mountain \sqcap vertebrate$ and $mountain \sqcap \neg vertebrate$ both as referring to nothing, despite considering *mountain* as containing the individual interpreting $Everest$.

Example 3: Suppose we have a database with the relation schema $flight$ (*airline*, *origin*, *destination*). It is also clear that a value *must* be present in columns *airline*, *origin* and *destination* in every tuple of an instance of relation $flight$; there are no flights with no airline, or no origin, or no destination in real life. Moreover, although it is permissible that a null value occurs in a row as a column value, its interpretation must refer to some value **any**, an *existent*, although *actually not known*, value (as standards like SQL, for example, mandates). Now, suppose we have a domain consisting of individuals that include flights, and we describe the schema of flights and its constraints through DL-Lite axioms as follows:

$flight \doteq airline\mathbf{:any}$; $flight \doteq origin\mathbf{:any}$; $flight \doteq destination\mathbf{:any}$;[1]
$airline\mathbf{:A}_j \sqsubseteq airline\mathbf{:any}$, for $\mathbf{A}_j$ an admissible airline;
$origin\mathbf{:O}_j \sqsubseteq origin\mathbf{:any}$, for $\mathbf{O}_j$ an admissible origin;
$destination\mathbf{:D}_j \sqsubseteq destination\mathbf{:any}$, for $\mathbf{D}_j$ an admissible destination;
$airline\mathbf{:any} \sqcap \neg\, airline\mathbf{:A}_j \sqcap ...$ ($\mathbf{A}_j$ an admissible airline) $... \sqsubseteq \bot$;
$origin\mathbf{:any} \sqcap \neg\, origin\mathbf{:O}_j \sqcap ...$ ($\mathbf{O}_j$ an admissible origin) $... \sqsubseteq \bot$;
$destination\mathbf{:any} \sqcap \neg\, destination\mathbf{:D}_j \sqcap ...$ ($\mathbf{D}_j$ an admissible destination) $... \sqsubseteq \bot$;
$origin\mathbf{:O}_j \sqcap destination\mathbf{:D}_{j'} \sqsubseteq \bot$, for $\mathbf{O}_j = \mathbf{D}_{j'}$;

Now, suppose we add the axiom $airline\mathbf{:KLM}(1502)$ asserting that there is a flight 1502 in KLM. We would expect of flight 1502 having a concrete origin and a concrete destination, according to the semantics of axioms. There is, however, no restriction on a flight to have only one origin in the knowledge base. Nonetheless, Philosopher P1 assumes the point, because there is no concept $origin\mathbf{:O}_1 \sqcap origin\mathbf{:O}_2$, for $\mathbf{O}_1 \neq \mathbf{O}_2$, in the knowledge base.

1.1 Contribution

In this paper, we present a family of description logics that semantically reflects the position of philosopher P1, and we call this family of logics STOC-DL-Lite (Strict Ontological-Committed DL-Lite). The family of logics we present enriches DL-Lite logics, allowing the user to specify (explicitly or implicitly) *ontological commitments* on concepts in a knowledge base K, heightening the status of each concept receiving a commitment (which we call generically K-*committed*) up to that of a universal entity: a distinct intension that characterizes a set of individuals in the described world. Semantically, the notion of entity in the logics is strict: the described world is composed solely of individuals receiving an ontological commitment in the knowledge base. Therefore, the semantics of any logic in the family prescribes that every individual in the domain of an interpretation *must* be a member of the interpretation of *at least one* K-committed concept, which *types* the individual as entity. We present here the *bool* fragment with no role hierarchies of the family (we use liberally the name STOC-DL-lite to design this fragment), and provides the reader with its syntax and semantics. Finally, we study the problem of deciding whether a STOC-DL-Lite knowledge base is satisfiable. We exhibit a sound and complete decision procedure for satisfiability, showing that, in terms of time complexity, the problem is tractable.

1.2 Related Work

The entailment of the existence of universal entities from the truth of statements made during a discourse is a problem focus of ample and controversial debate

---

[1] $C \doteq D$ standing for $C \sqsubseteq D$, $D \sqsubseteq C$.

in philosophy, epistemology, linguistics, and logic. Several positions have been fixed in the past regarding the issue [17]. Particularly, a clear distinction has been made among objects conforming a concept (or a mere idea), and objects characterizing a universal entity: objects having a common *identity* with (potential or real) distinct existence [16]. The use of description logics in modeling information systems or in providing semantics to Web data is nowadays almost a standard practice [15]. However, few logics clearly distinguish concepts from universal entities in their semantics. The subject, although not explicitly introduced in our terms, constitutes one goal of FCA (Formal Concept Analysis). There, ideal concepts are named *attributes* while reserving the name *concept* to denote a universal entity: an attribute or series of attributes with at least an individual asserting its existence in an FCA-context [9]. An attempt has been made to merge FCA-contexts with description logics [3]. The attempt falls short of goals, however: since first order Tarskian semantics is preserved, hard reasoning problems result from it. Our approach, differently, abandons classic semantics, and fully exploits the notion of implicational concept graph of FCA [19] as a mechanism for circumscribing precisely the set of universal entities that characterize the individuals in the described world, with exclusive base on the information provided in the knowledge base.

Our logic has a distinguished property: its satisfiability problem is tractable. This property occurs as a consequence of the enforcement of a light form of closeness in its semantics. Although partially non-monotonic w.r.t. reasoning, the semantics of our logic does not introduce the notion of a closed-world w.r.t. interpretation domains, as other approaches do [10,11]. Our logic closes the universe of possible combinations of concept literals to those explicitly combined in the axioms of the knowledge base. It behaves like a content controlled form of formula circumscription, different from other attempts to merge circumscription with description logics [7,14]. It is similar in scope, although different in realization, to the work in [4], where knowledge bases are translated into UML class diagrams, and finite satisfiability is decided on acyclic structures only. The same similarity in scope occurs with [18], where a syntax-driven tractable approximate decision procedure for satisfiability is presented.

2   STOC-DL-Lite Logics

Syntactically, a knowledge base K in STOC-DL-Lite is made of a TBox, an ABox, as any DL-Lite knowledge base [1], and a CBox, a set of global *ontological commitments*. An ontological commitment is an expression $\circ\, C$, where C a well formed concept s.t. $\bot$ is not part of the composition of C. The occurrence of an ontological commitment $\circ\, C$ as a member of the CBox of K explicitly asserts that C is a K-committed concept. Since K-commited concepts represent universal entities, the CBox of a knowledge base K acts as the set of universal entities *a priori* defined by K.

2.1   STOC-DL-Lite Concepts and Axioms

TBoxes and ABoxes in STOC-DL-Lite are similar to those corresponding to DL-Lite [1]. We have in STOC-DL-Lite a vocabulary composed by: a set $\mathcal{N}_\mathcal{I}$ of individual names, a set $\mathcal{N}_\mathcal{C}$ of concept names, and a set $\mathcal{N}_\mathcal{R}$ of role names. Role names give way to expressions of the form P and P^- named generically *roles*, for P a role name in $\mathcal{N}_\mathcal{R}$, and $\mathcal{R}$ denotes the set of roles occurring in K. Roles give way in turn to expressions of the form: $\geq q\,R$, R a role in $\mathcal{R}$, $q > 0$ a natural number, named generally *number restrictions*, and $\mathcal{R}_{nr}$ denotes the set of them occuring in K. Concept names and roles give way finally to well-formed concepts, and $\mathcal{C}$ denotes the set of them occurring in K. Concepts in $\mathcal{C}$ are classified as in DL-Lite into:

– *Global Concepts:* Expressions of the form $\bot$ or $\neg\bot$, their set denoted by $\mathcal{C}_{global}$;
– *Basic Concepts:* Concept names $A \in \mathcal{N}_\mathcal{C}$, or number restrictions $\geq q\,R$ in $\mathcal{R}_{nr}$, their set denoted by $\mathcal{C}_{base}$;
– *Concept Literals:* Expressions of the form A or $\neg A$, where A is a basic concept, their set denoted by $\mathcal{C}_{lit}$;
– *Complex Concepts:* Expressions of the for $A_1 \sqcap ... \sqcap A_n$, $n > 1$ a natural number, where $A_1, ..., A_n$ are concept literals, s.t. no A_j, A_i, $i \neq j$, involve the same basic concept.

Analogously with DL-Lite, TBoxes in STOC-DL-Lite are formed upon inclusion axioms of the form $C \sqsubseteq D$, where C and D are concepts in $\mathcal{C}$, $C \neq \bot$, $D \neq \neg\bot$, and ABoxes in are formed upon axioms of the form: $A(a)$, where A is a concept name, b is an individual name, or $P(a,b)$, where P is a role name, and a and b are individual names.

Notation: For simplicity, in what follows, for P a role name, expression $R^\sim$ stands for P^- whenever $R = P$, or P whenever $R = P^-$. In the case of concept literals, expression $\sim l$, l a concept literal, stands for concept $\neg A$ whenever $l = A$, and concept A whenever $l = \neg A$; expression $l_1 \parallel l_2$, which reads l_1 *clashes with* l_2, holds whenever $l_1 = \sim l_2$ and l_1 and l_2 involve the same concept name, and whenever $q' \leq q$, and $l_1 = \geq q\,R$ and $l_2 = \neg\geq q'\,R$, or vice versa.

We call any non-empty set of concept literals a (concept) *type*. In the case of non-global concepts, expression $\sqcap_{l \in L}\, l$, $L \neq \emptyset$, denote concept literal A whenever $L = \{A\}$, and inductively $(\sqcap_{l \in L \setminus \{A\}} l) \sqcap A$ whenever $\#L > 1$ ($\#L$ the size of set L) and $A \in L$. Since any non-global concept $C \in \mathcal{C}$ can be expressed as $\sqcap_{l \in L}\, l$, for some type L, concept $C = \sqcap_{l \in L}\, l$ will be represented in what follows by type L, and we will say that C is a concept with *type L*, consequently calling L the *type* for C. Let therefore L, L_1 and L_2 be concept types:

• we define $\mathcal{B}(L) = \{\, B \in \mathcal{C}_{base}\,|\, B \in L\, or\, \neg B \in L\}$, and
         we say that L is *clash-free* if and only if:
                 there not exists l_1 and $l_2 \in L$ s.t. $l_1 \parallel l_2$;

- we say that:
  - $L_1 \sqsubseteq_{(K)} L_2$ (L_1 is included into L_2 for K) if and only if:
    axiom $\sqcap_{l_1 \in L_1} l_1 \sqsubseteq \sqcap_{l_2 \in L_2} l_2$ is in K;
  - $L_1 \rightarrow_{(K)} L_2$ (L_1 implies L_2 in K), if and only if:
    $L_1 \sqsubseteq_{(K)} L_2$, or
    for $\geq q\, R, \geq q'\, R \in \mathcal{R}_{nr}$: $L_1 = \{\geq q\, R\}$ and $L_2 = \{\geq q'\, R\}$ and $q' < q$, or
    for $\neg \geq q\, R, \neg \geq q'\, R \in \mathcal{R}_{nr}$: $L_1 = \{\neg \geq q\, R\}$ and $L_2 = \{\neg \geq q'\, R\}$ and $q' > q$;

For $\gg$ a binary relation on concept types that we call generically *concept implication* and $S = dom(\gg) \cup ran(\gg)$, we define:

Definition 1. *Implication Closure: Let L be a clash-free concept type. The closure $L_{(\gg)}^+$ of L w.r.t. concept implication $\gg$ is the minimum set of concept literals that includes L and satisfies the following rule:*
- *for any pair L_1, $L_2 \in S$ s.t. $L_{(\gg)}^+ \supseteq L_1$, if $L_1 \gg L_2$ holds, then each literal $l_2 \in L_2$ is a member of $L_{(\gg)}^+$.*

Definition 2. *Consistency: Let $L_{(\gg)}^+$ be the closure of some clash-free type L w.r.t. relation $\gg$, and Φ a set of types. We say that set Φ is compatible with type L if and only if, for each $L' \in \Phi$, $L' \not\subseteq L_{(\gg)}^+$. We say that a type L is consistent w.r.t $\gg$ and Φ, if and only if $L_{(\gg)}^+$ is clash-free and Φ is compatible with L.*

Note that, even though a set of concept literals L was clash-free, L may be inconsistent w.r.t $\gg$ and Φ.

2.2   STOC-DL-Lite Semantics

In STOC-DL-Lite, universal entities are represented by K-committed concepts. Not every commitment should be given explicitly through an axiom in the CBox, however. Implicit commitments are also derived from the contents of the TBox and the ABox of K. The set of all K-committed concepts yields what we call the *strict commitment context* (SC-context, for short) for K, the set of types for all the K-committed concepts. The members of the SC-context for K are determined inductively through the application of a sound set of commitment rules on the contents of the CBox, the TBox and the ABox, as follows:

Definition 3. *Contexts: $\xi(K)$, and $\xi^-(K)$, the positive and negative pre-contexts for a knowledge base K, are the minimum sets of concept literals w.r.t. set inclusion satisfying the following:*

1- *if $\circ \sqcap_{l \in L} l$ is a member of the CBox of K, then L is a member of $\xi(K)$.*
2- *If l is a concept literal that occurs in K, then $\{l\}$ and $\{\sim l\}$ are members of $\xi(K)$.*
3- *if $\sqcap_{l_1 \in L_1} l_1 \sqsubseteq \sqcap_{l_2 \in L_2} l_2$, $\#L_1 > 0$, $\#L_2 > 0$, occurs in the TBox of K, then L_1 and L_2 are members of $\xi(K)$.*
4- *if $\sqcap_{l_1 \in L_1} l_1 \sqsubseteq \bot$, $\#L_1 > 0$, occurs in the TBox of K, and there is no $\sqcap_{l_2 \in L_2} l_2 \sqsubseteq \bot$, $L_1 \supset L_2$, in the TBox of K, then L_1 is a member of $\xi^-(K)$.*

5- *if* $\geq q\,R$ *occurs in* $\xi(K)$, *then* $\{\geq 1\,R\}$ *and* $\{\geq 1\,R^{\sim}\}$ *are members of* $\xi(K)$.

$\Xi^-(K)$, *the constraining context of knowledge base* K *is the maximal subset of* $\xi^-(K)$ *s.t. each member of* $\Xi^-(K)$ *is clash-free, their members called* $(K$-$)$ *constraints.* $\Xi(K)$, *the SC-context of knowledge base* K *is the maximal subset of* $\xi(K)$ *s.t. each member of* $\Xi(K)$ *is consistent w.r.t.* $\rightarrow_{(K)}$ *and* $\Xi^-(K)$.

The rationale of the rules above is easy to grasp. Any consistent concept literal occurring in K and any consistent concept explicitly mentioned in the knowledge base as the whole left or right part of a full concept inclusion is considered K-committed, and thus its type is considered a member of $\Xi(K)$, the SC-context for K. The type of the left right part of a concept inclusion with $\bot$ as its right part is considered a constraint, and is a member of set $\Xi^-(K)$ if it is minimal.

With the notion of SC-context in mind, we define the *implication* relation $\rightarrow_{\Xi(K)}$ among members of set $\Xi(K)$, as the restriction of relation $\rightarrow_{(K)}$ to members of $\Xi(K)$. We additionally define the *reachability* relation $\rightarrow^+_{\Xi(K)}$ as the transitive expansion of relation $\rightarrow_{\Xi(K)}$, and thus we read that $L_2 \in \Xi(K)$ is *reachable* from $L_1 \in \Xi(K)$ whenever we met $L_1 \rightarrow^+_{\Xi(K)} L_2$, and define $\sqsubseteq^+_{\Xi(K)}$, the *autonomy* relation as the transitive expansion of $\sqsubseteq_{(K)}$, restricted to members of $\Xi(K)$. Context $\Xi(K)$ and relation $\rightarrow_{\Xi(K)}$ induce together a directed graph:

Definition 4. *SC-Graph: The* SC-graph *for knowledge base* K *is a directed graph* $\mathcal{G}_{\Xi(K)} = (\mathcal{N}_{\Xi(K)}, \mathcal{A}_{\Xi(K)})$, *where each member of* $\Xi(K)$ *constitutes a node of* $\mathcal{N}_{\Xi(K)}$, *and each pair* L_1, L_2 *of nodes in* $\mathcal{N}_{\Xi(K)}$ *s.t.* $L_1 \rightarrow_{\Xi(K)} L_2$ *holds constitutes an arc* (L_1, L_2) *in* $\mathcal{A}_{\Xi(K)}$.

We can divide, through the reachability relation $\rightarrow^+_{(K)}$, the set of nodes $\mathcal{N}$ into a set of *classes of equivalence* w.r.t. $\rightarrow^+_{(K)}$, defining a class of equivalence of $\mathcal{N}$ w.r.t. $\rightarrow^+_{(K)}$ as a maximal size set of nodes in $\mathcal{N}$ s.t., for each pair $n_1 \neq n_2 \in \mathcal{N}$, n_1 and n_2 belongs to the same class if and only if both nodes are reachable from each other, i.e. the pairs (n_1, n_2) and (n_2, n_1) are members of $\rightarrow^+_{(K)}$. We consider the classes of equivalence induced by graph $\mathcal{G}$ as the genuine representatives of the committed concepts or "universal entities" in K. We thus define:

Definition 5. *Characteristic Graph: The* characteristic graph *for SC-cont-ext* $\Xi(K)$ *is a directed graph* $\mathcal{G}^{\leftrightarrow}_{\Xi(K)} = (\mathcal{N}^{\leftrightarrow}_{\Xi(K)}, \mathcal{A}^{\leftrightarrow}_{\Xi(K)})$, *where* $\mathcal{N}^{\leftrightarrow}_{\Xi(K)}$ *is the set of nodes* $n_c = \bigcup_{L \in c} L$, *for each class of equivalence* c *of* $\mathcal{N}$ *w.r.t.* $\rightarrow^+_{(K)}$, *and* $\mathcal{A}^{\leftrightarrow}_{\Xi(K)}$ *is the set of pairs* (n_{c_1}, n_{c_2}), $n_{c_1} \neq n_{c_2} \in \mathcal{N}^{\leftrightarrow}_{\Xi(K)}$, *such that there is an arc* $(n_1, n_2) \in \mathcal{A}_{\Xi(K)}$ *with* n_1 *a member of class* c_1 *and* n_2 *a member of class* c_2. *The reader should note that graph* $\mathcal{G}^{\leftrightarrow}_{\Xi(K)}$ *is acyclic.*

The SC-graphs for the cases in Examples 1, 2 and 3, are exhibited in Figs. 1a, b, and c, respectively. In Fig. 1d three classes of equivalence are depicted, their inner arcs depicted by dashed lines.

In STOC-DL-Lite, a knowledge base K is satisfied in *strictly committed interpretations* or SC-interpretations, for short. Differently from DL-Lite, an SC-interpretation is not defined directly through an interpretation function that

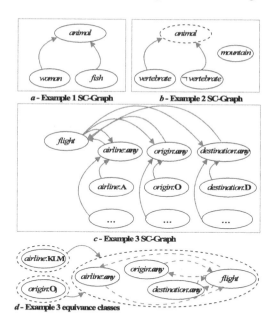

Fig. 1. Graphs for case examples

maps concepts to subsets of the interpretation domain; it is defined indirectly through a set of *licit types*, which are minimal consistent types that result from the closure of the nodes of the characteristic graph of K w.r.t. the concept implication induced by its arcs, and the application of the constraints in $\Xi^-(K)$. Accordingly, we define:

Definition 6. *Type Space: Let K be knowledge base, and let n be a member of $\mathcal{N}^{\leftrightarrow}_{\Xi(K)}$. We call a type $\Omega = n^+_{\mathcal{A}^{\leftrightarrow}_{\Xi(K)}}$ a prototype for K if and only if Ω is consistent w.r.t. $\mathcal{A}^{\leftrightarrow}_{\Xi(K)}$ and $\Xi^-(K)$. We say that a constraint $L \in \Xi^-(K)$ is applicable to a type L' if and only if $L \cap L' \neq \emptyset$, $L \setminus L' \neq \emptyset$, and $\mathcal{B}(L \setminus L') \cap \mathcal{B}(L') = \emptyset$. We call a type Λ closed for K if and only if Λ is consistent w.r.t. $\mathcal{A}^{\leftrightarrow}_{\Xi(K)}$ and $\Xi^-(K)$, and $\Lambda = W^{(k)}$, where k is the superscript of the last element of a sequence $(W^{(i)}, Z^{(i)})$ s.t. $W^{(1)} = \Omega$, Ω a prototype for K, and $Z^{(1)} = \emptyset$, and (if it exists) $W^{(i+1)} = (W^{(i)} \cup \{\sim l\})^+_{\mathcal{A}^{\leftrightarrow}_{\Xi(K)}}$, $l \in \Upsilon \setminus W^{(i)}$, for some $\Upsilon \in \Xi^-(K) \setminus Z^{(i)}$ applicable to $W^{(i)}$, and $Z^{(i+1)} = Z^{(i)} \cup \{\Upsilon\}$. For Λ a closed type for K s.t. set $Q = \{\neg \geq q\, R \in \mathcal{C}_{lit} \mid \geq q\, R, \neg \geq q\, R \notin \Lambda\}$ satisfies that $Q^+_{\mathcal{A}^{\leftrightarrow}_{\Xi(K)}} \subseteq \Lambda \cup Q$, we call type $\Theta = \Lambda \cup Q$ an SC-type for K, and we call it licit if and only if, additionally, there is no pair of SC-types Θ_1 and Θ_2 for K and basic concept $B \in \mathcal{C}_{base}$ s.t. $B \notin \Theta$, and $\Theta_1 = \Theta \cup \{B\}$ and $\Theta_2 = \Theta \cup \{\neg B\}$. We call $\mathcal{T}(K)$, the set of all licit types for K, the (SC-)type space of K.*

An SC-interpretation is defined with base on the type space K^τ, as follows.

Definition 7. *SC-Interpretation: An SC-interpretation $\mathcal{J}(K)$ for a knowledge base K is a tuple $(\Delta^{\mathcal{J}(K)}, \mathcal{T}(K), \theta^{\mathcal{T}(K)}, .^{\mathcal{J}(K)})$ where:*

- $\Delta^{\mathcal{J}(K)}$ *is the domain of interpretation $\mathcal{J}(K)$;*
- $\mathcal{T}(K)$ *is the type space of K;*
- $\theta^{\mathcal{T}(K)}$, *the typing function, is a mapping that assigns a licit type Θ in $\mathcal{T}(K)$ to each individual o in the domain $\Delta^{\mathcal{J}(K)}$;*
- $.^{\mathcal{J}(K)}$ *is a mapping that assigns a subset of $\Delta^{\mathcal{J}(K)}$ to each well-formed concept in the vocabulary of K, in a manner that conditions in Table 1 are satisfied.*

We say that a concept inclusion $C \sqsubseteq D$ holds in a SC-interpretation $\mathcal{J}(K)$, formally, $\mathcal{J}(K) \models_{SC} C \sqsubseteq D$, if and only if $C^{\mathcal{J}(K)} \subseteq D^{\mathcal{J}(K)}$. If a concept inclusion $C \sqsubseteq D$ holds in a SC-interpretation $\mathcal{J}(K)$, we say that $\mathcal{J}(K)$ is an SC-model of $C \sqsubseteq D$. We say that an assertion $A(a)$ holds in a SC-interpretation $\mathcal{J}(K)$, formally $\mathcal{J}(K) \models_{SC} A(a)$ if and only if $a^{\mathcal{J}(K)} \in A^{\mathcal{J}(K)}$. If an assertion $A(a)$ holds in an interpretation $\mathcal{J}(K)$, we say that $\mathcal{J}(K)$ is an SC-model of $A(a)$. A STOC-DL-Lite knowledge base K is said to be SC-satisfiable if and only if there exists an SC-model $\mathcal{J}(K)$ of all axioms in the TBox and ABox of K. If this is the case, we say that $\mathcal{J}(K)$ SC-satisfies K, formally, $\mathcal{J}(K) \models_{SC} K$, and that $\mathcal{J}(K)$ is an SC-model of K.

Let us study the impact SC-interpretations have in the example cases:

In the case of Example 1, set $\Xi(K)$ is made of concepts *woman*, *fish*, and *animal*. The licit types are: $\{woman, animal\}$, $\{fish, animal\}$, and $\{animal\}$. Notice that, in this case, $woman \sqcap fish$ cannot have non-empty interpretations, because there is no licit type Θ s.t. $\{woman, fish\} \subseteq \Theta$. The same occurs with $\{\neg woman, fish\}$, and $\{woman, \neg fish\}$. It is important to note that, if the axiom $\circ(woman \sqcap fish)$ had occurred in the CBox, concept $woman \sqcap fish$ would have been allowed to refer to a non-empty set, and at least one licit type would have been included concepts *woman* and *fish* together among its members.

In the case of Example 2, set $\Xi(K)$ is made of $\{vertebrate\}$, $\{\neg vertebrate\}$, $\{animal\}$ and $\{mountain\}$. The licit types are: $\{vertebrate, animal\}$, $\{\neg verte$-$brate, animal\}$, and $\{mountain\}$. Notice that $\{animal\}$ is a member of $\Xi(K)$ and therefore is a type, even though it cannot have proper autonomous individuals. It is not, however, a *licit* type, because $\{vertebrate, animal\}$ and $\{\neg vertebrate, animal\}$ are types and the minimality restriction for licit types does not hold. We highlight the fact, circling node *animal* with dashed lines in the graph presented in Fig. 1b.

In the case of Example 3, an important thing to remark, and depicted in Fig. 1c, is the presence of a "genuine" class of equivalence, a node that blend together the nodes $\{flight\}$, $\{airline\text{:}\mathbf{any}\}$, $\{origin\text{:}\mathbf{any}\}$, and $\{destination\text{:-}\mathbf{any}\}$ in the SC-graph, because of the double concept inclusions. Some of the classes, as nodes of the characteristic graph, are depicted in Fig. 1d by a dashed line surrounding the original nodes in the SC-graph. There is another important issue to remark here: the derivation of licit types involving disjunctions through the "application" of constraints. Let us consider the derivation of one licit type Θ originating in the class node for $airline$:**KLM**. Clearly the literals in the "blending node" $\{flight, airline$:**any**$, origin$:**any**$, destination$:**any**$\}$ are members of Θ, because they must be members of the prototype Ω_{KLM} generated by $airline$:**KLM** (the "blending" node is reachable from $airline$:**KLM**). Because of the occurrence of axiom $origin$:**any** $\sqcap \neg origin$:$\mathbf{O}_j \sqcap ... \sqsubseteq \bot$ in the knowledge base, and the consequent occurrence of constraint $\{origin$:**any**$, \neg origin$:$\mathbf{O}_j, ...\}$ in $\Xi^-(K)$, and that the intersection of that constraint and set $\{airline$:**KLM**$, flight, airline$:**any**$, origin$:**any**$,$ and $destination$:**any**$\}$ is set $\{origin$:**any**$\}$, one literal $origin$:$\mathbf{O}_j$, for some j, must be a member of Θ, because the constraint is "applicable" to Ω_{KLM}. The same argument follows for the constraint involving $destination$:$\mathbf{D}_{j'}$, and one literal $destination$:$\mathbf{D}_{j'}$, for some j', from the literals in constraint $\{destination$:**any**$, \neg destination$:$\mathbf{D}_{j'}, ... \}$, must occur in Θ. It is also noticeable that $\mathbf{O}_j \neq \mathbf{D}_{j'}$ must hold, because of the requisite that any SC-type must be consistent w.r.t. set $\Xi^-(K)$. The occurrence of constraints $\{origin$:$\mathbf{O}_j, destination$:$\mathbf{D}_{j'}\}$ in $\Xi^-(K)$, for $\mathbf{O}_j \neq \mathbf{D}_{j'}$, as a consequence of the occurrence of axioms: $origin$:$\mathbf{O}_j \sqcap destination$:$\mathbf{D}_{j'} \sqsubseteq \bot$, for $\mathbf{O}_j = \mathbf{D}_{j'}$ in the knowledge base, prevent the simultaneous occurrence of literals $origin$:$\mathbf{O}_j$ and $destination$:$\mathbf{D}'_j$, with $\mathbf{O}_j = \mathbf{D}_{j'}$, in a consistent type. The desired "completion" schema for the example is thus fulfilled.

Table 1. STOC-DL-Lite$_{bool}^{\mathcal{N}}$ Semantics

	Expression	Interpretation
1	$P \in \mathcal{R} \,\|\, P \in \mathcal{N}_\mathcal{R}$	$P^{\mathcal{J}(K)} \subseteq \Delta^{\mathcal{J}(K)} \times \Delta^{\mathcal{J}(K)}$
2	$P^- \in \mathcal{R} \,\|\, P \in \mathcal{N}_\mathcal{R}$	$(P^-)^{\mathcal{J}(K)} = \{(y,x)\| \, (x,y) \in P^{\mathcal{J}(K)}\}$
3	$\bot \in \mathcal{C}_{global}$	$\bot^{\mathcal{J}(K)} = \emptyset$
4	$\neg\bot \in \mathcal{C}_{global}$	$\neg\bot^{\mathcal{J}(K)} = \Delta^{\mathcal{J}(K)}$
5	$a \in \mathcal{N}_\mathcal{I}$	$a^{\mathcal{J}(K)} \in \Delta^{\mathcal{J}(K)}$
6	$A \in \mathcal{N}_\mathcal{C}$	$A^{\mathcal{J}(K)} = \{o \in \Delta^{\mathcal{J}(K)} \,\|\, A \in \theta^{\mathcal{T}(K)}(o)\}$
7	$\geq nR \in \mathcal{R}_{nr}$	$(\geq nR)^{\mathcal{J}(K)} = \{x \in \Delta^{\mathcal{J}(K)} \,\|\, \#\{y\|(x,y) \in R^{\mathcal{J}(K)}\} \geq n\}$
		$= \{o \in \Delta^{\mathcal{J}(K)} \,\| \geq nR \in \theta^{\mathcal{T}(K)}(o)\}$
8	$\neg A \,\|\, A \in \mathcal{N}_\mathcal{C}$	$(\neg A)^{\mathcal{J}(K)} = \{o \in \Delta^{\mathcal{J}(K)} \,\|\, \neg A \in \theta^{\mathcal{T}(K)}(o)\}$
9	$\neg \geq nR \in \mathcal{C}_{lit}$	$(\neg \geq nR)^{\mathcal{J}(K)} = \Delta^{\mathcal{J}(K)} \setminus (\geq nR)^{\mathcal{J}(K)}$
10	$\sqcap_{l \in L}l \,\|\, l \in \mathcal{C}_{lit}, \#L > 1$	$(\sqcap_{l \in L}l)^{\mathcal{J}(K)} = \{o \in \Delta^{\mathcal{J}(K)} \,\|\, L \subseteq \theta^{\mathcal{T}(K)}(o)\}$

3   Reasoning on STOC-DL-Lite Knowledge Bases

We have defined an SC-interpretation on grounds of type space $\mathcal{T}(K)$, the set of all licit types of the given knowledge base K. The computation of set $\mathcal{T}(K)$

is therefore a prerequisite for performing any reasoning task on the contents of K. We have thus devised a set of algorithms pursuing that goal:

a1: an algorithm that computes the contexts for K;
a2: an algorithm that computes $\mathcal{G}_{\Xi(K)}$, the SC-graph for K;
a3: an algorithm that computes $\mathcal{G}^{\leftrightarrow}_{\Xi(K)}$, the characteristic graph for K;
a4: an algorithm that computes $\mathcal{T}(K)$, the set of all licit types for K;

A crucial point for each of those algorithms in order to show good performance is the production of an efficient subroutine for computing closures of sets of attributes w.r.t. a set of implications. This is not a difficult point: it is well known that a linear time algorithm exists for the problem [6]. Since this algorithm does not always produce the best performance on real cases, we have adapted for the case the quadratic time algorithm CLOSURE from [5] instead, because it shows better time response in general over the one presented in [6]. In our case, we have parameterized CLOSURE with the implication relation given as a parameter. For creating sets $\Xi(K)$ and $\Xi^-(K)$, algorithm a1 performs a single pass on the axioms of K, building sets $\xi(K)$ and $\xi^-(K)$ and relations $\sqsubseteq_{(K)}$ and $\rightarrow_{(K)}$. Set $\Xi^-(K)$ is built immediately; then, for each candidate concept C with type L, performs a computation of the closure of L in order to decide consistency. Algorithm a2 for creating graph $\mathcal{G}_{\Xi(K)}$ is straightforward, since the set of nodes is set $\Xi(K)$, and it is already computed. The set of arcs $\mathcal{A}_{\Xi(K)}$ is computed, reducing the relation $\rightarrow_{(K)}$ formed in algorithm a1 to include only pairs $n_1, n_2 \in \Xi(K)$. The building of the characteristic graph in algorithm a3 is more involved: it requires the identification of the classes of equivalence induced by the arcs in the SC-graph. Algorithm a3 invokes a procedure SAME_CLASS that, for a node n and a set of nodes N not including n, returns a node $n' \in N$ s.t. n' is in the same class of n, if such a node n' exists. SAME_CLASS simply search for an n' that is reachable from n s.t. n is reachable form n'. SAME_CLASS compute closures from n and each $n' \in N$, until n' fulfills that n' belongs to the closure of n and n belongs to the closure of n'. N is initiated with the nodes in $\mathcal{N}_{\Xi(K)}$, and is reduced with each n' found. A failure of procedure SAME_CLASS closes a class. Since the intersection of sets of equivalent nodes is empty, the process can be repeated until there are no remaining nodes in N. It is easy to see, therefore, that determining the classes of equivalence of the SC-graph lies in Ptime. Then, with the set of nodes computed, a single pass on the set of arcs of the SC-graph suffices for determining the set of arcs of the characteristic graph.

Finally, algorithm a4 computes set $\mathcal{T}(K)$ from graph $\mathcal{G}^{\leftrightarrow}_{\Xi(K)}$. For determining the set of prototypes, a4 computes the closure of each node in $\mathcal{N}^{\leftrightarrow}_{\Xi(K)}$ w.r.t. the set of arcs $\mathcal{N}^{\leftrightarrow}_{\Xi(K)}$ (regarded as a relation), and checks for consistency. For deriving the SC-types, each prototype Θ is augmented in a cycle with one literal of each type $\Theta \setminus L$, L in $\Xi^-(K)$, and the resulting type check for consistency, until there is no applicable constraints left unregarded. Finally, a pruning step is performed, with base on the basic concepts that occur in $\Xi(K)$, for the restriction of minimality. It is not difficult to see that all the algorithms have a polynomial time asymptotic behavior. We thus have the following:

Theorem 1. *The computation of the set of all licit types for a STOC-DL-Lite knowledge base lies in Ptime.*

3.1 Deciding Satisfiability in STOC-DL-Lite

A procedure for deciding the satisfiability of a DL-Lite$^{\mathcal{N}}_{bool}$ knowledge base K is well known [1]. The method consists in translating the contents of knowledge base K into a first order sentence K^{τ} with at most one variable, and deciding whether the sentence has a Herbrand model. Assuming that Σ and $\mathcal{A}$ are respectively the TBox and ABox of K, $.^{\tau}$, the function that translates K, is defined as follows:

$$K^{\tau} = (\textstyle\bigwedge_{s \in \Sigma} s^{\tau}) \wedge (\textstyle\bigwedge_{R \in \mathcal{R}} \varepsilon(R) \wedge \delta^{\tau}(R)) \wedge \mathcal{A}^{\tau}, \text{ where}$$

- for $A \in \mathcal{N}_{\mathcal{C}}$, $B \in \mathcal{C}_{base}$, and $R \in \mathcal{R}$, the following terms are translated as:
  $(\perp)^{\tau} = \perp$, $A^{\tau} = A(x)$, $(\geq q\, R)^{\tau} = R_{\geq q}(x)$, $(\neg B)^{\tau} = \neg (B^{\tau})$,
- for $L \in 2^{\mathcal{C}_{lit}}$, $\#L > 1$, a complex concept is translated as:
  $(\sqcap_{l \in L}\, l)^{\tau} = \textstyle\bigwedge_{l \in L} l^{\tau}$;
- an axiom $s \in \Sigma$ of the form $\sqcap_{l \in L}\, l \sqsubseteq \sqcap_{l' \in L'}\, l'$, $\#L, \#L' > 0$, is translated as:
  $s^{\tau} = \forall x :(\sqcap_{l \in L}\, l_j)^{\tau} \Rightarrow (\sqcap_{l' \in L'}\, l')^{\tau}$;
- an axiom $s \in \Sigma$ of the form $\sqcap_{l \in L}\, l \sqsubseteq \perp$, $\#L > 0$, is translated as:
  $s^{\tau} = \forall x :(\sqcap_{l \in L}\, l)^{\tau} \Rightarrow \perp$;
- $\varepsilon(R)$ is the sentence: $\forall x : R_{\geq 1}(x) \Rightarrow R^{\sim}_{\geq 1}(sk_R)$;
- $\delta^{\tau}(R)$ is the sentence: $\forall x : R_{\geq q'}(x) \Rightarrow R_{\geq q}(x)$,
  = for each pair of naturals q and q' s.t. $q = 1$ or $\geq q\, R$ occurs in K,
  and $\geq q'\, R$ occurs in K, and $q < q'$,
  and there is no q'' s.t. $\geq q''\, R$ occurs in K and $q < q'' < q'$;
- $\mathcal{A}^{\tau}$ is the sentence: $\mathcal{A}^{\tau} = \bigwedge_{A(b) \in \mathcal{A}} A(b) \wedge \bigwedge_{R \in \mathcal{R}, a \in ob(\mathcal{A})} \geq q^{max}_{(R,a)} R(a)$,
  where $= ob(\mathcal{A})$ are all $a \in \mathcal{N}_{\mathcal{I}}$ that occur in $\mathcal{A}$, and
    $q^{max}_{(R,a)} => 0$, and $q^{max}_{(R,a)}$ is the maximum number from a set Q_R containing number 1 and all q for which a concept $\geq q\, R$ occurs in K, and
    there are $q^{max}_{(R,a)}$ many distict b with $P(a,b) \in \mathcal{A}$, for $R = P$, and
    there are $q^{max}_{(R,a)}$ many distict b with $P(b,a) \in \mathcal{A}$, for $R = P^{-}$.

Since sentence K^{τ} does not have any existentially quantified variable nor any function symbol, the Herbrand base $\mathcal{HB}(K^{\tau})$ for K^{τ} is made exclusively of the constants occurring in K^{τ}. Since we are searching for a Herbrand model, we must to search for a truth value assigment that turns **t** the conjunction of the propositional formulas $K^{\tau}_{x \mapsto c}$ resulting from substituting variable x with each constant c in $\mathcal{HB}(K^{\tau})$. This is a hard problem in general. Nonetheless, we are looking for SC-models instead of FO-models. Since each constant c is interpreted by a member o of the domain, and, according to the semantics of our logic, o has an associated licit type $\theta^{\mathcal{T}(K)}(o)$, we can systematically test the implication of sentence K^{τ} from each type in the type space $\mathcal{T}(K)$, and thus considering only the cases where value **t** is a concrete possibility. The decision procedure follows:

PROCEDURE SC-SAT$(K^\tau, \mathcal{T}(K))$:
Determine the Herbrand base $\mathcal{HB}(K)$ of K^τ, and sets $Ob(\mathcal{A})$ and $\mathcal{R}$ from K^τ
for each a in $\mathcal{HB}(K^\tau)$:
    for each Θ in $\mathcal{T}(K)$:
        if for each sub-sentence $s^\tau = \bigwedge_{l_1 \in L_1} l_1(x) \Rightarrow \bigwedge_{l_2 \in L_2} l_2(x)$:
            $L_1 \subseteq \Theta$ and $L_2 \subseteq \Theta$ hold, or $L_1 \not\subseteq \Theta$ holds, and
           for each sub-sentence $s^\tau = \bigwedge_{l \in L} l(x) \Rightarrow \bot$:
                $L \not\subseteq \Theta$ holds, and
          (there is no sub-sentence $B(a) \in \mathcal{A}^\tau$ or
           for each sub-sentence $B(a) \in \mathcal{A}^\tau$: $B \in \Theta$ holds);
        then $Type(a,\Theta).value = \mathbf{t}$, else $Type(a,\Theta).value = \mathbf{f}$;
        for each role $R \in \mathcal{R}$,
          if $\geq q\, R \in \Theta$ holds, for some $q > 0$,
          then $Type(a,\Theta).range_{R \neq \emptyset} = \mathbf{t}$, else $Type(a,\Theta).range_{R \neq \emptyset} = \mathbf{f}$;
          if $a = sk_R$ and $\geq q\, R \in \Theta$ holds, for some $q > 0$,
          then $Type(a,\Theta).dom_{R \neq \emptyset} = \mathbf{t}$, else $Type(a,\Theta).dom_{R \neq \emptyset} = \mathbf{f}$;
return $\bigwedge_{a \in \mathcal{HB}(K^\tau)} (\bigvee_{\Theta \in \mathcal{T}(K)} (Type(a,\Theta).value \wedge$
               $\bigwedge_{R \in \mathcal{R}} (Type(a,\Theta).range_{R \neq \emptyset} \Rightarrow \bigvee_{\Theta' \in \mathcal{T}(K)} Type(sk_{R\sim},\Theta').va$
                         $\wedge\ Type(sk_{R\sim},\Theta').dom_{R\sim \neq \emptyset})$

We have the following:

Theorem 2. *A STOC-DL-Lite knowledge base K has an SC-model if and only if procedure SC-SAT on sentence K^τ and set $\mathcal{T}(K)$, the set of all licit types for K returns $\mathbf{t}$.*

From the analysis of procedure SC-SAT we can distinguish two steps. In the first step, it cycles on the constants in the Herbrand base of sentence K^τ and on the types in $\mathcal{T}(K)$, filling with boolean values the component *value* and, for each role R, the components $range_{R \neq \emptyset}$ and $dom_{R \neq \emptyset}$, of the two-dimensional array $Type$. The final step decides the satisfiability, evaluating a boolean formula on the boolean contents of array $Type$, which involves a cycle on the constants in $\mathcal{HB}(K^\tau)$ with a double inner cycle on the licit types in $\mathcal{T}(K)$ and roles in $\mathcal{R}$. It is easy to see that procedure SC-SAT computes its value in a time proportional to $\# K^\tau \times \#_{lit} \mathcal{T}(K)$, $\#_{lit}$ designing in this case the number of (possibly repeated) literals occurring in the licit types. The size of sentence K^τ is linear in the size of knowledge base K. The size $\#_{lit}\mathcal{T}(K)$ is proportional to $\#\Xi(K) \times \#\Xi^-(K)$, therefore is polynomial in the size of knowledge base K. Since the computation of set $\mathcal{T}(K)$ is polynomial in the size of knowledge base K, we have the following:

Theorem 3. *Deciding if a given STOC-DL-Lite knowledge base K has an SC-model lies in Ptime.*

4   Conclusion

We have presented STOC-DL-Lite, an entity-oriented variant to logic DL-Lite$_{bool}^{\mathcal{N}}$, which allows the user to describe universal entities through ontological committed concepts given explicitly though a CBox, or derived from the contents of the TBox and ABox of a knowledge base. The semantics of the presented logic imposes a form of closeness to interpretations, ensuring that any individual in the domain of a model belongs to the interpretation of at least one committed concept. The restriction reduces the search space for models dramatically, thus turning the problem of satisfiability, which is NP-complete for the case of DL-Lite$_{bool}^{\mathcal{N}}$, tractable in the case of our variant.

References

1. Artale, A., Calvanese, D., Kontchakov, R., Zakharyaschev, M.: DL-lite in the light of first-order logic. In: 22nd AAAI Conference on Artificial Intelligence, pp. 361–366. AAAI Press (2007)
2. Artale, A., Calvanese, D., Kontchakov, R., Zakharyaschev, M.: The DL-lite family and relations. J. Artif. Intell. Res. (JAIR) **36**, 1–69 (2009)
3. Baader, F., Ganter, B., Sertkaya, B., Sattler, U.: Completing description logic knowledge bases using formal concept analysis. In: Veloso, M.M. (ed.) 20th IJCAI International Joint Conference on Artificial Intelligence, pp. 230–235 (2007)
4. Balaban, M., Maraee, A.: A UML-based method for deciding finite satisfiability in description logics. In: Baader, F., Lutz, C., Motik, B. (eds.) DL 2008. CEUR Workshop Proceedings, vol. 353. CEUR-WS.org (2008)
5. Bazhanov, K., Obiedkov, S.A.: Optimizations in computing the duquenne-guigues basis of implications. Ann. Math. Artif. Intell. **70**(1–2), 5–24 (2014)
6. Beeri, C., Bernstein, P.A.: Computational problems related to the design of normal form relational schemas. ACM Trans. Database Syst. **4**(1), 30–59 (1979)
7. Bonatti, P.A., Lutz, C., Wolter, F.: The complexity of circumscription in DLs. J. Artif. Intell. Res. (JAIR) **35**, 717–773 (2009)
8. Civili, C., Console, M., De Giacomo, G., Lembo, D., Lenzerini, M., Lepore, L., Mancini, R., Poggi, A., Rosati, R., Ruzzi, M., Santarelli, V., Savo, D.F.: MASTRO STUDIO: managing ontology-based data access applications. PVLDB **6**(12), 1314–1317 (2013)
9. Ganter, B., Stumme, G., Wille, R. (eds.): Formal Concept Analysis. LNCS (LNAI), vol. 3626. Springer, Heidelberg (2005)
10. Gries, O.: Generalized closed world reasoning in description logics with extended domain closure. In: Cuenca Grau, B., Horrocks, I., Motik, B., Sattler, U. (eds.) DL 2009. CEUR Workshop Proceedings, vol. 477. CEUR-WS.org (2009)
11. Grimm, S., Motik, B.: Closed world reasoning in the semantic web through epistemic operators. In: Cuenca Grau, B., Horrocks, I., Parsia, B., Patel-Schneider, P.F. (eds.) OWLED 2005. CEUR Workshop Proceedings, vol. 188. CEUR-WS.org (2005)
12. Khouri, S., Bellatreche, L.: DWOBS: data warehouse design from ontology-based sources. In: Yu, J.X., Kim, M.H., Unland, R. (eds.) DASFAA 2011, Part II. LNCS, vol. 6588, pp. 438–441. Springer, Heidelberg (2011)
13. Knublauch, H., Fergerson, R.W., Noy, N.F., Musen, M.A.: The protégé OWL plugin: an open development environment for semantic web applications. In: McIlraith, S.A., Plexousakis, D., van Harmelen, F. (eds.) ISWC 2004. LNCS, vol. 3298, pp. 229–243. Springer, Heidelberg (2004)

14. Krisnadhi, A.A., Sengupta, K., Hitzler, P.: Local closed world semantics: grounded circumscription for description logics. In: Rudolph, S., Gutierrez, C. (eds.) RR 2011. LNCS, vol. 6902, pp. 263–268. Springer, Heidelberg (2011)

15. Obrst, L., Grüninger, M., Baclawski, K., Bennett, M., Brickley, D., Berg-Cross, G., Hitzler, P., Janowicz, K., Kapp, C., Kutz, O., Lange, C., Levenchuk, A., Quattri, F., Rector, A., Schneider, T., Spero, S., Thessen, A., Vegetti, M., Vizedom, A., Westerinen, A., West, M., Yim, P.: Semantic web and big data meets applied ontology - the ontology summit 2014. Appl. Ontology **9**(2), 155–170 (2014)

16. Quine, W.V.: Ontology and ideology. Philos. Stud. **2**, 11–15 (1951)

17. Rayo, A.: Ontological commitment. Philos. Compass **2**(3), 428–444 (2007)

18. Ren, Y., Pan, J.Z., Zhao, Y.: Soundness preserving approximation for TBox reasoning. In: Fox, M., Poole, D. (eds.) 24th AAAI Conference on Artificial Intelligence. AAAI Press (2010)

19. Wille, R.: Implicational concept graphs. In: Wolff, K.E., Pfeiffer, H.D., Delugach, H.S. (eds.) ICCS 2004. LNCS (LNAI), vol. 3127, pp. 52–61. Springer, Heidelberg (2004)

SeeCOnt: A New Seeding-Based Clustering Approach for Ontology Matching

Alsayed Algergawy[1,2(✉)], Samira Babalou[3],
Mohammad J. Kargar[3], and S. Hashem Davarpanah[3]

[1] Institute of Computer Science, Friedrich Schiller University of Jena, Jena, Germany
alsayed.algergawy@uni-jena.de
[2] Department of Computer Engineering, Tanta University, Tanta, Egypt
[3] Department of Computer Engineering,
University of Science and Culture, Tehran, Iran

Abstract. Ontology matching plays a crucial role to resolve semantic heterogeneities within knowledge-based systems. However, ontologies contain a massive number of concepts, resulting in performance impediments during the ontology matching process. With the increasing number of ontology concepts, there is a growing need to focus more on large-scale matching problems. To this end, in this paper, we come up with a new partitioning-based matching approach, where a new clustering method for partitioning concepts of ontologies is introduced. The proposed method, called *SeeCOnt*, is a seeding-based clustering technique aiming to reduce the complexity of comparison by only using clusters' seed. In particular, *SeeCOnt* first identifies and determines the seeds of clusters based on the highest ranked concepts using a distribution condition, then the remaining concepts are placed into the proper cluster by defining and utilizing a membership function. The *SeeCOnt* method can improve the memory consuming problem in the large-scale matching problem, as well as it increases the matching quality. The experimental evaluation shows that *SeeCOnt*, compared with the top ten participant systems in OAEI, demonstrates acceptable results.

Keywords: Ontology matching · Clustering techniques · Large-scale matching

1 Introduction

Ontology is the main backbone of the Semantic Web, which provides facilities for integration, searching, and sharing of information on the web through making those information understandable for machines [14]. Despite this crucial role and due to the engineering of ontologies by different people or methods even if they are created for the same domain, there exist different sorts of heterogeneities. Semantic heterogeneity is a common and key problem in different knowledge-based systems [5,6]. To obtain meaningful interoperation, one needs a semantic mapping among ontologies. To cope with the semantic heterogeneity problem, a

© Springer International Publishing Switzerland 2015
T. Morzy et al. (Eds.): ADBIS 2015, LNCS 9282, pp. 245–258, 2015.
DOI: 10.1007/978-3-319-23135-8_17

set of correspondences that identify similar concepts across different ontologies have to be achieved through ontology matching. The construction of manual match is an error-prone and labor intensive task that requires complete knowledge of the semantics of the data in ontologies being matched. Solutions that try to provide some automatic support for ontology matching have received steady attention over the years [3,8,19,21].

Nowadays, there is a natural evolution of data, and consequently, there exist many complex and large-scale ontologies in the real domains. For example, the Foundational Model of Anatomy (FMA), SNOMED CT, and the National Cancer Institute Thesaurus (NCI) ontologies are semantically rich and contain tens of thousands of concepts[1]. However, to process these large-scale ontologies, the existing ontology matching tools have some problems, such as shortage of consumed memory and/or long time consumption [21]. For example, the OAEI campaign started in 2006 including the anatomy matching task as an evaluation criterion for large-scale matching. The anatomy matching task is to match the Adult Mouse Anatomy (2744 concepts) and the NCI Thesaurus (3304 concepts). However, in 2011, only 6 of 16 systems could process those ontologies. With the increasing number of concepts typical ontologies have, the OAEI campaign included a new match track, called *Large Biomedical Ontologies*. In 2014, only three matching systems could complete the total matching tasks in this track [22].

In order to enable matching of large-scale ontologies, dividing the ontologies into a set of partitions is a way which has been proposed so far via the methods such as divide and conquer [15], clustering [1], and modularization [23]. We argue that partitioning input ontologies plays a central role towards building an effective and efficient matching system. To this end, in this paper, we introduce a seeding-based clustering approach for partitioning ontologies, called *SeeCOnt*. In particular, input ontologies are parsed and represented as concept graphs. We then develop a *Ranker* function to rank ontology concepts exploiting concept graph features. The highest ranked concepts are then selected to constitute the cluster seeds. To assign remaining concepts to their proper clusters, we introduce a membership function. We demonstrate that *SeeCOnt* reduces the complexity of the comparisons by comparing concepts with only seeds instead of all the other concepts. Finally, we adapt the Falcon-AO matching system to apply our new approach. To validate the proposed approach, we conducted a set of experiments utilizing different data sets from OAEI. The experimental results display that *SeeCOnt* achieved acceptable performance compared with the top-ten matching systems participating recently in OAEI.

The rest of the paper is structured as follows. Related work is presented in Sect. 2. We describe the proposed approach Sect. 3. We report experiments conducted and analysis results in Sect. 4. Section 5 concludes the paper.

---

[1] http://www.cs.ox.ac.uk/isg/projects/SEALS/oaei/2014/.

2   Related Work

Because its importance, several approaches have been proposed to deal with the problem of matching two large ontologies [1,4,13,15,18,20,23]. Promising areas for large-scale matching lie in four main directions: reduction of search space for matching, parallel matching, self-tuning match workflows and reuse of previous match results [18]. In this section, we pay attention to the approaches that perform reduction of the search space. The standard approach of cross join evaluation for ontology matching reduces the matching quality and matching efficiency. In order to reduce the search space for matching, two methods can be used; early pruning of dissimilar element pairs and partition-based matching.

In general, partitioning-based matching has three main stages: (i) partitioning each input ontology into a set of sub-ontologies and determining similar partitions between two sets to form a matching task, (ii) applying a matching method to each matching task to produce a set of partial match results, and (iii) combining partial results to get the final match result. In the following we present a set of matching systems that follow this architecture.

Quick ontology matching (QOM) was one of the first approaches to implement the idea of early pruning of dissimilar element pairs [7]. It iteratively applies a sequence of matchers and can restrict the search space for every matcher. COMA++ was one of the first systems to support partition-based schema matching [4]. It depends on fragment matching which determines fragments of the two schemas and identifies the most similar ones.

Another matching system that supports partition-based matching is Falcon-AO [15]. It initially partitions the ontologies into relatively small disjoint blocks by using structural clustering. Then, matching is applied to the most similar blocks from the two ontologies. Dynamic partition-based matching is supported by AnchorFlood [20]. It avoids the a-priori partitioning of the ontologies by utilizing anchors (similar concept pairs). It takes them as a starting point to incrementally match elements in their structural neighborhood until no further matches are found or all elements are processed. Thus the partitions (segments) are located around the anchors.

Zhong et al. propose an unbalanced ontology matching approach, which concerns with matching a lightweight ontology with a more heavyweight one [25]. Algergawy et al. uses a clustering-based matching approach that is based on an agglomerative bottom-up hierarchical fashion [1]. The clustering scheme is performed based on the context-based structural node similarities. Then, a light weight linguistic technique is used to find similar partitions to match.

3   SeeCOnt

To cope with matching large ontologies, we present a new seed-based clustering approach, called *SeeCOnt*. As shown in Fig. 1, *SeeCOnt* consists of three components: *preprocessing*, *ranking*, and *clustering*. First, input ontologies are parsed and represented internally as labelled directed graphs, called *concept graphs.* During the preprocessing step, the number of cluster heads ($\mathcal{CH}$) is to be determined. To quantify the importance of a concept in the concept graph, we introduce a new function called *Ranker* exploiting the concept graph features. Finally, remaining concepts are placed into their corresponding clusters according to a proposed membership function. The outline of the *SeeCOnt* approach is shown in Algorithm 1. In the following sections, we portray the description of each phase of the algorithm.

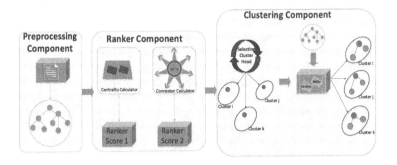

Fig. 1. Architecture of the *SeeCOnt* method.

3.1   Preprocessing

First input ontologies are parsed and inferred by Apache Jena[2] and then the concept graph is drawn by mapping the inferred result. We define *concept graph* $\mathcal{G} = (\mathcal{C}, \mathcal{R}, \mathcal{L})$ as a labeled directed graph. $\mathcal{C} = \{c_1, c_2, ..., c_n\}$ is a finite set of nodes presenting the concepts of the ontology. $\mathcal{R} = \{r_1, r_2, ..., r_m\}$ stands for a finite set of directed edges showing various relationships between concepts in an ontology $\mathcal{O}$, such that $r_k \in \mathcal{R}$ represents a directed relation between two adjacent concepts $c_i, c_j \in \mathcal{C}$. $\mathcal{L}$ is a finite set of labels of graph nodes defining the properties of each concept, such as the names of concepts. $n(= |\mathcal{C}|)$ and m are the number of nodes (concept) and edges (relationship) in $\mathcal{G}$, respectively. Given the number of concepts in an ontology, the number of cluster heads (i.e. $\mathcal{K}$) can be computed according to the following equation.

$$\mathcal{K} = \frac{|\mathcal{C}|}{\epsilon} \tag{1}$$

where ϵ is the maximum size of each cluster ($\epsilon < |\mathcal{C}|$) and should be set by an expert depending on the number of concepts.

Once each ontology is represented as a concept graph, the next step is to partition concepts, $\mathcal{C}$, of each graph into a set of separate (disjoint) clusters $\mathcal{T}_1, \mathcal{T}_2, ..., \mathcal{T}_k$ such that the cohesion of nodes in one cluster should be high, while the coupling of two clusters is low.

---

[2] https://jena.apache.org/.

Algorithm 1. Seeding-based clustering algorithm

Require: An ontology $\mathcal{O}$, a parameter ϵ limiting the maximum number of concepts in a cluster

Ensure: A set of clusters, $\mathcal{T} = \{\mathcal{T}_1, \mathcal{T}_2, ..., \mathcal{T}_k\}$

    {// **Phase 1: Preprocessing**}

1: $\mathcal{T} \Leftarrow \emptyset$;

2: $\mathcal{CG} \Leftarrow parse(\mathcal{O})$;

3: $\mathcal{K} \Leftarrow \frac{|\mathcal{CG}|}{\epsilon}$;

    {// **Phase 2: Concept Ranking**}

4: **for** $c_i \in C$ **do**

5:     $score_{c_i} \Leftarrow compute\_Ranker\_score(c_i)$;

6: **end for**

7: $\mathcal{CH} \Leftarrow select\_top\_rank(C)$;

    {//**Phase 3: Clustering**}

8: initialize each cluster with each $\mathcal{CH}$ node;

9: add each direct concept of $\mathcal{CH}$ to each cluster;

10: **for** $non - clustered\ c_i \in C$ **do**

11:     $max\_sim \Leftarrow 0$;

12:     **for** $c_j \in \mathcal{CH}_k$ **do**

13:         $sim_{ij} \Leftarrow MemFun(c_i, c_j)$;

14:         **if** $sim_{ij} \geq max\_sim$ **then**

15:             $max\_sim \Leftarrow sim_{ij}$;

16:             $concept\_place \Leftarrow j$;

17:         **end if**

18:     **end for**

19:     $Clust.put(C_i, \mathcal{CH}_j)$

20: **end for**

3.2 Concepts Ranking

The seeding-based clustering algorithm starts by selecting a set of nodes distinguishing as *important nodes*. These nodes are then selected to be cluster heads, $\mathcal{CH}$. In order to identify a node as an important one, we should first quantify its role in the concept graph. To this end, we introduce a new function, called *Ranker*. This function should be as simple as possible but effective. I.e. the *Ranker* function should not consume much time, however, correctly rank concepts inside an ontology, given that we deal with the large-scale matching problem.

Ranker Function. The importance of a node in a concept graph is understandable through the node itself and its surroundings [11]. This matter leads us to use graph-theoretic measures based on graph connections in the *Ranker* function. In the following, we present two different implementation of the *Ranker* function. The first is based on the *centrality* measure of a concept, while the second depends on the context of the concept.

First Rank Function. The definition of "centrality" measure of a concept in a concept graph is derived from the social network analysis [9]. Each person is given a score based on his or her position at the network showing the importance of each individual. To consider the effect of the concept itself through its edges, we use a set of centrality measures, as given below.

- Degree Centrality ($C1$): This measure is the simplest measure that calculates the number of connections of a node. In a directed graph, there is an in-degree and out-degree centrality that calculates the number of input and output links, respectively. The relationships between nodes can be considered as a power source during concept ranking; nodes with high degree of centrality are certainly more prominent than the others, since they receive a great deal of power [16].
- Closeness Centrality ($C2$): This measure shows the importance of the close nodes to the others in the graph. In this measure, reaching cost of a node to the others is measured [16].
- Betweenness Centrality ($C3$): This measure is considered the most relevant in that context. It consists in computing on each node the fraction of shortest paths that pass through it [10].
- EcCentrality ($C4$): This measure calculates the maximum distance between pairs of nodes. The intuition is that one node is the central if no node is far from it [12].
- Stress Centrality ($C5$): This measure calculates the absolute number of the shortest paths through a node [17].

A summary of these centrality measures and their descriptions are shown in Table 1.

Table 1. Different Centrality Measures.

No.	Name	Formula	Description
1	Degree Centrality	$C1(c_i) = degreeCentrality(c_i)$	-
2	Closeness Centrality	$C2(c_i) = \frac{1}{\sum_{c_j \in C} distance(c_i, c_j)}$	distance (c_i, c_j) function is the shortest path between i and j nodes in the graph
3	Betweeness Centrality	$C3(c_i) = \sum_{s,t \in c_i} \frac{\sigma_{s,t}(c_i)}{\sigma_{s,t}}$	$\sigma_{s,t}(c_i)$ is the number of shortest paths from s to t through c_i, and $\sigma_{s,t}$ is the total number of shortest paths from s to t
4	EcCentrality	$C4(c_i) = \frac{1}{\max_{c_j \in C} distance(c_i, c_j)}$	-
5	Stress Centrality	$C5(c_i) = \sum_{s,t \in c_i} \sigma_{s,t}(c_i)$	$\sigma_{s,t}(c_i)$ represents the number of the shortest paths from s to t via c_i

The arising question now is which centrality measure(s) should be used to implement our *Ranker* function. During the selection process we need to optimize between two criteria: an accurate and fast measure. To this end, and based on our experimental results shown later, we select the combination of the *Degree*, *C1* and *closeness*, *C2* centrality measures. As a result we can formulate the score of the first *Ranker* function for a given concept, c_i, as below:

$$Ranker\_score_1(c_i) = C1(c_i) + C2(c_i) \qquad (2)$$

Second Rank Function. During the employment of the first *Ranker* function, we observed that it is an effective measure but it requires a lot of time to rank concepts. This makes it unsuitable for matching large ontologies. Therefore, we propose another rank function that should be more applicable to large-scale matching. First, we introduce the definition of the concept *connexion set*, then show how to use this set to determine the importance of the concept.

Definition 1. *Given a concept graph* $\mathcal{G} = (\mathcal{C}, \mathcal{R}, \mathcal{L})$, *the connexion set of a concept* $c_i \in \mathcal{C}$ *is defined as:* $\Psi(c_i, d) = \{SubClass(c_i, d) \cup SuperClass(c_i, d)\}$

where $\Psi(c_i, d)$ is a set in which all the concepts with d levels that effect on c_i node. $SubClass(c_i, d)$ is the children of c_i with d hierarchical levels, and $SuperClass(c_i, d)$ is the parents of c_i with d hierarchical levels. It is evident that the importance of a concept increases as it has a larger number of surroundings. Based on this we propose the following score function that can be used to rank concepts of an ontology.

$$Ranker\_score_2(c_i) = |\Psi(c_i, d)| \qquad (3)$$

Determining Cluster Heads. Once computing the importance of concepts of a concept graph, the next step is to select which concepts represent cluster heads, $\mathcal{CH}$. If simply the nodes with the highest score are selected as the cluster heads, distribution would be disregarded. To avoid this problem, the distance between two cluster heads is measured, and among the highest score nodes, those with a minimum distance of d from each other are selected as the cluster heads. For this purpose, we adopt the *Connexion* set with d levels defined before.

3.3 Finalizing Clustering

At first, the *SeeCOnt* algorithm creates one cluster for each cluster head. Then, it places direct children in the corresponding cluster and finally, for remaining nodes, a membership function is used to determine the cluster of each node. In general, clustering is done through the following three steps:

- *Seeding*: Creating a cluster for each cluster head, *Algorithm 1, line 8*.
- *Direct Spread*: Assigning direct children of each cluster head to the corresponding cluster, *Algorithm 1, line 9*.
- *In-Direct Spread*: Calling a membership function for the remaining nodes, *Algorithm 1, lines 10–18*.

The *direct Spread step* reduces the time complexity, since the number of comparisons will be reduced as well as applying the membership function for all nodes is time consuming. While by placing the nodes via the call of the membership function, the same results would still be achieved.

Membership Function. Once determining cluster heads, ($\mathcal{CH}$), and assigning direct children to their proper heads, the next step is to place remaining concepts into their fitting cluster. To this end, we develop a membership function, *MemFun*. First, each concept is associated with a flag, $\mathcal{F}$, such that if the $\mathcal{F}$ of c concept is false, it means c is not assigned to any cluster and thus, the membership function is called for the concept c. In addition, the $\mathcal{F}$ flag can only have one value, i.e. each node can be placed in only one cluster so that no overlap is observed in clusters. The membership function determines that each concept $c_i \in \mathcal{C}$ should be placed in which $\mathcal{T}_i, i < \mathcal{K}$ cluster. For this, the similarity of c_i with all $\mathcal{CH}_i$ is calculated and then c_i is placed in a cluster with the maximum similarity value. Using the proposed membership function, each concept is compared with Cluster Heads, instead of comparing with all concepts like whatever was done in [1,15], which reduces the complexity of comparison.

In order to measure the membership of a concept to a cluster head, a linear weighted combination of the following structural and semantic similarity measures is calculated as in the following equation:

$$MemFun(c_i, \mathcal{CH}_k) = \alpha \times SNSim(c_i, \mathcal{CH}_k)$$
$$+ (1 - \alpha) \times SemSim(c_i, \mathcal{CH}_k) \quad (4)$$

where α is constant between 0 and 1 to reflect the importance of each similarity measure, $ShareNeighbors(SNSim)$ and semantic similarity $SemSim$ are two similarity measures that quantify the structural properties of the concept c_i, respectively.

Shared Neighbors. This measure considers the number of shared neighbors between c_i and $\mathcal{CH}_k$. The shared neighbour measure plays an important role in structural similarity, because similar concepts have similar neighbors [2,24]. The neighbors of a concept are the concept's children, concept's parents, concept's siblings and the concept itself. In our implementation, we determine the neighbors of the concept c_i and the neighbors of the cluster head $\mathcal{CH}_k$, then determine how many concepts are common between these two sets.

$$SNSim(c_i, \mathcal{CH}_k) = \frac{|SN_{c_i} \cap SN_{\mathcal{CH}_k}|}{|SN_{c_i} \cup SN_{\mathcal{CH}_k}|} \quad (5)$$

where SN_{c_i} and $SN_{\mathcal{CH}_k}$ are the neighbor sets of the concept c_i and the cluster head $\mathcal{CH}_k$, respectively.

Hierarchical Semantic Similarity. It is evident that a higher semantic similarity implies a stronger semantic connection, so we first calculate the semantic similarities between the concept c_i and the cluster head $\mathcal{CH}_k$. The most classic semantic similarity calculation is based on the concept hierarchy. The hierarchy semantic similarity between c_i and $\mathcal{CH}_k$ can be defined as below:

$$SemSim(c_i, \mathcal{CH}_k) = \frac{2 \times N_3}{N_1 + N_2 + 2 \times N_3} \tag{6}$$

where N_1 and N_2 are the numbers of sub-concept relations from c_i and $, \mathcal{CH}_k$ to their most specific common superconcept C, and N_3 is the number of sub-concept relations from C to the root of the concept hierarchy.

Matching. Once settling on the similar clusters of the two ontologies, the next step is to fully match similar clusters to obtain the correspondences between their elements. Each pair of the similar clusters represents an individual match task that is independently solved. Match results of these individual tasks are then combined in'to a single mapping, which represents the final match result. We adopt the Falcon matching system [15] to perform this task.

4   Experimental Evaluation

To develop *SeeCOnt*, the open source Falcon-AO system[3] was used. It was implemented in Java with Apache 2.0 license. Falcon-AO has some components including PBM (Partition Block Match) [24]. PBM is used for large-scale ontology matching which was replaced by the *SeeCOnt*. All the experiments were carried out on Intel core i5 with 4 GB internal memory on Windows 7 with Java compiler 1.7. Ontologies were parsed using Jena Apache, and the mapping functions were implemented by Alignment API[4].

4.1   Data Set

We tested with two common data sets from the OAEI[5]: Conference and Anatomy. The conference data set containing 16 ontologies is much used in ontology matching evaluation. The Anatomy data set contains two ontologies of human and mouse anatomy with 3306 and 2746 concepts, respectively.

4.2   Evaluation Criteria

In our implementation, we attempt to get answers to the following questions:

- Which centrality measure should be used to implement the first *Ranker* function?
- Which ranker function should be used to implement the ranker component?

---

[3] http://ws.nju.edu.cn/falcon-ao.
[4] http://alignapi.gforge.inria.fr.
[5] http://oaei.ontologymatching.org.

– What is the relative performance of the *SeeCOnt* approach w.r.t. recent matching approaches?

In order to answer these questions, we carried out sets of experimental evaluations. In the following, we report on the answers of these questions.

4.3   Experimental Results

Centrality Measure Evaluation. We conducted the first set of experiments to decide which centrality measure(s) should be used to implement the first *Ranker* function. To this end, we performed an evaluation using three different ontologies: *Linkling*, *MICRO*, and *cmt* from the Conference dataset. In this set, all 32 combinations of the five centrality measures were assessed. We asked a number of experts to select the top ten important concepts while we did not say anything about our criteria to them. Due to differences between important concepts by experts, we selected the most common shared important concepts. The results of 32 combinations of these criteria on Linkling ontology are shown in Fig. 2, where each bar is dedicated to one combination of different criteria, C1 refers to Degree Centrality, C2 is Clossness Centrality, C3 is Betweenness Centrality, C4 is EcCentrality, and C5 is Stress Centrality. In each test we use one combination of C1–C5 criteria and select top ten important concepts, we also examine how many of these criteria are similar to expert judge. The test examines which combination was more similar to the experts view. Based on our criteria, different sets of combinations could achieve these criteria, such as the combinations C1+C2, C1+C4, C1+C2+C5, C1+C2+C3+C4 are more similar to whatever experts think. From these combinations, we selected C1+C2 because it outperforms the other combinations.

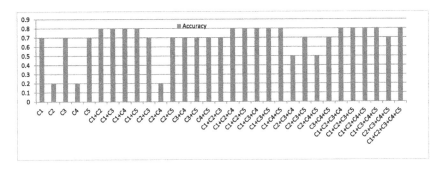

Fig. 2. Accuracy of combining centrality measures on *"Linkling"* ontology.

Selection of Ranker Function. In the previous experiments, we observed that the centrality measures effectively quantify the importance of ontology concepts. However, they consume much time, which makes using them for large-scale matching impractical. Therefore, we conducted another set of experiments to recognize which ranker function is suitable to the large matching problem. To this end, we compare the two ranker functions, implemented in Eqs. 2 and 3, respectively. We used the conference and anatomy ontologies for this test. Results are reported in Table 2. The table shows that the second ranker function outperforms the first one w.r.t. both matching quality and the time needed to complete the ranking process. This can be explained as the second ranker function exploits the concept connections which mostly reflects the importance of the concept without going into much details through computing the importance score as in the first ranker function. Therefore, we settle on the selection of the second ranker function to implement the ranker component.

Table 2. Comparing two ranker functions.

	ontology	First ranker	Second ranker
matching quality	conference	0.609	0.624
time (sec.)	anatomy/mouse	269.6	0.302
	anatomy/nci	355.8	0.57

***SeeCOnt* Quality.** This set of experiments has been conducted to validate the effectiveness and quality of the *SeeCOnt* approach. To this end, we use ontologies from the Conference and Anatomy data sets comparing *SeeCOnt* with Falcon-AO [15] and work done in [1]. Results are presented in Figs. 3 and 4. Figure 3 demonstrates that *SeeCOnt* produces higher precision, recall, and F-measure than the original Falcon-AO system. It could improves the quality of matching on the conference track by 7 % compared to the original Falcon system. However, as shown in Fig. 4, even if the *SeeCOnt* approach produces lower precision than the precision in both Algergawy' approach [1] and the original Falcon approach [15], however it achieves higher recall than the other two systems. This results in the F-measure produced by *SeeCOnt* is higher than the Algergawy' approach by 11 % and Falcon-AO by 14 %. These results demonstrate that our proposed

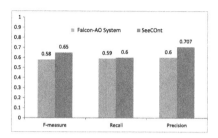

Fig. 3. Results for conference track.

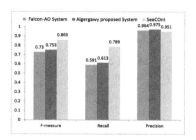

Fig. 4. Results for anatomy track.

seed-based clustering is capable of grouping similar concepts in one partition to be fully matched with another partition containing also similar concepts.

Figures 5 and 6 compare *SeeCOnt* with the top-ten matching systems, participating in OAEI competition held in 2011–2014, in the Conference test and in the Anatomy test, respectively. For simplicity of the chart, only F-Measure of each system is shown in Figs. 5 and 6. The horizontal axis shows the participating systems and the vertical axis shows F-measure. We see that *SeeCOnt* approach has comparable results with the others.

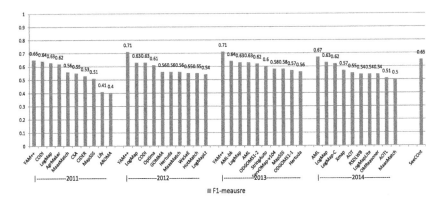

Fig. 5. Comparing *SeeCOnt* with top-ten systems Participating in OAEI Competitions in 2011–2014 in the Conference Test.

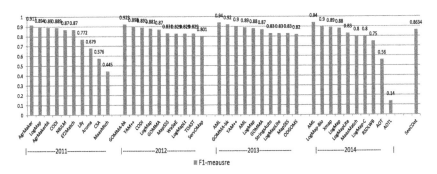

Fig. 6. Comparing *SeeCOnt* with top-ten Systems Participating in OAEI Competitions in 2011–2014 in the Anatomy Test.

5   Conclusions

In this paper, we introduced a new clustering approach, *SeeCOnt*, to be used within the context of matching large ontologies. *SeeCOnt* partitions a large-scale ontology to several disjoint sub-ontologies and the problem of large-scale ontology matching is converted into a set of small ontology matching tasks. Firstly, we represented input ontologies as concept graphs. We then introduced two different *Ranker* functions that can be used to quantify the importance of a concept in the graph. The highly important concepts are selected to be cluster heads. We further developed a new membership function that assign remaining concepts to their proper clusters. This membership function reduces the number of comparisons since it only compares each concepts to each cluster head. To validate *SeeCOnt*, we conducted an intensive set of experiments using the Conference and Anatomy data sets. We compared our proposed approach with recent matching systems participating in OAEI. Experimental results show that *SeeCOnt* presents acceptable performance. In the future, we plan to extend our work by looking for new strategies that determine and identify similar clusters and that match those similar partitions in parallel.

Acknowledgments. A. Algergawy'work is partly funded by DFG in the INFRA1 project of CRC AquaDiva.

References

1. Algergawy, A., Massmann, S., Rahm, E.: A clustering-based approach for large-scale ontology matching. In: Eder, J., Bielikova, M., Tjoa, A.M. (eds.) ADBIS 2011. LNCS, vol. 6909, pp. 415–428. Springer, Heidelberg (2011)
2. Algergawy, A., Nayak, R., Saake, G.: Element similarity measures in XML schema matching. Inf. Sci. **180**(24), 4975–4998 (2010)
3. Bellahsene, Z., Bonifati, A., Rahm, E.: Schema Matching and Mapping. Springer, Heidelberg (2011)
4. Do, H.H., Rahm, E.: Matching large schemas: approaches and evaluation. Inf. Syst. **32**(6), 857–885 (2007)
5. Doan, A., Halevy, A.: Semantic integration research in the database community: A brief survey. AAAI AI Mag. **25**(1), 83–94 (2005)
6. Doan, A., Halevy, A.Y., Ives, Z.G.: Principles of Data Integration. Morgan Kaufmann, USA (2012)
7. Ehrig, M., Staab, S.: QOM – quick ontology mapping. In: McIlraith, S.A., Plexousakis, D., van Harmelen, F. (eds.) ISWC 2004. LNCS, vol. 3298, pp. 683–697. Springer, Heidelberg (2004)
8. Euzenat, J., Shvaiko, P.: Ontology Matching, 2nd edn. Springer, Heidelberg (2013)
9. Freeman, L.C.: Centrality in social networks conceptual clarification. Soc. Netw. **1**(3), 215–239 (1979)
10. Freeman, L.C.: A set of measures of centrality based on betweenness. Sociometry **40**(1), 35–41 (1997)
11. Graves, A., Adali, S., Hendler, J.: A method to rank nodes in an RDF graph. In: 7th International Semantic Web Conference (Posters and Demos) (2008)
12. Hage, P., Harary, F.: Eccentricity and centrality in networks. Soc. Netw. **17**, 57–63 (1995)
13. Hamdi, F., Safar, B., Reynaud, C., Zargayouna, H.: Alignment-based partitioning of large-scale ontologies. In: Guillet, F., Ritschard, G., Zighed, D.A., Briand, H. (eds.) Advances in Knowledge Discovery and Management. SCI, vol. 292, pp. 251–269. Springer, Heidelberg (2010)

14. Hendler, J.: Agents and the semantic web. IEEE Intell. Syst. J. **16**, 30–37 (2001)
15. Hu, W., Qu, Y., Cheng, G.: Matching large ontologies: A divide-and-conquer approach. DKE **67**, 140–160 (2008)
16. Kermarrec, A.-M., Merrer, E.L., Sericola, B., Trdan, G.: Second order centrality: Distributed assessment of nodes criticity in complex networks. Comput. Commun. **34**, 619–628 (2011)
17. Koschützki, D., Lehmann, K.A., Peeters, L., Richter, S., Tenfelde-Podehl, D., Zlotowski, O.: Centrality indices. In: Brandes, U., Erlebach, T. (eds.) Network Analysis. LNCS, vol. 3418, pp. 16–61. Springer, Heidelberg (2005)
18. Rahm, E.: Towards large-scale schema and ontology matching. In: Bellahsene, Z., Bonifati, A., Rahm, E. (eds.) Data-Centric Systems and Applications, vol. 5258, pp. 3–27. Springer, Heidelberg (2011)
19. Rahm, E., Bernstein, P.A.: A survey of approaches to automatic schema matching. VLDB J. **10**(4), 334–350 (2001)
20. Seddiquia, M.H., Aono, M.: An efficient and scalable algorithm for segmented alignment of ontologies of arbitrary size. Web Semant. **7**(4), 344–356 (2009)
21. Shvaiko, P., Euzenat, J.: Ontology matching: State of the art and future challenges. IEEE Trans. Knowl. Data Eng. **25**(1), 158–176 (2013)
22. Shvaiko, P., Euzenat, J., Mao, M., Jimnez-Ruiz, E., Li, J., Ngonga, A.: editors. 9th International Workshop on Ontology Matching collocated with the 13th International Semantic Web Conference (ISWC 2014) (2014)
23. Wang, Z., Wang, Y., Zhang, S.-S., Shen, G., Du, T.: Matching large scale ontology effectively. In: Mizoguchi, R., Shi, Z.-Z., Giunchiglia, F. (eds.) ASWC 2006. LNCS, vol. 4185, pp. 99–105. Springer, Heidelberg (2006)
24. Hu, W., Zhao, Y., Qu, Y.: Partition-based block matching of large class hierarchies. In: Mizoguchi, R., Shi, Z.-Z., Giunchiglia, F. (eds.) ASWC 2006. LNCS, vol. 4185, pp. 72–83. Springer, Heidelberg (2006)
25. Zhong, Q., Li, H., Li, J., Xie, G.T., Tang, J., Zhou, L., Pan, Y.: A Gauss function based approach for unbalanced ontology matching. In: the ACM SIGMOD International Conference on Management of Data, (SIGMOD 2009), pp. 669–680 (2009)

Time Series Processing

ForCE: Is Estimation of Data Completeness Through Time Series Forecasts Feasible?

Gregor Endler[(✉)], Philipp Baumgärtel, Andreas M. Wahl, and Richard Lenz

Computer Science 6 (Data Management),
Friedrich-Alexander-Universität Erlangen-Nürnberg, Erlangen, Germany
{gregor.endler,philipp.baumgaertel,andreas.wahl,richard.lenz}@fau.de
https://www6.cs.fau.de

Abstract. Measuring the completeness of a data population often requires either expert knowledge or the presence of reference data. If neither is available, measuring population completeness becomes nontrivial. We present the ForCE approach (Forecasting for Completeness Estimation), a method to estimate the completeness of timestamped data using time series forecasting. We evaluate the method's feasibility using a medical domain real-world dataset, which we provide for download. The method is compared to three baselines. ForCE manages to surpass all three.

Keywords: Data quality · Population completeness · Time series · Forecasting

1 Introduction

Data quality is an important concern in all application domains of databases [9]. Quality of data is multidimensional: Many aspects are differentiated in the literature [1,18–20]. One of these dimensions is population completeness [15,19], which evaluates whether all entities of a population are represented within the database under scrutiny.

Measuring it is often difficult [6,13]. In rare cases, it can be guaranteed that every real-world entity has a related record within the database. If this is the case, and with the additional precondition that the semantics of occurring *null*-values [21] are known, population completeness can be measured by counting *null*-values. As an example, consider a school's information system: Manual entry and verification ensure that every pupil has a corresponding tuple in the database. The school year's final grades however are not entered at first and their column contains *null*-values. Before the end of the school year, these entries are not incomplete, as the final grades do not yet exist. After the school year however, *nulls* indicate missing values. By counting these, the completeness of the population of final grades can be measured.

The original version of this chapter was revised: The authors corrected errors in the figures appearing in Sect. 3.2 and the Appendix and adjusted the text referring to the figures. An erratum to this chapter can be found at DOI: 10.1007/978-3-319-23135-8_32

© Springer International Publishing Switzerland 2015
T. Morzy et al. (Eds.): ADBIS 2015, LNCS 9282, pp. 261–274, 2015.
DOI: 10.1007/978-3-319-23135-8_18

By contrast, many application scenarios do not satisfy these preconditions, meaning that the database cannot guarantee the existence of an entry for every real-world object. For instance, if a medical practitioner does not document a performed treatment, there may not be a corresponding tuple in the database. Thus, the data population of treatment entries is incomplete, even without the presence of *null*-values. In this case, strictly *measuring* completeness is only possible using reference data (also called master data in the literature) or expert knowledge about the number of real-world entities in question. In cases where neither is available, completeness needs to be estimated instead of measured. According to Dustdar et al., developing metrics for this is commonly regarded as highly nontrivial [4].

1.1 Motivation

Incompleteness is a frequent problem in many different application scenarios. Our work on population completeness is embedded in the project context of an ERP system[1] for German medical centers [5]. These centers are loose affiliations of medical practitioners, usually each with their own data management system. The center's manager requires accurate data about the benefits[2] the center's practitioners provide to their patients. This is because every practitioner operates on a budget mandated quarterly by the "Association of Statutory Health Insurance Physicians" (ASHIP). Once the budget is exhausted, all benefits a practitioner provides are remunerated only as a fraction of their actual value. At the end of each quarter, every practitioner receives an ASHIP invoice. This invoice contains a manually curated and completed record of all benefits over the quarter. At this point however, all accounts in the center have already been finalized. This means that financial controlling and steering, e.g. by transferring patients to doctors who have not yet exhausted their budget, is no longer possible.

While the ASHIP invoices constitute reference data for the center, they become available too late to detect problems with data completeness. Some of these cases may be caught with relatively straightforward data quality rules [10]. In many cases however, a data quality monitoring system [8] needs to offer more powerful mechanisms to estimate the completeness of data used for financial controlling.

The practices, and therefore the data sources, usually are independent of each other, and are not under the control of a central authority [7]. In many cases, this means that the practitioners in the center are not willing to replace their data management systems with a shared database. They may also be reluctant to change their data entry habits. To nevertheless create an overview over the data of the center, the practitioners' heterogeneous data sources send a daily summary of their data, which is integrated into a central database.

---

[1] enterprise resource planning system.

[2] A "benefit" is any creditable treatment, counseling, or similar action a practitioner performs.

There is no direct way of knowing whether a sent summary is complete – it may even be empty on some days. This does not necessarily imply an impairment of data completeness. Conversely, even if data is delivered, there may still be missing records.

1.2 Related Work

While our project context and test bed is specific to the German health system, many other applications also depend on complete information and suffer from similar problems [9]. There are many existing approaches to measuring population completeness, though most do use reference data or related concepts. Dugas et al. [3], for example, require that business processes define the necessary amount of data. Dersch-Mills et al. [2] use expert knowledge to choose the most likely complete population among several as the reference data for all other populations.

Several other approaches (Fan and Geerts [9], Naumann et al. [14], and Razniewski and Nutt [17], to name but a few) consider query completeness in incomplete or partially incomplete databases. All these solutions require either the presence of reference data or tables already annotated with information about their completeness. These annotations can only originate outside of the database according to Razniewski and Nutt [17]. In some cases, a closed world can be assumed [14].

To our best knowledge however, there is no commonly applicable method to measure completeness of data without using reference data, expert knowledge, or equivalent concepts. According to Fan and Geerts, "(...) effective algorithms and metrics are yet to be developed, to assess the completeness of information in our database for answering queries" [9, p. 93].

1.3 Contribution

We suggest an approach to measuring population completeness applicable to time-stamped data items. We maintain that time series modeling techniques can be used to classify parts of a data population as "complete" or "incomplete". Both technical and semantic timestamps are allowed, and the frequency with which data items are introduced to the population does not need to be constant. Our contributions are as follows:

1. An approach to classify parts of a data population as "complete" or "incomplete" respective of the data items they should contain (see Sect. 2).
2. An anonymized real-world dataset containing both dirty and cleaned data about a population of benefits data of 20 medical practitioners over the course of three months (see Sect. 3).
3. An evaluation of the feasibility of our approach (see Sect. 3).

1.4 Problem Description

As an example of our motivational use case, consider the data shown in Fig. 1. It shows the benefits a practitioner documents over a course of four weeks[3]. Each data point shows the change Δ of the population of this practitioner's overall benefits, i.e. the change of the number of entries in the practitioner's database. Looking at the data of the first three weeks, we make two assumptions:

1. There is a characteristic spike at the middle of each week.
2. Δ is around 0 over the weekends.

These assumptions form a rough description of the population's *volatility*. The volatility of a population denotes the behavior over time of the population's change Δ. It is noteworthy that volatility alone does not allow any statement about the quality of data, but rather is a characteristic of real-world entities or populations.

However, knowing the volatility of a population allows us to make statements about its expected growth. For example, judging from the data we expect the practitioner to do most of the work in the middle of the week. If there is a deviation from this expectation, we consider this to imply impaired completeness, e.g. on day 25: The "middle of the week spike" is missing. Instead, the same number of benefits as the day before is observed. If the practitioner actually did provide more benefits on day 25, then our data is incomplete.

Note that the data in this example has already been cleaned – there are a lot more errors, i.e. instances of incomplete data, in the original version.

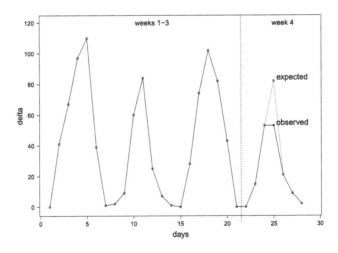

Fig. 1. Benefits per day of one practitioner

---

[3] Snippet of cleaned real-world data from a medical center. To exemplify our proposition, an artificial error has been introduced at data point 25.

2   Method

We call our approach ForCE (Forecasting for Completeness Estimation), and in the following we describe how it classifies population completeness.

2.1   The ForCE Approach

We assume the data in question to be in the form of a time series $\mathcal{T}$ with schema $\mathcal{T}(\Delta, point\_in\_time)$. The actual unit of measure of $point\_in\_time$ is not of consequence and may be of arbitrary granularity. Every change Δ of a population is considered. The identifying attribute of $\mathcal{T}$ is $point\_in\_time$ – a specific population has only one Δ at a specific time.

We use statistical time series methods (see Sect. 3.2) to model the change over time of the data population in question. As an estimate of the population's volatility, we fit a model to a training set composed of data from the time series up to a certain point in time. We can then forecast expected values of the series at future timestamps. If the observed values are different from the expected values, a problem with data quality may exist.

The classification process is shown as pseudocode in Algorithm 1. A portion of the dataset $data$ is chosen as a training set, to which a time series model is

---

input   : data (dataset to classify)
             start_size (starting size of training set)
             full_history (flag to use variable or fixed length training set)

output: predicted_class
             (completeness predictions for all entries in data starting from start_size+1)

for $i = start\_size;\ i < length(data);\ i{+}{+}$ **do**

    if *full_history* **then**

      | training_set = data[1:i]

    else

      | training_set = data[(1+i-start_size):i]

    end

    fit = model(training_set)

    expected_value = forecast(fit, horizon=1)

    observed_value = data[i+1]

    if *expected_value <= observed_value* **then**

      | predicted_class[i+1] = "complete"

    else

      | predicted_class[i+1] = "incomplete"

    end

end

return predicted_class

Algorithm 1: Classifying completeness with the ForCE approach fitted. According to the flag *full_history*, the training set either encompasses all values that have already been observed, or is limited to a window of fixed size over the observed values. More on this in Sect. 3.2. The resulting model is then used to forecast the expected value at a horizon of 1, i.e. one data point after the training set's end. If the expected value is smaller or equal to the actual value of the dataset, *data* is deemed "complete" at this point, else "incomplete". At each new iteration, the training set is extended to include the observed value, and the process is repeated until all data points in *data* have been assigned a completeness class.

Note that this way, the training set never includes the data point currently under scrutiny. Thus, the training and testing sets are disjoint, even though at the last step the whole dataset (excluding the last point) has been used in training the model.

2.2   Data Preprocessing

To be able to apply the approach to any timestamped data, the data first needs to be transformed into a time series. As a straightforward example in case of relational data with schema $\mathcal{R}(A, B, ..., timestamp)$, we build a time series by aggregating as follows:

```
Time series T :=
  SELECT COUNT(*) AS delta, bin(timestamp) AS point_in_time
  FROM R
  GROUP BY bin(timestamp)
  ORDER BY bin(timestamp)
```

bin() is a function that transforms the timestamp according to the desired frequency, for example by truncating hours, minutes and seconds in case the timestamp is of type `datetime` and the desired frequency is "daily". In case that the resulting relation does not contain all possible values of `bin(timestamp)`, we fill the gaps by inserting the missing tuples $(0 , < point\_in\_time >)$.

If there are several distinguishable populations within the same table that need to be separated, identifying columns should be selected and grouped by additionally.

3   Evaluation

We evaluate the ForCE approach on an aggregated dataset[4] based on 58130 benefits provided at a medical center participating in our project. Its schema is

$$BenefitData(ObservedBenefits, ActualBenefits, \underline{PractitionerId, Date})$$

The dataset comprises 1820 tuples: For each of the 20 distinct medical practitioners, *ObservedBenefits* shows the number of benefits they documented over

---

[4] available for download at www6.cs.fau.de/files/completeness_data.zip.

the course of one quarter (91 days), and *ActualBenefits* shows the actual number of benefits for each day taken from the ASHIP invoices (see Sect. 1.1). All values of *PractitionerId* are pseudonymized so they can be distinguished, but not related to actual persons. The number of benefits is 0 if no benefits were documented/invoiced that day.

As shown in Algorithm 1, we build a time series model over all values of T up to a point n to model the volatility of the population change of *BenefitData*. We then forecast Δ at time $n+1$ and in turn use this forecast to classify the completeness at $n+1$. The predicted classes are compared with the actual completeness of the data population at this point to obtain performance measures for the classifier.

3.1 Classification of Completeness

The classifier was implemented in the R environment[5] and makes use of the `forecast` package [11], which provides several functions to model time series. R has the advantage of offering all functions necessary for this feasibility study out of the box.

Performance Measures. For each day in the data, we classified the completeness of each practitioner's data. To evaluate the performance of our classifier, we calculate several common measures of classifier performance. We use the abbreviations tp = true positives, tn = true negatives, fp = false positives, fn = false negatives; "true positive" in this context means *correctly classified as complete*, "true negative" means *correctly classified as incomplete*. The measures are defined as follows:

- *acc*: Accuracy
  (overall percentage of correctly classified instances)

$$acc = \frac{tp+tn}{tp+tn+fp+fn} \tag{1}$$

- *mcc*: Matthews Correlation Coefficient
  (correlation between the actual completeness and the classifier's predictions)

$$mcc = \frac{tp*tn-fp*fn}{\sqrt{(tp+fp)(tp+fn)(tn+fp)(tn+fn)}} \tag{2}$$

- *ppv*: Positive Predictive Value / Precision
  (percentage of instances correctly classified as complete)

$$ppv = \frac{tp}{tp+fp} \tag{3}$$

- *npv*: Negative Predictive Value
  (percentage of instances correctly classified as incomplete)

$$npv = \frac{tn}{tn+fn} \tag{4}$$

---

[5] See `r-project.org`.

- *sens*: Sensitivity / Recall
  (percentage of actual complete instances classified as complete)

$$sens = \frac{tp}{tp+fn} \qquad (5)$$

- *spec*: Specificity
  (percentage of actual incomplete instances classified as incomplete)

$$spec = \frac{tn}{tn+fp} \qquad (6)$$

The mcc falls within the interval $[-1, 1]$, with -1 meaning "strongly negatively correlated", 0 meaning "no correlation", and 1 meaning "strongly positively correlated". All other measures take values from $[0, 1]$.

Baselines. To judge the feasibility of classifying completeness with our approach in the real-world example, we compare our classifier to three baselines:

- guess_p: only predict positives, i.e. always predict the data to be complete
- guess_n: only predict negatives, i.e. always predict data to be incomplete
- guess_r: predict the completeness randomly with $p = 0.5$ for each outcome

The approach guess_p is actually the de facto standard in medical centers: Since no reference data is available during a quarter, population completeness often gets ignored until the arrival of the ASHIP invoices.

We calculate the performance measures for the baselines dependent on the distribution of the classes. With $x = percentage\ of\ positives$, we get the following:

Table 1. Results of baselines

	tp	fp	tn	fn
guess_p	x	$1 - x$	0	0
guess_n	0	0	$1 - x$	x
guess_r	$0.5x$	$0.5(1-x)$	$0.5(1-x)$	$0.5x$

Applying these values to the measure formulas, we get:

Table 2. Performance measures of baselines

	acc	mcc	ppv	npv	sens	spec
guess_p	x	NaN	x	NaN	1	0
guess_n	$1-x$	NaN	NaN	$1 - x$	0	1
guess_r	0.5	$\frac{0}{0.25(x-x^2)}$	0.5	0.5	0.5	0.5

The NaN values are due to division by zero and denote that this measure is not applicable to the respective classifier. This is intuitively correct, e.g. a classifier that only predicts positives does not have a negative predictive value. The mcc also becomes incalculable in this case. Sensitivity trivially is 1 for guess_p[6], 0 for guess_n, and vice versa for specificity. For the random classifier guess_r, all performance measures except the mcc are 0.5. If there are only positives or only negatives, the mcc is NaN for guess_r, and zero in all other cases.

The mcc is the only one of the performance measures whose absolute value has expressivity since it is not sensitive to the cardinality of the classes "complete" respectively "incomplete" and does not lose expressivity if one class is significantly larger than the other. All other measures can only be assessed relatively to the cardinality of the classes. As an example, when all data is complete, a classifier that guesses "complete" for every instance reaches a perfect accuracy of 1. However, this classifier will be useless in situations where detection of negatives is critical. For this reason, we include the two "blind" classifiers guess_p and guess_n, as they depend directly on the number of positives and thus enable interpretation of the measures. In the remainder of this section, we mainly reason about accuracy and mcc, since together with the baselines they contain all the information necessary to evaluate the classifier's performance. For further information on results for the other measures, please see Fig. 6 in the appendix.

3.2   Results

In the following, we describe the results of our experiments. We start out by comparing the results of several forecasting methods, then move on to the effect of a fixed or variable length training set. We show differences in classifier performance between the 20 practitioners, and close with an overall summary of our method's performance compared to the three baselines. The results are visualized in box plots, showing median, 25 % and 75 % quantiles, 1.5 * interquartile range, and outliers. When discussing the results, we will use the term "box" to denote box, whiskers, and outliers of a single plot.

For each practitioner and forecasting method, the tests were repeated with training data of minimal size up to the maximum size the data could still accommodate, both with fixed and varying windows of training data. As an additional setting, when an observation was zero, the forecast was set to zero as well. This captures the assumption that critical failures leading to a total absence of data can be detected beforehand, e.g. through failed ping attempts to a data source.

Comparison of Forecasting Methods. We ran tests with the forecasting methods mentioned in the description of the *forecast* function [11, p. 27–29]. Figure 2 shows the mcc of the methods. Each box contains the results of all test datasets, i.e. of each practitioner. The last box, labeled "all", is a summary of all results, obtained by union of the result multisets of all other methods.

There is some noticeable variation from method to method, e.g. some produce more outliers than others (i.e. values that lie beyond 1.5 * interquartile range), and all quantiles vary slightly. None of the methods decidedly beats or falls short of the others. Therefore, to avoid cherry picking, we always use the summary of all methods in the following, since it shows the full range of results of all forecasting methods.

---

[6] If the classifier always guesses positive, all actual positives are caught.

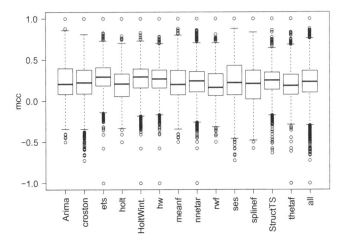

Fig. 2. Comparison of forecasting methods

Training Set: Full History vs. Fixed Length. Figure 3 shows the effect of the flag *full_history* (see Algorithm 1). Each box contains the results for all datasets and all forecasting methods. For the left box in each subplot, the full history was used as training data, for the right box, a window of fixed length was used. The results exhibit very little difference. This is in accordance to our expectations, since the forecasting functions give greater weight to current values than to older ones. This also means that the window size of the training data does little to influence the results.

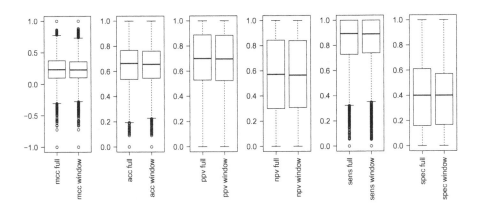

Fig. 3. Training set: full history vs. fixed length

Comparison of Datasets. To compare the performance dependent on dataset, Fig. 4 shows the results grouped by practitioners, ordered by median mcc. Their

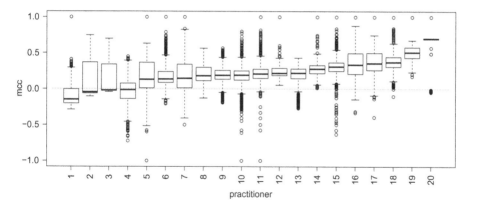

Fig. 4. Comparison of datasets

numbers correspond to their *PractitionerId* in the uploaded dataset. Each box contains the respective practitioner's results for all forecasting methods. The dotted line gives the result of guess_r. Since the mcc is sensitive to class cardinality, the results of guess_p and guess_n are accounted for in the values of the mcc.

The medians of practitioners 1 to 4 are below zero, meaning that on average their results are negatively correlated to the actual completeness. The four datasets do reach a reasonable mcc in some but not in the majority of their cases. This shows that the proposed method does not work well for these datasets. However, if a dataset is partitioned into subsets, the result of ForCE is largely stable between partitions. This means that "offending" datasets can be detected a priori.

All other datasets outperform guess_r in more than 75 % of their cases. Since all predictions except for practitioners 1 to 4 are positively correlated to the actual completeness in the majority of cases, the classifier on average is helpful in judging completeness.

Classifier Performance Compared to Baselines. Figure 5 compares the accuracy of the classifier to the baselines. Quantiles and 1.5 * interquartile range are labeled with their values, truncated to three decimal places for legibility's sake. The box labeled "ForCE" contains our classifier's results for all datasets and all forecasting methods. All other boxes contain the results depending on the number of positives in each dataset (see Table 2).

Practitioners 1 to 4 have been omitted from these results. This is valid since datasets for which ForCE does not work well can be detected a priori as discussed above. The baseline most difficult to beat here is guess_p, since the test data tends to be complete more often than incomplete. Both the classifier and guess_p reach perfect accuracy in some cases. The classifier's 25 % quantile is close to guess_p's median, meaning that almost 75 % of the classifier's results are better than the median of guess_p. Barring the 75 % quantile, the classifier's quantiles

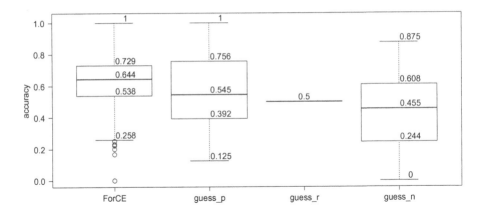

Fig. 5. Performance of ForCE compared to baselines

are higher than the respective quantiles of guess_p. Comparing the medians, the classifier surpasses the baselines by 9.9, 14.4 and 18.9 % points respectively.

4   Discussion and Future Work

The results show that time series forecasting methods can indeed be used to classify the completeness of populations of timestamped data. The ForCE approach outperforms naive approaches when tested on real-world data.

There are many ways in which one can improve the performance results established here: There is a wealth of different methods for time series analysis in statistics and machine learning. It would be interesting to see how the use of neural networks, hidden Markov models, support vector machines, Gaussian processes or the like influences the classifier's performance. Other nonlinear methods and advances in time series modeling [12] are applicable as well. Another subject of study is how well other approaches fare in this application, e.g. outlier detection or arrival processes. All incremental improvements to the employed techniques will also improve our method.

For this paper, we classified the completeness of populations. An extension of our approach may use the confidence intervals many forecasting methods deliver to give an estimate of the probability with which data items are missing.

Lastly, our approach needs to be tested with data from other domains. It should perform well in areas where established results show that the time series representation of the data in question can be forecast accurately. While our test data and therefore results are limited to the specific medical use case, the approach can easily be tested for other datasets. As an example, consider a sensor that only sends an update if its output changes significantly. In a heavily resource constrained environment [16], it may be necessary to omit heartbeats, making it impossible to reliably detect sensor outages short of manually checking the sensor. Applying our method may improve this situation.

As a perspective for future applications, ForCE can easily be extended to be a self-learning system. Its pool of forecasting and modeling functions can be expanded as described above. During operation, all of these can be evaluated on their performance measures, and the current best fit for a specific dataset can be chosen for future calculations. The choice can then be re-evaluated regularly to catch possible changes in the observed data's behavior. Since the performance measures can be calculated step-by-step for many different forecasting functions, adding functionality to choose one of these on the fly would only require minimal effort.

In the future, it will be interesting to test our approach by delivering the results of ForCE to practice managers, then quantifying its impact on the center's data quality.

Acknowledgements. Parts of this work are supported by the German Federal Ministry of Education and Research (BMBF), grant No. 13EX1013D.

Appendix

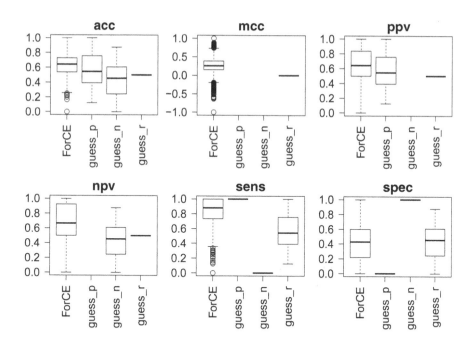

Fig. 6. Overview of all performance measures

References

1. Batini, C., Scannapieco, M.: Data Quality: Concepts Methodologies and Techniques. DCSA. Springer, Heidelberg (2006)

2. Dersch-Mills, D., Hugel, K., Nystrom, M.: Completeness of information sources used to prepare best possible medication histories for pediatric patients. Can. J. Hosp. Pharm. **64**, 10–15 (2011)

3. Dugas, M., Dugas-Breit, S.: A generic method to monitor completeness and speed of medical documentation processes. Methods Inf. Med. **51**(3), 252–257 (2012)

4. Dustdar, S., Pichler, R., Savenkov, V., Truong, H.L.: Quality-aware service-oriented data integration: requirements, state of the art and open challenges. SIGMOD rec. **41**(1), 11–19 (2012)

5. Endler, G.: Data quality and integration in collaborative environments. In: Proceedings of the SIGMOD/PODS 2012 PhD Symposium, PhD 2012, pp. 21–26. ACM, New York (2012)

6. Endler, G., Baumgärtel, P., Lenz, R.: Pay-as-you-go data quality improvement for medical centers. In: Ammenwerth, E., Hörbst, A., Hayn, D., Schreier, G. (eds.) Proceedings of the eHealth2013 (2013)

7. Endler, G., Langer, M., Purucker, J., Lenz, R.: An evolutionary approach to IT support for medical supply centers. In: Proceedings der 41. Jahrestagung der Gesellschaft für Informatik e.V. (GI) (2011)

8. Endler, G., Schwab, P.K., Wahl, A.M., Tenschert, J., Lenz, R.: An architecture for continuous data quality monitoring in medical centers. In: MEDINFO 2015 (2015)

9. Fan, W., Geerts, F.: Foundations of Data Quality Management. Morgan & Claypool Publishers, San Rafael (2012)

10. Gorupec, M., Endler, G.: ruleDQ: Ein Regelsystem zur Datenqualitätsverbesserung medizinischer Informationssysteme. In: Gesellschaft für Informatik (ed.) Lecture Notes in Informatics (LNI) Seminars 13 / Informatiktage 2014, pp. 37–40 (2014)

11. Hyndman, R.J.: R package 'forecast' - forecasting functions for time series and linear models. http://cran.r-project.org/web/packages/forecast/forecast.pdf (2015). Accessed on 14 April 2015

12. Kantz, H., Schreiber, T.: Nonlinear Time Series Analysis, vol. 7. Cambridge University Press, Cambridge (2004)

13. Miller, D.W., Yeast, J.D., Evans, R.L.: Missing prenatal records at a birth center: a communication problem quantified. In: AMIA Annual Symposium Proceedings of American Medical Informatics Association (2005)

14. Naumann, F., Freytag, J.C., Leser, U.: Completeness of integrated information sources. Inf. Syst. **29**(7), 583–615 (2004)

15. Pipino, L.L., Lee, Y.W., Wang, R.Y.: Data quality assessment. Commun. ACM **45**, 211–218 (2002)

16. Pollner, N., Steudtner, C., Meyer-Wegener, K.: Placement-safe operator-graph changes in distributed heterogeneous data stream systems. In: Datenbanksysteme für Business, Technologie und Web - Workshopband (2015)

17. Razniewski, S., Nutt, W.: Completeness of queries over incomplete databases. PVLDB **4**(11), 749–760 (2011)

18. Redman, T.C.: Data Quality: The Field Guide. Digital Press, Newton (2001)

19. Scannapieco, M., Missier, P., Batini, C.: Data quality at a glance. Datenbank-Spektrum **14**, 6–14 (2005)

20. Wang, R.Y., Ziad, M., Lee, Y.W.: Data Quality. ADS. Springer, New York (2002)

21. Zaniolo, C.: Database relations with null values. In: Proceedings of the 1st ACM SIGACT-SIGMOD Symposium on Principles of database systems, PODS 1982, pp. 27–33. ACM, New York (1982)

Best-Match Time Series Subsequence Search on the Intel Many Integrated Core Architecture

Mikhail Zymbler[(✉)]

South Ural State University, Chelyabinsk, Russia
mzym@susu.ru

Abstract. Subsequence similarity search is one of the basic problems of time series data mining. Nowadays Dynamic Time Warping (DTW) is considedered as the best similarity measure. However despite various existing software speedup techniques DTW is still computationally expensive. There are approaches to speed up DTW computation by means of parallel hardware (e.g. GPU and FPGA) but accelerators based on the Intel Many Integrated Core architecture have not been payed attention. The paper presents a parallel algorithm for best-match time series subsequence search based on DTW distance for the Intel Xeon Phi coprocessor. The experimental results on synthetic and real data sets confirm the efficiency of the algorithm.

1 Introduction

Subsequence similarity search is one of the basic problems of time series data mining and appears in various applications, e.g. climate modeling [1], medical monitoring [6], economic forecasting [5], etc. Best-match time series subsequence search assumes that a query sequence and a longer time series are given, and the task is to find a subsequence in the longer time series, whose distance from the query is the minimum among all the subsequences.

Nowadays the Dynamic Time Warping (DTW) [2] is considered as the best similarity measure in many time series applications [3]. DTW is computationally expensive and there are many software approaches that have been proposed to solve this problem, e.g. lower bounding [3], computation reusing [13], data indexing [9], early abandoning [11], etc. However, DTW is still very time-consuming and there are approaches to speed up DTW computation by means of parallel hardware, e.g. computer-cluster [15], multicore [14], FPGA and GPU [13,16] but none for the Intel Many Integrated Core [4] accelerators.

In this paper we present a parallel algorithm for best-match subsequence search based on DTW distance adapted for a central processor unit (CPU) accompanied with the Intel Xeon Phi many-core coprocessor. The remainder of the paper is organized as follows. Section 2 gives the formal definition of the problem and briefly considers the Intel Xeon Phi architecture and programming model and discusses related work. The suggested algorithm is described in Sect. 3. Experimental results evaluating the algorithm are presented in Sect. 4. Section 5 contains summary and directions of future work.

© Springer International Publishing Switzerland 2015
T. Morzy et al. (Eds.): ADBIS 2015, LNCS 9282, pp. 275–286, 2015.
DOI: 10.1007/978-3-319-23135-8_19

2   Background and Related Work

2.1   Problem Definition

A *time series* T is an ordered sequence $t_1, t_2, \ldots, t_N$ of real data points, measured chronologically, where N is a length of the sequence.

A *query* Q is a time series to be found in T; n is a length of the query, $n \ll N$.

A *subsequence* T_{im} of time series T is its continuous subset starting from i-th position and consisting of m data points, i.e. $T_{im} = t_i, t_{i+1}, \ldots, t_{i+m-1}$, where $1 \le i \le N$ and $i + m \le N$.

Best-match subsequence search aims to finding a subsequence T_{in} whose Dynamic Time Warping distance from Q is the minimum among all the subsequences, i.e. $DTW(T_{in}, Q) < DTW(T_{mn}, Q)$ for any m such that $1 \le m \le N - n$.

In this paper we do not consider *local-best-match search* [16], which aims to finding *all* the subsequences T_{in} whose distance from Q is the minimal among their neighboring subsequences whose distance from Q is under specified threshold.

Dynamic Time Warping (DTW) is a similarity measure between two time series X and Y, where $X = x_1, x_2, \ldots, x_N$ and $Y = y_1, y_2, \ldots, y_N$, is defined as follows.

$$DTW(X, Y) = d(N, N),$$

$$d(i, j) = |x_i - y_j| + min \begin{cases} d(i-1, j) \\ d(i, j-1) \\ d(i-1, j-1), \end{cases}$$

$$d(0, 0) = 0; d(i, 0) = d(0, j) = \infty; i = j = 1, 2, \ldots, N.$$

2.2   The Intel Xeon Phi Architecture and Programming Model

The Intel Xeon Phi coprocessor is an x86 many-core coprocessor of 61 cores, connected by a high-performance on-die bidirectional interconnect where each core supports 4× hyperthreading and contains 512-bit wide vector processor unit (VPU). Each core has two levels of cache memory: a 32 Kb L1 data cache, a 32 Kb L1 instruction cache, and a core-private 512 Kb unified L2 cache. The Intel Xeon Phi coprocessor is to be connected to a host computer via a PCI Express system interface. Being based on Intel x86 architecture, the Intel Xeon Phi coprocessor supports the same programming tools and models as a regular Intel Xeon processor.

There are three programming modes to deal with the Intel Xeon Phi coprocessor: native, offload and symmetric. In native mode the application runs independently, on the coprocessor only. In offload mode the application is running on the host and offloads computationally intensive part of work to the coprocessor. The symmetric mode allows the coprocessor to communicate with other devices by means of Message Passing Interface (MPI).

2.3   Related Work

Currently DTW is considered as best similarity measure for many applications [3], despite the fact that it is very time-consuming [7,15]. Research devoted to acceleration of DTW computation includes the following.

The SPRING algorithm [12] uses computation-reuse technique. However, this technique squeezes the algorithm's applications because data-reuse supposes non-normalized sequence. In [9] indexing technique to speed up the search was used, which need to specify the query length in advance. Authors of [8] suggested multiple index for various length queries. Lower bound technique was proposed in [7] and prunes off unpromising subsequences using the lower bound of DTW distance estimated in a cheap way. The UCR-DTW algorithm [11] integrates all the possible existing speedup techniques and most likely it is the fastest of the existing subsequence matching algorithms.

All the aforementioned algorithms aim to decrease the calling times of DTW computation, not accelerating DTW itself. However, because of its complexity, DTW still takes a large part of the total application runtime. That is why there are researches exploiting the effectiveness of parallel hardware by means of allocation of DTW calculation of different subsequences into different processing elements.

In [14] subsequences starting from different positions of the time series are sent to different Intel Xeon processors, and each processor computes DTW. In [15] different queries are distributed onto different cores, and each subsequence is sent to different cores to be compared with different queries. GPU implementation [17] parallelize the generation of the warping matrix but still process the path search serially. GPU implementation proposed in [13] utilizes the same ideas as in [14]. FPGA implementation described in [13] focuses on the naive subsequence similarity search, and do not exploit any pre-processing techniques. It is generated by a C-to-VHDL tool and due to lack of insight into the FPGA can not be applied in big-scale tasks. To address these problems in [16] a stream oriented framework was proposed. It implements coarse-grained parallelism by reusing data of different DTW computations and uses a two-phase precision reduction technique to guarantee accuracy while reducing resource cost.

In this work a parallel algorithm of the time series subsequence DTW-based similarity search on the Intel Xeon Phi many-core coprocessor is presented where the UCR-DTW serial algorithm is used as a basis.

3   Best-Match Subsequence Search on the Intel Xeon Phi

Development of the best-match subsequence search algorithm consists of the following steps, which will be discussed in detail further.

At first, we developed a parallel version of the UCR-DTW serial algorithm [11] using OpenMP technology. However, experiments have shown that, despite the one-order speedup of parallel algorithm, it works slower on the Intel Xeon Phi coprocessor in *native* mode than on CPU. This results from low operational intensity of our algorithm, i.e. insufficient FLOPs (floating point operations) per byte of data to be effectively processed on the Intel Xeon Phi coprocessor.

Next, we modified our algorithm combining CPU and coprocessor to process time series, i.e. CPU and the Intel Xeon Phi run parallel algorithm developed at the previous step. Here we used *offload* mode to transfer code and data to the coprocessor. Experiments, where we varied the portion size of data to be transferred to the coprocessor, show results similar to those obtained at the first step due to the same reason.

Finally, we developed an advanced version of the algorithm where the coprocessor is exploited only for DTW computations whereas CPU performs pruning and supports a queue of subsequences for the coprocessor. This significantly increased the operational intensity of the computations on the coprocessor and experiments shown acceptable performance of the algorithm.

3.1   Parallel Algorithm for CPU

The UCR-DTW algorithm proposed in [11] is one of the fastest existing subsequence matching algorithms. This algorithm (Fig. 1) uses a cascade estimation of the lower bound of DTW distance. If the lower bound has exceeded some threshold, the DTW distance also exceeds the threshold, so the subsequence can be pruned off. Here the `bsf` (best-so-far) variable stores the distance to the most similar subsequence.

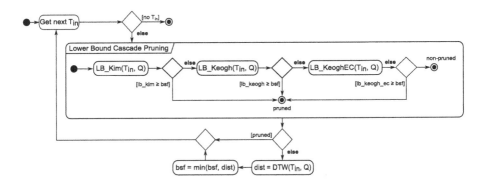

Fig. 1. Serial algorithm

A parallel version of the UCR-DTW algorithm is depicted in Fig. 2. We parallelize the original algorithm using the OpenMP technology. The time series is splitted into equal-length portions and each portion is processed by a separate OpenMP-thread. Let P denotes a number of OpenMP-threads, then a portion assigned for processing to the k-th thread, $0 \le k \le P - 1$, is defined as a subsequence T_{sl}, where

$$s = \begin{cases} 1 & , k = 0 \\ k \cdot \lfloor \frac{N}{P} \rfloor - n + 2 & , else \end{cases}$$

$$l = \begin{cases} \lfloor \frac{N}{P} \rfloor & , k = 0 \\ \lfloor \frac{N}{P} \rfloor + n - 1 + (N \bmod P) & , k = P - 1 \\ \lfloor \frac{N}{P} \rfloor + n - 1 & , else \end{cases}$$

It means that the head part of every portion except first overlaps with the tail part of previous portion in $n - 1$ data points. This permits to keep possible resulting subsequences from the junctions of portions.

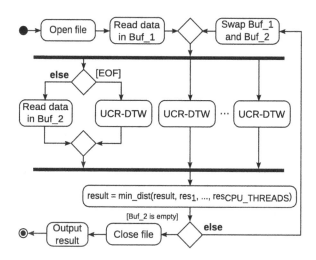

Fig. 2. Parallel algorithm for CPU

Here UCR-DTW is a subroutine that implements the original serial algorithm. In contrast with the serial version the bsf variable is shared among the threads. This allows each thread to prune off unpromising subsequence using lower bounding. Master thread reads new data from a file simultaneously with processing of data that have been read.

The obtained algorithm is ready to run on the Intel Xeon Phi in *native* mode but experiments have shown that (Fig. 6) although parallel algorithm expectedly surpasses the serial algorithm it works slower on the coprocessor than on CPU. This was a result of low operational intensity of our algorithm, i.e. insufficient FLOPs per byte of data to be effectively processed on the Intel Xeon Phi coprocessor.

3.2   Naïve Parallel Algorithm for CPU and the Intel Xeon Phi

Figure 3 depicts the modified version of the algorithm. This version is called "naïve", because in comparison with the previous version it only distributes

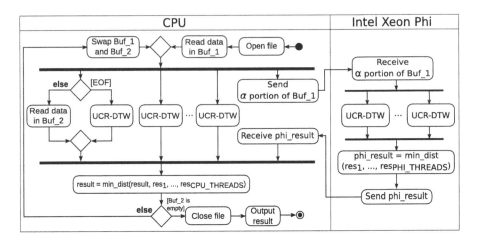

Fig. 3. Naïve parallel algorithm for CPU and the Intel Xeon Phi

work among CPU and the coprocessor. Here α is a parameter that determines a proportion of data to be transferred to the coprocessor. We use *offload* mode to organize data exchange between CPU and the coprocessor. The min_dist subroutine chooses a subsequence with minimal value of DTW.

As well as in the previous step we evaluated the obtained algorithm and experiments have shown that regardless of the α value the algorithm has worse performance in comparison with the parallel algorithm for CPU.

This is because we still have not increased operational intensity of calculations on the coprocessor. Additionally, bsf shared variable can not be synchronized between the CPU and the coprocessor while offloading is performed (the synchronization is possible at the beginning and at the end of offload section). That is why we have more non-pruned subsequences to compute DTW.

3.3 Advanced Parallel Algorithm for CPU and the Intel Xeon Phi

The advanced version of the algorithm obtained at the previous step is depicted in Fig. 4.

The algorithm is based on the following two ideas. First, the coprocessor should be exploited only for DTW computation whereas CPU prunes unpromising subsequences and computes DTW in case if it really does not have another job. Second, CPU should support a queue of subsequences that are candidates to be offloaded to the coprocessor to compute DTW for each candidate subsequence.

To reduce amount of data transferred to the coprocessor the following technique has been used. Queue does not store each candidate subsequence T_{in} but stores its corresponding tuple (i, A), where A is an n-element array containing LB_{Keogh} lower bounds for each position of the subsequence which is used for early abandoning of DTW [11]. CPU offloads current part of the time series once whereas queue is offloaded each time it is full.

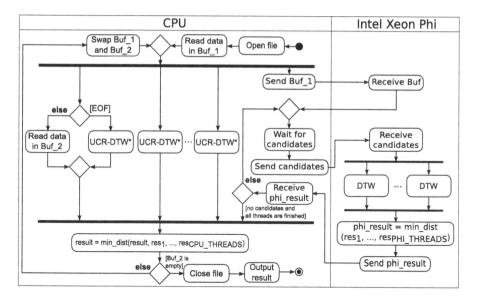

Fig. 4. Advanced parallel algorithm for CPU and the Intel Xeon Phi

The number of elements in the queue is calculated as $C \cdot h \cdot W$, where C is a number of cores of the coprocessor, h is a hyperthreading factor of the coprocessor and W is a number of candidates to be processed by a coprocessor's thread (i.e. W is a parameter of the algorithm).

One of the CPU threads is declared as a master and the rest as workers. At start master sends a buffer with the current portion of the time series to the coprocessor. If queue is full then master offloads it to the coprocessor to perform DTW computation for the corresponding subsequences by the coprocessor's threads.

A worker's behavior is depicted in the Fig. 5. Worker computes cascade estimates for the current subsequence. If it is dissimilar to the query then the worker

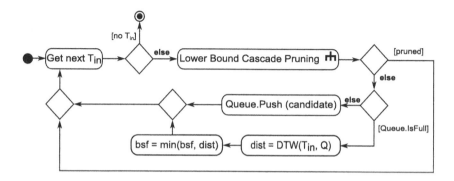

Fig. 5. UCR-DTW* subroutine

prunes it off otherwise worker pushes this subsequence to the queue. If the queue is full and data previously transferred to the coprocessor have not been processed yet, the worker computes DTW by itself.

At the end of offload section the information about most similar subsequence found on the coprocessor is transferred to the CPU. The final result is computed among the most similar subsequence found on the CPU and same that found on the coprocessor.

4   Experimental Results

To evaluate the developed algorithm we performed experiments on the Tornado SUSU supercomputer's node (Table 1 contains its specifications).

Table 1. Specifications of the Tornado SUSU supercomputer's node

Specifications	Processor	Coprocessor
Model	Intel Xeon X5680	Intel Xeon Phi SE10X
Cores	6	61
Frequency, GHz	3.33	1.1
Threads per core	2	4
Peak performance, TFLOPS	0.371	1.076

We measured search runtime while varying query length. Experiments have been performed on synthetic and real time series. We also investigated the impact of queue size on the speedup and compared performance of the algorithm with analogues for GPU and FPGA.

4.1   Performance

In the first experiment we used synthetic time series generated by one-dimensional random walk [10] comprising of 100 million data points. Experimental results (Fig. 6a) show that our algorithm is more effective for longer queries. In case of shorter queries the algorithm has the same performance as parallel algorithm for CPU only.

The second experiment investigates the algorithm's performance on real electrocardiographic (ECG) data with about 20 million data points (approximately 22 hours of ECG sampled at 250 Hz). Our algorithm shows (Fig. 6b) a three times higher performance than the parallel algorithm for CPU only.

4.2   Impact of Queue Size

Results of experiments investigating the impact of queue size on performance are depicted in Fig. 7. In the current experimental environment, i.e. hyperthreading

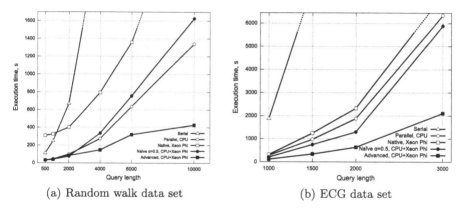

(a) Random walk data set   (b) ECG data set

Fig. 6. Performance of the algorithm

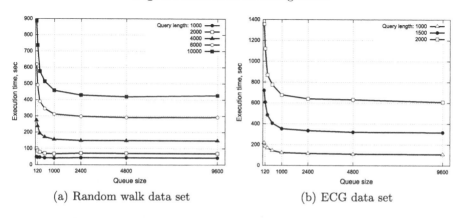

(a) Random walk data set   (b) ECG data set

Fig. 7. Impact of queue size on the speedup

factor of the coprocessor h is 4, number of cores of the coprocessor C is 60^{1}, optimal number of candidates to be processed by a coprocessor's thread W is 10, so optimal number of the elements in the queue is 2400. Experimental results described in Sect. 4.1 have been achieved with this queue size.

4.3   Comparison with Analogues

We compared the performance of our algorithm with analogues for GPU and FPGA developed in [13]. We repeated the experiments presented in that paper using the same data set and query length.

The results of the experiments are depicted in Fig. 8, here percentage on the top of the bar indicates a proportion of subsequences that have not been pruned

---

[1] One of the coprocessor's cores is not involved in computations as it is recommended by the Intel Xeon Phi programmer's manual.

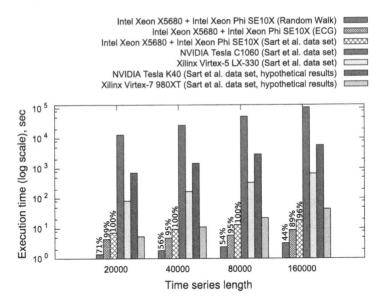

Fig. 8. Comparison of performance

and subjected to the DTW calculation in our experiments. We also add to the chart results of experiments on random walk and ECG data sets.

We took into account that the peak performance of the hardware we used is significantly greater than its counterparts of that paper, i.e. overall peak performance of our hardware was 1.44 TFLOPS whereas GPU as NVIDIA Tesla C1060 had 77.8 GFLOPS and FPGA as Xilinx Virtex-5 LX-330 had 65 GFLOPS.

To provide more "fair" comparison we added to the chart *hypothetical* results for modern NVIDIA Tesla K40 (1.43 TFLOPS) and Xilinx Virtex-7 980XT (0.99 TFLOPS) multiplying real results of NVIDIA Tesla C1060 and Xilinx Virtex-5 LX-330 by a respective scaling factor. As we can see our algorithm does not concede to analogous on performance.

5   Conclusion

In this paper an approach to best-match time series subsequence search under DTW distance on the Intel Many Integrated Core architecture has been presented. The parallel algorithm combines capabilities of CPU and the Intel Xeon Phi many-core coprocessor. The coprocessor is exploited only for DTW computations whereas CPU performs lower bounding, prepares subsequences for the coprocessor and computes DTW as a last resort. CPU supports a queue of candidate subsequences to be offloaded to the coprocessor to compute DTW. Experiments on synthetic and real data sets have shown that our algorithm does not concede to analogous algorithms for GPU and FPGA on performance.

As future work we plan to extend our research in the following directions: implement our algorithm for the cases of several coprocessors and cluster system based on nodes equipped with the Intel Xeon Phi coprocessor(s) and apply our approach to the task of local-best-match time series subsequence search.

Acknowledgment. This work was financially supported by the Ministry of education and science of the Russian Federation ("Research and development on priority directions of scientific-technological complex of Russia for 2014–2020" Federal Program, contract No. 14.574.21.0035).

References

1. Abdullaev, S., Lenskaya, O., Gayazova, A., Sobolev, D., Noskov, A., Ivanova, O., Radchenko, G.: Short-range forecasting algorithms using radar data: translation estimate and life-cycle composite display. Bull. S. Ural State Univ. Ser. Comput. Math. Soft. Eng. **3**(1), 17–32 (2014)
2. Berndt, D.J., Clifford, J.: Using dynamic time warping to find patterns in time series. In: Fayyad, U.M., Uthurusamy, R. (eds.) KDD Workshop, pp. 359–370. AAAI Press (1994)
3. Ding, H., Trajcevski, G., Scheuermann, P., Wang, X., Keogh, E.J.: Querying and mining of time series data: experimental comparison of representations and distance measures. PVLDB **1**(2), 1542–1552 (2008)
4. Duran, A., Klemm, M.: The intel many integrated core architecture. In: Smari, W.W., Zeljkovic, V. (eds.) HPCS, pp. 365–366. IEEE (2012)
5. Dyshaev, M., Sokolinskaya, I.: Representation of trading signals based on kaufman adaptive moving average as a system of linear inequalities. Bull. S. Ural State Univ. Ser.: Comput. Math. Soft. Eng. **2**(4), 103–108 (2013)
6. Epishev, V., Isaev, A., Miniakhmetov, R., Movchan, A., Smirnov, A., Sokolinsky, L., Zymbler, M., Ehrlich, V.: Physiological data mining system for elite sports. Bull. S. Ural State Univ. Ser.: Comput. Math. Soft. Eng. **2**(1), 44–54 (2013)
7. Fu, A.W.C., Keogh, E.J., Lau, L.Y.H., Ratanamahatana, C.A.: Scaling and time warping in time series querying. In: Böhm, K., Jensen, C.S., Haas, L.M., Kersten, M.L., Larson, P., Ooi, B.C. (eds.) Proceedings of the 31st International Conference on Very Large Data Bases, pp. 649–660. ACM, Trondheim, Norway, 30 August – 2 September 2005 (2005)
8. Keogh, E.J., Wei, L., Xi, X., Vlachos, M., Lee, S.-H., Protopapas, P.: Supporting exact indexing of arbitrarily rotated shapes and periodic time series under euclidean and warping distance measures. VLDB J. **18**(3), 611–630 (2009)
9. Lim, S.-H., Park, H.-J., Kim, S.-W.: Using multiple indexes for efficient subsequence matching in time-series databases. In: Li Lee, M., Tan, K.-L., Wuwongse, V. (eds.) DASFAA 2006. LNCS, vol. 3882, pp. 65–79. Springer, Heidelberg (2006)
10. Pearson, K.: The problem of the random walk. Nat. **72**(1865), 294 (1905)
11. Rakthanmanon, T., Campana, B.J.L., Mueen, A., Batista, G.E.A.P.A., Westover, M.B., Zhu, Q., Zakaria, J., Keogh, E.J.: Searching and mining trillions of time series subsequences under dynamic time warping. In: Yang, Q., Agarwal, D., Pei, J. (eds.) KDD, pp. 262–270. ACM (2012)
12. Sakurai, Y., Faloutsos, C., Yamamuro, M.: Stream monitoring under the time warping distance. In: Chirkova, R., Dogac, A., Tamer Özsu, M., Sellis, T.K. (eds.) Proceedings of the 23rd International Conference on Data Engineering, ICDE 2007, pp. 1046–1055. IEEE, The Marmara Hotel, Istanbul, Turkey, 15–20 April 2007 (2007)

13. Sart, D., Mueen, A., Najjar, W.A., Keogh, E.J., Niennattrakul, V.: Accelerating dynamic time warping subsequence search with gpus and fpgas. In: Webb, G.I., Liu, B., Zhang, C., Gunopulos, D., Wu, X. (eds.) ICDM, pp. 1001–1006. IEEE Computer Society (2010)

14. Srikanthan, S., Kumar, A., Gupta, R.: Implementing the dynamic time warping algorithm in multithreaded environments for real time and unsupervised pattern discovery. In: Department of Computer Science and Motial Nehru National Institute of Technology Engineering, ICCCT, pp. 394–398. IEEE Computer Society (2011)

15. Takahashi, N., Yoshihisa, T., Sakurai, Y., Kanazawa, M.: A parallelized data stream processing system using dynamic time warping distance. In: Barolli, L., Xhafa, F., Hsu, H.H. (eds.) CISIS, pp. 1100–1105. IEEE Computer Society (2009)

16. Wang, Z., Huang, S., Wang, L., Li, H., Wang, Y., Yang, H.: Accelerating subsequence similarity search based on dynamic time warping distance with FPGA. In: Hutchings, B.L., Betz, V. (eds.) The 2013 ACM/SIGDA International Symposium on Field Programmable Gate Arrays, FPGA 2013, pp. 53–62. ACM, Monterey, 11–13 February 2013 (2013)

17. Zhang, Y., Adl, K., Glass, J.R.: Fast spoken query detection using lower-bound dynamic time warping on graphical processing units. In: 2012 IEEE International Conference on Acoustics, Speech and Signal Processing, ICASSP 2012, pp. 5173–5176. IEEE, Kyoto, Japan, 25–30 March 2012 (2012)

Feedback Based Continuous Skyline Queries Over a Distributed Framework

Ahmed Khan Leghari[1][(✉)], Jianneng Cao[2], and Yongluan Zhou[1]

[1] Institute of Mathematics and Computer Science (IMADA),
University of Southern Denmark, Odense, Denmark
{ahmedkhan,zhou}@imada.sdu.dk

[2] Department of Data Analytics, Institute for Infocomm Research,
A*, Singapore, Singapore
caojn@i2r.a-star.edu.sg

Abstract. Continuous skyline query processing is becoming wide spread. Most of the work done in this field is focused to process skyline queries on a single machine. Our focus is to process continuous skyline queries over data streams, where data is arriving at server in the form of continuous updates from multiple distributed input sources. A single machine solution to run continuous skyline queries over streaming data is not very scalable. Moreover, streaming data arriving from multiple sources can overwhelm server's computing power, specially if the skyline queries are involved to compute high quality multidimensional skyline points. We propose a three layer solution to compute continuous skyline points. A bottom layer in our approach sends the local skyline points to the middle layer, which after receiving feedback from the server filters the false-positives, and produces the semi-global skyline points to be sent to the server for global skyline. Our approach being scalable distributes the workloads across the network on multiple machines and reduces the number of unnecessary data points to be sent to the server, allowing it to produce qualitative skyline points.

1 Introduction

A skyline query [1,2] running over a multi-dimensional data set produces a subset of data points which are not dominated by others. In a multi-dimensional data set, a data point is said to be dominant over the others if it is not worse than the others in all dimensions and better in at least any single dimension. Skyline queries are widely used in many decision making applications where one dimension can be contrary to others, such as finding a cheap car online with luxury features, powerful engine, impressive fuel economy, and cheap maintenance cost.

In a single machine persistent data setup the execution of a skyline query incurs in just CPU and memory cost, but challenges are different while processing streaming data. It is possible to process streaming data and run skyline queries on a single machine, but running continuous queries over multiple streams result in performance degradation caused by huge processing load, increased response

© Springer International Publishing Switzerland 2015
T. Morzy et al. (Eds.): ADBIS 2015, LNCS 9282, pp. 287–301, 2015.
DOI: 10.1007/978-3-319-23135-8_20

time and limited scalability. Since, in distributed stream processing applications a time bounded response is of very high significance, and plays a decisive role for the reliability and effectiveness of the application. This time bounded response is inversely proportional to the processing load on the machine, therefore, running continuous skyline queries to process multiple streams adversely affect the performance of the applications. Moreover, due to the hardware constraints the number of streams that can be processed in parallel on a single machine are also limited. Therefore, to achieve a better response time and greater scalability a more preferable option is to reduce the processing load on single server by distributing the processing of skyline queries among multiple machines.

In this paper we propose a distributed processing model for processing continuous skyline queries over data streams as shown in Fig. 1, originating from multiple sources and from different geographical locations. Our technique extends the concept proposed by Lu et al. [1]. In [1] a two layer model of distributed continuous skyline computing over data stream was proposed. In that model multiple data sites continuously and directly send local skyline points to the server, even though many of the local skyline points do not turn into actual global skyline points. This leads to increased processing time and load on the server.

Moreover, as the streaming data is processed by a single server, hence it is not very scalable and can process a limited number of data streams. To reduce the processing load and to achieve better scalability we extend this approach to three layers. The top layer sends periodic feedbacks to the middle layer that enables middle layer to perform a local filtering of unnecessary skyline points to

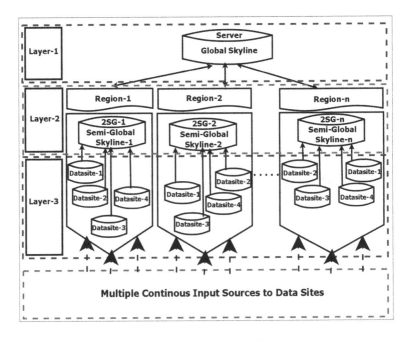

Fig. 1. Three tier architecture of the system

reduce the amount of data to be sent to the server allowing it to compute high quality skyline points.

The contributions of this paper are as follows: *(i)* An improved technique to process continuous skyline queries over data streams in a distributed setup *(ii)* A feedback based approach to minimize the false-positive skyline points to be sent to the server. *(iii)* A three layer continuous skyline query processing architecture that is scalable and computes high-quality multidimensional data points.

The remainder of this paper is organized as follows: Sect. 2 provides an insight into the past research related to our present work. Section 3 presents problem description, illustrates the terminology used in the paper and system model. Section 4 presents the fundamentals of the approach. Section 5 presents the steps required to generate the skyline points in our approach. Section 6 explains the adjustment of skyline points and early filtration of data using the feedback mechanism. Section 7 describes the experiments and obtained results, and Sect. 8 presents the conclusion of the study.

2 Related Work

Skyline operator was first proposed by Borzsony et al. [2] as SKYLINE OF clause in SQL's SELECT statement, and since then the skyline queries have been studied by many [3–29]. The work that is most relevant to ours is done by [1,18].

Wu et al. [18] proposed a distributed progressive skyline query processing over multiple machines. Their solution is based on data partitioning across share-nothing architecture. The focus of their work is to achieve scalability through parallelization of one time skyline queries on persistent data, whereas we focus to distribute and parallelize the processing load to achieve better scalability while running continuous skyline queries over multiple streams.

The work done by Lu et al. [1] provides the foundations for our work. They proposed a two tier architecture to compute skyline points over dynamic data arriving from multiple sources. Their approach is based on initialization and maintenance phase, in the first phase the results in the form of local skylines are obtained from the remote data sites, and in the maintenance phase the local skylines are merged to obtain a global skyline. The global skyline is then continuously updated on the basis of the local skylines. This approach performs really well, the data points before being sent to global server are locally filtered, but even after performing a local filtering many of the false-positive skyline points are still processed by the global server. The number of these useless data points are directly proportional to the frequency of updates performed at the local data sites, and the number of streams involved in the process of computing skyline points. This approach is not very scalable and frequent updates from multiple data streams can overwhelm the computing power of a single server.

In this paper we extend the prior work done by Lu et al. [1] to incorporate a middle layer in the system to make the solution more efficient by providing a feedback mechanism to reduce the number of false-positive skylines processed by the global server.

3   Background

3.1   Problem Description

We want to reduce the processing load on the server that can be achieved through early pruning of the useless data points and by distributing the processing load among multiple machines. The problem can be formulated as:

Let Q be a continuous skyline query registered at the centralized server. N is the number of data sites, each maintaining a relation R of local skyline points, minimize the number of unnecessary data points to be sent to the server, not involved in the final computing of global skyline points.

To achieve this objective we perform a two-stage pruning of useless data points described in the forthcoming sections.

3.2   Basic Definitions

(i) **Dominance Relationship:** A skyline query retrieves data points which are not dominated by others in a multi-dimensional data set. A data point P is said to be dominant over another point Q represented as $P \succ Q$, if P is not worse than Q in any dimension and at-least better in one dimension. The terms "better" and "worse" are general and have different meanings in different contexts.

(ii) **Data Site:** Refer to Fig. 1, a data site is a processing unit at the bottom layer of our model that receives streaming data from multiple streams.

(iii) **Local Skyline Points:** Skyline points which are computed from the streaming data arriving at each data site.

(iv) **Semi-Global Skyline Generator (2SG):** A 2SG is a processing unit in the middle layer of our model as shown in Fig. 1 that receives local skyline points from the data sites connected to it.

(v) **Semi-Global Skyline (SGS) Points:** SGS points are computed at each 2SG, from the local skyline points sent by concerned data sites.

(vi) **Global Skyline Points:** Refer to Fig. 1, global skyline points are computed at the top layer of our model. A server after receiving SGS points from 2SGs generates global skyline points.

3.3   System Model

Refer to Fig. 1, there is a hierarchy of three processing layers. On the bottom layer, there are n data sites. Each data site is located at different geographical location, and receives continuous inputs in the form of data streams. Streaming data can have many forms such as sensor data, web click stream, log streams or stock streams. In this paper we use the stock streams as our streaming data, but our proposed model is general and can be used with any form of streaming data.

Table 1. Local skylines generated at a data site

Data site	ID	name	ask_price	bid_price	change	volume
A	Tuple_1	ABCD	44.45	42.60	-0.04 -0.09 %	1790802
B	Tuple_4	SMSG	551.60	548.14	+0.10 +0.12 %	1886825
C	Tuple_3	AFCD	82.72	80.72	-0.01 -0.07 %	1088307
D	Tuple_2	MCDL	66.30	64.32	+0.06 +0.10 %	1355313

In our model stock streams arrive from multiple stock markets. A data site after receiving data tuples from multiple stock streams performs local filtering and produces local skylines as depicted in Table 1. These local skylines are then sent to the corresponding regional Semi-Global Skyline Generators (2SGs) in the middle layer. The number of 2SG machines can be adjusted according to the volume and pace of data received from multiple data sites, and is an important factor to satisfy the need of time critical applications. Each 2SG machine after receiving multiple local skyline performs a filtering and generates a *Semi-Global Skyline* (SGS). The generation of SGS from multiple local skylines reduces the volume of data to be sent to the server by reducing the false-positive skyline points. The SGS points are then sent to the top layer. This layer corresponds to a centralized server. The server provides a skyline query interface, and generates a global skyline from the SGS points.

4   Base of the Approach

Suppose a continuous skyline query is submitted at server to fetch stock commodities with the largest difference between `ask_price` and `bid_price`, maximum positive `change` per share and the highest `volume` of traded shares. In response to that query each data site will generate local skyline points as shown in Table 1 and send them to the corresponding 2SG. A 2SG after receiving local skyline points from its connected data sites discards the data points dominated by others such as Tuple_1 from data site A shown in Table 1, and computes SGS points. These SGS points are then sent to server for generation of global skyline points as depicted in Table 2.

We assume that each 2SG would send their SGS points to the server at random interval and independent to any other 2SG. Refer to Fig. 1 in Sect. 1 and Fig. 2, the very first SGS points received by server from a 2SG-1 will be fixed as a *Skyline_Threshold*.

The server would then inform all 2SGs (except 2SG-1) that any future SGS points computed at 2SGs must not be sent to server unless they are dominant over the Skyline_Threshold. As 2SG-1 would not receive any feedback in a certain time, it would determine that the SGS points it sent are set as the Skyline_Threshold. Later on, if SGS points at 2SG-1 change then the new SGS points will immediately be sent to the server. The SGS points fixed as Skyline_Threshold

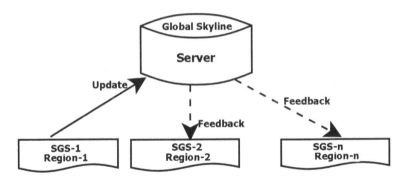

Fig. 2. Adjustment of SGS points on the basis of the server feedback

will continuously be updated to represent the most recent dominant SGS points
sent by one or more 2SGs.

It is possible that multiple 2SGs send their SGS points at the same time,
or before receiving any Skyline_Threshold. In that situation server will compute global skyline from all the SGS points it received and then set a Sky-
line_Threshold. Suppose Table 2 shows SGS points received by server at the
same time. From these SGS points the Server will compute a Skyline_Threshold
based on the dominant dimensions of each tuple as shown in Table 3.

Table 2. SGS points received by Server

2SG	Data site	ID	name	ask_price	bid_price	change(%)	volume
2	B	Tuple_4	SMSG	551.60	548.14	+0.10 +0.11	1086825
4	C	Tuple_3	AFCD	82.72	80.10	+0.01 +0.07	1888307
3	D	Tuple_2	MCDL	66.30	65.32	+0.06 +0.20	1355313

This Skyline_Threshold will be sent to all those 2SGs which have not yet
sent their SGS point to server. A 2SG after receiving threshold information
from server compares its present SGS points with the threshold information and
if it finds them better or dominant in any dimension then it sends them to the
server, otherwise it just discards them.

While processing streaming data the most important and valuable tuples
are the most recent ones, therefore, each tuple that arrives in the system has a
certain life span. An old tuple from a source S is discarded on a data site when
a new tuple from S arrives. A tuple at a data site is also discarded when it fails
to be part of the local skyline sent to a 2SG. A tuple on a 2SG is discarded

Table 3. Skyline_Threshold set by server

max. diff.=(ask_pirce-bid_price)	max. change per share(%)	max. volume traded
3.46	0.14	1888307

being false-positive if it is not part of SGS points to be sent to the server. In the same way a tuple expires if it is not part of global skyline points generated at the server.

4.1   Importance of the Feedback

Feedback plays a very important role to reduce the processing load on the server. Refer to Fig. 2 that shows the feedback messages sent from server to 2SGs. After receiving SGS points from a 2SG, server sets a Skyline_Threshold and sends feedback to all other 2SGs except one to update them about the current dominant skyline points computed at server. This feedback reduces the number of false-positive skyline points to be sent to server. The number of messages sent as a feedback can be obtained through the Eq. 1 as follows.

$$Feedback = (N - 1) * T/C \qquad (1)$$

Here, N represents the total number of 2SGs in the setup subtracting the one who sends most recent tuple to server, T represents the total number of tuples in the stream and C represents the change in Skyline_Threshold caused by the dominance relationship between tuples in the stream. Therefore, in a setup of one server and three 2SGs, if there are 100 tuples in the stream and each 4th tuple in the stream modifies the Skyline_Threshold, then the total number of feedback messages sent by server would be as shown in Eq. 2.

$$50 = (3 - 1) * 100/4 \qquad (2)$$

The dominance relationship among tuples arriving in the stream affects the number of feedback messages sent by the server.

5   Generating Skyline

5.1   Generating Local Skyline

Refer to the Algorithm 1, every data site maintains a local relation that contains the most recent input tuples received from a source. After the arrival of a new input tuple the values of the old tuple that belongs to the same source are removed from the local relation. After each update in the local relation, local skyline points are computed.

For the sake of simplicity and efficiency we assume that the updates are sent periodically by the sources and when there is any significant change in the price of any stock commodity that exceeds a certain threshold. The updated local skyline points are then sent to the respective 2SG.

Algorithm 1. Maintaing a local relation, and local skyline points at a data site

1: **Input:** $tuple_j$, is the most recent input received by a data site DS_k, sent by stock market S_m.
2: **do**
3: **if** $tuple_j$ has been received **then**
4:     copy $tuple_j$ into the memory
5:     **if** some old $tuple_i$ from S_m already exists in the relation **then**
6:         discrad $tuple_i$
7:         update the local relation R with $tuple_j$
8:         regenerate the local skyline
9:     **else**
10:         include the $tuple_j$ as the first update received from S_m
11:         generate the local skyline
12:     **end if**
13: **end if**
14: **while(true)**

5.2   Generating Semi-Global Skyline

Semi-Global Skyline (SGS) points are computed from the local skyline points received from different data sites. Like a data site, a 2SG also maintains a relation that consists of local skyline points sent by the data sites connected to a 2SG. After receiving a local skyline from a data site the old skyline points of the same data site are replaced by the new skyline points, and a dominance check is performed with any existing local skyline points sent by other data sites. SGS points are computed after updating the local relation and performing a dominance check. Every tuple in the relation maintained by a 2SG is a local skyline point, but not necessarily a SGS point. SGS points generated at each 2SG machine are subset of its local relation consisting of local skyline points received from different data sites.

5.3   Generating Global Skyline

After a skyline query is registered at the server, the server sends the query requests to all 2SGs. The 2SGs interpret the query requests and send them to the data sites for obtaining local skylines. Each data site in return generates and sends its local skyline to its corresponding 2SG. Each 2SG generates SGS points from the local skylines and sends them to the server. A tuple arrived at a data site is gradually promoted to be part of global skyline at server iff it is part of local skyline points generated at a data site, it is not a false-positive SGS point at a 2SG, it is a part of SGS points generated at a 2SG and it is not part of false-positive global skyline points at the server. A false-positive skyline point is a skyline point that is part of local skyline at a data site, but is not part of SGS point at a 2SG, or it is a part of SGS point at a 2SG, but not a part of global skyline points generated at server.

Algorithm 2. Seting Skyline_Threshold and Generating global skyline at the Server

---

1: **Input:** SGS_k are the most recent semi-global skyline points received by server.
2: **do**
3: **if** SGS_k has been received **then**
4:     copy SGS_k into the memory
5:     **if** some other SGS_j points already exist in the memory **then**
6:         compare SGS_k with SGS_j
7:         **if** $SGS_k \succ SGS_j$ **then**
8:             discard SGS_j
9:             update the $Skyline\_Threshold$
10:            send new $Skyline\_Threshold$ information to concerned 2SGs
11:            Merge SGS_k with existing global skyline points
12:        **else**
13:            update the $Skyline\_Threshold$ accordingly
14:            send new $Skyline\_Threshold$ information to concerned 2SGs
15:            Merge SGS_k with existing global skyline points
16:        **end if**
17:    **else**
18:        mark SGS_k as the first SGS points
19:        set the initial $Skyline\_Threshold$
20:        send the initial $Skyline\_Threshold$ information to concerned 2SGs
21:    **end if**
22: **end if**
23: **while(true)**

Server receives SGS points and performs a dominance check and if newly received SGS points qualify to be part of global skyline points then they are merged with the existing points as sated in line 8–11,13–15,18–20 in Algorithm 2 and the $Skyline\_Threshold$ is also updated accordingly. The global skyline points computed at a given point in time t_1 expire after the generation of global skyline points computed at a later point in time t_2, here $t_2 > t_1$.

6   Adjusting Skyline

6.1   Adjusting the Semi-Global Skyline

Refer to the Algorithm 3, every time local skyline points are received from a data site, a 2SG regenerates semi-global skyline points.

If a 2SG has already received a threshold feedback from the server, it performs a dominance check against the threshold. After the dominance check, new SGS points will be sent to the server for generation of global skyline points, iff they found to be better in at least one dimension than the present threshold.

If the newly generated SGS points are not better than the threshold, then they would simply be removed.

If a 2SG has not yet received any threshold information, it means that the server has not received SGS point from any 2SG, in that case a 2SG would immediately send its SGS points to the server. The two way communication between server and 2SGs takes place as depicted in Fig. 2.

Algorithm 3. Adjusting the SGS at 2SG, after reciving feedback from server

---
1: **Input:** Let FB_i be the recent feedback received from server to a 2SG at time t_1. Let SGS_j be a semi-global skyline to be sent to the server at time t_2.
2: **do**
3: **if** FB_i has been received **then**
4:    copy FB_i into the memory
5:    compare SGS_j to the threshold information in FB_i
6:    **if** $FB_i \succ SGS_j$ **then**
7:       discard SGS_j
8:    **else**
9:       send SGS_j to the server
10:   **end if**
11: **end if**
12: **while(true)**

6.2 Adjusting the Global Skyline

After receiving SGS points from a 2SG the server performs actions according to the scenarios as described bellow:

(i) The newly received SGS points are better in all dimensions than the present Refer to lines 3–11 in Algorithm 2, in this situation the present Skyline_Threshold would be updated to the new SGS points, and the new SGS points would also be added into the relation of global skyline points updating any past global skyline points if already existed.

(ii) The received SGS points are the very first SGS points received by the server. Refer to lines 18–20 in Algorithm 2, these SGS points would be set as a Skyline_Threshold, and a feedback is sent to all 2SGs to send their respective SGS points iff the SGS points are better than the Skyline_Threshold in any dimension. A 2SG whose SGS is set as a Skyline_Threshold would not receive any feedback from the server, and in future if it computes any updated SGS, then the updated SGS points are sent immediately to the server.

(iii) The newly received SGS points are better in one or multiple dimensions. Refer to lines 13–15 in Algorithm 2, the Skyline_Threshold at server would be updated according to the new dimensions and contain the dominant points of the newly arrived SGS and the old threshold. The SGS points would be added into the relation of global skyline points.

7 Experimental Evaluation

As our work extends the prior work done by Lu et al. [1], therefore we developed a system based on the techniques described in this paper as well as borrowed some of the concepts from [1]. The approach presented in [1] is simple and therefore, in the forthcoming text we refer to it as the näive approach. The efficiency of näive approach and feedback based approach is measured in the number of tuples processed by server and the required processing time.

7.1 Experimental Setup

We performed multiple experiments. Our setup was based on a single server, three 2SGs where each 2SG represented a distinct region and was connected to four data sites in it's proximity, and there were twelve data sites in total. Each data site sent its local skyline points to corresponding 2SG, and all 2SGs were sending their SGS points to the server. A continuous skyline query as the example in Sect. 4 is used to process streams consisted of three million and six million tuples, based on the historical stock quotes of the NASDAQ [30]. All our experiments were involved in 2, 4 and 6 dimensional data tuples.

7.2 Results and Discussion

Refer to Fig. 3 that shows the number of tuples processed by each approach. In the first round of experiment the stream consisted of 3 million tuples (3MT) and the data dimensionality were gradually set as 2, 4 and 6. The server in Feedback based approach processed 980668 (33 %), 1532321 (51 %) and 1546483 (52 %) of the total tuples respectively while increasing the dimensionality from 2 to 6. While the server in näive approach processed all 3 million tuples (3MT) that it received.

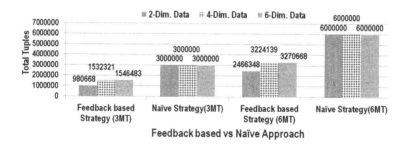

Fig. 3. Tuples processed by server in Feedback based vs. Näive approach

Second experiment consisted of 6 million tuples (6MT). Server in Feedback based approach processed 2466348 (41 %), 3224139 (54 %) and 3270668 (55 %) tuples, while the näive approach processed all 6 million tuples (6MT). The results in Fig. 3 show a considerable reduction in tuples processed by Feedback based strategy.

Refer to Fig. 4 showing the number of feedback messages sent by server to 2SGs. While computing two dimensional skyline points from 3 million and 6 million tuples, server sent 653778 and 1644232 feedback messages respectively.

While computing 4-dimensional data points there were 1021547 and 2149426 feedback messages and the stream consisted of 3 million and 6 million tuples respectively.

In the same way while computing 6-dimensional data points there were 1030988 feedback messages when the stream consisted of 3 million tuples, and 2180445 feedback messages were sent when the stream consisted of 6 million tuples. The number of feedback messages increased with the growing number of input tuples. Likewise, increasing the data dimensionality also affected the dominance relationship among tuples as described in Sect. 4.1 and resulted in an increase in the feedback.

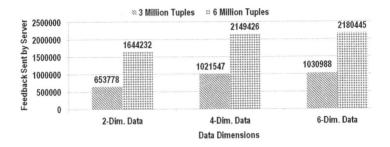

Fig. 4. The affect of data dimensionality over feedback

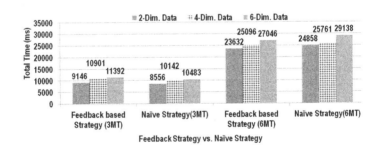

Fig. 5. Processing time of Feedback based vs. Näive approach

Refer to Fig. 5 showing the comparison of processing time between Feedback based and näive approach. While processing stream of 3 million tuples (3MT) with growing number of data dimensionality, both of the approaches have shown almost the same performance. Feedback based approach performed slightly better while processing 6 million tuples (6MT). The results show that even though the feedback based approach is involved in two-stage computation of skyline points it still showed performance nearly identical to the näive approach.

The time spent in computing SGS points at 2SGs is a trade-off to reduce the processing load on server, but as the server has to process considerably reduced amount of data, thus the time saved at server neutralizes the affect.

Figure 6 shows the number of useless tuples discarded by 2SGs while processing the streams. These tuples were discarded to reduce the processing load on the server. While processing stream of 3 million and 6 million tuples and computing 2-dimensional SGS points a total of 2019332 (67 %) and 3533652 (59 %) tuples were discarded respectively. In the same way while computing 4-dimensional SGS points 1467679 (49 %) and 2775861 (46 %) tuples were discarded.

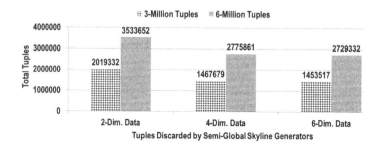

Fig. 6. Early prunning of unwanted tuples by 2SGs

Similarly, 1453517 (48 %) and 2729332 (45 %) tuples were discarded while computing 6-dimensional SGS points over stream of 3 million and 6 million tuples.

Our experiments showed that the approach used by [1], in which server directly receives local skyline points from data sites leads to higher processing load on the server, and wastes precious resources. Feedback based approach considerably reduces the amount of data as well as the false-positives to be processed by the server, allowing it to process continuous skyline queries over data stream, and generate the global skylines in a timely manner. Moreover, the middle layer in feedback based approach provides greater flexibility, and scalability to handle multiple streams. An increase/decrease in number of input streams can be handled by adjusting the number of 2SGs in the middle layer.

Even though the feedback based approach is better than the näive approach in many ways, it also has a trade-off like any other approach that the number of feedbacks depends on the dominance relationship among tuples in the stream as discussed in Sect. 4.1, which can be affected by the growing data dimensionality. Many real life skyline applications do not involve in so many dimensions therefore, the feedback based approach can perform well in a wide variety of fields.

8   Conclusion

We presented an approach to process continuous skyline queries over data streams in a distributed framework. Past strategies to execute skyline queries and process multiple streams can overload the computing resources of the single server. This could lead to a potential bottleneck and affect the applications that require a time bounded response. In our approach, we introduce a feedback based mechanism in a three layer architecture to process continuous skyline queries over data stream. Through extensive experiments we proved that feedback based strategy dramatically reduces the processing load on server, while exhibiting the time requirements identical to other approaches.

References

1. Lu, H., Zhou, Y., Haustad, J.: Continuous skyline monitoring over distributed data streams. In: Gertz, M., Ludäscher, B. (eds.) SSDBM 2010. LNCS, vol. 6187, pp. 565–583. Springer, Heidelberg (2010)
2. Borzsony, S., Kossmann, D., Stocker, K.: The skyline operator. In: Proceedings of the 17th International Conference on Data Engineering, pp. 421–430. IEEE (2001)
3. Endres, M., Roocks, P., Kießling, W.: Scalagon: an efficient skyline algorithm for all seasons. In: Renz, M., Shahabi, C., Zhou, X., Chemma, M.A. (eds.) DASFAA 2015. LNCS, vol. 9050, pp. 292–308. Springer, Heidelberg (2015)
4. Liknes, S., Vlachou, A., Doulkeridis, C., Nørvåg, K.: APSkyline: improved skyline computation for multicore architectures. In: Bhowmick, S.S., Dyreson, C.E., Jensen, C.S., Lee, M.L., Muliantara, A., Thalheim, B. (eds.) DASFAA 2014, Part I. LNCS, vol. 8421, pp. 312–326. Springer, Heidelberg (2014)
5. Endres, M., Kießling, W.: High parallel skyline computation over low-cardinality domains. In: Manolopoulos, Y., Trajcevski, G., Kon-Popovska, M. (eds.) ADBIS 2014. LNCS, vol. 8716, pp. 97–111. Springer, Heidelberg (2014)
6. Chester, S., Sidlauskas, D., Assent, I.: Bøgh, K.S.: Scalable parallelization of skyline computation for multi-core processors (2015)
7. Chomicki, J., Godfrey, P., Gryz, J., Liang, D.: Skyline with presorting. In: ICDE, vol. 3, pp. 717–719 (2003)
8. Tan, K.L., Eng, P.K., Ooi, B.C., et al.: Efficient progressive skyline computation. In: VLDB, vol. 1, pp. 301–310 (2001)
9. Kossmann, D., Ramsak, F., Rost, S.: Shooting stars in the sky: an online algorithm for skyline queries. In: Proceedings of the 28th International Conference on Very Large Data Bases, VLDB Endowment, pp. 275–286 (2002)
10. Papadias, D., Tao, Y., Fu, G., Seeger, B.: An optimal and progressive algorithm for skyline queries. In: Proceedings of the 2003 ACM SIGMOD International Conference on Management of Data, pp. 467–478. ACM (2003)
11. Yiu, M.L., Mamoulis, N.: Efficient processing of top-k dominating queries on multidimensional data. In: Proceedings of the 33rd International Conference on Very Large Data Bases, VLDB Endowment, pp. 483–494 (2007)
12. Papadias, D., Tao, Y., Fu, G., Seeger, B.: Progressive skyline computation in database systems. ACM Trans. Database Syst. (TODS) 30(1), 41–82 (2005)
13. Dellis, E., Seeger, B.: Efficient computation of reverse skyline queries. In: Proceedings of the 33rd International Conference on Very Large Data Bases, VLDB Endowment, pp. 291–302 (2007)
14. Morse, M., Patel, J.M., Jagadish, H.: Efficient skyline computation over low-cardinality domains. In: Proceedings of the 33rd International Conference on Very Large Data Bases, VLDB Endowment, pp. 267–278 (2007)
15. Lee, K.C., Zheng, B., Li, H., Lee, W.C.: Approaching the skyline in Z order. In: Proceedings of the 33rd International Conference on Very Large Data Bases, VLDB Endowment, pp. 279–290 (2007)

16. Pei, J., Jiang, B., Lin, X., Yuan, Y.: Probabilistic skylines on uncertain data. In: Proceedings of the 33rd International Conference on Very Large Data Bases, VLDB Endowment, pp. 15–26 (2007)

17. Balke, W.-T., Güntzer, U., Zheng, J.X.: Efficient distributed skylining for web information systems. In: Bertino, E., Christodoulakis, S., Plexousakis, D., Christophides, V., Koubarakis, M., Böhm, K. (eds.) EDBT 2004. LNCS, vol. 2992, pp. 256–273. Springer, Heidelberg (2004)

18. Wu, P., Zhang, C., Feng, Y., Zhao, B.Y., Agrawal, D.P., El Abbadi, A.: Parallelizing skyline queries for scalable distribution. In: Ioannidis, Y., Scholl, M.H., Schmidt, J.W., Matthes, F., Hatzopoulos, M., Böhm, K., Kemper, A., Grust, T., Böhm, C. (eds.) EDBT 2006. LNCS, vol. 3896, pp. 112–130. Springer, Heidelberg (2006)

19. Huang, Z., Jensen, C.S., Lu, H., Ooi, B.C.: Skyline queries against mobile light-weight devices in manets. In: Proceedings of the 22nd International Conference on Data Engineering, ICDE 2006 p. 66. IEEE (2006)

20. Zhu, L., Tao, Y., Zhou, S.: Distributed skyline retrieval with low bandwidth consumption. IEEE Trans. Knowl. Data Eng. **21**(3), 384–400 (2009)

21. Huang, Z., Lu, H., Ooi, B.C., Tung, A.: Continuous skyline queries for moving objects. IEEE Trans. Knowl. Data Eng. **18**(12), 1645–1658 (2006)

22. Lin, X., Yuan, Y., Wang, W., Lu, H.: Stabbing the sky: Efficient skyline computation over sliding windows. In: Proceedings 21st International Conference on Data Engineering, ICDE 2005, pp. 502–513. IEEE (2005)

23. Tao, Y., Papadias, D.: Maintaining sliding window skylines on data streams. IEEE Trans. Knowl. Data Eng. **18**(3), 377–391 (2006)

24. Wu, P., Agrawal, D., Egecioglu, O., El Abbadi, A.: Deltasky: Optimal maintenance of skyline deletions without exclusive dominance region generation. In: IEEE 23rd International Conference on Data Engineering, ICDE 2007, pp. 486–495. IEEE (2007)

25. Zhang, Z., Cheng, R., Papadias, D., Tung, A.K.: Minimizing the communication cost for continuous skyline maintenance. In: Proceedings of the 2009 ACM SIGMOD International Conference on Management of data, pp. 495–508. ACM (2009)

26. Mouratidis, K., Papadias, D., Hadjieleftheriou, M.: Conceptual partitioning: an efficient method for continuous nearest neighbor monitoring. In: Proceedings of the 2005 ACM SIGMOD International Conference on Management of Data, pp. 634–645. ACM (2005)

27. Mullesgaard, K., Pedersen, J.L., Lu, H., Zhou, Y.: Efficient skyline computation in mapreduce. In: 17th International Conference on Extending Database Technology (EDBT), pp. 37–48 (2014)

28. Lu, H., Zhou, Y., Haustad, J.: Efficient and scalable continuous skyline monitoring in two-tier streaming settings. Inf. Syst. **38**(1), 68–81 (2013)

29. Cui, B., Lu, H., Xu, Q., Chen, L., Dai, Y., Zhou, Y.: Parallel distributed processing of constrained skyline queries by filtering. In: IEEE 24th International Conference on Data Engineering, ICDE 2008, pp. 546–555. IEEE (2008)

30. NASDAQ. http://www.infochimps.com/. Accessed 03 December 2014

Performance and Tuning

Partitioning Templates for RDF

Rebeca Schroeder[1]([⊠]) and Carmem S. Hara[2]

[1] Universidade Do Estado de Santa Catarina - UDESC,
Joinville, SC 89.219-710, Brazil
`rebeca.schroeder@udesc.br`
[2] Universidade Federal Do Paraná- UFPR, Curitiba, PR 81531-990, Brazil
`carmem@inf.ufpr.br`

Abstract. In this paper, we present an RDF data distribution approach which overcomes the shortcomings of the current solutions in order to scale RDF storage both with the volume of data and query requests. We apply a workload-aware method that identifies frequent patterns accessed by queries in order to keep related data in the same partition. In order to avoid exhaustive analysis on large datasets, a summarized view of the datasets is considered to deploy our reasoning through partitioning templates for data items in an RDF structure. An experimental study shows that our method scales well and is effective to improve the overall performance by decreasing the amount of message passing among servers, compared to alternative data distribution approaches for RDF.

1 Introduction

We have witnessed an ever-increasing amount of RDF data made available in different application domains. The DBpedia dataset[1] has now reached a size of 2.46 billion RDF triples extracted from Wikipedia. According to the W3C, some commercial datasets may be even bigger reaching the score of 1 trillion triples[2]. The envisioned architecture to manage these huge datasets is based on elastic cloud-based datastores supported by parallel techniques for querying massive amounts of data [5]. In order to scale RDF storage, datasets must be partitioned across multiple commodity servers. By placing partitions on different servers, it is possible to speedup query processing when each server can scan its partitions in parallel. On the other hand, message passing among servers can be required at query time when related data is spread among arbitrary partitions. These rounds of communication over the network can become a performance bottleneck, leading to high query latencies. Therefore, the scalability of query processing depends on how data is partitioned or replicated across multiple servers.

RDF data are represented by triples given by `subject-predicate-object` (`s, p, o`) statements. In an RDF dataset, triples are related to each other representing a graph. Thus, the RDF partitioning problem has been addressed as a graph cut problem [5,15]. Likewise the general problem, partitioning a

---

[1] http://wiki.dbpedia.org/Datasets.
[2] http://www.w3.org/wiki/LargeTripleStores.

© Springer International Publishing Switzerland 2015
T. Morzy et al. (Eds.): ADBIS 2015, LNCS 9282, pp. 305–319, 2015.
DOI: 10.1007/978-3-319-23135-8_21

distributed database is known to be NP-hard [8] and, therefore, heuristic-based approaches become more attractive. In general, the heuristics applied by current methods are solely based on the RDF graph structure, generating partitions that do not express query patterns of the workload. As result, the query performance decreases when data required by the same query pattern is distributed over different servers. Besides the workload-oblivious reasoning, most of the current approaches apply a graph partitioner algorithm on the whole RDF graph. However, large graphs are hard to partition.

In this paper, we introduce a data partitioning approach which overcomes the shortcomings of current solutions by reasoning over a set of query patterns assumed as the expected workload. The contribution of this approach is twofold. First, partitions are extracted from clusters of data accessed together by frequent query patterns. Such coverage of query patterns provides scalability for query processing by reducing the amount of message passing among machines at query time. Second, we are able to define how data items must be clustered solely based on the structure of query patterns. The query patterns are formulated over a summarization schema that represents the data structures for an RDF dataset. Thus, we define partitioning *templates* as the partitioning strategy to be applied to instances of an RDF structure. By doing so, we avoid exhaustive analyses on the whole data graph for defining data partitioning.

Despite the fact that most RDF datasets are schema-free, the lack of a schema makes it harder to formulate queries on RDF graphs and define suitable strategies for indexing and clustering. In fact RDF datasets range from structured data (e.g. DBLP) to unstructured data (e.g. Wikipedia). However, there is a bit of regularity in RDF data [9] and it is relatively easy to recover large part of the implicit class structure underlying data stored in RDF triples as demonstrated in [7]. In our approach, RDF structures are applied to identify the query patterns in order to partition datasets. By following such a workload-agnostic approach, we are able to efficiently handle the most frequent queries. Likewise in traditional design approaches and the so-called 20–80 rule, we favor the important 20 % of queries which corresponds to 80 % of the total database load.

The rest of the paper is organized as follows. Section 2 introduces the partitioning problem. Our workload characterization method is presented in Sect. 3. In Sects. 4 and 5, we describe our partitioning method involving data fragmentation and allocation. In Sect. 6, we experimentally investigate the impact of our method and compare to related approach. We discuss related work in Sect. 7 and conclude in Sect. 8.

2 Preliminaries and Partitioning Objective

RDF data can be defined as a finite set of triples composed of `subject`, `property` and `object` (`s`, `p`, `o`). Assume there are pairwise disjoint infinite sets $\mathcal{U}$ and $\mathcal{L}$, where $\mathcal{U}$ are URIs denoting Web resources, and $\mathcal{L}$ are literals. Thus, an RDF triple (`s`, `p`, `o`) $\in (\mathcal{U} \times \mathcal{U} \times \{\mathcal{U} \cup \mathcal{L}\})$. RDF follows a data model in which triples are related to each other, which can be represented as a directed graph. We denote an RDF graph as D. That is, D is a set of triples which denote facts where the

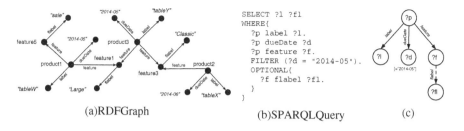

Fig. 1. RDF graph and a SPARQL query example

subject is the origin node of a property labelled edge directed to its object node. As an example, the subject `product1` is related to the object `feature1` through the property `feature` in Fig. 1a.

SPARQL is the W3C Recommendation language for querying RDF datasets. The SPARQL core syntax is based on a set of triple patterns like RDF triples except that subjects, properties and objects may be defined as variables. In our work, pattern graphs represent the conjunctive fragment of SPARQL queries. We assume the existence of a set $\mathcal{V}$ of variables that is disjoint of the sets $\mathcal{U}$ and $\mathcal{L}$. Variables in $\mathcal{V}$ are denoted by a question mark (?) prefix.

Definition 1 *(Pattern Graph): A pattern graph is denoted by $G = (V, E, r)$ where: (1) $V \subseteq \{\mathcal{V} \cup \mathcal{U} \cup \mathcal{L}\}$; (2) $E \subseteq (V \times \mathcal{U} \times V)$, where for each edge $(\hat{s}, \hat{p}, \hat{o}) \in E$, $\hat{s}$ is the source of the edge, $\hat{p}$ is the property, $\hat{o}$ is the target of the edge; and (3) r is a set of filter expressions for variable nodes in G. A filter is expressed in the form $?x \; \theta \; c$, where $?x \in \mathcal{V}$, $c \in \{\mathcal{U} \cup \mathcal{L}\}$ and $\theta \in \{=, >, \leqslant, <, \geqslant\}$. Hereafter, we use $V(G)$ and $E(G)$ to denote the set of vertices and the set of edges of a pattern graph, respectively.*

An example of pattern graph is given in Fig. 1c where variable nodes are annotated with the associated filter expressions. The conjunctive fragment of SPARQL queries involving operators AND, FILTER, OPTIONAL and UNION can be represented as graph patterns as follows. Pattern triples are represented by connected nodes denoting operators *AND* (solid edges) and *OPTIONAL* (dashed edges). To simplify, we represent pattern graphs connected by the *UNION* operator as independent graphs. Figure 1b shows a SPARQL query that retrieves data for products and features associated with product where the *due-Date* is "2014-05". The equivalent representation for the pattern graph is shown in Fig. 1c. Observe that although in the example the query is represented as a tree, cycles are admitted by the pattern graph definition.

The workload is defined as pattern graphs representing a set of SPARQL queries Q. Given that SPARQL is a graph-matching language, processing a query against RDF graphs consists of a subgraph matching problem which can be computed by graph homomorphism [17]. The subgraphs shown in Fig. 2a correspond to matches of the pattern graph of Fig. 1c applied to the RDF graph in Fig. 1a. We use $B(q) = \{b_1, ..., b_n\}$ to denote the result of a query q, where b_i is a subgraph of an RDF graph D, i.e., $b_i \subseteq D$.

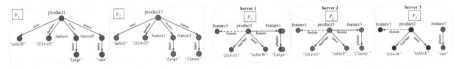

(a) SPARQL Query Results   (b) Query Segmentation on Partitioned Datasets

Fig. 2. SPARQL query results on partitioned data

Consider now processing the same query over a partitioned dataset. Figure 2b illustrates the graph in Fig. 1a partitioned across 3 server. When the query is issued, it is processed in parallel in all servers. Ideally, each subgraph in a result should be stored in a single server. However, in our example, subgraphs b_1 and b_2 are segmented across two servers. Retrieving b_1 requires *Server1* and *Server3* to be accessed, while *Server1* and *Server2* are needed to retrieve b_2. In order to avoid this message passing among servers, the main goal of our approach is to partition data so that query can be processed in parallel without inter-server communication whenever it is possible. More formally, we are interested in generating a partitioning $\mathcal{P} = \{P_1, ..., P_m\}$, for an RDF graph denoted by D across m servers, where the amount of partitions required to retrieve each subgraph in a query result $B(q)$ is minimized. To this end, we define the segmentation of the subgraphs in $B(q)$ with respect to a partitioning $\mathcal{P}$ and a query q as follows:

Definition 2 *(Query Segmentation): Given a partitioning $\mathcal{P}$ of an RDF graph D, the query segmentation measure $\hat{P}$ of $\mathcal{P}$ with respect to q is defined as:*

$$\hat{P}(q, \mathcal{P}) = \left| \{(b, P) \in (B(q) \times \mathcal{P}) | b \cap P \neq \varnothing\} \right| - \left| B(q) \right| \tag{1}$$

In this equation, the minuend determines how many partitions (or servers) have to be accessed to retrieve all triples in each subgraph result. That is, given a subgraph result $b \in B(q)$ and a partition P, a pair (b, P) is in the minuend set whenever P contains a triple in b. Ideally, no subgraph should be segmented. That is, the size of the minuend should be equal to the number of subgraphs in the result $B(q)$, which leads to $\hat{P} = 0$. Intuitively, $\hat{P}$ measures the amount of inter-server communication to compute a query result. Given that a workload consists not only of a single query, but a set of queries Q, the overall objective of our partitioning strategy is to minimize $\hat{P}$ for the set Q. To this end, we assume that each query q in the set is associated with its expected frequency in a period of time, which is denoted by $f(q)$. Thus, we can formally define our problem as to find a partitioning $\mathcal{P}$ that minimizes the following equation:

$$min \sum_{q \in Q} f(q).\hat{P}(q, \mathcal{P}) \tag{2}$$

Observe that frequent queries have a higher impact on the equation than infrequent ones. Intuitively, our strategy is based on favoring the most frequent queries in the workload. To achieve our goal, we characterize the workload for

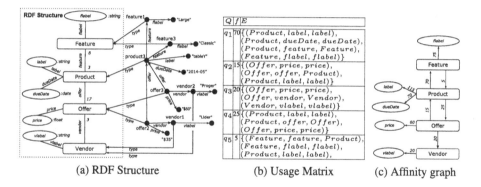

(a) RDF Structure   (b) Usage Matrix   (c) Affinity graph

Fig. 3. Workload data

examining the paths traversed by the queries and their frequencies in order to quantify the affinity between pairs of nodes. Such affinity measure is the basis for our partitioning reasoning.

3   Workload Characterization

In this section we present a method for representing workload information. The core of this method is based on identifying and measuring affinity relations among RDF nodes. We start by defining an RDF Structure, containing both the structure of the RDF graph and the expected size of its instances. Although RDF can define a schema-free model, in general an RDF graph represents both schema and instances. Most datasets define the `type` property connecting entities to their respective classes. In Fig. 3a, the RDF Structure is illustrated in the dashed shape containing classes as well as relationships among them. An RDF Structure is an undirected cyclic graph defined as a 6-tuple $S = (C, L, l, A, s, o)$, where (1) C is a set of labelled nodes representing RDF classes; (2) L is a set of labelled nodes denoting class properties with literal values; (3) l assigns a data type to each node in L; (4) A is a set of undirected edges $(n_1, n_2) \in (C \times \{C \cup L\})$ which corresponds to associations between nodes; (5) s is a function that assigns the expected size for the instances of nodes in $\{C \cup L\}$; and (6) o gives the expected cardinality of associations between two nodes; that is, it is a function that maps a pair in $(C \times \{C \cup L\})$ to an integer that defines for each node $n_1 \in C$ the expected number of occurrences of associations to a node $n_2 \in \{C \cup L\}$.

Figure 3a shows an RDF Structure. In the example, $o(\mathtt{Product}, \mathtt{Feature}) = 8$ because the average number of occurrences of `Feature` associated to an instance of `Product` is 8. Similarly, an instance of `Feature` is related to 3 instances of `Product` in average. That is, $o(\mathtt{Feature}, \mathtt{Product}) = 3$. Besides, there are multi-valued relationships between $(\mathtt{Product}, \mathtt{Offer})$ and $(\mathtt{Vendor}, \mathtt{Offer})$. We assume that for the remaining associations relating any other nodes n_1 and n_2 in the example, $o(n_1, n_2) = 1$. The size of a node n is not depicted in the example. If n is a literal node, $s(n)$ is the number of bytes needed for storing its value.

For class nodes, on the other hand, the size corresponds to the size required to store their property structures. To simplify the example, we consider that for any node n, $s(n) = 1$.

Given a representation of an RDF Structure, we now turn to the workload characterization. We define a workload as a set of queries Q represented as pattern graphs and a function f that defines the expected frequency of each query in Q. The workload can be represented as a usage matrix as depicted in Fig. 3b. According to the example, q_1 is expected to be executed 70 times and involves the literal nodes `label`, `dueDate`, `flabel` and the classes `Product` and `Feature`.

Given a workload on an RDF Structure, the affinity of two nodes n_i and n_j in an RDF Structure as the frequency they are accessed together by any query in the workload. Towards this goal, an affinity function $aff(n_i, n_j)$ takes as input a set of queries Q and computes the sum of frequencies of queries that involve both n_i and n_j by a path in a specific direction, i.e., n_i is the source node and n_j is the target node. More formally, we define $Q_{ij} = \{q \in Q \mid (n_i, p_{ij}, n_j) \in q\}$, and $aff(n_i, n_j) = \sum f(q)$, $q \in Q_{ij}$. As an example, consider the workload given in Fig. 3b. The affinity between *Product* and *label* consists of the sum of frequencies of queries q_1, q_2, q_4 and q_5. Thus, $aff(\texttt{Product}, \texttt{label}) = f(q_1) + f(q_2) + f(q_4) + f(q_5) = 115$. The affinity function can be used to label edges in a directed graph involving all nodes in an RDF Structure, as depicted in Fig. 3c. We refer to this graph as an affinity graph, which is defined as a tuple $\mathcal{A} = (N, \hat{E}, \texttt{aff})$, where N is the set of nodes in the RDF Structure and $\hat{E}$ is a set of edges which relates two nodes n_i and n_j by an affinity value $(\texttt{aff}(n_i, n_j))$.

We present our partitioning technique in two steps. The first consists of data fragmentation. That is, determining how to cut an RDF Structure in order to keep closely related data by affinity relations in a storage unit. The second concerns data clustering thus, it relates to the problem of allocating related fragments in the same server.

4   RDF Fragmentation

Distributed query processing performance is not only affected by the amount of message passing, but also by the size of the messages. A suitable size for messages motivated us to adopt a storage threshold as the basis for our partitioning technique. We refer to this storage threshold as Γ. Intuitively, our goal is to partition nodes of an RDF Structure, such that partitions contain as many correlated nodes as possible that can fit in a given storage size. In what follows, we introduce the RDF fragmentation problem and our proposal for solving it.

Given an RDF Structure $S = (C, L, l, A, s, o)$ and an affinity graph $\mathcal{A} = (N, \hat{E}, aff)$, we are interested in obtaining a fragmentation template $T = \{t_1, ..., t_m\}$, $m \geq 1$, such that t_i is a subgraph of S, $\bigcup_{i=1}^{m}(t_i) = (N, E')$, where $E' \subseteq E$ and each t_i is defined with disjoint sets of nodes. Figure 4a presents an example of a fragmentation template for the RDF Structure depicted in Fig. 3a. Instances of template t_1 extracted from an RDF graph according to this fragmentation template are illustrated in Fig. 4.

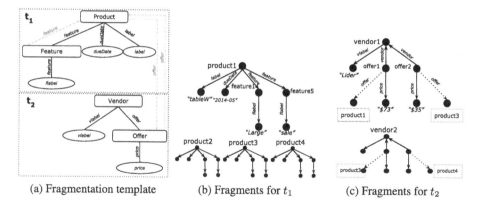

(a) Fragmentation template   (b) Fragments for t_1 (c) Fragments for t_2

Fig. 4. Templates and fragments

Given that the fragmentation process is based on a storage threshold, we also need the notion of the size of a fragmentation template $t_i \in T$. The size of t_i is given by the sum of the expected number of occurrences of nodes multiplied by their sizes. The tree composition of fragmentation templates requires us to measure the node occurrence in the nested structure. The function $occ(n)$ maps each node in a template t_i to its expected number of occurrences in an instance of t_i. It is recursively defined as follows: $occ(n) = 1$ if n is the root node of t_i, and $occ(n) = occ(p) \times o(p, n)$ where p is a parent node of n in t_i. The size of t_i is denoted by $size(t_i) = \sum_{n \in t_i} (occ(n) \times s(n))$.

In order to formally state our problem, we need the notion of a strongly correlated set scs for a node in the affinity graph, defined as follows: $scs(n) = \{n' | aff(n, n') \geq aff(n', n'')$ for every node n'' directly connected to $n'\}$. Intuitively, scs determines which nodes have stronger affinity with n than with any other in the graph. We denote by scs^+ the transitive closure of the scs relation.

We can now state our fragmentation problem: Find T such that the following conditions are satisfied: (1) $size(t_i) \leq \Gamma$ for every $t_i \in T$; and (2) if n_1 and n_2 are nodes in the same fragment then $n_2 \in scs^+(n_1)$. The first condition defines that all fragments in T must fit in Γ and the second generates fragments that are related by affinity values higher than the values with nodes in other fragments.

As an example, consider $\Gamma = 20$ and the affinity graph depicted in Fig. 3c. The fragmentation template in Fig. 4a satisfies our conditions because (1) the size of templates fits in the storage threshold, that is $size(t_1) = 19$ and $size(t_2) = 4$; and (2) the affinity between any node in t_1 with any node in t_2 is lower than the affinity between any pair of nodes in the same fragment, for example, $aff(\texttt{Offer}, \texttt{Product}) < aff(\texttt{Offer}, \texttt{Vendor})$.

We propose a fragmentation algorithm based on RDF Structures and workload. The Algorithm *affFrag* takes as input an RDF Structure S with information on node sizes and number of occurrences, an affinity graph $\mathcal{A}$ and a storage threshold Γ. The algorithm computes templates of fragments based on strongly correlated sets of nodes if their sizes lie within Γ.

Algorithm 1. Algorithm affFrag

Input: RDF Structure $S = (C, L, l, A, s, o)$, Affinity Graph $\mathcal{A} = (N, E, aff)$ and Γ
Output: T fragmentation template

```
1  T ← {};
2  allNodes ← N;
3  allEdges ← E;
4  repeat
5  |    (n₁, n_b) ←edge in allEdges with highest affinity;
6  |    tNodes ← {n₁};
7  |    tEdges ← {};
8  |    tSize ← s(n₁);
9  |    Occ(n₁) ← 1;
10 |    border ← {(n₁, n_b)|n_b ∈ allNodes};
11 |    allNodes ← allNodes − {n₁};
12 |    while tSize < Γ and border! = {} do
13 |    |    (n₁, n_b) ← extract edge from border with highest affinity, where n₁ ∈ tNodes and n_b ∉ tNodes;
14 |    |    n_bEdges ← {(n_b, n) ∈ allEdges|n ∈ allNodes};
15 |    |    if for all edges e ∈ n_bEdges: aff(e) ≤ aff(n₁, n_b) then
16 |    |    |    Occ(n_b) ← Occ(n₁) × o(n_b);
17 |    |    |    if s(n_b) × Occ(n_b) + tSize ≤ Γ then
18 |    |    |    |    tNodes ← tNodes ∪ {n_b};
19 |    |    |    |    tEdges ← tEdges ∪ {(n₁, n_b)};
20 |    |    |    |    border ← border ∪ n_bEdges;
21 |    |    |    |    allNodes ← allNodes − {n_b};
22 |    |    |    |    tSize ← tSize + s(n_b) × Occ(n_b);
23 |    |    |    end
24 |    |    end
25 |    end
26 |    T ← T ∪ {(tNodes, tEdges)};
27 |    allEdges ← allEdges − tEdges;
28 until allNodes = {};
29 output T;
```

The algorithm processes the edges in $\mathcal{A}$ in descending order of affinity. Given an edge (n_1, n_b), the primary goal is to compute $scs(n_1)$. The node n_1 is set to be the root of the fragment being computed because it is the source node of the edge with the highest affinity. A new fragment is generated by processing edges (n_1, n_b) in *border* as follows: n_b is only considered to be inserted in the current fragment if it is related with higher affinity to some element in the current fragment than to any other outside the fragment (Lines 14-15). According to Line 13, the candidate nodes are processed in descending order of affinity in order to fill up the fragment with those with highest affinity. At the end, all nodes have been assigned to some fragment. However, before inserting new nodes in the *tNodes* we check whether it is possible to do so within the size of Γ given the size and occurrence of the node to be included (Line 16-17).

As an example, consider the affinity graph of Fig. 3c and $\Gamma = 20$ as the input to *affFrag*. The first edge to be processed is the one with highest affinity involving nodes **Product** and **label**. **Product** is inserted into a fragment t_1 as the root node. The size of t_1 is initially set to 1, given our assumption that all nodes have size 1. Since this is below the threshold, we keep inserting nodes to t_1 among those connected to **Product** which are kept in *border*. The one with highest affinity is **label**. Such node is inserted in t_1, since it is not connected to any other node with higher affinity and this insertion does not exceed the value of Γ. The same happens for inserting nodes **dueDate**, **Feature** and **flabel** into t_1. At this point, $tSize = 19$ given the simple occurrence of **dueData** and **label** with the multiple occurrence of **Feature** and **flabel**. The next edges in *border* to be considered relates **Product** to **Offer** and **price**. **Offer** should not be inserted in the fragment because its affinity is higher with nodes that are not in the current fragment. Thus, the first fragment is created with nodes

Product, label, dueDate, Feature and flabel. A similar process creates the second fragment with Offer, price, Vendor and vlabel. The final fragmentation template generated is the one depicted in Fig. 4a.

The fragmentation template defines how to partition instances of an RDF Structure, i.e., an RDF graph. Thus, a fragment is generated for each instance of the *root* node according to the fragmentation template of $t_i \in T$. In the example, t_1 must generate fragments for each *product* instance. According to the RDF graph of Fig. 3a, the fragment generated for product instances may be represented by the trees in Fig. 4b.

5   Clustering Fragments

Given our approach for the fragmentation problem, we now turn to the allocation problem. That is, given that a fragment is our storage unit, we are now interested in determining which fragments should be allocated in the same server. Although our fragmentation algorithm cuts the affinity graph based on affinity relations, nodes in distinct fragments may still keep strong affinity relations. This is because the fragmentation process has been designed to satisfy a storage threshold. Since there may be several template elements connected by affinity relations, we choose to group the ones with stronger affinities. More specifically, consider a fragmentation template $T = \{t_1, ..., t_m\}$ defined based on an affinity graph $\mathcal{A} = (N, \hat{E}, \mathtt{aff})$. Let $E_T \subseteq \hat{E}$ be the set of edges connecting a node in a fragment t_i to the root of a fragment t_j. Observe that it is possible that $i \neq j$ as well as $i = j$. By connecting *templates* through a *root* node, we are able to extend their tree structures to define a nesting arrangement among related data. We define a clustering template as $G = \{g_1, ..., g_n\}, n \leq m$, such that G is a forest of linked fragmentation templates. Similar to the *affFrag* algorithm, groups in G are built considering edges in E_T in descending order of affinity values. Although we do not define a threshold for the group size, it is limited by the storage capacity of the server.

According to the fragmentation template in Fig. 4a, the dashed arrows denote unprocessed edges in the fragmentation process. As discussed before, only edges directed to root nodes in template elements are considered to define clusters of fragments. Here, the edges (Feature, Product) and (Offer, Product) meet this requirement. Given that both edges are directed to Product, we choose only one of them in order to nest Product and keep the tree structure among the template elements. To do so, we choose the one with the highest affinity. Figure 5a presents a clustering template that relates t_1 and t_2 through the edge (Offer, Product) with the highest affinity. Instances of this cluster template are presented in Fig. 5b.

We apply a clustering template to an RDF graph in order to extract fragments and cluster them properly. Some issues can arise in this process. First, a fragment should be generated for each of the root classes in the fragmentation template. However, it is possible that more fragments are required given by the variability of the size of the nodes and the number of instances for multi-valued relationships in the RDF graph. It is important to remind that both the size

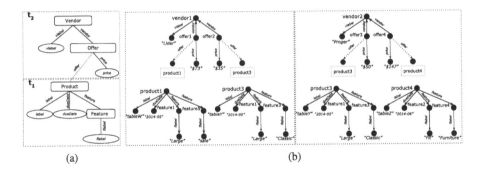

(a)   (b)

Fig. 5. Clustering templates and fragments

and the instances considered in the RDF structure correspond to average values provided as the expected workload. These values are applied to predict the size of fragments in order to define fragmentation templates. In addition, we create edges to represent edges unprocessed by the fragmentation process. To do so, edges are created in the fragments that contain their source nodes. As an example, notice that the edge (Offer, Product) denotes the cut between the fragmentation templates t_1 and t_2 in Fig. 4a. However, the edges among Offer and Product instances are created in the instances of t_2 in order to keep the connection among fragments as depicted in Fig. 5a.

The tree structure created by clustering and fragmentation templates may produce some data redundancy of nested data related to multi-valued relationships. However, we control the amount of replicas by applying a threshold to the amount of replicated data allowed. Due to space limitations, we omit a detailed discussion here.

6   Experimental Study

We have developed *ClusterRDF*, a system to deploy our approach based on an architecture where RDF data is partitioned across a set of servers over a distributed in-memory key-value store. We use the key-value datastore Scalaris [12] as a scalable system to leverage scalability and content locality in order to support our clustering solution. We have conducted an experimental study for determining the effect of our approach on the performance of query data retrieval. We compare *ClusterRDF* with its closest related approaches: the one introduced by Huang et al. [5] and Trinity.RDF [16] using the Berlin SPARQL Benchmark (BSBM).

Huang et al. applies the METIS [1] partitioner on an RDF graph, followed by a replication step to overlap data across partitions according to an *n-hop* guarantee. We refer to this approach as *METIS-2hops* because we have implemented the undirected 2-hop guarantee version of this method. Although Trinity.RDF is focused on providing a query engine for RDF data, this system considers a hash partitioning of RDF nodes and the power law distribution of node degrees to cluster data.

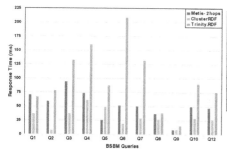

Dataset	#Triples	Size	Triple Overhead	
			ClusterRDF	Metis-2hops
BSBM_1	40405	10.2MB	14141	27071
BSBM_2	75620	19.2MB	22686	44615
BSBM_3	191650	48.9MB	67329	120739
BSBM_4	375163	96MB	105045	213842
BSBM_5	3567636	922.3MB	891909	1748141
BSBM_6	35300350	9.97GB	7766077	15532154
BSBM_7	100399052	27GB	20079810	40159620

(b) Statistics of datasets

(a) Response Time - 8 servers and BSBM_5

Fig. 6. Response time and statistics

BSBM provides a workload with 12 queries and a data generator that supports the creation of arbitrarily large datasets using the number of products as scale factor. Among the 12 queries defined for the benchmark, we have chosen 11, because the remaining one does not satisfy our definition of a pattern graph. For a specific dataset size and workload provided by BSBM, we have generated data clusters according to *ClusterRDF*, *METIS-2hops* and *Trinity.RDF*. Figure 6b summarizes the statistics of the datasets used in this study. As expected, *ClusterRDF* and *Metis-2hop* produce space overhead in terms of triple replication. However, *Metis-2hop* produces twice as many triples compared to our method.

The goal of the experiments reported in this section is to determine the effect of our clustering method on the system performance, and compare it with both *Metis-2hops* and *Trinity.RDF*. The comparison is based on the response time required to retrieve query data from the datastore.

First, we compare the clustering approaches on a cluster of 8 servers and BSBM_5 dataset. The results are shown in Figs. 6a and 7b. The reported times in milliseconds are the average values computed over multiple runs of the experiment and represent the cost of retrieving query data in parallel on a distributed datastore. Each server in the distributed system starts a thread and performs an arbitrary number of local or cross-server requests to retrieve the query data. In such a parallel retrieval, the thread that executes the highest number of cross-server requests determines the query response time. We have collected both the maximum number of distributed requests issued by a single server as well as the total number of distributed requests for all threads in Fig. 7a. Observe that the total number of distributed requests corresponds to the query segmentation denoted by the $\hat{P}$ measure (Definition 2). In addition, we have collected the total number of requests (local and distributed) in Fig. 7b. Observe that the latter corresponds to the size of query results. That is, it is a measure of the total number of fragments retrieved.

Cross-server Requests. As expected, there is a direct correspondence between the number of distributed requests and the response time. That is, a high number of cross-server requests induces a high cost to retrieve data spread among

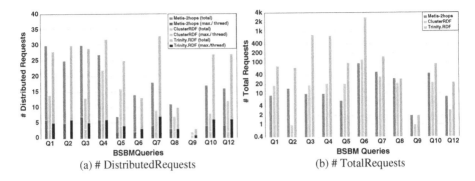

(a) # DistributedRequests

(b) # TotalRequests

Fig. 7. Number of requests- 8 servers and BSBM_5

distributed servers. Indeed, observe that the execution of $Q1$ on *ClusterRDF* requires at most 4 servers accesses per thread, which takes 37.27 ms. The execution of the same query on the *Metis-2hops* and *Trinity.RDF* almost doubles the number of requests and has the same effect on the response time (70.94 ms and 67.52, respectively).

Intuitively, the number of cross-server requests required to retrieve query data measures the effectiveness of the partitioning methods. The difference between the results for the approaches can be explained by the coverage that each method provides in terms of the query patterns. We may say that *Metis-2hops* assures a 2-hop coverage for any pattern graph. However, a 2-hop guarantee is not enough to cover the whole pattern of the majority of queries in the BSBM workload.

Trinity.RDF provides a simple pattern graph coverage in most cases given its fine-grained storage unit based on RDF nodes. This explains why *Trinity.RDF* presents the worst results among the three. *ClusterRDF* provides a complete coverage for queries $Q2$ and $Q6$, given that requests are issued to only one server. For the remaining queries, *ClusterRDF* does not avoid cross-server requests. However, it reduces the number of servers to be accessed if compared to the two other alternatives. The results reported in Fig. 6a show that *ClusterRDF* outperforms *Metis-2hops* and *Trinity.RDF* for most queries, except for $Q5$ and $Q9$. This is because *ClusterRDF* assigns data to clusters according to the access pattern of the most frequent queries of the workload.

Total Requests. The size of query results is reported by the quantity of total requests in Fig. 7b. This measure represents the total amount of fragments (storage units) retrieved. Scalaris provides a functionality for packing a set of requests for the same server into a single message for minimizing the cost of message passing. We have observed that the cost of these packed message can be ignored when the amount of requests is up to 10 requests per server. This measure is also related to the amount of irrelevant data in the fragments being retrieved. Notice that *ClusterRDF* requires a lower number of server requests than *Metis-2hops* in $Q6$, however *ClusterRDF* achieves a higher number of fragment requests. This can be explained by the fact that the requested data are in the same cluster but

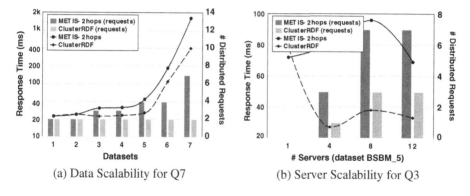

(a) Data Scalability for Q7   (b) Server Scalability for Q3

Fig. 8. Data and server scalability

probably not in the same fragment. In *Trinity.RDF*, this amount is even bigger for all queries because of the its fine-grained storage model.

Data Scalability. We test the methods running on a cluster of 8 servers on 7 datasets (BSBM_1 to BSBM_7) of increasing sizes. The results are shown in Fig. 8a for query 7(in logarithmic scale). In general, the results of these queries increase as the size of the dataset increases. The increase of the dataset size leads to a higher number of distributed requests in most cases. This may be explained by a higher degree of the RDF nodes which requires to balance the load among servers. However, this only happens when the whole set of query data items is not set to be clustered.

Server Scalability. We have deployed the systems in clusters with varying number of servers, and test its performance on dataset BSBM_5. The results are shown in Fig. 8b for query 3. In general, the increase on the number of servers brings the benefits of the parallel processing and reduces the load of servers. However, this increase can also lead data to be distributed among servers when query data items are not set to be clustered. We believe that the high number of requests being performed by each thread in parallel increases the competition for resources and impacts the system performance. The worst effect of this competition is observed in *Q3* on a cluster of 8 server for *METIS-2hops*, where each thread requires to access all servers. Notice that the effect of the parallel processing only reduces the response time when system capacity is increased to 12 servers and the number of server requests remains stable.

7 Related Work

Similar to our work, there are several graph-based approaches focused on database partitioning. However, they differ on the data model and the heuristics applied. A similar heuristic is used in the traditional algorithm *MakePartition* [6] proposed for relational databases. However, the number of fragments generated for a given dataset tends to be larger given that they do not focus on the storage capacity of the fragments. Affinity-based solutions have also been applied to XML fragmentation [3,11,13]. Our approach targets the RDF model and provides an extended coverage of such affinity-based approaches by clustering affinity fragments.

Our approach to generate fragmentation templates is similar to traditional vertical fragmentation techniques. Here, each instance of a template root node produces a fragment with its adjacent nodes. It is also similar to the hierarchical data model applied by Google F1 [14]. Clustering templates may also be associated to horizontal partitioning of traditional databases. In this paper we have compared *ClusterRDF* to other methods based on RDF graphs. As pointed out in Sect. 6, Huang et al. [5] assigns an RDF graph to a traditional graph partitioner and replicates cross-partition nodes in order to improve the query coverage. However, they only consider the associations of RDF vertexes and not the query patterns in order to provide an approximated coverage. *Trinity.RDF*[16] applies a simplest heuristic on RDF graph. In this case, high-degree nodes are identified to be clustered together with their adjacent nodes. We have demonstrated through a benchmark use case that a clustering approach based on workload analysis achieves a better approximation in terms of the coverage of frequent query patterns.

8   Conclusion Remarks

We have proposed an approach for partitioning RDF data according to an application workload defined on the structure of RDF graphs. This work makes contributions in the context of highly distributed databases, where communication costs must be reduced to provide a scalable service. In particular, *ClusterRDF* is able to reduce communication costs for distributed query evaluation by providing a suitable partition for datasets. Our experiments show that *ClusterRDF* can improve the query performance by roughly 27 % to 86 %, compared to *METIS-2hops*[5], a closely related approach for RDF partitioning. We have also reported that *ClusterRDF* can perform up to 10 times faster then the hash-partitioning introduced by *Trinity.RDF*. Although *ClusterRDF* and *METIS-2hops* replicates RDF data in order to provide better results, *ClusterRDF* reduces by 50 % the replication storage overhead produced by *METIS-2hops*.

Recent works evidence both the feasibility of such methods [2,10] as well as the availability of workload data [4]. In *ClusterRDF*, both the query patterns as well as the partitioning strategy are formulated over a summarization schema that represents the data structures for an RDF dataset. By doing so, the same partitioning *template* for a query workload may be continually applied to new data. However, considering dynamicity of query patterns is a topic for future work. In addition, we plan to investigate metadata management, indexing structures and query optimization strategies.

Acknowledgments. This work was partially supported by CAPES, CNPq, Fundação Araucária and by AWS in Education.

References

1. METIS (2013). http://glaros.dtc.umn.edu/gkhome/views/metis
2. Aluc, G., Özsu, M.T., Daudjee, K.: Workload matters: why RDF databases need a new design. PVLDB **7**(10), 837–840 (2014)
3. Bordawekar, R., Shmueli, O.: An algorithm for partitioning trees augmented with sibling edges. Inf. Process. Lett. **108**(3), 136–142 (2008)
4. Curino, C., Jones, E., Zhang, Y., Madden, S.: Schism: a workload-driven approach to database replication and partitioning. VLDB Endow. **3**(1–2), 48–57 (2010)
5. Huang, J., Abadi, D.J.: Scalable SPARQL querying of large RDF graphs. PVLDB **4**(11), 1123–1134 (2011)
6. Navathe, S., Ra, M.: Vertical partitioning for database design: a graphical algorithm. In: ACM SIGMOD International Conference on Management of Data, vol. 18, pp. 440–450 (1989)
7. Neumann, T., Moerkotte, G.: Characteristic sets: accurate cardinality estimation for RDF queries with multiple joins. In: ICDE, pp. 984–994 (2011)
8. Ozsu, M.T., Valduriez, P.: Principles of Distributed Database Systems. Prentice-Hall, Inc, Upper Saddle River (1991)
9. Pham, M.: Self-organizing structured RDF in MonetDB. In: IEEE International Conference on Data Engineering Workshops, pp. 310–313 (2013)
10. Quamar, A., Kumar, K.A., Deshpande, A.: SWORD: scalable workload-aware data placement for transactional workloads. In: EDBT, pp. 430–441 (2013)
11. Schroeder, R., Mello, R., Hara, C.: Affinity-based XML fragmentation. In: International Workshop on the Web and Databases (WebDB), Scottsdale (2012)
12. Schütt, T., Schintke, F., Reinefeld, A.: Scalaris: reliable transactional p2p key-/value store. In: ACM SIGPLAN Workshop on ERLANG, pp. 41–48 (2008)
13. Shnaiderman, L., Shmueli, O.: iPIXSAR: incremental clustering of indexed XML data. In: International Conference on Extending Database Technology - Workshops, pp. 74–84 (2009)
14. Shute, J., Whipkey, C., Menestrina, D., et al.: F1: a distributed SQL database that scales. VLDB Endow. **6**(11), 1068–1079 (2013)
15. Yang, T., Chen, J., Wang, X., Chen, Y., Du, X.: Efficient SPARQL query evaluation via automatic data partitioning. In: Meng, W., Feng, L., Bressan, S., Winiwarter, W., Song, W. (eds.) DASFAA 2013, Part II. LNCS, vol. 7826, pp. 244–258. Springer, Heidelberg (2013)
16. Zeng, K., Yang, J., Wang, H., Shao, B., Wang, Z.: A distributed graph engine for web scale RDF data. VLDB Endow. **6**(4), 265–276 (2013)
17. Zou, L., Mo, J., Chen, L., Özsu, M.T., Zhao, D.: gStore: answering SPARQL queries via subgraph matching. VLDB Endow. **4**(8), 482–493 (2011)

Efficient Computation of Parsimonious Temporal Aggregation

Giovanni Mahlknecht, Anton Dignös$^{(\boxtimes)}$, and Johann Gamper

Free University of Bozen-Bolzano, Bolzano, Italy
giovanni.mahlknecht@gmail.com, {dignoes,gamper}@inf.unibz.it

Abstract. Parsimonious temporal aggregation (PTA) has been introduced to overcome limitations of previous temporal aggregation operators, namely to provide a concise yet data sensitive summary of temporal
data. The basic idea of PTA is to first compute instant temporal aggregation (ITA) as an intermediate result and then to merge similar adjacent
tuples in order to reduce the final result size. The best known algorithm
to compute a correct PTA result is based on dynamic programming (DP)
and requires $\mathcal{O}(n^2)$ space to store a so-called split point matrix, where
n is the size of the intermediate data. The matrix stores the split points
between which the intermediate tuples are merged.

In this paper, we propose two optimizations of the DP algorithm for
PTA queries. The first optimization is termed *diagonal pruning* and identifies regions of the matrix that need not to be computed. This reduces
the runtime complexity. The second optimization addresses the space
complexity. We observed that only a subset of the elements in the split
point matrix are actually needed. Therefore, we propose to replace the
split point matrix by a so-called *split point graph*, which stores only those
split points that are needed to restore the optimal PTA solution. This
step reduces the memory consumption. An empirical evaluation shows
the effectiveness of the two optimizations both in terms of runtime and
memory consumption.

1 Introduction

In a temporal database [2,8,12,13], each tuple is associated with a time interval
that represents the time period when the represented fact is true in the modeled
reality. An important operation in temporal databases is aggregation, which has
been studied in various flavors. In instant temporal aggregation (ITA) [1,7,9,16,
17,19], the aggregate value at a time instant t is computed from the set of all
tuples whose timestamp contains t. Result tuples at consecutive time instants
with identical aggregate values are then coalesced into result tuples over maximal
time intervals during which the aggregate results are constant. By aggregating at
the finest granularity level, ITA is the most precise form of temporal aggregation
and considers the distribution of the data, however the size of the result relation
might become up to twice as large as the input relation [1]. In contrast, span
temporal aggregation (STA) [1,7,14] allows an application to specify the time

© Springer International Publishing Switzerland 2015
T. Morzy et al. (Eds.): ADBIS 2015, LNCS 9282, pp. 320–333, 2015.
DOI: 10.1007/978-3-319-23135-8_22

intervals for which to report result tuples, e.g., for each month in 2014. For each of these intervals a result tuple is computed by aggregating over all input tuples that overlap with the interval. While the result size is predictable, STA may fail to provide good summaries of the data since the user-specified intervals do not consider the distribution of the data.

To overcome the shortcomings of ITA and STA, the work in [5] introduced parsimonious temporal aggregation (PTA), which combines the best features of ITA and STA. By merging similar adjacent tuples in the ITA result, the PTA operator remains data-sensitive, yet allows the user to control the result size. PTA computes compact aggregation summaries that reflect the most significant changes in the data over time.

Example 1. As a running example we use relation *patients* in Fig. 1a, which stores costs of hospital stays. Each tuple records a patient P, the department D he/she is admitted, the daily costs C and the time period T of the hospital stay. Consider the query *"What is the sum of the daily costs for each department over time?"* This can be answered by a ITA query as shown in Fig. 1b. In order to obtain a more concise summary that reflects only the most important changes in the data, the PTA operator should be used instead. Figure 1c shows the PTA result with a result size of four tuples, where the ITA tuples s_2, s_3 and s_4, s_5 are merged into the PTA tuples z_2 and z_3, respectively.

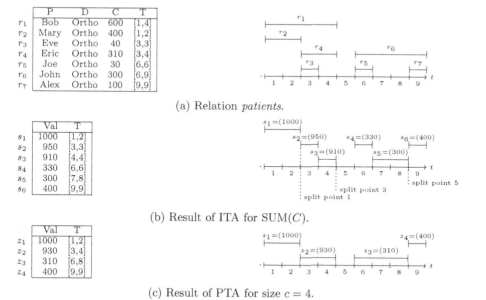

(a) Relation *patients*.

(b) Result of ITA for SUM(C).

(c) Result of PTA for size $c = 4$.

Fig. 1. Input relation *patients* with ITA and PTA result.

Gordevičius et al. [5] propose an exact algorithm, named PTAc, to compute PTA queries with a given result size c. It is based on dynamic programming and has a computational complexity of $\mathcal{O}(n^2 c)$ and a space complexity of $\mathcal{O}(n^2)$, where n is the size of the ITA result. The algorithm computes two $c \times n$ matrices: an error matrix, which stores the error introduced when tuples are merged, and a split point matrix, which tells us where to split the intermediate ITA result in order to obtain the compression with the minimal error.

Figure 2 shows the two matrices for our running example, where we reduce the final result to $c = 4$ tuples. The columns $i = 1, \ldots, n$ represent the chronologically ordered ITA tuples, and the rows k the PTA result size, ranging from 1 to c. An element $\mathbf{E}_{k,i}$ represents the smallest error of reducing the first i ITA tuples to k tuples. An element $\mathbf{J}_{k,i}$ represents the index of the ITA tuple (i.e., split point), where the ITA relation must be split to obtain the minimal error $\mathbf{E}_{k,i}$. In Fig. 2, the minimal error for a PTA result of size $c = 4$ is $\mathbf{E}_{4,6} = 1400$ (value in the lower right corner of the error matrix). The optimal split points can be retrieved from the split point matrix. The first split point (from right to left) is the value in the lower right corner of the matrix, i.e., $\mathbf{J}_{4,6} = 5$. We use this value as column index in the previous line of the matrix, which gives the next split point 3, etc. The three split points are in boldface in Fig. 2b. The matrices $\mathbf{E}$ and $\mathbf{J}$ are filled row-wise for all $k = 1, \ldots, c$, and for each k from left to right.

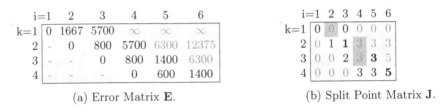

(a) Error Matrix $\mathbf{E}$.   (b) Split Point Matrix $\mathbf{J}$.

Fig. 2. Matrices used by the PTAc algorithm for $n = 6$ and $c = 4$ (Color figure online).

In this paper, we propose two optimizations of the PTAc algorithm [5]. The first optimization is *diagonal pruning*, which identifies regions of the error matrix $\mathbf{E}$ and split point matrix $\mathbf{J}$ that need not to be computed (see light gray cells in Fig. 2). This reduces the number of computations and thus the runtime. The second optimization is to replace the split point matrix by a *split point graph*. By analyzing the split point matrix, we discovered that many elements in the matrix become obsolete during the computation, that is they are not needed in order to construct the optimal solution once the DP scheme has completed. Such split points could be eliminated during the incremental computation of the matrix, yielding a sparse matrix. In Fig. 2b, the split points marked in dark gray become obsolete when the next level of the matrix is computed. As a more memory efficient solution we propose a split point graph, which stores only those matrix elements that need to be kept until the DP scheme terminates.

To summarize, the main contributions of this paper are as follows:

- We show how *diagonal pruning* identifies regions of the error and split point matrices that need not to be computed; this reduces the runtime.
- We replace the split point matrix by a memory-efficient *split point graph*, storing only those split points that are needed to restore the optimal solution.
- We conduct an experimental evaluation that shows the effectiveness of our optimization techniques.

The rest of the paper is structured as follows. Section 2 discusses related work. Section 3 describes how to prune the search space of the DP scheme. Section 4 describes the split point graph as a memory efficient alternative to the split point matrix. In Sect. 5 we present the results of an experimental evaluation.

2   Related Work

Various forms of temporal aggregation have been studied in the past (for an overview see [3]). They differ mainly in how the time line is partitioned. Instant temporal aggregation (ITA) [7,9,17] operates at the smallest granularity level. Conceptually, it computes for each time instant t an aggregate over all tuples that hold at t, followed by a coalescing phase that combines consecutive time instants with identical aggregate values into tuples over time intervals. Moving-window (or cumulative) temporal aggregation (MWTA) [11,14,19] extends ITA and computes for each time instant t the aggregate values over all tuples that hold in a window that is anchored at t. Though ITA and MWTA work at the most detailed level and provide data sensitive summaries, the main drawback is that the result size might become up to twice as large as the input relation. Span temporal aggregation (STA) [1,7,14] allows the user to control the result size by partitioning the time line into intervals that are specified in the query. For each such interval, a result tuple is computed over all argument tuples that overlap with that interval. STA does not consider the distribution of the data, and most approaches consider only regular time spans expressed in terms of granularities, e.g., years or months.

The work in [1,17] provides a uniform framework that generalizes previous temporal aggregation operators and allows the comparison of different temporal aggregation variants.

Tao et al. [15] were the first to introduce an approximate temporal aggregation technique. For a given time interval, it finds an approximate aggregation result from all tuples that overlap with that interval, thus approximating span temporal aggregation. By approximating STA, the method is not data-sensitive.

Another approximation operator is parsimonious temporal aggregation [4,5], which aims to overcome limitations of previous temporal aggregation operators. It is an approximation of ITA, hence it is data sensitive yet allows to control the size of the result relation. Thus, the operator combines the best features of ITA and STA. In this paper, we provide two optimization techniques for the dynamic programming algorithm in [5] that computes correct PTA results.

3 Diagonal Pruning

The PTAc algorithm uses a dynamic programming technique to compute two matrices, the error matrix $\mathbf{E}_{c \times n}$ and the split point matrix $\mathbf{J}_{c \times n}$, each with c columns and n rows, where c is the size of the reduction (i.e., the size of the PTA result) and n is the number of tuples in the input relation.

In this section we propose an approach, called *diagonal pruning* that omits unnecessary computations by reducing the search space of the dynamic programming algorithm.

Consider the cell $\mathbf{E}_{k,i}$ of the error matrix, i.e., the cell of the kth row and i^{th} column in $\mathbf{E}$. For this cell the PTAc algorithm computes the minimum cost to merge the first i ($1 \leq i \leq n$) input tuples $\langle s_1, \ldots, s_i \rangle$ into k ($1 \leq k \leq c$) result tuples $\langle z_1, \ldots, z_k \rangle$. The idea of diagonal pruning is to stop the computation of matrix elements when i becomes too large, that is, when not enough input tuples $(n - i)$ are left to produce $c - k$ result tuples.

Lemma 1 (Diagonal Pruning). *For the computation of the error matrix $\mathbf{E}$ and split point matrix $\mathbf{J}$, given input size n, result size c, row k and column variable i, i can be upper-bounded by $i \leq n - (c - k)$.*

Proof. Recall that cell $\mathbf{E}_{k,i}$ of the error matrix stores the cost to merge the first i of the n input tuples into k result tuples. The cost to merge a given number of tuples into c result tuples is always lower than to merge it into $< c$ result tuples, since value-equivalent an adjacent tuples do not exits in the result of temporal aggregation. Thus, to prove the lemma we need to show that, when $i > n - (c - k)$, there are not enough input tuples left to produce c result tuples. Consider $i = n - (c - k) + 1$. The first $n - (c - k) + 1$ input tuples are merged into k result tuples. In total we need c result tuples, but there are only $n - (n - (c - k) + 1) = n - n + c - k - 1 = c - k - 1$ input tuples left to be merged into $c - k$, which is not possible. The same arguments apply for the split point matrix $\mathbf{J}$.

Example 2. Consider our running example with 6 input tuples and $c = 4$ in Fig. 2. The computation of $\mathbf{E}_{3,6}$ is unnecessary, since it is the cost of merging the first 6 input tuples into 3 result tuples, but there are no input tuples left to merge the input into 4. As such this result can never be used to compute $\mathbf{E}_{4,6}$. Similar for $\mathbf{E}_{2,5}$, $\mathbf{E}_{2,6}$, $\mathbf{E}_{1,4}$, $\mathbf{E}_{1,5}$ and $\mathbf{E}_{1,6}$. In Fig. 2a elements that need not to be computed are marked in light gray.

The number of cells in the matrix that can be pruned by diagonal pruning depends on the reduction size c. In the k-th row of the matrix $(c - k)$ cells can be pruned. For c rows this yields $\sum_{k=1}^{c} (c - k) = \frac{c^2 - c}{2}$ pruned cells. Diagonal pruning does not only reduce the number of computations, but it also reduces the space consumption, since pruned cells do not have to be stored (see next section).

The upper bound i defined by diagonal pruning can also be used in combination with the upper bound i_{max} provided in the PTAc algorithm [5], which considers temporal gaps in the data. By taking the minimum of the two upper bounds we can obtain an even more effective pruning. Notice, however, that the pruning strategy in [5] works only if the data contain gaps, whereas our pruning strategy is independent of the data.

4 Space-Efficient PTA Computation

In this section, we address the quadratic space complexity of the PTAc algorithm.

4.1 Overview and Approach

Recall that the PTAc algorithm uses two matrices: an error matrix $\mathbf{E}$, which stores the error introduced when tuples are merged, and a split point matrix $\mathbf{J}$, which tells us where to split the intermediate ITA result in order to obtain the compression with the minimal error.

The error matrix falls into a category of dynamic programming problems, where a straightforward space reduction can be applied [6,10]. Since the computation of row k in $\mathbf{E}$ requires only the result of row $k-1$, only the last two rows need to be kept in memory, yielding linear space complexity. This optimization is already implemented in PTAc.

Unfortunately, this is not true for the split point matrix $\mathbf{J}$. The optimal split points that minimize the merging error can only be retrieved from the split point matrix after the dynamic programming scheme terminates, i.e., when the elements $\mathbf{E}_{c,n}$ and $\mathbf{J}_{c,n}$ in the lower right corner of the error and split point matrix, respectively, are computed. Therefore, the PTAc algorithm has to store the complete matrix, yielding $O(n^2)$ memory complexity.

By analyzing the split point matrix we observed that some split points in the matrix become obsolete during the computation of the DP scheme, hence they can be removed and need not to be kept until the end. More specifically, after computing a row k of the split point matrix, it is possible to identify split points in row $k-1$, which cannot become part of any optimal solution anymore. This yields a sparse matrix, where only a small number of elements need to be stored until the end. In Fig. 2b, all matrix elements with a dark gray background could be removed.

In the remainder of this section we introduce the *split point graph*, which provides an efficient representation of the sparse split point matrix.

4.2 Split Point Graph

The split point graph is an alternative and memory-efficient structure to store the same information as in the matrix $\mathbf{J}$. In contrast to the matrix, the cells that are not computed thanks to diagonal pruning are not stored in the graph. Moreover, during the computation nodes can be eliminated once we are guaranteed that they cannot become part of any solution with minimum error.

The split point graph is organized into levels. Each level k corresponds to a row of the split point matrix. A node in the split point graph has a label representing a split point, and at most one outgoing edge. We denote by $N_{k,i}$ the node in the split point graph that corresponds to the cell $\mathbf{J}_{k,i}$ of the split point matrix. Node $N_{k,i}$ at level k has label i and points to a parent node at level $k-1$; nodes on level 1 have no outgoing edges. By following a path from a node $N_{k,i}$ up to level 1 we obtain the *split path*, i.e., sequence of split points, that incurs the minimum error when reducing the first i tuples to size k. Consequently, the split path for the PTA result relation of size c can be retrieved by following the path from node $N_{c,n}$ upwards until level $k = 1$ is reached.

Example 3. The split point graph for our running example is shown in Fig. 3d. The split points for the reduction of 6 input tuples into 4 result tuples is obtained from the labels, following the edges starting from node $N_{4,6}$ up to a terminating node. The split path for our running example is $[5, 3, 1]$ (compare to Fig. 1b).

Graph Construction. The split point graph is constructed level-wise for $k = 1, \ldots, c$. A node is inserted for each $i = k, \ldots, (n - (c - 1))$, hence unnecessary nodes that are not computed thanks to diagonal pruning are not inserted. For each new node at level k, an edge to a parent node in level $k - 1$ is inserted. The parent node is the node whose label corresponds to the split point element that is computed by the PTAc algorithm. To find the parent node efficiently, we keep a hash map on the node labels of the preceding row. Figure 3 shows the level-wise construction of the split point graph for our running example.

Path Pruning. A node $N_{k,i}$ of the split point graph can have multiple incoming edges if the merging of tuples s_i and s_{i+1} is either not possible due to a gap or would result in a higher cost. If some nodes have more than one incoming edge, some other nodes must have no incoming edge since the number of nodes at each level is the same. We call such nodes *orphan nodes*. Orphan nodes are guaranteed not to be part of the final optimal split path, since they are not reachable from node $N_{c,n}$, where the optimal split path starts.

Orphan nodes at level k can be detected once the nodes at level $k + 1$ have been computed, since only nodes at level $k + 1$ can have edges to nodes at level k. In the split point matrix orphan nodes correspond to the split points that are not reachable from the next row. For instance, in Fig. 3 node $N_{2,4}$ is identified as an orphan node, since level 3 has been computed and the node has no incoming edge. The node corresponds to the entry $\mathbf{J}_{2,4}$ in the split point matrix shown in Fig. 2b. This element is not reachable, as there is no cell with value 4 in row 3.

To efficiently detect orphan nodes during graph construction, we store an edge counter in each node, which counts the number of incoming edges. The counter is incremented whenever a new edge pointing to that node is inserted. When a row k is completed, we scan through the map of row $k - 1$ to find orphan nodes, i.e., nodes with an edge count of 0. If an orphan node is encountered, it is deleted from the graph, and the edge counter of its parent node is decremented.

If the parent node becomes an orphan node due to the deletion of a child node, we recursively apply this deletion procedure. This process is called *path pruning*, and it is applied each time a new row has been completed. For instance, when we delete node $N_{2,4}$ in Fig. 3c, node $N_{1,3}$ becomes also orphan and can be deleted.

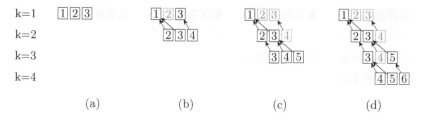

Fig. 3. Construction of split point graph. Gray boxes are never computed nor stored due to diagonal pruning. Gray framed boxes are eliminated due to path pruning (Color figure online).

Example 4. Figure 3 shows the step-by-step construction of the split point graph. In Fig. 3a, three root nodes are created. The labels of these nodes correspond to the index of the chronologically ordered input tuples $\langle s_1, \ldots, s_n \rangle$. For example, the third node represents that in order to reduce the first three tuples to one tuple, the split point is at tuple s_3, that is all tuples are merged together. Figure 3b illustrates the computation of the second row, which inserts again three new nodes in the graph. The nodes 2 and 3 point to parent 1. This represents that in order to compress the first two or three tuples to two tuples, we have to split at tuple s_1. Similarly, if the tuples $s_1, \ldots, s_4$ are reduced to two tuples, the split point is at tuple s_3. Since the node 2 at level 1 has no incoming edge, path pruning can be applied and the node is removed from the graph. Figure 3c shows the computation of the third level of the split point graph. At this level, path pruning removes node 4 at level two and node 3 at level one. Figure 3d shows the final split point graph. The split path of the optimal solution can be retrieved by starting from node 6 (right most leave node) and following the path up to the root node 1. The labels of the intermediate nodes along this path are the split points, i.e., $\{5, 3, 1\}$.

4.3 Analysis

When using the split point graph instead of the split point matrix, split points that are not computed thanks to diagonal pruning are not stored in the graph. The number of elements that need to be computed despite diagonal pruning is $(n - c + 1) \cdot c$. Depending on c, the maximum number of nodes of the split graph with diagonal pruning reaches the value $\frac{n^2 + 2n + 1}{4}$ for $c = \frac{n+1}{2}$.

Additionally to diagonal pruning, we apply path pruning, which eliminates nodes that cannot be part of the final optimal path. Path pruning can be applied when nodes during computation have no incoming edges, due to other nodes that have a smaller cost or due to gaps between tuples that are always a split point. In our experiments we observed that on average path pruning removes 65 % of the nodes in the split point graph.

Figure 4a shows the number of elements of the split point matrix computed by PTAc (i.e., $n \cdot c$), the number of nodes in the split point graph with diagonal pruning but without path pruning (i.e., $(n - c + 1) \cdot c$), and the number of nodes in the split point graph with path pruning when 65 % are removed on average.

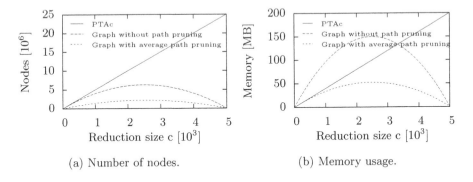

(a) Number of nodes.   (b) Memory usage.

Fig. 4. Impact of diagonal pruning and path pruning ($n = 5000$).

The effect of path pruning depends on the reduction size c. For a small reduction size c, more paths can be pruned than for large values of c. For large values of c, almost all nodes are part of a possible split path, because almost between any input tuple a split point is placed. This is the reason why the graph with path pruning is right-skewed in practice, as we will see in the experimental evaluation in Fig. 9.

When using a split point graph instead of a matrix, nodes can be pruned, but more space is required for single elements, since nodes need to be connected by pointers and we additionally need an (incoming) edge counter per node for path pruning. An element in the matrix requires 64 bits in our implementation, whereas for a node we need 192 bits (64 bit each for label, pointer and counter). Figure 4 illustrates the number of nodes and the corresponding memory usage for the matrix, the graph without path pruning and the graph with path pruning.

5   Experimental Evaluation

In this section, we compare our proposed optimizations with the original PTAc algorithm. We show that diagonal pruning reduces the runtime of PTAc and that our split point graph successfully reduces the space requirements.

5.1   Setup and Data

For the experiments we used an Intel 1.7 GHz Core i5 machine with 8 GB main memory running Mac OS X. The algorithms were implemented in C++ and run on a single core.

We used two datasets: (1) ETDS, a synthetic employee data set donated by F. Wang [18], which records the evolution of employees in a company, and (2) SYNTH, a synthetic dataset. The values of the aggregation attribute are uniformly distributed in the range [1;1000], and the duration of the tuples' timestamp is uniformly distributed in the range [1;40].

The input of the PTAc algorithm is the result of an instant temporal aggregation (ITA) query shown in Table 1. The runtime to create the ITA result is not included in the measurements.

Table 1. ITA queries used as input for the experiments.

Name	Grouping attributes	Aggregation function	ITA size	c_{min}
ETDS	deptno	avg(salary)	57,408	9
SYNTH	none	avg(value)	500,000	1

In the experiments we compare the original PTAc algorithm with a matrix implementation that extends the PTAc with diagonal pruning (MP) and the graph implementation with path pruning and diagonal pruning (GP). An overview is given in Table 2.

Table 2. Algorithm configuration used for experimental evaluation.

Label	Split point implementation	Diagonal pruning
PTAc	matrix	no
MP	matrix	yes
GP	graph	yes

5.2 Diagonal Pruning

In the first experiment we show the impact of diagonal pruning on the runtime performance of PTAc. We use 5000 records each from SYNTH and ETDS as input and vary the reduction size c from 100 to 4600. Fig. 5 shows the runtime performance of the original PTAc algorithm (PTAc) and the matrix implementation with diagonal pruning (MP). As expected MP always outperforms the original PTAc algorithm thanks to the pruning strategy. The improvement of the runtime grows with an increasing reduction size c, since with a larger c more cells of the matrix can be pruned (cf. Lemma 1).

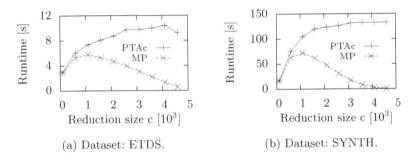

(a) Dataset: ETDS.   (b) Dataset: SYNTH.

Fig. 5. Runtime for varying reduction size c, with and without diagonal pruning.

5.3 Graph Implementation

In the next experiment we compare the overhead of the split point graph implementation (GP) with respect to the matrix implementation (MP), using the same setting as in the previous experiment. The result is shown in Fig. 6. The overhead by GP in terms of runtime compared to MP is very small. It is mainly due to the overhead of the dynamic graph structure and path pruning.

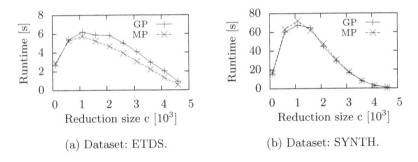

(a) Dataset: ETDS.   (b) Dataset: SYNTH.

Fig. 6. Runtime for varying reduction size c, matrix versus graph.

In the next experiment we choose small values of c, where diagonal pruning has little effect. We vary the input cardinality and compare the graph implementation GP with the original PTAc algorithm. Figure 7 shows the comparison for $c = 1\,\%$, $5\,\%$ and $10\,\%$ of the input cardinality. We can see that even for small values of c, GP has a better or comparable runtime performance.

5.4 Space Efficiency

We now compare the space requirement of the GP algorithm that uses the split point graph and the original PTAc algorithm that uses a matrix. We focus only on the split point implementation of the two approaches. The size of a node in

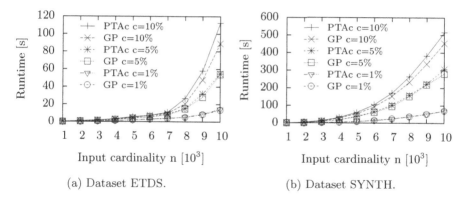

(a) Dataset ETDS.   (b) Dataset SYNTH.

Fig. 7. Runtime for varying input cardinality n, matrix versus graph.

the split point graph is 192 bits (64 bit label, 64 bit incoming edge counter for path pruning and 64 bit pointer), whereas the size of a cell in the split point matrix is 64 bits. We use both data sets, ETDS and SYNTH, and vary the cardinality of the input. Figure 8 shows the memory usage in MB for c values of 1%, 5% and 10% of the input cardinality. We can observe that GP uses much less space than the matrix implementation. The average values over all input cardinalities are given in Table 3, where for instance -86.8% means that GP uses 86.8% less memory than PTAc.

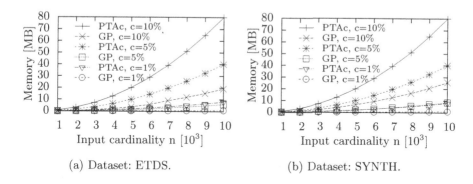

(a) Dataset: ETDS.   (b) Dataset: SYNTH.

Fig. 8. Memory usage for varying input cardinality n, matrix versus graph.

Next, we analyze the space reduction of our optimizations when varying c. We use 5000 records of our data sets and compare the original matrix implementation (PTAc) with the graph implementation with path pruning (GP) and the graph implementation without path pruning. The result is shown in Fig. 9.

The experiments confirm our previous analysis and show that the split point graph implementation with path pruning (GP) successfully reduces the memory requirements for the computation of PTA. The matrix implementation (PTAc)

Table 3. Average memory usage of GP compared to PTAc.

Dataset	Reduction size in % of the input			
	1 %	2 %	5 %	10 %
ETDS	-86.8 %	-89.2	-85.5 %	-77.6 %
SYNTH	-85.9 %	-85.8 %	-78.6 %	-67.9 %

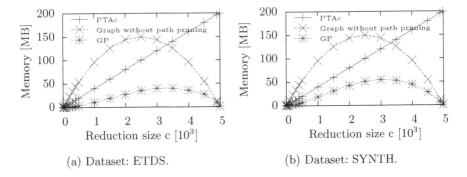

(a) Dataset: ETDS.   (b) Dataset: SYNTH.

Fig. 9. Memory usage for varying reduction size c, matrix versus graph.

and the graph implementation without path pruning are independent of the data distribution; therefore, they use the same amount of memory for both datasets. GP has a huge pruning effect. It prunes approximately 2/3 of the graph. Path pruning is more effective for the ETDS data set, since more dominant nodes in the graph exist that attract more incoming edges, which results in more orphan nodes that can be pruned.

6   Conclusion

In this paper, we introduced two optimization techniques for the PTAc algorithm to compute correct parsimonious temporal aggregation queries. The first optimization decreases the runtime of the algorithm by reducing the search space of the dynamic programming scheme adopted by PTAc. The second optimization regards the memory consumption, which for PTAc is quadratic in the input data. We proposed to replace the split point matrix by a split point graph, which stores only those split points that are needed to restore the optimal PTA solution. Experiments showed that the two optimizations reduce the memory requirements to about one third of the memory required by the original PTAc algorithm. The effectiveness of both optimizations depends on the compression rate. The best memory reduction is achieved for a reduction size up to 10 % or greater than 80 % of the input relation. Runtime improvements are present for all reduction sizes. The maximum effect is achieved when the reduction size is close to the size of the input relation.

As part of future work we will study whether the proposed optimization techniques can be generalized to other dynamic programming problems and algorithms.

References

1. Böhlen, M.H., Gamper, J., Jensen, C.S.: Multi-dimensional aggregation for temporal data. In: Ioannidis, Y., Scholl, M.H., Schmidt, J.W., Matthes, F., Hatzopoulos, M., Böhm, K., Kemper, A., Grust, T., Böhm, C. (eds.) EDBT 2006. LNCS, vol. 3896, pp. 257–275. Springer, Heidelberg (2006)
2. Dignös, A., Böhlen, M.H., Gamper, J.: Temporal alignment. In: SIGMOD, pp. 433–444 (2012)
3. Gamper, J., Böhlen, M.H., Jensen, C.S.: Temporal aggregation. In: Liu, L., Özsu, M.T. (eds.) Encyclopedia of Database Systems, pp. 2924–2929. Springer, Heidelberg (2009)
4. Gordevicius, J., Gamper, J., Böhlen, M.H.: Parsimonious temporal aggregation. In: EDBT, pp. 1006–1017 (2009)
5. Gordevicius, J., Gamper, J., Böhlen, M.H.: Parsimonious temporal aggregation. VLDB J. **21**(3), 309–332 (2012)
6. Hirschberg, D.S.: Algorithms for the longest common subsequence problem. J. ACM **24**(4), 664–675 (1977)
7. Kline, N., Snodgrass, R.T.: Computing temporal aggregates. In: ICDE, pp. 222–231 (1995)
8. Lorentzos, N.A.: Period-stamped temporal models. In: Liu, L., Özsu, M.T. (eds.) Encyclopedia of Database Systems, pp. 2094–2098. Springer, Heidelberg (2009)
9. Moon, B., Lopez, I.F.V., Immanuel, V.: Efficient algorithms for large-scale temporal aggregation. IEEE Trans. Knowl. Data Eng. **15**(3), 744–759 (2003)
10. Myers, E.W., Miller, W.: Optimal alignments in linear space. Comput. Appl. Biosci. **4**(1), 11–17 (1988)
11. Navathe, S.B., Ahmed, R.: A temporal relational model and a query language. Inf. Sci. **49**(1–3), 147–175 (1989)
12. Snodgrass, R.T. (ed.): The TSQL2 Temporal Query Language. Kluwer, Norwell (1995)
13. Snodgrass, R.T.: Developing Time-Oriented Database Applications in SQL. Morgan Kaufmann, San Francisco (1999)
14. Snodgrass, R.T., Gomez, S., McKenzie, L.E.: Aggregates in the temporal query language TQuel. IEEE Trans. Knowl. Data Eng. **5**(5), 826–842 (1993)
15. Tao, Y., Papadias, D., Faloutsos, C.: Approximate temporal aggregation. In: ICDE, pp. 190–201 (2004)
16. Tuma, P.: Implementing Historical Aggregates in TempIS. Ph.D. thesis, Wayne State University, Detroit, Michigan (1992)
17. Lopez, V.I.F., Snodgrass, R.T., Moon, B.: Spatiotemporal aggregate computation: A survey. IEEE Trans. Knowl. Data. Eng. **17**(2), 271–286 (2005)
18. Wang, F.: Employee temporal data set (2009). http://timecenter.cs.aau.dk/
19. Yang, J., Widom, J.: Incremental computation and maintenance of temporal aggregates. VLDB J. **12**(3), 262–283 (2003)

TDQMed: Managing Collections
of Complex Test Data

Johannes Held$^{(\boxtimes)}$ and Richard Lenz

Computer Science 6 (Data Management),
Friedrich-Alexander Universität Erlangen-Nürnberg, Erlangen, Germany
{johannes.held,richard.lenz}@fau.de

Abstract. Medical devices like Medical Linear Accelerators (LINAC)
are extensively tested before they are used in routine practice. Such systems typically interact with multiple other systems that produce complex input data, like medical images annotated with extensive metadata.
Before such a system is actually used in a hospital with real patients
it has to be tested with test data as realistic as possible. Suitable test
data, however, cannot be easily generated. For this reason vendors typically accumulate large collections of patient files over the years to have
them available for various test scenarios. In the TDQMed project we
have developed methods and tools that enable a tester to estimate both
the quality of a test data collection and its applicability for a particular test goal. A prototype system has been implemented to demonstrate
the feasibility of measuring specific test data related quality criteria like
coverage of test space and closeness to reality. An evaluation with professional testers indicates that the overall approach is promising.

Keywords: Test-data quality · Knowledge discovery · Information
extraction and integration

1 Introduction

The system test of data-intensive and data-driven systems – in our case medical
devices – bears some impediments. In our project, test data are patient files
composed of DICOM[1] images from modalities like CT or MRT, treatment plans
like DICOM-RT[2] for e.g. cancer therapy, and other medical documents. Every
patient is unique and the device has to be capable to process the corresponding
patient data flawlessly. Only for very special purposes, these test data items
may be generated despite their complexity. For example, test data generation
can be used for stress tests, as is it unlikely to find a real patient file containing
e.g. a lot of images. Generated test data with intentional errors can be used to
test the system's overall reaction to corrupted input. Yet, these errors are still
artificial and do not resemble reality and complexity of a live system or a patient,

---

[1] *Digital Imaging and Communications in Medicine* [9,10].

[2] DICOM-RT is an extension suited for radio therapy [8].

© Springer International Publishing Switzerland 2015
T. Morzy et al. (Eds.): ADBIS 2015, LNCS 9282, pp. 334–347, 2015.
DOI: 10.1007/978-3-319-23135-8_23

somewhat a requirement for proper system testing. Therefore, manufacturers of medical devices collect patient files in big test data collections to resemble the huge variety needed for system testing. To adapt these collections to new medical devices and scenarios, new test data has to be added and the collection evolves (and degrades) over time and suffers from low accessibility. Availability of test data impacts the test planning and execution and testers need to know to what extent the collected test data supports the actual test goals. As the software testing of new medical devices is pending, manufactures need to know how good their available test data covers the needed variety of scenarios to ensure a proper test planning and acquisition of test data.

We propose a concept to measure test-data quality and introduce metrics to gauge a test data collection's applicability for system testing data-intensive systems like medical devices. Therefore, we build an index on test data items based on test scenarios (classes of input) and analyse their distribution in the test space. We assess test-data quality by different indicators to assist test planning. One indicator is the early identification of non-testable test cases due to missing test data as this early knowledge can enables a timely acquisition of test data. Another indicator is the assessment of the test data collection's closeness to reality, estimating to what extent test cases can be executed with realistic input data. Another indicator is an estimation of the test data collection's diversity. This can be used to detect clusters and potentially redundant test data items, which can be removed from the test data collection to reduce its complexity and overall size.

Background and Method. We had a close cooperation with our project partner and one validation partner. Our project partner is specialized in testing medical devices and develops software solutions for model-based testing. The results of our research are going to be implemented in their software to add a test-data quality component. The validation partner is a major manufacturer of medical devices and two divisions, planning and executing system tests, were available for rounds of talks and validation meetings. Due to our partner's alignment, we decided on the test phase system test. During the project's lifespan, we met every month to ensure a target-aimed research. We interviewed domain experts and test experts about their requirements for test data collections and asked for their notion of quality. Based on that, we developed generic quality criteria and analysis methods for test data collections.

Structure. In Sect. 2 we give an overview on related work. We then describe our concepts of *test space* and *scenario*, and our test-data quality indicators and data model in Sect. 3. Section 4 describes in detail proposed analyses to gauge test-data quality. We specify calibration techniques in Sect. 5 and present implementation details of our prototype in Sect. 6. The evaluation, based on questionnaires on our proposed methods, is given in Sect. 7 and is followed by an overall discussion in Sect. 8. Section 9 summarizes our contribution.

2   Related Work

Data quality is an interdisciplinary research topic which is discussed in different research communities from different viewpoints. Different approaches for generally applicable frameworks for data quality have been proposed (e.g.: [1,13,17–19]). A common understanding is that data quality is multi-dimensional. Most general approaches to data quality try to identify and classify typical dimensions of data quality along with typical indicators that can measure such a particular data quality aspect. Examples for frequently mentioned data quality dimensions are *correctness, accuracy, completeness, consistency* and *currency*. However, the definitions for these dimensions differ greatly. Moreover, each of these dimensions typically can be broken down to more specific criteria, which might be relevant or irrelevant in a particular context. Juran [7] coined the term 'fitness for use' as a definition for quality in general and data quality in particular. This definition also underpins the view that 'data quality lies in the eyes of the beholder.' Thus, it is not surprising that data quality dimensions need to be redefined, reinterpreted or at least adjusted to the particular context of test data management. *Correctness* of data, for example, is usually understood as 'absence of errors.' This is an important quality dimension for a typical production context, where data is used to control processes or support decisions. Test data, however, have a different use, namely to determine whether a system can deal with a potential input or workload. It might even be an explicit test goal to check whether the system can deal with incorrect data. Obviously, general classifications of data quality criteria will not deliver a ready-to-use framework for measuring test-data quality, but they can help to systematically investigate what data quality in the context of test data actually means. Once, the relevant dimensions for test-data quality have been identified, we must find indicators and metrics that can contribute to measuring these criteria.

Test data are of good quality if they are suitable to support the particular test goals. Testing in general is aimed at reducing the number of hidden errors as early as possible. For the type of systems we are looking at, a complete test of all imaginable circumstances is impossible, so test case creation and selection is of utmost importance. Regarding black-box and system testing, a lot of researchers have proposed methods for adequate test case prioritization and selection. Di Nardo et al. surveyed four techniques of test case prioritization for regression tests, focussing on various coverage criteria for e.g. code coverage and examined the influence of coverage granularity on the results [6]. Thomas et al. proposed a black-box technique that does not depend on source code or access to the system under test [16]. They analyse 'linguistic' properties (like identifier name, comments, etc.) of test case descriptions and use text analysis algorithms to identify a test case's function.

3   Model, Mapping and Metrics

We propose two criteria to describe quality of test data, valued as a test data collection's applicability for test goals.

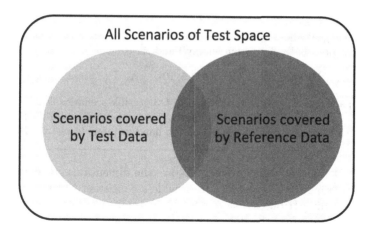

Fig. 1. At a glance: *coverage of test space* and *closeness to reality*.

Coverage of Test Space. The quality of a test data collection is indicated by the ratio of the size of scenarios contained in the test data items to the size of scenarios in the test space.

Closeness to Reality. The quality of a test data collection is indicated by its degree of similarity to a reference data collection. The similarity measure is based on the coverage of scenarios and the distribution of both data collections in the test space.

Test Space and Scenario. According to Ostrand and Balcer [11] a test space S of a test goal g contains all possible scenarios, that are classes of input for the system under test for that test goal. For the sake of brevity, we consider only one test goal (and test space) for the remainder of this paper. A scenario is influenced by parameters and environment, which can be described as *test-relevant attributes*. Such an attribute A is defined by its range, which is a set of nominal, ordinal or numerical values.

$$A = \{a_1, a_2, \ldots, a_k\} \tag{1}$$

Domain experts and test experts need to identify these test-relevant attributes for every test goal. Using our *domain specific language* (DSL), they create a configuration $\mathcal{A}$ that describes all test-relevant attributes (see Sect. 6). The crossproduct of all test-relevant attributes $A_i \in \mathcal{A}$ for the test goal yields all possible scenarios, thus building the n-dimensional test space S.

$$S = \prod_{\mathcal{A}} = A_1 \times A_2 \times \cdots \times A_n \tag{2}$$

A scenario $s \in S = (v_1, v_2, \ldots, v_n)$ resembles a class of input and is the minimum unit of the test space S. Test data items are indexed by their contained scenarios.

Assume a test goal whose scenarios are constituted by three test-relevant attributes A_1 ('patient position'), A_2 ('beam energy') and A_3 ('region').

$$A_1 = \{\text{HFS}, \text{HFP}, \text{FFS}\}, \quad A_2 = \{1, 2, 5, 10\}, \quad A_3 = \{\text{Throat}, \text{Lung}, \text{Hand}\}$$

The test space $S = A_1 \times A_2 \times A_3$, contains $3 \cdot 4 \cdot 3$ possible scenarios. E.g. a scenario (HFS, 5, Hand) reads as 'Use a treatment plan for a patient in position HFS, being treated with beams of level 5 targeted on the hand as input to test the SUT.'

As the test-relevant attributes $A_i \in \mathcal{A}$ are the dimensions spanning the test space S, their ranges must be finite to provide a consistent test space. For non-finite test-relevant attributes (like *free text*), a mapping to finite (e.g. categorical) values can be defined by the domain experts in the configuration. Nonetheless, we provide an evolutionary approach and accept and incorporate new values, as such encounters bear a lot of information (see Sect. 4).

Mapping. Before we are able to evaluate test-data quality or perform any analyses on the test data collection, we have to index each test data item and import the information about its contained scenarios into our *test-data quality database* (TDQ-DB). Because of its complexity, a test data item may provide multiple scenarios. With D being a set of test data, the function map maps a test data item $d \in D$ to the power set of S (2^S).

$$\text{map} : D \rightarrow 2^S \tag{3}$$

Likewise, a test data collection (or any other set of data, e.g. reference data) is described as union of all scenarios provided by its contained data items.

$$\mathcal{D} = \bigcup_{d \in D} \text{map}(d) \tag{4}$$

Featuring data safety and loose coupling, the mapping, highlighted in Fig. 2, has to be realized externally to the *test-data quality system* (TDQ-Sys) by manufacturers. For this very reason, our approach to test-data quality is independent from the actual test data model and the use case, as the TDQ-Sys has (and needs) no access to the actual data and does not need to be able to understand the original data format. Mapping and provenance information - not the actual test data - is stored in the TDQ-DB and used as input for our proposed quality measurement and the provided search feature.

Mapping reference data (data or statistics exported from live systems or gathered during studies, etc.) is done in same way. However, it is not trivial to obtain reference data and sometimes impossible due to data security concerns. The mapping component can permit the usage of statistical data, as it can e.g. map 'fake' data items to scenarios as encountered in the statistical data.

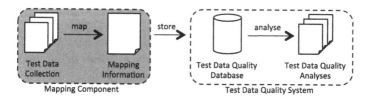

Fig. 2. Information flow in our test-data quality system. The mapping is not part of our system.

Metrics. coverage and closeness fit into the data quality dimension *completeness*: The more complete the test data collection covers the test space or the reference data, the better. To describe the completeness of a data set, Pipino et al. introduced the metric *population completeness* as the size of actual data in a collection divided by the size of the population [12]. In our case, we define the population as the test space S and the actual data as the set of all scenarios covered by data items. We define a function pc, very similar to *population completeness*, that allows to measure *coverage of test space* and *closeness to reality* (see Sect. 4). It takes two subsets α, β of S and returns the ratio of the intersection's size to the size of the second set, the population.

$$\text{pc} : 2^S \times 2^S \to \mathbb{Q}$$

$$\text{pc}(\alpha, \beta) = \frac{|\alpha \cap \beta|}{|\beta|} \tag{5}$$

Due to the *empty space* phenomenon [3], the result of $\text{pc}(\alpha, \beta)$ will be near 0 if $|\beta| \gg |\alpha|$. We took care of this problem and offer three powerful calibration techniques to reduce and shape the test space in Sect. 5.

The *coverage of test space* of the test data collection D, is measured by coverage(D).

$$\text{coverage}(D) = pc(D, S) = \frac{|D \cap S|}{|S|} = \frac{|D|}{|S|} \tag{6}$$

Let $R \subseteq S$ be the set of all scenarios contained in reference data. The *closeness to reality* of the test data collection (described by D) is measured by closeness(D, R) using R as population.

$$\text{closeness}(D, R) = pc(D, R) = \frac{|D \cap R|}{|R|} \tag{7}$$

4   Gauging Test-Data Quality

The mapping information can be utilized in many ways during the test process.

Analyzing the Mapping Step

By itself, the mapping step is an important part of test-data quality, as a data item containing undefined values for test-relevant attributes contains an important piece of advice. Two cases have to be distinguished:

- If the value is recognized as a standardized value, then the domain experts and test experts forgot this value during the definition of the test-relevant attribute. As a consequence, this value and therefore some scenarios would not have been considered for testing at all.
- If the origin of this value remains unknown, valuable insights can be gathered from the data item's provenance. Being from a reference data collection, this find contributes to the overall knowledge of the heterogeneity and the compliance to standards in the live system.

A smooth mapping step increases the overall confidence in the identified test space and therefore the knowledge about the system under test and its anticipated use cases.

Helpful during test phase *Test Analysis and Design* **and for** *Test Data Management.*

Analyzing Coverage and Closeness

A part from the values calculated by `coverage` and `closeness`, more knowledge can be lifted. A list of scenarios that are not covered by test data is easy to obtain: $S \setminus \mathcal{D}$. Defining hazard scenarios $H \subset S$, it is equally easy to identify which of these hazards cannot be tested, due to missing test data: $H \setminus \mathcal{D}$. Early enough in the test phase *test analysis and design*, this information can be used to start a directed acquisition of test data.

`closeness` can be used pre-test to evaluate how many and which realistic scenarios cannot be tested due to missing input data: $\mathcal{R} \setminus \mathcal{D}$. `closeness` can also be used post test: Taking actual tested scenarios as δ and scenarios, gathered from studies or beta tests, as ω, `closeness` (δ, ω) then evaluates as an overall quality indicator of the test process' real-world anticipation.

Helpful during test phases *Test Analysis and Design* **and** *Test Execution.*

Analyzing Distribution and Diversity

Evaluating the test data collection's distribution in the test space leads to a good overview. This is especially helpful because it contains a lot of test data items and requires a lot of storage. Suitable visualization techniques can depict whether this great number of test data items are evenly distributed in test space or if there are fields of attention. At the moment, we use scatter plots and 'Parallel Sets' [2] to visualize the distribution but many more visualization techniques can be used. Clusters of scenarios contained in test data items can be taken as hints on an unrestricted growth of the test data collection. This information can be a valuable input for the test data management and start a potential data cleansing. Clusters of scenarios contained in reference data can be treated as an indicator for frequent real-world use cases. These point to scenarios that should be tested thoroughly. However, these clusters can stem from provenance: Little exported data from unaltered export hubs (clinics, etc.) can be a reason for clusters. During our project, we noted that most of the test-relevant attributes are of categorical type. This has an impact on the visualization and can be the cause of clusters or correlations.

Besides this variety of visualizations to evaluate a test data collection's distribution, its diversity should also be evaluated. The function $s : D \times D \rightarrow \mathbb{R}$ measures the similarity of two test data items $d_i, d_j \in D$. Many possible implementations for set similarity metrics are given by Deza et al [5] or Stahl [15]. These measures can be helpful to identify similar and therefore potentially redundant test data items. It is important to know the test data's diversity to better control the growth of the test data collection.

Helpful during test phase *Test Analysis and Design* **and for** *Test Data Management.*

Supporting Test Case Selection

The mapping information stored in the TDQ-DB can be evaluated to support existing methods for test case selection. Each test case is prioritized according to whether its contained scenario is *executable, realistic* or *realistic and executable.* A *realistic* test case is a test case for which its scenarios are contained in reference data. A *non-executable* test case is a test case for which no test data item contains the needed scenarios to act as input data. Non-executable test cases can then be substituted early in time, lowering a testers failure quota during test execution searching for non-existent test data items as input data. Next, *realistic* test cases can be ranked according to their scenario's relative frequency in the reference data, thus prioritizing common real world use cases.

Helpful during test phase *Test Analysis and Design.*

Supporting Test Data Selection

Based on the n-dimensional mapping information for each test data item, we provide a search alongside the test-relevant attributes spanning the n-dimensional test space. For a given test case, a tester can query the database for test data items that are suitable as input. The query does not depend on fully qualified search criteria for each dimension and by using similarity measurements, even fuzzy searches are possible. However, fuzzy results may not be applicable for every domain. The list of suitable test data items can be ranked according to their *closeness to reality* or other quality means like *reputation* or *provenance.*

Helpful during test phase *Test Execution.*

5   Calibration Techniques

The size of the test space S is much greater than the number of scenarios $\mathcal{D}$ covered by test data. Due to $|S| \gg |\mathcal{D}|$, the *empty space* phenomenon [3] becomes a real problem as e.g. *coverage* tends to be zero, the list of missing scenarios is long and a search often finds zero if few results. We provide three calibration techniques to shape the test space and to reduce its size. All three techniques can be freely combined to shape the test space. The tester must be aware of his tweaks, to understand and evaluate the results of his analyses upon the shaped test space.

Equivalence Classes. The first way to reduce the test space is to group values into equivalence classes, a common technique in software testing [14]. For each value of a class, the SUT shows similar behaviour. This arrangement can be based on specifications or other functional classification, e.g. storage footprint, and can be nested to create hierarchies of equivalence classes.

Analyse Subspaces. Another technique to reduce the test space is to perform the analyses on subspaces. Based on actual tasks or analysis queries, testers select $\mathcal{A}^* \subset \mathcal{A}$ to span the subspace S^*. E.g. if some task is independent of A_3 (*region*), a subspace $S^* = A_1 \times A_2$ can be used for analyses. The TDQ-Sys transparently maps all stored mapping information into the new subspace and takes care of the correct assignment and aggregation. Now, testers can perform their queries and analyses on a reduced space and expect more meaningful answers. The results need to be combined by the testers to reconstruct a holistic view on the test data collection.

Mark Unwanted Scenarios. As the test space S is built via the crossproduct of all test-relevant attributes $\mathcal{A}$, all possible combinations of values are treated as scenarios, describing an input for the SUT. Because of that, some scenarios are impossible in the live system, out of range for an actual configuration of the SUT, or simply no in the actual test focus. These unwanted scenarios can be marked via rules. Applying these rules during analyses, the TDQ-Sys automatically hides these unwanted scenarios.

6 Prototype

This section describes the most important parts of our prototype. The whole system is written as a web application using the programming language 'Ruby'. The system configures itself at runtime using the configuration $\mathcal{A}$.

Data Model. Data items are stored as pointers into the used test data management system and provide the necessary provenance information and are enriched with additional descriptive information, like timestamps. We store the mapping information about these data items in structure similar to a *data warehouse* (DWH), see Fig. 3. This permits querying the database with powerful OLAP queries and supports the creation of subspaces, as they are equal to the `slice&dice` operator for data warehouses [4]. The qualifying dimensions, in our case the test-relevant attributes, are stored according to the *star* schema, thus enabling third-party tools to access the hierarchies of equivalence classes. Instead of storing numerical values as quantifying facts, we store a list of data items in order to enable the search feature and the evaluation of test data diversity. To allow third-party tools to take full advantage of the stored data, we make all configuration information accessible via separate tables.

```
1   describe 'PatientPosition' do
2       range 'HFP', 'HFS', 'HFDR', 'HFDL', 'FFP', 'FFS', 'FFDR', ↩
            'FFDL'
3       type  NOMINAL
4       query 'dicom://0018,5100'
5   end

7   describe 'SetupType' do
8       range 'Relative', 'Absolute'
9       type  NOMINAL
10      query 'dicom://300a,0128', 'dicom://300a,0129', ↩
            'dicom://300a,012a'
11      separator '^^'
12      assign ->(vertical, longitudinal, lateral) {
13          if(vertical == '0' && longitudinal == '0' && lateral == ↩
                '0')
14              'Relative'
15          else
16              'Absolute'
17          end
18      }
19  end
```

Listing 1.1. Definition of two test-relevant attributes.

Domain Specific Language. The description of the test-relevant attributes is a vital part of assessing test-data quality as it sets the stage: the test space. Like Ostrand and Balcer [11], we rely on experienced testers to identify and describe the test-relevant attributes. We profit from their valuable expertise and their knowledge of attributes that are not present in standard documents or specifications. To support this task, we designed a DSL and let testers describe the test-relevant attributes with plain text that serves as configuration file for the mapping component and the TDQ-Sys. An exemplary definition is shown in Listing 1.1.

7   Evaluation

With the aid of our prototype, we showed that our proposed concepts and methods for assessing a test data collection's data quality can be implemented with justifiable expenditure. Moreover, the generic implementation emphasizes our concept's independence of domain and application. We synchronized our concepts and methods with the tester's demands and their requirements for test-data quality, following the 'fitness for use' principle. We verified this alignment with an evaluation, carried out with six domain and test experts from our validation partner. However, we could only evaluate the compliance and acceptance of our concepts and methods. We were not able to measure real-world benefits as a proof that the methods actually increase the likelihood that errors will be detected earlier or that the methods help to simplify test case selection can be performed only by an intervention. The reason behind this is that the validation

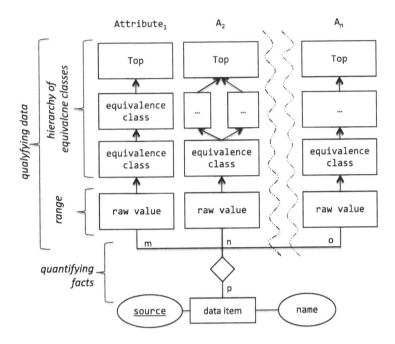

Fig. 3. Data warehouse like schema to store the mapping information

partner had no budget to duplicate his testing process and was not willing to use the prototypic implementation in production.

For the evaluation, we presented our concepts and methods as well as our prototype. Using questionnaires, we polled the alignment of the tester's demands and requirements with our concept, asked if they would take advantages utilizing our methods and if they want to use the methods in their daily routine. Figure 4 shows the results of the evaluation.[3] The interviewees state that they get additional information on the test (Fig. 4a) and note that this is helpful to evaluate the test-data quality of the test data collection (Fig. 4b). They rated their ability to assess test-data quality as partly good (Fig. 4c). Although the assessment itself was hindered by the prototypical implementation (Fig. 4d), they want to have such a tool (Fig. 4e) and value its integration in their daily routine as manageable (Fig. 4f). The masking of unwanted scenarios is reported as a good possibility to shape the test space (Fig. 4g) and the interviewees approve our concept of test data as sets of scenarios (Figs. 4h and i). The initial costs of preparation and identifying test-relevant attributes are estimated as high (Fig. 4j), telling us that these steps need to be prepared well and supervised by experts.

8   Discussion

Treating test data as sets of scenarios, n-tuples of values of its test-relevant attributes, comes along with loosing information. For example, the frequency of

---

[3] Detailed information and results upon request.

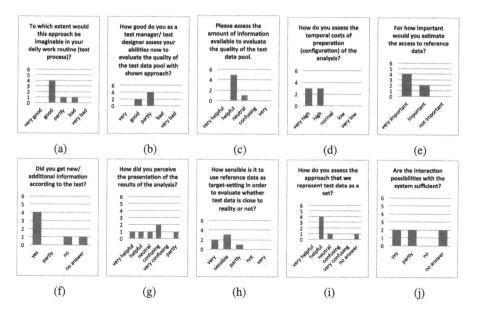

Fig. 4. Evaluation results with six participants.

scenarios within a test data item can be a valuable information for test data selection. But it is not used for test-data quality and therefore is is not tracked.

Another point is the used level of abstraction. Defining a test data item as a complete patient file, involves a loss of associations between scenarios. After mapping, it is not possible to detect which scenarios are covered by which part of the patient file. For a finer granularity of mapping, it is required to lower the level of abstraction and treat e.g. a single document inside a patient file as test data item. However, the granularity depends on the actual use cases and operational sequence for testing, being the domain of the manufacturer and his responsibility.

Although the test-relevant attributes are identified and described at the beginning, the test space can be populated with new values for test-relevant attributes, as stated earlier. Also, it is always possible to add or modify test-relevant attributes, enforcing a complete re-mapping of all test data items, as all mapping information in the TDQ-DB must be updated. Of course, it is always possible to add new test data items.

There are costs associated with the usage of test-data quality. A priori, there are the running costs of the TDQ-Sys, consisting of a database server and a web server. First, depending on the complexity of the test data type, the implementation costs of the mapping component cannot be neglected. Other costs are then time and labor for user training and for the required process to identify and describe all test-relevant attributes.

9 Contribution

We introduced a concept to assess a test data collection's quality as an indicator for its applicability for test goals. We showed some benefits during the test process that can be realized by utilizing our concept. First, the necessity to identify test-relevant attributes and their values to build the test spaces for test goals forces test experts to reflect on and revive their domain knowledge, inducing positive effects on test planning. Next, analyzing the mapping step builds confidence about the identified values and may discover unknown scenarios or use cases which can be used for a more thorough test. The evaluation of *coverage of test space* identifies untestable scenarios due to missing test data, triggering a target-aimed acquisition in sufficient time. Analyzing the test data collection's distribution in the test space enables test data management to better govern further growth. The similarity measurement can reveal possible redundant test data. Shrinkage the test data collection relieves test data management and eases searching for test data items. The n-dimensional and fuzzy searches help to find matching test data for test cases and can be used to link test data to test cases early on.

Although our research had a medical background, our concept is domain independent, because of the decoupling done via the externalized mapping. However, they are best suited to be used for the system test, as the complexity of test data and the size of the test space are relatively low for unit testing or integration testing. We implemented a generic prototype that is configured at runtime by configuration files written in our domain specific language, and explained calibration techniques to adapt our methods in a demand-driven fashion.

Acknowledgements. This project is supported by the German Federal Ministry of Education and Research (BMBF), project grant No. 01EX1013G.

References

1. Batini, C., Cappiello, C., Francalanci, C., Maurino, A.: Methodologies for data quality assessment and improvement. ACM Comput. Surv. **41**(3), 1–52 (2009)
2. Bendix, F., Kosara, R., Hauser, H.: Parallel sets: visual analysis of categorical data. In: IEEE Symposium on Information Visualization, INFOVIS 2005, pp. 133–140, No. 1. IEEE (2005)
3. Carreira-Perpiñán, M.: A review of dimension reduction techniques. Technical report, University of Sheffield, Sheffield (1997)
4. Chaudhuri, S., Dayal, U.: An overview of data warehousing and OLAP technology. ACM SIGMOD Rec. **26**(1), 65–74 (1997)
5. Deza, E., Deza, M.M.: Encyclopedia of Distances. Springer, Heidelberg (2009)
6. Di Nardo, D., Alshahwan, N., Briand, L., Labiche, Y.: Coverage-based test case prioritisation: an industrial case study. In: 2013 IEEE Sixth International Conference on Software Testing, Verification and Validation, March 2013
7. Juran, J.M.: Juran on Planning for Quality. Free Press, New York (1988)

8. Law, M.Y., Liu, B.: Informatics in radiology: DICOM-RT and its utilization in radiation therapy. Radiographics Rev. Publ. Radiol. Soc. North Am. Inc. **29**(3), 655–667 (2011)

9. Mildenberger, P., Eichelberg, M., Martin, E.: Introduction to the DICOM standard. Eur. Radiol. **12**(4), 920–927 (2002)

10. Mustra, M., Delac, K., Grgic, M.: Overview of the DICOM standard. In: 50th International Symposium, ELMAR, pp. 10–12. IEEE, Zadar, September 2008

11. Ostrand, T.J., Balcer, M.J.: The category-partition method for specifying and generating functional tests. Commun. ACM **31**(6), 676–686 (1988)

12. Pipino, L.L., Lee, Y.W., Wang, R.Y.: Data quality assessment. Commun. ACM **45**(4), 211–218 (2002)

13. Redman, T.C.: Data Quality: The Field Guide, data manag edn. Digital Press, Newton (2001)

14. Sommerville, I.: Software Engineering. International Computer Science Series, 8th edn. Addison-Wesley, Reading (2007)

15. Stahl, H.: Clusteranalyse großer Objektmengen mit problemorientierten Distanzmaßen. Verlag Harri Deutsch, Thun, Frankfurt am Main, reihe wirtschaftswissenschaften edn. (1985)

16. Thomas, S.W., Hemmati, H., Hassan, A.E., Blostein, D.: Static test case prioritization using topic models. Empirical Softw. Eng. **19**(1), 182–212 (2014)

17. Wand, Y., Wang, R.Y.: Anchoring data quality dimensions in ontological foundations. Commun. ACM **39**(11), 86–95 (1996)

18. Wang, R.Y., Strong, D.M.: Beyond accuracy: what data quality means to data consumers. J. Manage. Inf. Syst. **12**(4), 5–34 (1996)

19. Wang, R.Y., Ziad, M., Lee, Y.W.: Data Quality. Springer - Kluwer Academic, Boston (2002)

Advanced Query Processing

A Self-tuning Framework for Cloud Storage Clusters

Siba Mohammad$^{(\boxtimes)}$, Eike Schallehn, and Gunter Saake

Institute of Technical and Business Information Systems,
Otto-von-Guericke-University of Magdeburg, Building 29,
Universitätsplatz 2, 39106 Magdeburg, Germany
{smohamma,eike,saake}@iti.cs.uni-magdeburg.de

Abstract. The well-known problems of tuning and self-tuning of data management systems are amplified in the context of Cloud environments that promise self management along with properties like elasticity and scalability. The intricate criteria of Cloud storage systems such as their modular, distributed, and multi-layered architecture add to the complexity of the tuning and self-tuning process. In this paper, we provide an architecture for a self-tuning framework for Cloud data storage clusters. The framework consists of components to observe and model certain performance criteria and a decision model to adjust tuning parameters according to specified requirements. As part of its implementation, we provide an overview on benchmarking and performance modeling components along with experimental results.

Keywords: Cloud storage clusters · Self-tuning · Performance modelling · Regression analytic · Benchmarking

1 Introduction

Although, conventional database systems are used for Cloud applications where strict consistency and transactional processing are needed, properties of the Cloud environment (multi tenancy, component failure, etc.) and the needs of its application (scalability, availability, and fault tolerance, etc.) resulted in a new breed of data storage systems. These systems were primarily developed for internal use by companies such as Google, Amazon, Facebook, etc. For Cloud-based and big data applications, Cloud storage systems are the storage systems of choice to meet the mentioned requirements.

From the architectural point of view, these systems have a modular, multi-layered architecture. According to application needs, multiple component systems are combined together to provide needed functionalities. As a basic component, a distributed file system (e.g. Google and Hadoop file systems) supports scalable, fault tolerant data storage and access. On top of it, typically lays a structured-data storage system (e.g. Bigtable [4], Cassandra [10]). Systems of this layer structure data in non-relational data models; key-value model

© Springer International Publishing Switzerland 2015
T. Morzy et al. (Eds.): ADBIS 2015, LNCS 9282, pp. 351–364, 2015.
DOI: 10.1007/978-3-319-23135-8_24

being the dominant. They also provide API access and SQL-like query languages (e.g. CQL). Because of the non-relational nature of the underlying data model, RDBMS-style aggregations and joins were typically not supported. To perform such operations and more complex data analytic, a distributed processing system is used on top of the previous layers; Map Reduce framework being the dominant. Though later versions of Cloud storage systems support joins and aggregations, these operations are internally transformed into Map Reduce jobs (e.g. Hive and Pig)[1]. A more detailed architectural overview and a classification of cloud storage systems can be found in [6].

Though Cloud storage systems were developed to be self-managing regarding many aspects, e.g. dynamically adding or removing resources, there are still numerous decisions to be made to actually fit the requirements of given applications to provide suitable performance. The contributions of this paper are as follows:

- As a precondition for the proposed framework, we relate tasks of (self-)tuning to layers and sub-clusters within a typical Cloud storage architecture.
- We describe the top-level view of our framework applicable to various optimisation goals and parameters describing the application or the configuration.
- Based on measured and/or modelled performance of applications, we describe a decision model suitable for self-tuning of configuration parameters.
- More specifically we address the common problem of adjusting the size of (sub-)clusters in an experimental evaluation.

The rest of the paper is organized as follows. First, we provide a motivational scenario in Sect. 2. In Sect. 3, we give an overview of the framework. After that, we provide more detailed description of benchmarking and modelling components in Sect. 4. Then, we discuss the current experimental results in Sect. 5. After that, we provide an overview of related work in Sect. 6. The paper ends with conclusion and future work in Sect. 7.

2   Motivation

The tuning and self-tuning of Cloud storage systems has gained more attention by both industrial and academic research [20,22,24–26]. Because of the typical shared nothing architectures with data partitioning and replication, some performance aspects can be easily addressed for the overall system. Nevertheless, the typical multi-layered distributed architecture of several component systems adds complexity to the tuning tasks. Moreover, if there are several applications with different and possibly changing requirements, using the same data storage cluster, there is little chance to tune for a specific application.

Based on this assumption our approach applies the concept of creating dedicated sub-clusters for single applications/workloads or groups of workloads with similar requirements as shown Fig. 1. Here, application requirements can be

---

[1] https://cwiki.apache.org/confluence/display/Hive/LanguageManual+Joins.

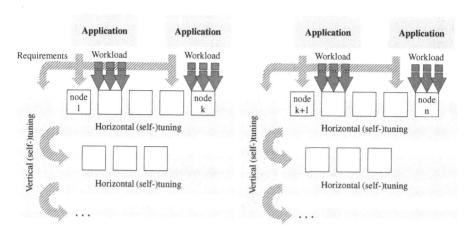

Fig. 1. (Self-)tuning for cloud storage systems

mapped to different tuning knobs to achieve applications optimisation goals. In the multi-layered, modular architecture, these requirements can now be handled on two dimensions. The first one, we call horizontal (self-)tuning, which takes place within layer. The horizontal (self-)tuning includes aspects such as partitioning, load balancing, replication, update strategies, and automatic scaling, etc. Problems on this dimension are better supported because of the homogeneous processes of a single component type within one layer. The second one, we call vertical (self-)tuning, which is carried out across layers. Vertical (self-)tuning includes the mapping of application requirements expressed as optimization goals, service levels, etc. to specific tuning knobs on each level of the storage architecture. For the remainder of the paper, we focus on aspects of horizontal self-tuning.

For illustration purposes, consider the example shown in Fig. 2, which is based on data gathered from experiments described, in more details, in Sect. 4. As shown in Fig. 2a, there are three different workloads, being read-heavy (r90w10), evenly mixed (r50w50), and write-heavy (r10w90) showing very different performance characteristics measured for different cluster sizes (overall latency for entire workload, average over 5 independent runs). Decisions that can be made based on this data include:

- finding the best cluster size for a single workload, e.g. indicated by a global mimimum within resource restrictions, or
- creating an optimal setup of sub-clusters for all workloads.

Based on our overall approach, the latter is of great importance, and the results of optimal sub-cluster configurations given different node constraints are shown in Fig. 2b. For reasons of simplicity, the sum of the overall latency was used as an optimisation goal, though different aggregation functions are conceivable. The optimisation of this given problem can easily be done by brute force algorithms, because it is linear and discrete in the number of nodes and only exponential

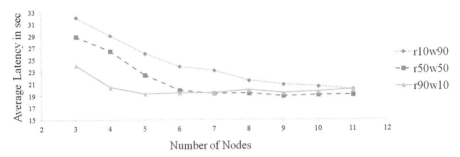

(a) Measured latency characteristics for different workloads

(b) Optimal allocation of nodes to workloads

Number of Nodes	9	10	11	12	13	14	15	16	17	18	19	20	21	22	23
Nodes for r10w90	3	3	4	3	4	5	5	6	6	8	8	9	9	10	11
Nodes for r50w50	3	3	3	5	5	5	6	6	6	6	6	6	7	7	7
Nodes for r90w10	3	4	4	4	4	4	4	4	5	4	5	5	5	5	5
Latency in sec		85	81.4	78.3	74.9	71.9	68.9	66.3	64.2	63.1	61.8	60.8	60.1	59.6	59.2	58.7

Fig. 2. Optimal allocation of nodes for three workloads 10 % read and 90 % write (r10w90), 50 % read and 50 % write (r50w50), and 90 % read and 10 % write (r90w10) for different cluster sizes

in the number of workloads. More sophisticated approaches may be required for non-discrete cases and those involving more complex parameter combinations. Furthermore, the general framework presented in this paper will have to deal with the fact, that measurements gathered from monitoring the system or test runs are incomplete within the huge space of possible parameter combinations. To predict the performance, a model of it needs to be derived from the available data.

3   A Framework for Tuning Cloud Data Storage Cluster

In this section, we discuss our approach of addressing the aforementioned problem scenario. After formalizing the problem, we illustrate different components of our infrastructure and their functionalities.

3.1   Problem Statement and Solution Approach

For our framework, we define the general optimization approach as follows: the optimisation goal opt is to find a cluster configuration c out of a set of possible configurations CC that minimizes (assuming a standard form of the problem) the costs for all workloads w of a set of workloads WL that need to be supported by the overall cluster.

$$opt = \underset{c \in CC}{\text{minimize}} \quad \Gamma_{w \in WL} cost(c, w)$$

Here Γ represents some aggregation function suitable to the given cost components considered, e.g. sum for energy consumption or average or maximum for response time. Constraints can be defined on the cluster configuration as discussed below:

Cluster Configurations in CC**.** These independent variables are controlled variables and represent the actual knobs that can be used to achieve the optimisation goal. Typical configuration aspects are for instance the cluster size, hardware being used, replication factor and other database parameters, etc. Formally, c can be described as an n-tuple that holds relevant parameters as components, e.g. $c_1 = \{cn = 10, rf = 3\}$ for a cluster of 10 nodes and a replication factor of 3.

Workload Characteristics in WL**.** These independent variables describe the application, but are not controlled by the systems administrators or developers, i.e. though they may be highly dynamic, they can not be changed deliberately to achieve an optimisation goal. These include for instance workload characteristics, access frequencies, user numbers, data volume and schema, etc., which, again, can be modelled as an n-tuple, e.g. $wl_1 = \{r = 90, w = 10, sf = 10, nc = 5\}$ for a workload having 90 % read operations performing on a 10GB database of 5 column families.

Optimisation Goal opt**.** The dependent variables used in prediction models for system optimisation are typically those, for which an optimal value should be achieved. For Cloud storage, these may include variables such as throughput, latency, energy consumption, resource utilisation, consistency, etc. The optimisation task, for which the model is being used, may be multi-objective, requiring specific techniques not discussed in this paper.

Not all of the possible parameters describing a workload or a cluster configuration may be relevant or desirable to consider in a given application scenario. Furthermore, there might be strong correlations between some of the independent variables, which can be used to simplify the models creation and application. While we discuss techniques to create a performance model in this paper, here it is not our intention to investigate the complex space of variables and their dependencies in its entirety, but rather focus on – in our opinion – a most relevant subspace to discuss the modelling and prediction techniques, namely finding the optimal size of sub-clusters for a given set of workloads. To achieve this, we express the relation between performance metrics of a workload w with cluster configuration c as a cost function where N is the total number of nodes in the infrastructure:

$$opt = \sum_{w \in WL} cost(w, n_w)) \rightarrow min$$

subject to

$$\sum_{w \in WL} n_w \leq N$$

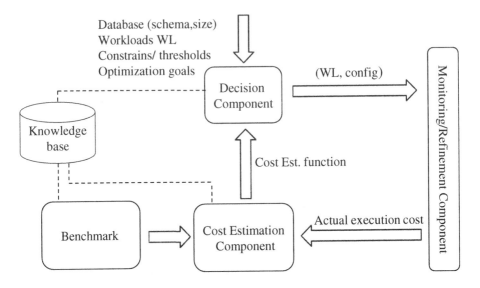

Fig. 3. Self-tuning framework for cloud data management systems

3.2 Framework Architecture

As we illustrate in the Fig. 3, our framework is composed from the following components:

Benchmark. The purpose of this component is to generate the training data needed to model the performance of the data storage cluster for a certain workload with different cluster sizes or possibly different configurations.

Cost Estimation Component. This component uses statistical-based data-driven modeling to build performance models as mathematical functions. These functions are derived by regression techniques done on statistical data gathered from the benchmarking phase.

Decision Component. Tuning knobs are expressed as different independent variables during the modeling process. Based on conditions derived from workload thresholds, this component performs a filtering process on the value-space of the independent variables. Then it solves the optimization problem of the cost models, based on the optimisation goals of different workloads, to find preferable values of the tuning knobs.

Monitoring/Refinement Component. This components is responsible for adding measurement to the knowledge-base and initiating a re-modeling process if

$$|Actual - Prediction| \geq threshold$$

Knowledge-base. Stores information for reuse by the framework. Information includes workloads description (i.e. schema and data access pattern) and cost models.

4   Benchmarking and Performance Modeling

In this section, we provide our current theoretical and empirical results in implementing the tuning framework. As first steps in this direction, we developed a benchmark and a cost estimation component. Our approach for predicting the performance of a Cloud database cluster is to benchmark the cluster based on several runs of workloads. Then, build a cost(performance) model using regression analytic techniques. We provide more details in the following subsections.

4.1   Benchmarking Cloud Storage Clusters

The purpose of our benchmark is to generate data to model the performance of a Cloud storage cluster for a certain workload with different cluster sizes. Many benchmarks [1,2,4,7] exist for comparing the performance of several database systems for certain workloads or even checking the performance of one database systems with different workloads to identify bottlenecks. Our implementation supports the typical requirements of a benchmark such as allowing workload configuration: read/write ratio, data size, throughput, etc. and it automates the testing process for an increasing number of database system cluster size. We provide the architecture of our benchmark in Fig. 4. The essential component of the benchmark is the benchmark manager which is responsible for starting the benchmark instances. It also acts as data storage system cluster controller, performing operations of:

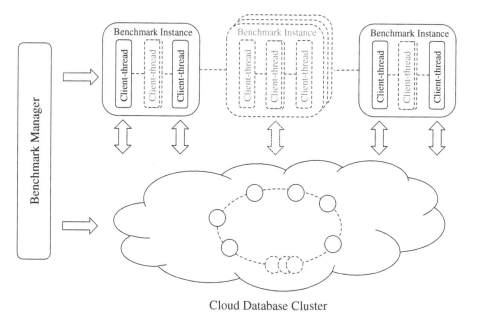

Cloud Database Cluster

Fig. 4. Benchmark architecture

- Preparing and starting the database cluster for certain number of nodes.
- Creating the database schema, generating and loading data before the actual workload, if needed.
- Rebooting database and operating system to flush the file system caches, main memory and CPU caches in the case of a cold run.

The benchmark instances are responsible for starting the workload and collecting measurements of performance. We designed the benchmark to allow specifying the following workload characteristics:

- Database schema (table, number of columns), record size and replication factor.
- Data access specifications: read/write ratio, number of rows to be read or written, throughput (number of concurrent access).

As specified by the workload setting and based on the replication factor and the maximum number of nodes intended for the data storage cluster, several phases of the benchmark are performed. Each phase is defined by the number of nodes in the cluster (cluster size). The cluster size varies between the data replication factor and the maximum number of nodes available. In each phase, multiple remote benchmark-instances are started by the benchmark-manager using SSH (Secure Socket Shell). Each benchmark-instance starts multiple client-threads depending on the number of the CPU cores and the memory size of the host machine. After the workload ends, statistical data describing the performance are retrieved from all client-generator machines and combined in one output file. Within one phase, the benchmark automatically repeats the experiment a number of times defined by the user and the average measurement is used for the modeling process. After the experiment is done for the current number of nodes, the benchmark starts again for next number of nodes.

4.2   Performance Modelling

This step includes analyzing the collected data to discover the underlying model. There are several machine learning techniques used for modeling. These include clustering, tree-based, genetic evolutionary algorithms and neural networks [9]. Regression analytic techniques are considered one of the simplest techniques for predictive modeling. Their process relies on statistical and regression analysis to find a formula or mathematical model to represent the relationship between a dependent variable being the measured performance aspect (e.g. latency) and one or more application-specific requirement such as workload criteria (e.g. read/write ratio) or system configuration aspects (e.g. database cluster size).

As stated by Mark Kotanchek et al. [8], there is an infinite number of predictive models that fit a finite data set. Our goal is to find a model that fits the data and has a relatively small error. To achieve this, we use different regression analysis techniques and measure the error rate. The dependent variable in our implementation is the response time or latency and the independent variables are the number of nodes in the database cluster (cluster size) and the read/write ratio.

Table 1. Software and hardware configuration

OS Ubuntu: 13.04 kernel		Version: Linux 3.8.0-35-generic-pae	
CPU: Intel(R) Xeon(R) E5-2650 0 2.00GHz		2 Cores Cache size: 20480 KB	
Disk:	90.18 GB 7200RPM	Memory:	8 GB
Network	100 MBits	Java Version	1.7.0_25
Cassandra Version 1.2.13		Virtual machines for cluster deployment: 11	
Replication factor: 3		Virtual machines for generating workload: 3	

5   Experiment and Evaluation

For our experiment, we choose Cassandra [10]. Our approach is database agnostic, and Cassandra was chosen only as an example of Cloud storage systems. Cassandra was designed for internal use by Facebook and was later adopted by Apache. Large clusters of Cassandra are being used by systems like Netflix, Spotify, and eBay[2], etc. Cassandra provides scalable structured data storage, supporting tune-able consistency, column family data model and a SQL-like query language called CQL. We deploy Cassandra on a network of virtual machines in our labs. Configuration of the testbed for this experiment is illustrated in Table 1. For the deployment of the benchmark, we dedicate another set of virtual machines in the same network with the same configuration. With the goal of modeling the performance of the database cluster with different cluster sizes and different workloads, the workloads we tested vary in the read/write percentage. Other workload criteria such as the schema, consistency level, row size, and goal throughput (concurrent accesses) is kept the same. Each workload operates on one column family. Each read or write operation touches one row. Write operations insert randomly generated strings. Read operations are select point queries; the whole row is retrieved. The percentage of read/write operations in the tested workloads are: 0, 10, 30, 50, 70, 90, and 100. Each workload was tested with Cassandra cluster of sizes that vary between 3 (replication factor) and 11 (the maximum number of virtual machines dedicated for the database in our infrastructure). Each experiment is repeated 5 times and the average value is used for the modelling process.

The result from our experiment is illustrated by the surface in Fig. 5 which represents how the cluster behaves (its latency in ms) with different workloads (characterized by their read/write percentage) and different cluster sizes. We test several regression analytic techniques: simple linear regression, polynomial regression with several degrees, and exponential regression. As a result from the regression process using the cubic regression gives the best residual standard deviation among the tested techniques, with a slight difference from the

---

[2] Usecase higlights for Cassandra are found on http://planetcassandra.org and http://www.datastax.com/customers.

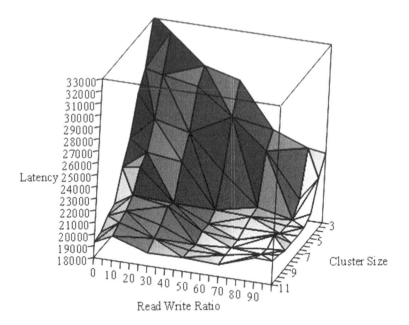

Fig. 5. Benchmarking results

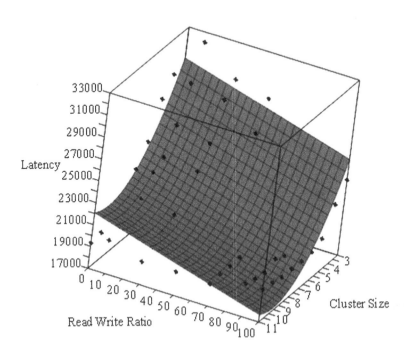

Fig. 6. Regression analysis results vs. input measurements

quadratic regression. Figure 6 illustrates the surface representing the resulted cubic model versus the points representing the input measurements.

To validate the result from the regression process, we test the model prediction power against new workloads and calculate the mean absolute error percentage. The cubic model gives high prediction accuracy of 96.4 %.

The result from our experiment and evaluation shows that the cubic model has the best residual standard deviation and characterizes the performance with high prediction accuracy. Such a model (even with the low number of independent parameters) can be beneficial to avoid allocating resources to the database cluster that will obtain insignificant benefit from them. Our benchmark allows specifying several workloads parameters, which allows extending the model. However, more experiments must be done to generate the statistical (training data) that is required for creating extended models.

6   Related Work

The work described in this paper is based on three areas of research: benchmarking, performance modeling, and (self-)tuning. In the next paragraphs, we provide a short overview of related work of these fields in the context of cloud storage systems.

Related work on benchmarking Cloud storage systems includes general purpose benchmarks [1,2,4,7] which measure latency for different systems and focus on providing details about selecting workloads and benchmark architecture that corresponds to the Cloud environment and its applications. Several studies [3,12] build on these benchmarks. Another group of benchmarks focus on how a system performance changes with different technical, or platform choices. An example of that can be found in [11] which focuses on analyzing the performance of Cassandra on two platforms using HDD or Flash memory. Another example [12] provides read/write and structured query benchmark which investigates how different implementation techniques of different systems affect the performance. There is also the work of Rabl et al. [5] which provides an overview of the performance impact of different storage architectures. The last group of benchmarks examine specific properties of Cloud storage systems such as replication, consistency, and elasticity. An example of such work can found in [3] which uses different replication strategies and consistency levels and measures their effect on latency and throughput.

Since virtualization is a major technique for cloud environment, a large part of work [14,17,18] is dedicated to performance modeling for cloud application in virtualized environments. The work of Noorshams et al. [14] investigates performance modeling for virtualized storage systems. Another example is the work of Kraft et al. [18] in which they present a simple model for predicting the degradation in performance that results from storage devices contention in virtualized environments. Other aspects such as modeling the scalability behavior of network/CPU intensive applications can be found in [17]. Different techniques for building models exists. A part of the research efforts uses machine learning techniques such as [19], which uses Kernel Canonical Correlation Analysis to model the execution time of MapReduce jobs. Another approach can be found in the work of [15] where they present an analytical model of the Spotify storage architecture that allows to estimate the distribution of response time of storage system.

Related work on (self-)tuning for Cloud storage systems falls in two parts. The first one includes tuning database systems for specific workload, optimization goal, or execution environment. Examples of such efforts include work of [26] which aims to reduce energy consumption and thus cooling costs by applying resource aware data placement and migration strategy. A part of work in this category falls under scheduling. Chi et al. perform cost aware scheduling of queries based on service level agreements [22] whereas Polo et al. perform Map Reduce jobs scheduling to maximize resource utilization [20]. The second part of the (self-)tuning efforts for Cloud storage systems is external to the database system and includes tuning the underlying resources to achieve the optimization goals of the database workload. Example of this is the work of [25] which focuses on partitioning the CPU capacity of physical machines among different database appliances. A more general example is the work of Xiong et al. [21] where they perform cost aware resource management. Herodotou et al. developed a self-tuning framework, starfish [24], for big data analytics. An interesting approach incorporates different DBMSs within one system (called DBMS+) and depends on the query optimizer, of the incorporated systems, to perform the tuning process [23]. Then the tuning process selects the appropriate execution plan for each request.

7   Conclusion and Future Work

While Cloud storage systems are good at self-management, e.g. automatically mapping to available resources, there are still many open issues to actually make them self-tuning, i.e. adjust their parameters to application-specific requirements. In this paper, we presented an approach how such self-tuning functionality can be integrated with an according system, by observing, modelling, predicting the performance, and adjusting the configuration depending on a described decision model. The practical evaluation applied Cassandra and focused on the problem of assigning an optimal number of nodes to various workloads running on sub-clusters to achieve the best possible overall latency.

As discussed throughout this paper, many open questions remain. From our point of view, the most important ones are to be related to multi-objective optimisation, which is the standard case for most real-life systems, where some controlled trade-off between related or even contradicting goals has to be found. Furthermore, we currently investigate the effect of heterogeneous environments and their effect on predictability and resource assignment.

References

1. Curino, C., Difallah, D.E., Pavlo, A., Cudre-Mauroux, P.: Benchmarking OLTP/web databases in the cloud: the OLTP bench framework. In: Proceedings of the 4th International Workshop on Cloud Data Management, pp. 17–20. ACM, New York (2012)
2. Binnig, B., Kossmann, D., Kraska, T., Loesing, S.: How is the weather tomorrow? towards a benchmark for the cloud. In: Proceedings of the 2nd International Workshop on Testing Database Systems, pp. 9:1–9:6. ACM, New York (2009)
3. Wang, H., Li, J., Zhang, H., Zhou, Y.: Benchmarking replication and consistencystrategies in cloud serving databases: HBase and Cassandra. In: Big Data Benchmarks, Performance Optimization, and Emerging Hardware 4th and 5th Workshops, pp. 71–82. Springer, Switzerland (2014)
4. Chang, F., Dean, J., Ghemawat, S., Hsieh, W.C., Wallach, D.A., Burrows, M., Chandra, T., Fikes, A., Gruber, R.E.: Bigtable: a distributed storage system for structured data. ACM Trans. Comput. Syst. **26**, 4:1–4:26 (2008)
5. Rabl, T., Gómez-Villamor, S., Sadoghi, M., Muntés-Mulero, V., Jacobsen, H.A., Mankovskii, S.: Solving big data challenges for enterprise application performance management. PVLDB **5**(12), 1724–1735 (2012)
6. Mohammad, S., Breß, S., Schallehn, E.: Cloud data management: a short overview and comparison of current approaches. In: 24th GI-Workshop on Foundations of Database, pp. 41–46. CEUR-WS, Aachen (2012)
7. Cooper, B. F., Silberstein, A., Tam, E., Ramakrishnan, R., Sears, R.: Benchmarking cloud serving systems with YCSB. In: 1st ACM Symposium on Cloud Computing, pp. 143–154. ACM Press, New York (2010)
8. Kotanchek, M., Smits, G., Vladislavleva, E.: Trustable symbolic regression models: using ensembles, interval arithmetic and pareto fronts to develop robust and trust-aware models. In: Riolo, R., Soule, T., Worzel, B. (eds.) Genetic Programming Theory and Practice V. Genetic and Evolutionary Computation Series, pp. 201–220. Springer, US (2008)
9. Matsunaga, A., Fortes, J.A.B.: On the use of machine learning to predict the time and resources consumed by applications. In: 10th International Conference on Cluster, Cloud and Grid Computing, pp. 495–504. IEEE Press, New York (2010)
10. Lakshman, A., Malik, P.: Cassandra - a decentralized structured storage system. Operating Syst. Rev. **44**(2), 35–40 (2010)
11. Aplin, P.: Benchmarking Cassandra on Violin - Violin Memory. Technical report, Violin Memory (2013)
12. Shi, Y., Meng, X., Zhao, J., Hu, X., Liu, B., Wang, H.: Benchmarking cloud-based data management systems. In: 2nd International Workshop on Cloud Data Management, pp. 47–54. ACM Press, New York (2010)
13. Piao, J.T., Yan, J.: Computing resource prediction for MapReduce applications using decision tree. In: Sheng, Q.Z., Wang, G., Jensen, C.S., Xu, G. (eds.) APWeb 2012. LNCS, vol. 7235, pp. 570–577. Springer, Heidelberg (2012)
14. Noorshams, Q., Bruhn, D., Kounev, S., Reussner, R.: Predictive performance modeling of virtualized storage systems using optimized statistical regression techniques. In: Proceedings of the 4th International Conference on Performance Engineering, pp. 283–294. ACM, New York (2013)
15. Yanggratoke, R., Kreitz, G., Goldmann, M., Stadler, R.: Predicting response times for the spotify backend. In: Proceedings of the 8th International Conference on Network and Service Management, pp. 117–125. International Federation for Information Processing, Laxenburg (2013)

16. Kundu, S., Rangaswami, R., Dutta, K., Zhao, M.: Application performance modeling in virtualized environment. In: Proceedings of the 16th International Symposium on High Performance Computer Architecture (HPCA), pp. 1–10. IEEE (2010)
17. Chen, X., Ho, C.P., Osman, R., Harrison, P.G., Knottenbelt, W.J.: Understanding, modeling and improving the performance of web applications in multicore virtualised environments. In: Proceedings of the 5th International Conference on Performance Engineering, pp. 197–207. ACM, New York (2014)
18. Kraft, S., Casale, G., Krishnamurthy, D., Greer, D., Kilpatrick, P.: Performance models of storage contention in cloud environments. Softw. Syst. Model. **12**, 681–704 (2013)
19. Ganapathi, A., Chen, Y., Fox, A., H. Katz, R., Patterson, D.A.: Statistics-driven workload modeling for the cloud. In: Workshops Proceedings of the 26th International Conference on Data Engineering (ICDE), pp. 87–92. IEEE, California (2010)
20. Polo, J., Castillo, C., Carrera, D., Becerra, Y., Whalley, I., Steinder, M., Torres, J., Ayguadé, E.: Resource-aware adaptive scheduling for MapReduce clusters. In: Kon, F., Kermarrec, A.-M. (eds.) Middleware 2011. LNCS, vol. 7049, pp. 187–207. Springer, Heidelberg (2011)
21. Xiong, P., Chi, Y., Shenghuo, Z., Moon, H.J., Calton, P., Hacigümüş, H.: Intelligent management of virtualized resources for database systems in cloud environment. In: Proceedings of the 27th International Conference on Data Engineering (ICDE), pp. 87–98. IEEE, Washington (2011)
22. Chi, Y., Moon, H.J., Hacigümüş, H.: iCBS: incremental cost-based scheduling under piecewise linear SLAs. Proc. VLDB Endow. **4**, 563–574 (2011)
23. Lim, H., Han, Y., Babu, S.: How to fit when no one size fits. In: Proceeding of the 6th Biennial Conference on Innovative Data Systems Research (CIDR), Online Proceedings (2013)
24. Herodotou, H., Lim, H., Luo, G., Borisov, N., Dong, F., Bilgen, F., Babu, S.: Starfish: a self-tuning system for big data analytics. In: Proceedings of the 5th Biennial Conference on Innovative Data Systems Research (CIDR), pp. 261–272. Online Proceedings (2011)
25. Aboulnaga, A., Salem, K., Soror, A.A., Minhas, U.F., Kokosielis, P., Kamath, S.: Deploying database appliances in the cloud. IEEE Data Eng. Bull. **32**, 13–20 (2009)
26. Goiri, I., Le, K., Nguyen, T.D., Guitart, J., Torres, J., Bianchini, R.: GreenHadoop: leveraging green energy in data-processing frameworks. In: The 7th European Conference on Computer Systems, pp. 5770. ACM, New York (2012)

Optimizing Sort in Hadoop
Using Replacement Selection

Pedro Martins Dusso[1,2]([✉]), Caetano Sauer[1], and Theo Härder[1]

[1] Technische Universität Kaiserslautern, Kaiserlautern, Germany
pmdusso@gmail.com
[2] Universidade Federal do Rio Grande do Sul, Porto Alegre, Brazil

Abstract. This paper presents and evaluates an alternative sorting component for Hadoop based on the replacement selection algorithm. In comparison with the default quicksort-based implementation, replacement selection generates runs which are in average twice as large. This makes the merge phase more efficient, since the amount of data that can be merged in one pass increases in average by a factor of two. For almost-sorted inputs, replacement selection is often capable of sorting an arbitrarily large file in a single pass, eliminating the need for a merge phase. This paper evaluates an implementation of replacement selection for MapReduce computations in the Hadoop framework. We show that the performance is comparable to quicksort for random inputs, but with substantial gains for inputs which are either almost sorted or require two merge passes in quicksort.

Keywords: Sorting · Quicksort · Replacement selection · Hadoop

1 Introduction

This work implements and evaluates an alternative sorting component for Hadoop based on the replacement selection algorithm [9]. Sorting is used in Hadoop to group the map outputs by key and deliver them to the reduce function. Because of Hadoops big data nature, this sorting procedure usually is an external sorting. The original implementation is based on the quicksort algorithm, which is simple to implement and efficient in terms of RAM and CPU.

Sorting performance is critical in MapReduce, because it is not trivially parallelizable as map and reduce tasks. The data is parallelized by partitions on the reduce key value, but this requires a lot of data movement. The sort stage of a MapReduce job is network- and disk-intensive, and often reading a page from the hard disk takes longer than the time to process it. Thus, CPU instructions stop being the unit to measure the cost in the context of external sorting, and we replace it by the number of disk accesses—or I/O operations—performed. This difference makes algorithms designed only to minimize CPU instructions not so efficient when analyzed from the I/O point of view. This means that the superiority of quicksort for in-memory processing may not be directly manifested in this scenario.

© Springer International Publishing Switzerland 2015
T. Morzy et al. (Eds.): ADBIS 2015, LNCS 9282, pp. 365–379, 2015.
DOI: 10.1007/978-3-319-23135-8_25

Our goal in this paper is to assess replacement selection for sorting inside Hadoop jobs. To the best of our knowledge, this is the first approach in that direction, both in academia and in the open-source community. We observe that our implementation performs better for almost-sorted inputs and for inputs that are considerably bigger than the available main memory. Furthermore, it exploits multiple hard drives for better I/O utilization.

The remainder of this work is organized as follows. Section 2 reviews related work and motivates the use of replacement selection. Section 3 reviews algorithms for disk-based sorting, focusing on the replacement selection algorithm. In Sect. 4, we discuss implementation details. First, we present the internals of Hadoop, focusing on the algorithms and data structures used to implement external sorting using quicksort. Second, we present a custom memory manager used to manage the space in the sort buffer in main memory efficiently. Third, we present optimizations in the key comparison during the sort and finally a custom heap with a byte array as the placeholder for the heap entries. In Sect. 5, we compare our replacement selection method against the original quicksort. Finally, Sect. 6 concludes this paper, providing a brief overview of the pros and cons of our solution, as well as discussing open challenges for future research.

2   Background

Data management research has shown that replacement selection may deliver higher I/O performance for large datasets [10,11] in the context of traditional DBMS applications. Replacement selection is a desirable alternative for run generation for two main reasons: first, it generates longer runs than a standard external-sort run-generation procedure. As Knuth remarks in [9], "the time required for external merge sorting is largely governed by the number of runs produced by the initial distribution phase". The fact that the lesser number of runs created by replacement selection leads to a faster merger phase is also reported in [7]. Thus, to decrease the time required for the merge phase, we increase the size of the runs.

As a second advantage, this algorithm performs reads and writes in a continuous record-by-record process and, hence, it can be carried out in parallel. This is particularly advantageous if a different disk device is used for writing runs because heap operations can be interleaved with I/O reads and writes asynchronously [7]. These optimizations are not possible with quicksort, which operates in a strict read-process-write cycle of the entire input.

A potential disadvantage of replacement selection compared to quicksort is that it requires memory management for variable-length records. Replacement selection is based on the premise that when we select the smallest record from the sort buffer in main memory, we can replace it with the incoming record. In practice, this premise does not hold because variable-length records are the rule and not the exception. When a record is removed from the sort workspace, the free space it leaves must be tracked for use by following input records. If a new input record does not fit in the free space left by the last selection, more records must be selected until there is enough space. This leads to another problem, namely the fragmentation of memory space. Thus, managing the space in the sort buffer efficiently when records are of variable length becomes a necessity. To address this issue, we implemented a memory manager based on the design proposed by P. Larson in [10]. These characteristics will be evaluated empirically in Sect. 4.

Run generation in quicksort is a simple, nevertheless effective, strategy to create the sorted subfiles. First the records from the input are read into main memory until the available main memory buffer is full. These records are then sorted in-place using quicksort. Finally, they are written into a new temporary file (run). If we assume fixed-length records such that m records fit in main memory, this process is repeated $\frac{s}{m}$ times, resulting in $r := \frac{s}{m}$ runs of size m stored in the disk.

Another advantage of quicksort is the vast research effort dedicated to it during the last decades [1,2,7,14]. We call attention to the work in AlphaSort [13], which is a sort component that enhances quicksort with CPU cache optimizations. Two of these techniques include the minimization of cache misses and the sorting of only pointers to records rather than whole records, which minimizes data movements in memory. The reason behind the adoption of quicksort in AlphaSort is "it is faster because it is simpler, makes fewer exchanges on average, and has superior address locality to exploit the processor caching" [13]. These techniques are also employed by Hadoop in its quicksort implementation. In Sect. 4.3 we discuss these techniques in the context of replacement selection.

3   Replacement Selection Sort

Sort algorithms can be classified into two broad categories. When the input set fits in main memory, an *internal* sort can be executed. If the input set is larger than main memory, an *external* sort is required. Memory devices slower than main memory (e.g., hard disk) must work together to bring the records in the desired ordering. We call a sorted subfile a *run*. The combination of sorting runs in memory followed by an external merge process is described in two phases: *run generation*, where intermediary sorted subfiles (i.e., runs) are produced, and *merge*, where multiple runs are merged into a single ordered one.

A *record* is a basic unit for grouping and handling data. A *key* is a particular field (or a subset of fields) used as criterion for the sort order. A *value* is the data associated with a particular key, i.e., the rest of the record. Whether key and value are disjoint or one is a subset of the other is irrelevant for our discussion.

3.1   Run Generation

The replacement selection technique is of particular interest because the expected length of the runs produced is in average two times the size of available main memory. This estimation was first proposed by E.H. Friend in [5] and later described by E.F. Moore in [12], and is also described in [9]. In real-world applications, input data often exhibits some degree of *pre-sortedness* (i.e., there is a correlation between the input and output orders). For instance, the order in which products are ordered from a retail warehouse is closely correlated to the order in which they are delivered. Thus, re-ordering a dataset from one criterion to the other would require only small dislocations in the position of each record. Replacement selection exploits this fact by trying to perform such movements within an in-memory data structure. In such cases, the runs generated by replacement selection tend to contain even more than $2m$ records. In fact, for the best case scenario, namely when all records can be ordered by dislocating no more than m positions, where m is the number of records that fit in main memory, replacement selection produces *only one* run. This means that an arbitrarily large file can be sorted in a single pass.

Table 1. Run generation with replacement selection

Step	Memory contents				Output
1	503	087	512	061	061
2	503	087	512	908	087
3	503	170	512	908	170
4	503	897	512	908	503
5	(275)	897	512	908	512
6	(275)	897	653	908	653
7	(275)	897	(426)	908	897
8	(275)	(154)	(426)	908	908
9	(275)	(154)	(426)	(509)	(end of run)
10	275	154	426	509	154
11	275	612	426	509	275

Assume a set of tuples ⟨*record, status*⟩, where *record* is a record read from the unsorted input and *status* is a Boolean flag indicating whether the record is *active* or *inactive*. Active records are candidates for the current run while inactive records are saved for the next run. The idea behind the algorithm is as follows: assuming a main memory of size m, we read m records from the unsorted input data, setting its status to active. Then, the active tuple with the smallest key is selected and moved to an output file. When a tuple is moved to the output (*selection*), its place is occupied by another tuple from the input data (*replacement*). If the record recently read is smaller than the one just written, its status is set to inactive, which means it will be written to the next run. Once all tuples are in the inactive state, the current run file is closed, a new output file is created, and the status of all tuples is reset to active.

We introduce an example from Knuth [9] in Table 1 to explain in detail the replacement selection algorithm. Assume an input dataset consisting of twelve records with the following key values: *061, 512, 087, 503, 908, 170, 897, 275, 653, 426, 154, 509* and *612*. We represent the inactive records in parentheses. To select the smallest current record, Knuth advises in [9] to make this selection by comparing the records against each other (in $r - 1$ comparisons) only for a small number of records. When the number of records is larger than a certain threshold, the smallest record can be determined with fewer comparisons using a *selection tree*. In this case, a heap data structure can be employed. With a selection tree, we need only $log\ (r)$ comparisons, i.e., the height of the tree.

In Step 1, we load the first four records from the input data into the memory. We select 061 as the lowest key value, move it to the output and replace the whole record with a new record with key value 908. The lowest key value then becomes 087, which is moved to the output and replaced with 170. The just-added record is also the smallest in Step 3, so we move it out and replace it with 897. Now we have an interesting situation: when the record 503 is replaced, the record read from the input is 275, which is *lower* than 503. Thus, since we cannot output 275 in the current run, it is set as inactive—a state that will be kept until the end of the current run. Steps 6, 7, and 8 normally proceed until we move out record 908, which is replaced by 509. At this point, in Step 9, all records in memory are inactive. We close the current run (with twice the size of the available memory), revert the status of all records to active, and continue the algorithm normally.

3.2   Merge

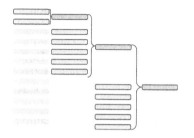

Fig. 1. Merging twelve runs into one with merging factor of six.

The goal of the second phase of external sorting, namely the *merge phase*, is to create a final sorted file from the existing runs. A heap data structure is used to select the smallest record among all runs, in the exact same way as done in run generation with replacement selection. A second improvement over the nave procedure is to take advantage of read and write buffers. Given r runs and memory of size m, a read buffer of size $\frac{m}{r+1}$ can be used for each input run. The size of the write buffer is then $m - (\frac{m}{r+1})r$.

Assume that the minimum buffer size is b. If the first phase of the algorithm produces more than $\frac{m}{b} - 1$ runs, then we cannot merge these runs in a single step. A natural solution for this limitation is to repeat the merging procedure on the merged runs, producing a *merge tree*. Figure 1 shows an example of merge tree. At each iteration, $\frac{m}{b} - 1$ runs are merged into a new sorted run. The result is $r - \frac{m}{b} + 2$ runs for the next iteration—the total number of runs minus the merged runs in this turn plus the new merged run. Several heuristics exist to merge runs in such a way that the resulting tree yields minimum I/O cost, such as cascade and polyphase merges, and can be found in [6,9].

4   Implementation

In this section, we discuss details of our implementation, in which the open-source Hadoop framework was extended with a new sort component. First, we present the internals of Hadoop, focusing on the algorithms and data structures used to implement external sorting using quicksort. Second, we present a custom memory manager used to manage the space in the sort buffer efficiently. Third, we present optimizations in the key comparison during the sort as well as a memory-efficient customized heap data structure.

4.1 Hadoop Internals

In this section, we provide a brief review of the internal components involved in a MapReduce computation in Hadoop. A primary goal of our design is to reuse this infrastructure as much as possible, supporting the replacement selection algorithm in a pluggable way. A detailed analysis of Hadoop's architecture and the components involved during the execution of a MapReduce job in Hadoop can be found in [3].

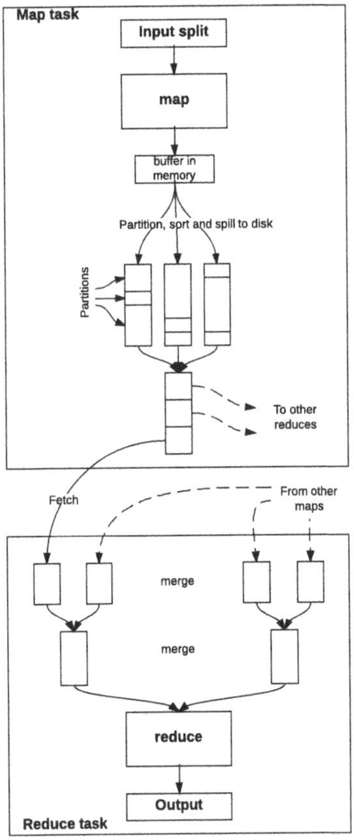

Fig. 2. M/R tasks in detail

The following process happens in a pipelined fashion, i.e., as soon as one step finishes, the next can start using the output emitted by the former. Figure 2 illustrates the process. The map function emits records (key-value pairs) while its input partition is processed, and these records are separated into partitions corresponding to the reducers that they will ultimately be sent to. However, the map task does not directly write the intermediary results to disk. The records stay in a memory buffer until they accumulate up to a certain minimum threshold, measured as the amount of occupied space in a buffer called kvbuffer; this threshold is by default 80 % the size of kvbuffer. Hadoop keeps track of the records in the key-value buffer in two metadata buffers called kvindices and kvoffsets. When the buffer reaches the threshold, the map task sorts and flushes the records to disk. When sorting kvoffsets, quicksort's *compare* function determines the ordering of the records accessing directly the partition value in kvindices through index arithmetic. But quicksort's *swap* function only moves data in the kvoffsets buffer. This corresponds to the pointer sort technique to be discussed in Sect. 4.3.

When the records are sorted, the map task finally writes them to a run file (or *spill file*, in the Hadoop nomenclature). Every time the memory buffer reaches the threshold, the map task flushes it to the disk and creates a new spill. When the map function completes processing the input partition and finishes emitting the key-value pairs, one last spill is executed to flush the buffers. Because the input split normally is larger than the memory buffer, when the map task has written the last key-value, several runs could be present. The map task then must externally merge these spill files into a final output file that becomes available for the reducers. Just like the spill files, this final output file is ordered by partition and, within each partition, by key.

After the completion of a map task, each reduce task (possibly on a different node) copies its assigned slice into its local memory. However, as long as this *copy* phase is not finished, the reduce function may not be executed since it must wait for the complete map output. Typically, the copied portion does not fit into the reduce task's local memory, and it must be written to disk. Once all map outputs are copied, a cluster-wide *merge* phase begins. As we noted in Sect. 3.2, if we have more than $m - 1$ map task outputs, the reduce cannot merge the intermediary results of all maps at the same time. The natural solution is to merge these spills iteratively. Hadoop implements an iterative merging procedure, where the property *io.sort.factor* specifies how many of those spill files can merged into one file at a time. The details underlying Hadoop's iterative merging procedure can be found in [3].

Hadoop's sort component not only has to take care of sorting in-memory keys and partitions but also merge these multiples sorted spills into one single, locally-ordered file. It has to consider data structures carefully to store the records and algorithms to manage and reorder these records. It should be clear that this merge is only a local merge (performed by each map task). A second, cluster-wide merge performed on the reduce side will merge the locally-ordered files that each map task has processed. This global merge is beyond the scope of this work because its performance is dictated only by the size and number of runs generated, and it is thus independent of the in-memory sort algorithm. Therefore, our analysis will consider a single merge phase, regardless of whether it is local or global.

4.2 Memory Management

As introduced in Sect. 2, replacement selection needs a memory manager to manage the space in the buffer efficiently when records are of variable length. It is important to emphasize that memory management is not an issue in Hadoop's quicksort strategy because it only requires swapping records into main memory. This means that a set of records is loaded into main memory and sorted *in-place* without requiring additional space.

Our naïve implementation simply uses Java's PriorityQueue class to implement the selection tree. Both memory management and heap implementation are reused from Java's standard library. However, this approach is inefficient due to the JVM's garbage collection overhead. To eliminate this overhead, we implemented a custom memory manager based on the *first-fit* design proposed by Larson in [10]. Other alternatives like the *best-fit* approach proposed in [11] exist, but the evaluation of its efficacy is left for future work. The performance gains of the customized memory manager are shown in the experiment of Fig. 3, where run generation is performed for an input of 9 GB and a buffer of 16 MB.

The optimized implementation is approximately 20 % faster than the nave, Java-based implementation.

In our implementation, the sort buffer is divided into *extents*, and the extents are divided into *blocks* of a predetermined size. The block sizes are spaced 32 bytes apart, which results in blocks of 32, 64, 96, 128, and so on. The extent size is the largest possible block size, which is 8KB in our implementation.

For each block size, we keep a list containing all free blocks of that particular size. The number of free lists is given by the extent size divided by the block size, thus $8 \times 1024/32 = 256$. The memory manager provides two main methods: one to *allocate* a memory block big enough to hold a record of a given size, and one to *free* a block that is not in use anymore (i.e., a block just selected for output).

Allocating a memory block big enough to hold a given record means to locate the smallest

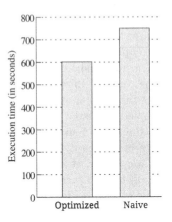

Fig. 3. Comparion of buffer implementations

(we want to avoid waste at maximum) free block larger than the record size. The allocate method works as follows: round up the record size to next multiple of 32 ($\lceil recordSize/blockSize \rceil * blockSize$). Find the index of the resulting rounded size in the free lists ($roundedSize/blockSize - 1$). Check if the list at the calculated index has a free block: if it does, return it. Otherwise, increment the index and look in the next list (which will be 32 bytes larger). If no block with the rounded record size is found, a larger block is taken and carved to the appropriate size, returning the excess as a smaller free block on its appropriate list. For instance, in the initial case where there is only one free block of 8192 bytes (the extent size), suppose the memory manager must allocate a block for a record of size 170. The rounded size of 170 is 192; because all other lists are empty, the manager gets the 8192 block from its list. To avoid a major wasting, the 192 first bytes of the 8192 block are returned, and the other 8000 are placed in its appropriate list. When the record is spilled from the buffer and its memory block becomes free, we return the block to the appropriate list. We illustrate this process in the example of Fig. 4, which shows a possible buffer state after a sequence of allocate and free invocations.

All lists start initially empty except the last one, which points to blocks of maximum size. As the blocks are allocated and freed, the lists are populated with smaller blocks. In the example of Fig. 4, we have 10 blocks of variable sizes. The block sizes are shown inside the blocks in the main-memory buffer (lower part of the figure) and on top of the free list they belong to (upper part of the figure). As smaller records are freed, the allocation process becomes faster, as fewer lists have to be searched to find smaller blocks.

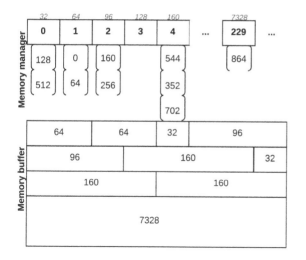

Fig. 4. A possible state of the memory manager and its free lists

However, this can lead to the following situation. Without loss of generality, imagine that the memory manager is continuously being asked for records of 32 bytes. After 256 block requests—the amount of blocks with 32 bytes in an 8KB extent—assume 4 contiguous (i.e., physically adjacent) blocks are freed. At this point, the memory manager has 128 bytes of free memory fragmented into four blocks of 32 bytes. If this stream of small records is interrupted by a larger record with 96 bytes, the memory manager will not find any block sufficiently large for that record—despite having enough free memory to answer the request. To remedy this situation, adjacent free blocks must be detected and *coalesced* into a single free block of total size equal to the sum of their sizes. For example, a block of 192 bytes being freed next to a free block of 64 bytes can be coalesced into a block of 256 bytes. Such coalescence also requires updating the free lists accordingly.

Detecting adjacent free blocks requires special free/occupied markers at the beginning and end of blocks. When a block is freed, the markers of the neighboring blocks are verified, and coalescence occurs if either neighbor has a free marker. Because implementing this technique is not a trivial task, especially in a memory-managed language like Java, we chose a simpler implementation without block coalescence. Instead, we perform a global defragmentation operation when large records cannot be allocated. As we show in Sect. 5, our implementation still delivers superior results than quicksort for the targeted cases, despite the defragmentation penalty.

4.3   Pointer Sort

As introduced in Sect. 2, one of the main advantages of quicksort is the *pointer sort* technique used to move fewer data. However, Nyberg et al. [13] state that "pointer sort has poor reference-locality because it accesses records to resolve key comparisons". In an ideal scenario, the whole selection tree should fit on the CPU data cache. But, in practice, the keys used to resolve record comparisons may be too large to fit all at the same time in the data cache. Nyberg et al. suggest the use of a prefix of the key rather than the full key to minimize cache misses. The idea is that a small prefix of the sort key (e.g., 2 bytes) is usually enough to resolve the vast majority of comparisons [7]. The complete record only has to be accessed in the rare occasions in which the prefixes are equal. Further techniques for key normalization and key reordering exist as in [8] and should be evaluated in future work.

We implement this technique of *pointer sorting with key prefixes* by storing only a pointer and a key prefix in the heap data structure. The pointer refers to the block allocated for the record in the memory manager, as discussed above. Since the entries in the heap are of fixed size, we optimized the algorithm even further by implementing a custom heap instead of using Java's PriorityQueue.

4.4   Custom Heap

Despite the customized memory manager for records, the JVM is still in charge of managing entries in the heap data structure (i.e., the selection tree). Our objective is to eliminate as far as possible the creation of objects in JVM's heap space during runtime, and allocate every needed array or object as soon as possible. One of the main advantages of custom managed memory buffer was the serialization of keys and values in a byte buffer of fixed size. To achieve the same result but for the selection tree, we implemented a custom heap which employs a byte array as the placeholder for the heap entries. We illustrate the idea in Fig. 5, which shows the format of entries in the optimized heap. We use Java's `ByteBuffer` class to wrap the byte array, which provides methods such *putInt* and *getInt*, as well similar methods to set and get other data types in arbitrary positions in the byte array. When we add a record to the sort buffer, we add its metadata "heap entry object" by directly writing the run, partition, etc., into the custom heap byte array. With this design, we can directly control how much memory the selection tree will consume, pre-allocating the heap space in a single contiguous block.

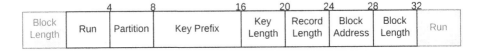

Fig. 5. The format of entries in our optimized heap

5   Experiments

This section evaluates the performance of replacement selection in the context of actual Hadoop jobs. First, we evaluate the run generation process, i.e., in-memory sorting using quicksort vs. replacement selection. Our goal is to show that (i) replacement selection indeed generates less runs, and (ii) the efficiency of replacement selection is not much worse than quicksort, i.e., the gains made at the merge phase are not wasted in a slower run generation phase. In fact, when able to exploit the continuous run generation characteristic of replacement selection, described in [7], where reads and writes overlap as the input is consumed and the output is produced, replacement selection outperforms quicksort. Second, we take a look at the special case of inputs with a certain degree of presortedness, which is where replacement selection is preferred to quickdort. These experiments are executed as micro-benchmarks to isolate sort performance on a single machine. Finally, in the third experiment set, we execute a full Hadoop job in a cluster comparing the run time with both sorting algorithms.

5.1   Run Generation

To confirm the prediction that replacement selection generates less runs empirically, we ran an experiment with the *lineitem* table from the TPC-H benchmark [15]. To randomize the sort order on the input, we consider a lineitem table sorted by *comment*, and them used the column *shipdate* as sorting key. Since the comment field is generated randomly by the benchmark, no correlation to the ship date is expected.

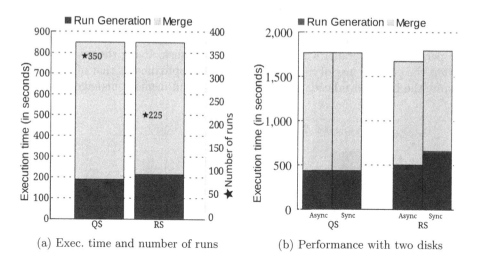

(a) Exec. time and number of runs   (b) Performance with two disks

Fig. 6. Experiment results for random inputs

Results for this experiment are presented in Fig. 6a. The buffer size used in these experiments is 50 MB, of which 10 % are overheads of auxiliary data structures (e.g., key prefixes). The table size is 700 MB which yields a ratio of 14 between input and buffer size. As shown in the graph, the run generation phase in replacement selection takes a little longer, but it produces approximately two thirds of the number of runs in quicksort. Note that the merge time is approximately the same, despite the substantial difference in the number of runs. This is expected because, in both cases, one merge pass was enough to produce a single sorted output. In this case, replacement selection yields only marginal gains in terms of CPU overhead in the merge phase, which are due to the smaller size of the heap used to merge the inputs. Nevertheless, the goal of this experiment is to show that replacement selection delivers comparable performance to quicksort when inputs are randomly ordered, which is clearly shown in the results. It substantially outperforms quicksort when inputs have a certain degree of pre-sortedness or, similarly, when multiple passes are required in quicksort. These cases are analyzed in Sects. 5.2 and 5.3 below.

The quicksort algorithm exhibits a fixed read-process-write cycle that does not allow I/O overlapping. One of the advantages of replacement selection is the continuous run generation process, alternately consuming input records and producing run files [7]. To demonstrate this fact empirically, we extended Hadoop with an asynchronous writer following the producer-consumer pattern. The idea is to place sorted blocks of data into a circular buffer instead of writing them directly to disk. Then, an asynchronous writer thread consumes blocks from this buffer and performs the blocking I/O write. While it waits, the sorting thread can sort other blocks of data in parallel. The results of our experiment are shown in Fig. 6b, where we compare the elapsed time of run generation with two hard disks—one for input and one for output. As predicted, quicksort delivers the same performance regardless of whether the writer is synchronous or asynchronous, whereas replacement selection benefits from writing asynchronously, performing run generation approximately 30 % faster. This is because reads from one disk are performed in parallel with writes on another disk. Using the asynchronous writer, the performance of replacement selection approximates that of quicksort, despite the higher overheads of heap operations and memory management.

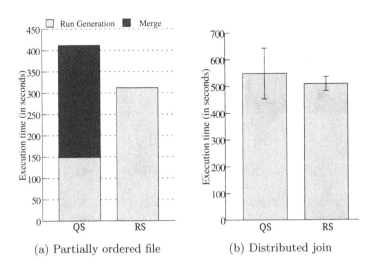

(a) Partially ordered file   (b) Distributed join

Fig. 7. Experiment results for pre-sorted input and join computation

5.2   Exploiting Pre-sortedness

One of the major advantages of replacement selection is that it can exploit pre-sortedness on the input file. Estivill-Castro and Wood showed mathematically in [4] that the length of the runs created by replacement selection increases as the order in the input file increases. To confirm this prediction empirically, we ran an experiment with the *lineitem* table again. We took as input the lineitem table sorted by *shipdate*, using the column *receiptdate* as new sorting key. We prepared the table by sorting the file by *shipdate* beforehand, but in practice this scenario could also occur if there is a clustered index or materialized view on *shipdate*. Since there is a strong correlation between the dates on which orders are shipped and received, this constitutes a good example of pre-sorted input.

The input dataset used in this experiment is the same as on the experiment of Fig. 6a, but with a different pre-sort order. As shown, the run generation phase takes considerably longer with replacement selection, but it produces only a single run at the end, meaning that no merge phase is required. As predicted, replacement selection acts as a sort sliding-window in this case. Note that quicksort finishes run generation earlier, but an additional pass over the whole input is required in the merge phase, requiring about 35 % more time in total.

5.3   Distributed Join

To conclude the experiments, we created a test scenario where a distributed join of two TPC-H tables is performed. Joins are a common operation in data management systems, and in MapReduce both inputs must be sorted by the join key (i.e., a sort-merge join algorithm). The joined tables are *lineitem*, which has about 1 GB, and *orders*, with 600 MB. This experiment was performed in a small cluster running Hadoop 2.4.0 with six nodes and we measured the total execution time of the jobs. The buffer size was 16 MB, which yields a 62.5 ratio with the lineitem table and a 37.5 ratio with the order table. Note that despite the tables being relatively small for typical Hadoop scales, the determinant factor for performance is actually the ratio between input and sort buffer size. A real-world large scale scenario would probably deal with sizes up to 1000× larger, i.e., tables of 1 TB and 600 GB, and a sort buffer of 16 GB. In such situations, the relative performance difference between the two algorithms would be very close to what is observed in our experiment, because the ratio is the same. The results in Fig. 7b confirm our expectation that replacement selection is faster, because less runs are generated. Furthermore, it seems that it is also more robust in terms of performance prediction, given the lower standard deviation.

6   Conclusion

This work described the implementation and evaluation of an alternative sort component for Hadoop based on the replacement selection algorithm. The original implementation, based on quicksort, is simple to implement and efficient in terms of RAM and CPU. However, we demonstrated that under certain conditions, such as pre-sorted inputs and large ratio between input and memory size, replacement selection is faster due to the lower number of runs to be merged. For the remaining cases, we showed that the performance is very close to that of quicksort, meaning that the average long-term gain in a practical scenario is in favor of replacement selection.

Despite the demonstrated advantages of replacement selection, we believe the implementation has the potential to outperform quicksort even further with certain optimizations. The main task in that direction is to optimize the main memory management component (e.g., by implementing block coalescence [10]) and the heap data structure (e.g., by minimizing the number of comparisons with a tree-of-losers approach or by optimizing key encoding [7–9]). Our work has been published as an open source pluggable module for Hadoop[1]. We hope to implement the mentioned optimizations and improve our code to integrate it with the official Hadoop distribution.

Acknowledgements. We thank Renata Galante for her helpful comments and suggestions on earlier revisions of this paper.

References

1. Bentley, J.L., Sedgewick, R.: Fast algorithms for sorting and searching strings. In: Proceedings of the Eighth Annual ACM-SIAM Symposium on Discrete Algorithms. SODA 1997, pp. 360–369. SIAM, Philadelphia, PA, USA (1997)
2. Cormen, T.H., Leiserson, C.E., Rivest, R.L., Stein, C.: Introduction to Algorithms, 3rd edn. The MIT Press, Cambridge (2009)
3. Dusso, P.M.: Optimizing Sort in Hadoop using Replacement Selection. Master thesis, University of Kaiserslautern (2014)
4. Estivill-Castro, V., Wood, D.: Foundations for faster external sorting (extended abstract). In: Thiagarajan, P.S. (ed.) FSTTCS. LNCS, vol. 880, pp. 414–425. Springer, Heidelberg (1994)
5. Friend, E.H.: Sorting on electronic computer systems. J. ACM **3**(3), 134–168 (1956)
6. Graefe, G.: Query evaluation techniques for large databases. ACM Comput. Surv. **25**(2), 73–169 (1993)
7. Graefe, G.: Implementing sorting in database systems. ACM Comput. Surv. **38**(3) (2006)
8. Härder, T.: A scan-driven sort facility for a relational database system. In: Proceedings of VLDB, pp. 236–244 (1977)
9. Knuth, D.E.: The Art of Computer Programming. Sorting and Searching, vol. 3, 2nd edn. Addison Wesley Longman Publishing Co. Inc., Redwood City (1998)
10. Larson, P.A.: External sorting: run formation revisited. IEEE Trans. Knowl. Data Eng. **15**(4), 961–972 (2003)
11. Larson, P.A., Graefe, G.: Memory management during run generation in external sorting. In: Proceedings of the 1998 ACM SIGMOD International Conference on Management of Data, pp. 472–483. SIGMOD 1998. ACM, New York, NY, USA (1998)

---

[1] http://bitbucket.org/pmdusso/hadoop-replacement-selection-sort.

12. Moore, E.: Sorting method and apparatus, 9 May 1961. http://www.google.com. br/patents/US2983904

13. Nyberg, C., Barclay, T., Cvetanovic, Z.: AlphaSort: a RISC machine sort. In: Proceedings of SIGMOD, pp. 233–242 (1994)

14. Skiena, S.S.: The Algorithm Design Manual. Springer, London (1998)

15. Transaction Processing Performance Council: TPC Benchmark H (Decision Support) Standard Specification. http://www.tpc.org/tpch/. Accessed 10 January 2014

Distributed Sequence Pattern Detection Over Multiple Data Streams

Ahmed Khan Leghari[1]([✉]), Jianneng Cao[2], and Yongluan Zhou[1]

[1] Department of Mathematics and Computer Science (IMADA),
University of Southern Denmark, Odense, Denmark
{ahmedkhan,zhou}@imada.sdu.dk
[2] Department of Data Analytics, Institute for Infocomm Research,
A*, Singapore, Singapore
caojn@i2r.a-star.edu.sg

Abstract. Sequence pattern detection over streaming data has many real world applications. Most of the present work is aimed to process sequence queries over single data stream. Situations where streaming data arrive from multiple sources have not been explored much. In traditional approaches a single centralized machine handles and processes sequence queries over multiple data streams. While running sequence queries on a single server, even though many of the events in data streams do not lead to successful pattern detection they are still handled and processed by the server. This consumes precious network bandwidth, server's computing resources and precious time. In this paper we focus on sequence pattern detection, where patterns are defined on chains of events that arrive from multiple distributed data streams. We propose a three layer distributed framework to avoid unnecessary event processing by the server, and to efficiently process sequence queries to detect sequence patterns relying upon chains of events. The bottom layer of data sources sends continuous data streams to the middle layer, which then performs pattern detection locally, and on the basis of the feedback received from the top layer of global server, sends events to the global server to detect complete patterns. Our present work is aimed to detect sequence patterns over multiple data streams, but, our proposed model can be extended to many other areas of distributed stream processing.

Keywords: Stream processing · Pattern detection · Chain-dependency

1 Introduction

Stream processing has gained immense popularity in scientific community still, complex event processing over multiple distributed data streams is relatively less explored area and is focus of the present research. In a typical distributed stream processing environment a single machine provides the interface for query submission, query execution, processing of multiple data streams and for the production of the result. Due to the possible importance and relevance of each

© Springer International Publishing Switzerland 2015
T. Morzy et al. (Eds.): ADBIS 2015, LNCS 9282, pp. 380–394, 2015.
DOI: 10.1007/978-3-319-23135-8_26

event in a pattern, the server handles and processes each event in the data stream, even though many of the events do not lead to successful pattern detection. These unwanted events consume precious network bandwidth, and valuable resources at server. Situation can be worse when streaming data arrive from multiple sources and a time critical response is required for the correctness of result, and for the reliability of applications. Therefore, in a single machine distributed stream processing setup [1], the time wasted in processing unwanted events can be a decisive factor in many time critical applications.

In complex event processing applications, the events arriving from various distant sources can be correlated and interdependent on the arrival order and time. Consider the scenario of an international stock market, where many of the traders wait for the drop or rise of stock commodities in a particular order. Prices of certain stock commodities can have affect on some other commodities in international markets. This can be described as: if the price of commodity A falls in stock market X at time t_1, which caused the of commodity B to fall in stock market Y at time t_2, here $t_1 \leq t_2$. Then buy the shares of commodity P, as this commodity is going to be very profitable in near future. This type of chain-dependency of stock events can be detected by running sequence queries over stock streams arriving from distant stock markets around the world.

Moreover, stock streams carry information about hundreds of commodities, but, majority of the stock traders just focus on particular shares, and specific market trends as mentioned above, thus, sending continuous data streams containing many unwanted and useless events directly to server is not an optimal method for many reasons, and needs a considerable improvement. To tackle this problem we propose a three layer architecture as depicted in Fig. 1 in Sect. 4 to handle streaming data and process sequence queries without sending unwanted events to the server.

The contribution of this paper are as follows.

(i) A three layer architecture to detect sequence patterns involving chains of events, arriving from multiple distributed data streams. (ii) A coordination mechanism between machines in three layer architecture that efficiently utilizes the network bandwidth by discarding the unwanted events to be sent to the server. (iii) Inclusion of a middle layer in the traditional distributed stream processing model to minimize the computing load, reduce processing time, and conserve precious resources on the server, by evaluating the sequence patterns in distributed fashion.

In remainder of this paper Sect. 2 presents background, Sect. 3 presents related work. Section 4 explains our approach. Section 5 presents experiments and results and Sect. 6 presents the conclusion of our work.

2 Background

2.1 Basic Terminology

Data Stream. We consider data stream as a massive sequence of events. Each event is modelled as a $(d+1)$-dimensional tuple $(ID, ts, A_1, A_2, \ldots, A_d)$, where

ID is the stream's unique identifier, ts is the timestamp showing when the event is generated, and A_i's are the attributes for $i = 1, 2, \ldots, d$. Consider a simplified example of stock market. Suppose that the stock trading stream is of structure (ID, ts, name, ask price, bid price, volume), which represents stock's identity, the timestamp of the event, name, price asked by the seller, price bid by the buyer, and the number of shares that changed hands.

Sequence Pattern. A sequence pattern is an ordered sequence of events, which may come from multiple data streams. We define a sequence pattern as ($\{e_1, e_2, \ldots, e_k\}$, window constraints, event correlations), where

- event e_i must happen before event e_j (i.e., $e_i.ts \leq e_j.ts$) for $1 \leq i < j \leq k$,
- window constraints are the time window, in which all the events should happen,
- and event correlations specify the correlations between events.

For example, e_1(ask price, Shanghai, ts) and e_2(ask price, Hongkong, ts) are the ask prices of Industrial and Commercial Bank of China (ICBC) in China mainland and Hongkong stock markets, respectively. Suppose that a user is interested in the case that: (a) if the ask price of in Shanghai increases than x and the difference of ask price between the two stock markets is larger than 1 Chinese Yuan, and (b) the time difference of the price asking is within 30 s. Then, the window constraint is "WITHIN 30 s", and the event correlations are

$$\left[(e_1.\texttt{askprice} > \texttt{x}) - (e_2.\texttt{askprice})\right] > 1 \text{ and } |e_1.ts - e_2.ts| \leq 30 \text{ Sec}.$$

Sequence Query. A sequence query is to detect a sequence pattern. Consider the example of stock market again. The SQL-like sequence query to detect the sequence pattern of price difference of ICBC is.

SELECT e_1, e_2
FROM   Shanghai AND Hongkong
WHERE $\left[(e_1.\texttt{ask price>x}) - (e_2.\texttt{ask price})\right] > 1$ AND $|e_1.ts - e_2.ts| \leq 30$ Sec

The example query runs over two different streams and can be processed by a single server, but in a single machine setup a query involving multiple streams can pose serious scalability challenges and processing issues due to the limited resources.

Refer to Fig. 1 in Sect. 4 that depicts the distributed three layered architecture of our approach and the example sequence query. When a sequence query detecting a pattern over multiple streams is submitted to the server (Layer-1), it is decomposed into multiple sub-queries, and each sub-query is then placed on a relevant Semi-Global Server (SGS) (Layer-2). An SGS is a regional server that detects a sub-pattern running a sub-query on the streams connected to it (Layer-3). After successful detection of the sub-pattern the detailed information is sent to the global server, which after receiving the information from all the concerned SGSs combines it and marks the detection of sequence successful.

Chain-Dependency. Refer to the example query, if the ask price in Shanghai increases than x, then it would be useful to evaluate the ask price in Hongkong and subsequently calculate the difference of two ask prices. Otherwise, there is no use to evaluate the ask price in Hongkong. The evaluation of event e_2 from Hongkong stock stream is dependent on the evaluation of event e_1 from Shanghai stock stream. This relationship of events where one of the event in a stream must be evaluated before the other event from a different stream is termed as *Chain-Dependency*.

Local-Chain Dependency. Refer to Fig. 1, suppose that events in the query submitted to the global server are chain-dependent on each other, then an event such as e_b (Layer-2) is said to be *Local-Chain Dependent* on event e_a, as they are processed by the same SGS in their dependency order.

Global-Chain Dependency. Refer to Fig. 1, when chain-dependent events such as e_i that depends on e_b arrive from event streams not connected through same SGS (Layer-2), the relationship among them is termed as *Global-Chain Dependency*.

3   Related Work

Event stream processing have been studied by many [2–26]. Ramakrishnan et al. [7] discussed importance of sequence queries and suggested a model for sequence data processing. Law et al. [18] discussed the limitations of relational algebra and SQL. They suggested changes in the data model to support sequence and other stream queries. Seshadri et al. [15] presented techniques and query evaluation plans to efficiently process sequence queries. Wu et al. [19] presented SASE (Stream based and shared event processing) to generate query plans and arrange query operators to efficiently handle sequence queries and detect patterns.

The work that is most relevant to ours is done by Mei et al. [21], Balkesen et al. [22] and Hirzel [23].

Mei et al. [21] presented ZStream to detect sequential patterns, and other stream operations. Their approach considers many tree based plans and according to the cost model selects the best possible plan to detect a pattern. Their approach is also based on single machine and chooses an optimal plan for pattern detection. While the focus of our work is to detect sequence patterns involving chain dependent events arriving from multiple event streams, and to minimize the number of unwanted events processed by server to reduce the processing load and time on server.

Balkesen et al. [22] proposed an intra query parallelism scheme for scalable pattern recognition. Their approach partitions the input stream and distributes the input events among multiple cores for parallel processing. Their work targets the multi-core architecture of a single machine, but a single machine implementation can easily lead to performance bottleneck while handling multiple fast data streams.

Hirzel proposed [23] a partitioning scheme for distributed pattern matching based on partition keys. His approach showed high throughput on multi-core machines and on clusters. To achieve high throughput Hirzel's scheme requires a partitioning key, while in our approach multiple streams are handled in distributed fashion without any partitioning.

4 Sequence Pattern Detection

The sequence queries when submitted to the global server are grouped as follows.

(1). Queries that Detect Sequence Patterns Involving Local-Chain Dependent Events (Local Queries). Some of the sequence patterns can be successfully evaluated by processing data streams connected through the same SGS. As the events are local-chain dependent in these sequence patterns therefore, there is no need for query decomposition.

(2). Queries that Detect Sequence Patterns Involving Global-Chain Dependent Events (Global Queries). Queries in which the detection of a complete sequence pattern requires involvement of more than one SGS are termed as Global Queries. Refer to Fig. 1, in which global server (Layer-1) receives a query that is further decomposed into three sub-queries. A sub-query such as (e_a, e_b) consists of those events that arrive from streams connected though the same SGS. Each sub-query runs over relevant streams and detects a sub-pattern, or partial sequence.

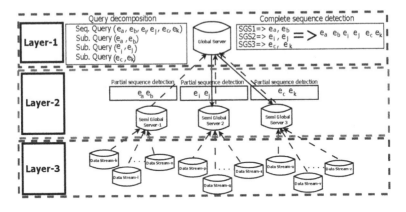

Fig. 1. The architecture of distributed pattern detection

While detecting patterns involving global-chain dependent events the feedback from global server plays a very important role.

Importance of the Feedback. A feedback is a two way communication that takes place between SGSs and global server to coordinate and complete the detection of a sequence pattern.

Example: Fig. 1 shows that global server has received a query involving global-chain dependent events (global query). Starting from the right in chain dependency order, events e_c, e_k would be processed by SGS3, which are chain dependent on events e_i, e_j processed by SGS2, which themselves are chain dependent on e_a, e_b. As events in the sequence pattern would be processed by multiple SGSs, hence they are forming a global-chain. To detect a sequence pattern involving global-chain dependent events, the query submitted to global server would be decomposed among SGSs participating in the detection process. Therefore, to detect a sequence pattern such as $(e_a, e_b, e_i, e_j, e_c, e_k)$ a query and it's associated time constraints would be decomposed into three sub-queries as $(e_a, e_b), (e_i, e_j)$ and (e_c, e_k) as mentioned in Fig. 1. Each sub-query would be placed on an SGS that receives the relevant events from the connected streams. An SGS would then detects a sub-pattern using it's sub-query in it's assigned sub-window.

In the Fig. 1, SGS-1 after detecting the sub-pattern e_a, e_b sends a notification that a sub-pattern has been detected, upon feedback received from SGS-1, the global sever then directs SGS-2 to detect sub-pattern e_i, e_j and so on. A feedback sent to global server contains the Seq_ID, sub-sequence_ID, the time when the first and the last event of sub-sequence was detected in the sub-window, and when the notification was sent by a particular SGS.

4.1   Global Query Processing

When a sequence query is submitted to the global server, it inspects the query and places it into the categories according to the characteristics of the chain dependency of the events. The next step is to decompose the query into multiple sub-queries. A sub-query is then placed on a particular SGS that receives a data stream on which the sub-query is required to be run as shown in Fig. 1. Events in global queries are processed according to the order of their chain-dependency, therefore, these queries are further categorized according to the processing orders of the events.

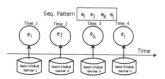

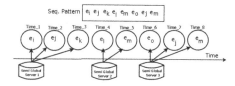

Fig. 2. A simple case of global-chain dependent events in sequence

Fig. 3.   Non-interleaved events in a sequence pattern

Cat-(1). One Event Arriving from One SGS. Refer to Fig. 2, the sequence pattern consists of 4 events: e_i, e_j, e_k, e_l, each would arrive at a different SGS from any of the stream connected to it. An event such as e_j that arrives at time t_2 on SGS2 is chain dependent on another event e_i arrived at time t_1 on SGS1 and so on, here $t_2 > t_1$. This is the simplest case of global-chain dependency, as the sequence pattern consists of multiple events, each from a single SGS. To avoid any processing and communication delay, an SGS after detecting the corresponding event would send it to the global server, which then evaluates the sequence pattern using the query placed on it within the respective time-window as described in the example query in Sect. 2.1. A time window is generated as follows.

$$Sub\_window_i = Time\_window/(T_{avg} * \# \ of \ events) \tag{1}$$

Here, T_{avg} denotes the average time required to evaluate an event multiplied by the number of consecutive events to be evaluated at an SGS. Each sub-query and associated sub-window is then provided to an SGS to detect a sub-sequence pattern. Every SGS receives a sub-query relevant to the event streams connected to it.

(Cat-2). Non-interleaved Events Arriving from Multiple SGS. If the processing order of events in a sequence pattern is non-interleaved as depicted in Fig. 3, then each SGS receives a sub-query to detect relevant sub-pattern at corresponding SGS along with it's respective sub-window calculated using the Eq. 1. Therefore, Events e_i, e_j, e_k would be processed by SGS1, events e_l, e_m would be processed by SGS2, and SGS3 would process events e_o, e_j, e_m. All the SGSs would process their respective events following the chain dependency order.

It is possible that events required to evaluate a sub-sequence would appear multiple times in a sub-window, then each time a sub-sequence is detected by an SGS, a notification is sent to global server that the partial evaluation of the sequence is successful. If any of the relevant event in sub-sequence does not arrive

---

Algorithm 1. Evaluating a sequence pattern at Global Server

1: **Input:** N_i is the most recent notification received at server, from an SGS_j
2: **do**
3: **if** N_i is received **then**
4:    copy N_i into the memory
5:    **if** N_i notifies that a sequence Seq_k has been successfully evaluated at SGS_j **then**
6:        Mark the evaluation of Seq_k successful
7:    **else**
8:        **if** N_i notifies that a sequence Seq_k has been partially evaluated at SGS_j **then**
9:            evaluate Seq_k using information in N_i
10:           **if** evaluation of Seq_k completes using information in N_i **then**
11:               Mark the evaluation of Seq_k successful
12:           **else**
13:               inform concerned SGS to send information required for further evaluation of Seq_k
14:           **end if**
15:       **end if**
16:   **end if**
17: **end if**
18: **while(true)**

---

within the specified sub-window then the sub-query and any relevant information is removed from the memory of an SGS. If the global server does not receive a notification after the time assigned as a sub-window it marks the evaluation of the sequence pattern unsuccessful, and removes the relevant information from it's memory. In the same way all other SGS awaiting to hear from global server also discard the relevant information from their memories.

Each time the global server receives a notification from an SGS, such as SGS1, denoting the successful evaluation of a sub-sequence, it inspects it's sequence_ID and sub-sequence_ID to determine the query associated to it, and if required then directs the next SGS such as SGS2, that should further continue the process of sequence pattern detection. The SGS2 after receiving a signal from global server starts detecting events in it's assigned sub-window, and each time it detects a sub-sequence it notifies it to the global sever. The global server after receiving notification denoting the successful evaluation of a sub-sequence at SGS2, directs SGS3 to start and continue the process of sub-sequence detection. The notification based communication is described in Algorithm 1.

(Cat-3). Interleaved Events Arriving from Multiple SGSs. Figure 4 shows a sequence pattern $(e_i, e_j, e_k, e_l, e_m, e_n)$ in which events making up the sequence are required to be processed in interleaved order. The events e_i, e_k, e_m would be processed by SGS1, and events e_j, e_l, e_n would be processed by SGS2. Each event is dependent on the detection of other event except the first event e_i. Hence, the processing of event e_j at SGS2 requires the detection of event e_i by SGS1, and SGS1 requires the detection of event e_j at SGS2 before it continues the detection of event e_k and so on. To process interleaved events in a pattern as shown in Fig. 4, a straight forward process is that when an SGS detects an event it should send a notification to the global server, who should direct the relevant SGS to detect the next event in the pattern. But, as each successive event in the pattern would arrive at alternating SGS, hence each SGS has to send multiple notifications for events that it detects. The global server also has to communicate back and forth with both SGSs to direct them to carry on the detection process. This would decrease the overall efficiency of the approach due to the higher communication cost, and potential delay while evaluating the sequence pattern.

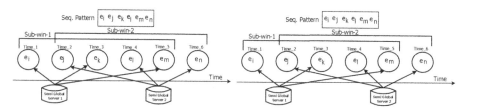

Fig. 4. Interleaved events in a sequence pattern

Fig. 5. Multiple interleaved events in a sequence pattern

In global queries with interleaved events, size of sub-window would be calculated according the *Degree of Proximity* (DoP) of events (Algorithm 2).

The DoP is the number of interleaved events to be arrived between two successive events at a single SGS. In Fig. 4 the DoP between events e_i, e_k and e_j arriving at SGS1 is one, and the DoP between events arriving at SGS2 is also one.

$$Sub\_window_i = Time\_window/[T_{avg} * (\# \ of \ events + DoP)] \tag{2}$$

A careful selection of DoP is important to calculate a suitable sub-window. This can result in simultaneous evaluation of events on multiple SGSs, and reduce back and forth communication between global server and SGSs. For all the events evaluated in a single sub-window, a single notification carrying the relevant information is sent to the global server.

However, choosing large overlapping sub-windows can waste precious resources. Refer to Fig. 5, as an event e_n at $time\_6$ is global-chain dependent on another event arriving at $time\_5$, therefore, assigning a large sub-window at SGS2 to evaluate interleaved event e_j arriving at time $time\_2$ in the same sub-window is poor strategy. Because before e_n there are multiple chain-dependent events after e_j which must be evaluated in the arrival order at SGS1, and if any of the chain-dependent event before e_n and after e_j does not arrive, then the arrival of the remaining events have no importance. Therefore, a balanced

---

Algorithm 2. Query decomposition involving interleaved events

---

1: **Input:** Q_i is a sequence query submitted to global server
2: **do**
3:   examin the query
4:   **if** events in Q_i are interleaved **then**
5:     calculate the Degree of Proximity
6:     decompose Q_i into sub-sequence queries
7:     for each sub-sequence query set the sub-window size according to the DoP
8:     distibute sub-queries to repective SGSs
9:   **end if**
10: **while(true)**

---

strategy is required while calculating the size of a sub-window using DoP. In case of multiple interleaved events in a sequence pattern a good idea is to generate multiple sub-windows at each SGS for these interleaved events.

The processing in Cat-3 queries would take place as follows. Refer to Fig. 4, at the time of query decomposition the sub-windows are calculated as described in Eq. 2. The sub-windows assigned to SGS1 and SGS2 both are overlapped. SGS1 would start detecting the events at time_1 and when SGS1 would detect the first event e_i of the sub-sequence, it will send a signal to the global server which then direct SGS2 to start detecting the events relevant to the sub-query assigned to it. The SGS2 would start detecting the relevant events in it's sub-window, that possibly would start at time_2 creating an overlapping sub-window, and stop at time_6. If SGS1 would successfully evaluate sub_sequence pattern using the sub-query assigned to it, then it would send a notification at $time\_5$ to the global server like in the non-interleaved order of events. The global server after receiving a notification from SGS1 waits for a notification to be received from SGS2 at $time\_6$. After receiving a notification from both of the SGSs denoting the successful partial evaluation of the sequence pattern, it then performs a global evaluation and marks the detection of a sequence pattern successful.

4.2 Local Query Processing

At the system startup time, each SGS receives it's individual set of local queries as well as another subset of sub-sequence queries from global server to detect sequence patterns which can be partially detected by events arriving on that SGS, but their successful evaluation depends on events arriving at multiple SGS. An SGS detects sequence patterns running sequence queries over multiple streams. None of the SGS sends events to the global server unless an event whose successful evaluation requires global-chain dependent events from multiple SGS. Events which are not relevant to any sequence pattern when arrive at an SGS are simply discarded. Detection of a relevant event at an SGS would lead to the steps described below.

(i). If an event e_i detected is the very first event of a sequence pattern, then it is checked that, are there other relevant events (making up the pattern sequence) which are local-chain dependent to the very first event. All the relevant events arriving in a time-window are examined for various combinations that can lead to the successful detection of a sequence pattern. If events making up a sequence pattern are local-chain dependent to the SGS, then the present and upcoming events (if there is any) would be evaluated locally by SGS, and a notification is sent to the global server that a sequence pattern has been successfully detected. If events in a sequence pattern are global-chain dependent, and the sequence is partially evaluated at an SGS using a sub-query, then a notification is sent to the server specifying the detail required to carry on further evaluation.

(ii). If an event just detected is not the first event in a sequence pattern, then it would be examined whether it belongs to a sequence pattern being evaluated locally, or is a global-chain dependent event. In the latter case, the event can be required to continue the partial evaluation of a sequence pattern. In the former case an event can be the last event required for a successful evaluation of a sequence pattern, or an event that belongs to the middle of a sequence pattern and evaluated accordingly, and if required global server would be notified. On the basis of the information received the global server performs some basic arithmetic operations, and marks the evaluation of a particular sequence successful.

5 Experimental Evaluation

5.1 Experimental Setup

In the experimental system there was one global server and three semi-global servers. Each SGS was receiving three event streams. An event stream represented stock events from a distinct stock market consisted of NASDAQ's [27] historical stock quotes. A single SGS received 1 million events in total from three stock streams connected to it. There were 9 input streams in total, 3 at each SGS, and the entire system received 3 million events. The sequence queries were generated randomly and submitted to the global server after a random interval. Each sequence query consisted of a variable set of stock events. The length of a pattern in the sequence query was ranging between 3 to 6. The total number of queries generated in the system was evenly divided among four categories of the queries as per their order of chain-dependent events. Therefore, 25 percent local queries were detecting patterns with local-chain dependent events on multiple SGSs, while rest of the 75 percent global queries consisted of 3 different categories as mentioned in Sect. 4.

5.2   Results and Discussion

Refer to Fig. 6a, in the first experiment a number of 48063 local queries and 144189 global queries were used to detect various stock patterns from nine incoming event streams. The total number of sequence pattern detected locally i.e. at three SGSs were 72, and there was 458 patterns detected globally (Fig. 6b). 151 patterns were detected in Cat-1 queries involving global-chain dependent events. Similarly 145, and 162 patterns were detected in the Cat-2 and Cat-3 sequence queries. Figure 6c shows that in 144189 global queries, 769008 events were required to successfully detect all the patterns. In a traditional approach the global server would have processed 3 million events to detect these patterns, but, in our approach the total number of events received by the global server is 2458, that is 0.0819 % of the total data.

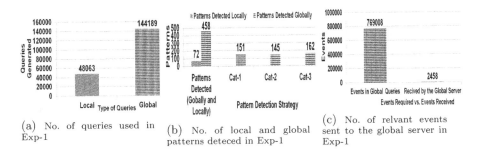

(a) No. of queries used in Exp-1   (b) No. of local and global patterns deteced in Exp-1   (c) No. of relvant events sent to the global server in Exp-1

Fig. 6. Configuration and results of experiment-1

The small number of events received by the global server is due to the reasons that SGSs discarded the events which were not relevant to the sequence patterns, or did not arrive in the time-window mentioned in the sequence query. All the events which were not arriving in the order of chain-dependency were are also discarded by the SGSs. Moreover, queries detecting patterns that relied upon event streams from nearby regions were also processed at SGSs, these all factors

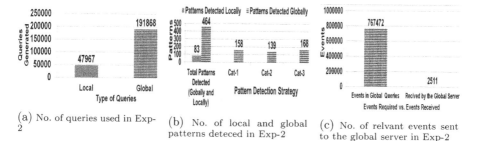

(a) No. of queries used in Exp-2

(b) No. of local and global patterns deteced in Exp-2

(c) No. of relvant events sent to the global server in Exp-2

Fig. 7. Configuration and results of experiment-2

contributed to minimize the number of events sent to the global server and thus, conserved precious bandwidth and reduced considerable processing load on the server.

In Experiment-2 (Fig. 7a), there were 47967 local queries and 191868 global queries. The total number of patterns detected locally were 83, and there was 464 patterns involving different categories of global-chain dependent events (Fig. 7b). Similarly, (Fig. 7c) a total of 767472 events were required in certain order to detect the patterns by global queries, but the global server just received 2511 events from multiple SGSs. These 2511 events were the actual events of interest and were evaluated successfully.

This experiment also shows a considerable gain in terms of bandwidth consumption, as just 0.0837 % events out of 3 million events were sent to the global server. Reducing 99 % of data rate and corresponding processing load on the global server.

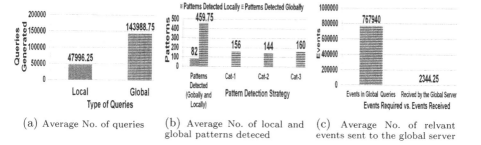

(a) Average No. of queries

(b) Average No. of local and global patterns deteced

(c) Average No. of relvant events sent to the global server

Fig. 8. Average results of multiple experiments

We repeated experiments many times. The average number of local and global queries used in the experiment were 47996.25, and 143988.75 (Fig. 8a). The average number of patterns detected locally were 82, and there were 459.75 patterns involving events received from multiple SGSs (Fig. 8b). There were 767940 total events in sequence queries required to evaluate the patterns, and the global server just received 2344.25 events (Fig. 8c).

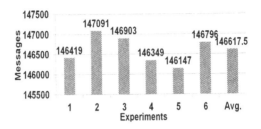

Fig. 9. No. of messges exchanged as a feedback

Refer to Fig. 9, in each experiment we also kept the record of total number of messages, exchanged as a feedback between global server and SGSs while processing global queries. The number of messages exchanged while processing local queries are not included, as there was just two messages exchanged between each SGS and global server for each query that successfully detected a sequence pattern. First, when the local query is placed on the SGS, and the Second, when the SGS sent a notification about the successful detection of a pattern.

If a local query does not detect a pattern then no message is sent to global server. It can be clearly seen that the number of messages exchanged as a feedback are much smaller than the number of events that would have been received by the global server without removal of unwanted events at SGSs. Thus by adopting a three layer architecture to process queries in which events are chain-dependent we have on average saved 95 % of the bandwidth (99 % excluding the feedback) that would have been consumed by useless events.

6   Conclusion

In this paper we introduced sequence patterns in which events are chain-dependent. Our experimental study shows that by establishing a feedback mechanism between two layers of servers to evaluate such patterns and pruning the unwanted events at SGSs can result up to 99 % cutback in the data sent to the global server, and can reduce the overall processing load on server.

References

1. Lu, H., Zhou, Y., Haustad, J.: Continuous skyline monitoring over distributed data streams. In: Gertz, M., Ludäscher, B. (eds.) SSDBM 2010. LNCS, vol. 6187, pp. 565–583. Springer, Heidelberg (2010)
2. Brenna, L., Gehrke, J., Hong, M., Johansen, D.: Distributed event stream processing with non-deterministic finite automata. In: Proceedings of the Third ACM International Conference on Distributed Event-Based Systems, p. 3. ACM (2009)
3. Golab, L., Özsu, M.T.: Issues in data stream management. ACM Sigmod Rec. **32**(2), 5–14 (2003)

4. Liu, M., Li, M., Golovnya, D., Rundensteiner, E.A., Claypool, K.: Sequence pattern query processing over out-of-order event streams. In: 2009 IEEE 25th International Conference on Data Engineering. ICDE 2009, pp. 784–795. IEEE (2009)
5. Stonebraker, M., Çetintemel, U., Zdonik, S.: The 8 requirements of real-time stream processing. ACM SIGMOD Rec. **34**(4), 42–47 (2005)
6. Kawashima, H., Kitagawa, H., Li, X.: Complex event processing over uncertain data streams. In: 2010 International Conference on P2P, Parallel, Grid, Cloud and Internet Computing (3PGCIC), pp. 521–526. IEEE (2010)
7. Ramakrishnan, R., Cheng, M., Livny, M., Seshadri, P.: What's next? sequence queries. In: Proceedings of International Conference Management of Data. Citeseer (1994)
8. Wang, Y., Cao, K., Zhang, X.: Complex event processing over distributed probabilistic event streams. Comput. Math. Appl. **66**(10), 1808–1821 (2013)
9. Jiang, Q., Chakravarthy, S.: Scheduling strategies for processing continuous queries over streams. In: Williams, H., MacKinnon, L.M. (eds.) BNCOD 2004. LNCS, vol. 3112, pp. 16–30. Springer, Heidelberg (2004)
10. Sharaf, M.A., Labrinidis, A., Chrysanthis, P.K.: Scheduling continuous queries in data stream management systems. Proc. VLDB Endowment **1**(2), 1526–1527 (2008)
11. Mani, M.: Efficient event stream processing: handling ambiguous events and patterns with negation. In: Xu, J., Yu, G., Zhou, S., Unland, R. (eds.) DASFAA Workshops 2011. LNCS, vol. 6637, pp. 415–426. Springer, Heidelberg (2011)
12. Schultz-Møller, N.P., Migliavacca, M., Pietzuch, P.: Distributed complex event processing with query rewriting. In: Proceedings of the Third ACM International Conference on Distributed Event-Based Systems, p. 4. ACM (2009)
13. Agrawal, J., Diao, Y., Gyllstrom, D., Immerman, N.: Efficient pattern matching over event streams. In: Proceedings of the 2008 ACM SIGMOD International Conference on Management of Data, pp. 147–160. ACM (2008)
14. Babcock, B., Babu, S., Datar, M., Motwani, R., Thomas, D.: Operator scheduling in data stream systems. VLDB J. Int. J. Very Large Data Bases **13**(4), 333–353 (2004)
15. Seshadri, P., Livny, M., Ramakrishnan, R.: Sequence query processing. In: ACM SIGMOD Record, vol. 23, pp. 430–441. ACM (1994)
16. Wu, J., Tan, K.-L., Zhou, Y.: QoS-oriented multi-query scheduling over data streams. In: Zhou, X., Yokota, H., Deng, K., Liu, Q. (eds.) DASFAA 2009. LNCS, vol. 5463, pp. 215–229. Springer, Heidelberg (2009)
17. Diao, Y., Immerman, N., Gyllstrom, D.: Sase+: An Agile Language for Kleene Closure Over Event Streams. ACM Press, New York (2007)
18. Law, Y.N., Wang, H., Zaniolo, C.: Query languages and data models for database sequences and data streams. In: Proceedings of the Thirtieth International Conference on Very Large Data Bases-Volume 30, VLDB Endowment, pp. 492–503 (2004)
19. Wu, E., Diao, Y., Rizvi, S.: High-performance complex event processing over streams. In: Proceedings of the 2006 ACM SIGMOD International Conference on Management of Data, pp. 407–418. ACM (2006)
20. Sadoghi, M., Singh, H., Jacobsen, H.A.: Towards highly parallel event processing through reconfigurable hardware. In: Proceedings of the Seventh International Workshop on Data Management on New Hardware, pp. 27–32. ACM (2011)
21. Mei, Y., Madden, S.: Zstream: a cost-based query processor for adaptively detecting composite events. In: Proceedings of the 2009 ACM SIGMOD International Conference on Management of Data, pp. 193–206. ACM (2009)

22. Balkesen, C., Dindar, N., Wetter, M., Tatbul, N.: Rip: Run-based intra-query parallelism for scalable complex event processing. In: Proceedings of the 7th ACM International Conference on Distributed Event-Based Systems, pp. 3–14. ACM (2013)

23. Hirzel, M.: Partition and compose: Parallel complex event processing. In: Proceedings of the 6th ACM International Conference on Distributed Event-Based Systems, pp. 191–200. ACM (2012)

24. Zhou, Y., Ma, C., Guo, Q., Shou, L., Chen, G.: Sequence pattern matching over time-series data with temporal uncertainty. In: EDBT, pp. 205–216 (2014)

25. Leghari, A.K., Wolf, M., Zhou, Y.: Efficient pattern detection over a distributed framework. In: Castellanos, M., Dayal, U., Pedersen, T.B., Tatbul, N. (eds.) BIRTE 2013 and 2014. LNBIP, vol. 206, pp. 133–149. Springer, Heidelberg (2015)

26. Wu, J., Zhou, Y., Aberer, K., Tan, K.L.: Towards integrated and efficient scientific sensor data processing: a database approach. In: Proceedings of the 12th International Conference on Extending Database Technology: Advances in Database Technology, pp. 922–933. ACM (2009)

27. http://www.infochimps.com/. 03 December 2014

Approximation and Skyline

Space-Bounded Query Approximation

Boris Cule, Floris Geerts$^{(\boxtimes)}$, and Reuben Ndindi

University of Antwerp, Antwerp, Belgium
{boris.cule,floris.geerts,reuben.ndindi}@uantwerpen.be

Abstract. When dealing with large amounts of data, exact query answering is not always feasible. We propose a query approximation method that, given an upper bound on the amount of data that can be used (*i.e.*, for which query evaluation is still feasible), identifies a part C of the data D that (i) fits in the available space budget; and (ii) provides accurate query results. That is, for a given query Q, the query result $Q(C)$ is close to the exact answer $Q(D)$. In this paper, we present the theoretical framework underlying our query approximation method and provide an experimental validation of the approach.

Keywords: Big data query processing · Query approximation · Data reduction

1 Introduction

Traditional query processing has primarily focused on the efficient computation of exact answers to queries. In applications with huge amounts of data, however, even simple queries that require a single scan over the entire database cannot be answered within an acceptable time bound. To accommodate for this, one can either try to leverage parallelism and distributed computation, settle for approximate query answering, rely on data-reduction techniques, or combinations thereof. In this paper we consider approximate query answering, or AQA for short.

Motivated by the need for big data analytics, recent work on AQA mainly concentrates on the efficient and accurate evaluation of simple aggregate queries. A recent proposal in this context is the BlinkDB system [1]. In a nutshell, BlinkDB addresses the following question: Given a query workload $\mathcal{Q} = \{Q_1, \ldots, Q_\ell\}$ consisting of aggregate queries Q_i, each equipped with an importance weight p_i, for $i \in \{1, \ldots, \ell\}$, a database D and given a storage capacity $\mathbb{B}$, what is the best set of samples $\mathcal{S} = \{S_1, \ldots, S_\ell\}$ of D that one should materialise such that (i) the samples fit in the available storage, i.e., $|S_1 \cup \cdots \cup S_\ell| \leqslant \mathbb{B}$; and (ii) evaluating the queries on samples in $\mathcal{S}$ provides an accurate estimate of the exact query answer. In addition, the most important queries (*i.e.*, those with high weight) should be approximated more accurately than the less important queries (*i.e.*, those with low weight).

In this paper, we consider a similar setting as in BlinkDB but for *non-aggregate queries*. That is, we are interested in finding the best set C of tuples

© Springer International Publishing Switzerland 2015
T. Morzy et al. (Eds.): ADBIS 2015, LNCS 9282, pp. 397–414, 2015.
DOI: 10.1007/978-3-319-23135-8_27

in the database D that one should store within the available storage capacity $\mathbb{B}$ such that $Q_i(C)$ is "close" to $Q_i(D)$ for any $i \in \{1, \ldots, \ell\}$. For this purpose, we equip databases with distance functions to measure the closeness between two databases (Sect. 2), replace the samples used in BlinkDB by so-called coverings C of the data, $i.e.$, sets of tuples that are within a certain distance from the original database, and introduce the valid selection covering problem (Sect. 3). We then show how to estimate the size of coverings (Sect. 4) and how the distance between coverings and the original data propagates through the queries in the workload (Sect. 5). Finally, similar to BlinkDB, we identify the desired coverings of the data by means of a mixed integer linear program (Sect. 6). An experimental validation of our AQA framework (Sect. 7) concludes the paper.

2   Preliminaries

▷ *Databases.* Let $\mathcal{R} = (R_1, \ldots, R_n)$ be a relational schema consisting of n relations R_i, each having a fixed arity m_i. Let $R(A_1, \ldots, A_k)$ be a relation in $\mathcal{R}$. We assume that each relation carries a distinct set of attributes. Furthermore, each attribute A_i in R comes equipped with a domain $dom(A_i) \subseteq \mathbb{U}$, where $\mathbb{U}$ is a countably infinite set. A tuple t of $R(A_1, \ldots, A_k)$ is simply an element of $dom(A_1) \times \cdots \times dom(A_k)$. A database of $\mathcal{R}$ is given by $D = (I_1, \ldots, I_n)$, where I_i is a finite set of tuples of R_i, for $i \in \{1, \ldots, n\}$. The active domain of D, denoted by $adom(D)$, is the set of elements from $\mathbb{U}$ present in D. Finally, the size $|D|$ of D refers to the number of tuples in D.

▷ *Distances and Metric Databases.* We further assume the presence of distance functions $\mathsf{d}_{A_i} : dom(A_i) \times dom(A_i) \to \mathbb{R}$, one for each attribute A_i in $\mathcal{R}$. A *metric database* (D, d) simply consists of a database D over $\mathcal{R}$ together with a collection of distance functions d_{A_i} for attributes A_i in $\mathcal{R}$.

To compare the distance between tuples on arbitrary sets X of attributes, we define $\mathsf{d}_X(s, t) = \max\{\mathsf{d}_{A_i}(s[A_i], t[A_i]) \mid A_i \in X\}$, provided, of course, that s and t are defined over a set Y of attributes such that $X \subseteq Y$. For example, when dealing with numerical attributes A_i for which $d_{A_i}(s, t) = |s[A_i] - t[A_i]|$, we have that $\mathsf{d}_X(s, t) = \max\{|t[A_i] - s[A_i]| \mid A_i \in X\}$. We also need to lift distance functions to sets of tuples, $i.e.$, database instances. Given two sets C and D of tuples, we define for $s \in C$, $\mathsf{d}_X(s, D) := \min\{\mathsf{d}_X(s, t) \mid t \in D\}$ and $\mathsf{d}_X(C, D) = \max\{\max_{s \in C} \mathsf{d}_X(s, D), \max_{t \in D} \mathsf{d}_X(t, C)\}$. In addition, we define $\mathsf{diam}_X(D) := \max\{\mathsf{d}_X(s, t) \mid s, t \in D\}$. We denote the diameter with $\mathsf{diam}(D)$ when X is the set of all attributes in D.

Example 1. Consider a part D of the Lineitem table from the TPCH benchmark as shown in Fig. 1 with attributes line number (LN), quantity (QT), extended price (EP), line status (LS), ship date (SD) and ship mode (SM). To turn D into a metric database (D, d) we equip each of the attributes with a distance function. For example, on the numerical attributes extended price and quantity we can use the absolute difference between values, on ship date one can use a date-specific distance function, and on the remaining categorical attributes the discrete distance function can be used. Consider tuples t_1 and t_2 in D. Their distance on the

	LN	QT	EP	LS	SD	SM
t_1:	1	39	50634.87	O	1997-04-12	REG AIR
t_2:	2	8	11379.84	F	1992-10-23	AIR
t_3:	3	32	53079.36	F	1994-04-23	RAIL
t_4:	4	12	22341.12	F	1993-08-11	REG AIR
t_5:	5	27	29542.86	F	1992-10-28	TRUCK
t_6:	6	11	16350.18	O	1997-11-28	REG AIR
t_7:	7	3	4065.99	O	1996-10-07	RAIL
t_8:	8	12	18102.24	O	1996-04-30	RAIL
t_9:	9	37	56625.91	O	1997-09-12	TRUCK
t_{10}:	10	22	39112.26	F	1992-12-24	RAIL

Fig. 1. An instance D of the `Lineitem` relation.

extended price attribute is then given by $d_{EP}(t_1, t_2) = |50634.87 - 11379.84| = 39255.03$. Now if we include the quantity attribute when comparing these two tuples, then $d_{\{EP,QT\}}(t_1, t_2) = \max\{|50634.87 - 11379.84|, |39 - 8|\} = 39255.03$. In this case the distance remains unchanged since the EP attribute dominates the distance value. As another example, using the discrete distance function on the line status attribute we have that $d_{LS}(t_1, t_2) = 1$ whereas $d_{LS}(t_2, t_3) = 0$.

To illustrate how the distance between sets of tuples is measured, consider two subsets S_1 and S_2 of the instance D given by $S_1 = \{t_1, t_3\}$ and $S_2 = \{t_5, t_6\}$. For simplicity, we only use the quantity attribute. First, note that the distance between t_1 and S_2 is given by $d_{QT}(t_1, S_2) = \min\{d_{QT}(t_1, t_5), d_{QT}(t_1, t_6)\} = \min\{12, 28\} = 12$. Similarly, one can verify that $d_{QT}(t_3, S_2) = 5$, $d_{QT}(t_5, S_1) = 5$ and $d_{QT}(t_6, S_1) = 21$. Then, the distance between S_1 and S_2 is given by $d_{QT}(S_1, S_2) = \max\{\max\{12, 5\}, \max\{5, 21\}\} = 21$. □

▷ *Conjunctive Queries.* We consider conjunctive queries (CQ) specified by

$$Q_i(Y) = \pi_Y \sigma_{F_i}(R'_1 \times \cdots \times R'_k),$$

where each R'_j is a renaming $\rho_j(R_j)$ of a relation R_j in $\mathcal{R}$, Y is a set of attributes and F_i is a selection predicate consisting of equality conditions of the form $A_i = A_j$ or $A_i = c$ for attributes A_i and A_j and constant $c \in dom(A_i)$.

3   The Valid Covering Selection Problem

In this section we first define the notion of a covering of a metric database relative to a set of attributes as a way of *approximating the data*. Next, we use these coverings to *approximate the results of queries* in some workload with a given budget on the available space to store the coverings of the data.

▷ *Data Approximation by Coverings.* Consider two metric databases (D, d) and (C, d) over $\mathcal{R}$ using the same distance functions d_{A_i}. Let X be a set of attributes and let $\epsilon \geq 0$. We say that (C, d) is an (X, ϵ)-*covering* of (D, d) if for any tuple

$s \in D$ there exists a tuple $t \in C$ such that $\mathsf{d}_X(s,t) \leqslant \epsilon$, indicating that any tuple in D is close to a tuple in C relative to the set X of attributes. Furthermore, given a space budget constraint $\mathbb{B}$, we say that an (X,ϵ)-covering (C,d) of (D,d) is *valid relative* to $\mathbb{B}$, or simply *valid*, if $|C| \leqslant \mathbb{B}$, *i.e.*, the covering fits in the available space. A collection of coverings $(C_1,\mathsf{d}),\ldots,(C_\ell,\mathsf{d})$ of (D,d) is valid if $|C_1 \cup \cdots \cup C_\ell| \leqslant \mathbb{B}$. Clearly, when valid coverings are concerned, the budget $\mathbb{B}$ imposes constraints on ϵ, and vice versa. For example, suppose that $\mathbb{B} = 1$ then $\epsilon = \mathsf{diam}(D)$; if $\mathbb{B} = |D|$ then ϵ can be taken to be zero.

▷ *Query Approximation by Coverings.* We want to use coverings to approximate query answers. More specifically, we are given a space budget $\mathbb{B}$ and a query workload $\mathcal{Q} = \{Q_1,\ldots,Q_\ell\}$ consisting of CQ queries. In addition, the frequency or importance of query Q_i in the workload is given by a parameter p_i. Then, given a metric database (D,d) we want to find the best valid collection of ℓ coverings (C_i,d), $i \in \{1,\ldots,\ell\}$, of (D,d) that can be used to approximate each query Q_i in $\mathcal{Q}$ relative to a user-defined set Z_i of attributes in the result schema of the query and accuracy bounds δ_i. Formally:

Valid Covering Selection Problem. *Given a metric database (D,d), query workload $\mathcal{Q} = \{Q_1(Y_1),\ldots,Q_\ell(Y_\ell)\}$, sets of attributes $Z_i \subseteq Y_i$, weights p_i, accuracy bounds δ_i, for $i \in \{1,\ldots,\ell\}$, and a budget constraint $\mathbb{B}$, find for each query Q_i a covering (C_i,d) of (D,d) such that $\max_{i \in \{1,\ldots,\ell\}} p_i \cdot |\delta_i - \mathsf{d}_{Z_i}(Q_i(D), Q_i(C_i))|$ is minimised and in addition, the collection $(C_1,\mathsf{d}),\ldots,(C_\ell,\mathsf{d})$ is valid relative to $\mathbb{B}$.* □

That is, this problem asks which coverings one should store in the available space as to "best" approximate the queries in the query workload. Here, with "best" we mean that $Q(C_i)$ approximates $Q(D)$ on the given attributes Z_i as close as possible to the user-defined accuracy bound δ_i. A naive approach for solving this problem is to just try all possible coverings and select the best ones. Not only is this exhaustive enumeration of coverings undesirable, it also requires the identification of coverings that are valid. Clearly, one cannot compute all such coverings efficiently. Furthermore, to select the best set of coverings one needs to compute $Q_i(D)$ (and also $Q_i(C)$ for that matter). Recall that we want to speed-up query evaluation by considering approximations. The exact computation of $Q_i(D)$ to find out the best way to approximate $Q_i(D)$ is clearly not an option! To make the valid covering selection problem feasible, one therefore needs to address the following two challenges.

Size Estimation: We are looking for valid coverings. This implies that one must be able to determine the sizes of (X,ϵ)-coverings to identify those coverings that are valid, *i.e.*, fit into the budget. For this purpose we extend the catalog of the DBMS with information about valid coverings. We show in the next section how this can be efficiently implemented on top of a DBMS.

Error Propagation: What can we say about $\mathsf{d}_Z(Q(D), Q(C))$ *without storing* (C,d) and *without evaluating* $Q(D)$ and $Q(C)$? That is, how does the accuracy bound on the data affect the accuracy bound on the query results? We will show in Sect. 5 that one can estimate $\mathsf{d}_Z(Q(D), Q(C))$ solely based on the structure of the query Q and the knowledge that (C,d) is an ϵ-covering.

4   Size Estimation of Coverings

In this section we consider how to estimate the size of an (X, ϵ)-covering of a metric database (D, d) for a given set X of attributes and accuracy value ϵ. The size of a minimum (X, ϵ)-covering is often referred to as the (X, ϵ)-*covering number* and will be denoted by $N(D, X, \epsilon)$. Not surprisingly, it is infeasible to compute $N(D, X, \epsilon)$ in practice. Indeed, one can verify that the computation of $N(D, X, \epsilon)$ corresponds to finding a solution to the VERTEX COVER problem, which is known to be NP-complete [2]. Although algorithms exist that approximate $N(X, D, \epsilon)$ (*e.g.*, based on [3]), they rely on efficient methods that, given a set S of tuples in D find a tuple t that maximises $d_X(t, S)$. Unfortunately, most database systems do not adequately support the indexing of tuples relative to arbitrary distance functions; a crucial feature for finding farthest removed tuples. It is outside the scope of this paper to bring database systems up-to-date with recent advances in metric indexing techniques as reported in [4].

Instead we aim to expand the DBMS's catalog with quantitative information on (X, ϵ)-coverings for various sets X of attributes and accuracy values ϵ. We particularly want that this information is easy to compute and maintain within the DBMS. As a first attempt, one can use $\tilde{N}(D, X, \epsilon) = \mathsf{diam}_X(D)/\epsilon$ as a trivial upper bound on $N(D, X, \epsilon)$. Intuitively, this upper bound assumes uniform distribution of values in X. A more sensible upper bound is given by $\min\{\tilde{N}(D, X, \epsilon), |\pi_X(D)|\}$ since $|\pi_X(D)|$ also provides an upper bound on $N(D, X, \epsilon)$.

To obtain a more fine-grained, yet efficient-to-compute upper bound for $N(D, X, \epsilon)$, we further assume that the domain values of attributes in X can be (*e.g.*, lexicographically) sorted. We can then obtain an estimate for $N(D, X, \epsilon)$ by counting the number of non-empty buckets in an equi-width histogram $H(D, X, \epsilon)$. We denote this estimate by $\hat{H}(D, X, \epsilon)$ for a given histogram $H(D, X, \epsilon)$.

Recall that an equi-width histogram $H(D, X, \epsilon)$ consists of k tuples $t_1, \ldots, t_k$ such that (i) $t_1[X] < t_2[X] < \cdots < t_k[X]$; (ii) for each $i \in \{1, \ldots, k-1\}$, $d_X(t_i, t_{i+1}) = \epsilon$; and finally (iii) for $D_i = \{t \in D \mid t_i[X] \leqslant t[X] < t_{i+1}[X]\}$ we have $D = D_1 \cup \cdots \cup D_k$. That is, $H(D, X, \epsilon)$ partitions the data into "buckets" D_i of diameter ϵ.

We next describe a procedure, referred to as Cover_Estim, for computing $\hat{H}(D, X, \epsilon)$. The pseudo-code of this procedure is shown in Fig. 2. In a nutshell, Cover_Estim (D, X, q, ϵ) recursively processes the attribute list X (line 5). When the current attribute is A_i, the algorithm has already computed a histogram $H(D, \langle A_1, \ldots, A_{i-1}\rangle, \epsilon)$ consisting of $\hat{H}_{i-1}$ non-empty buckets and now further refines each of these buckets B in $H(D, \langle A_1, \ldots, A_{i-1}\rangle, \epsilon)$ by means of the sub-procedure bucket(B, A_i, ϵ). The result is a histogram $H(D, \langle A_1, \ldots, A_i\rangle, \epsilon)$ consisting of $\hat{H}_i$ buckets, where $\hat{H}_i$ is obtained from $\hat{H}_{i-1}$ by adding for each bucket B in $H(D, \langle A_1, \ldots, A_{i-1}\rangle, \epsilon)$ the number of buckets returned by bucket(B, A_i, ϵ). The recursive procedure halts when either an empty bucket is considered (line 2) or when enough attributes have been processed (line 7), at which point we return the trivial upper bound $\min\{\mathsf{diam}_{X'}(D)/\epsilon, |\pi_{X'}(D)|\}$ for the remaining

---

Procedure Cover_Estim

Input: A database D, list of attributes $X = \langle A_1, A_2, \ldots, A_p \rangle$, number of attributes to be processed by the non-naive method q, accuracy threshold ϵ.

Output: Number $\hat{H}(X,D,\epsilon)$ of non-empty buckets in equi-width histogram $H(X,D,\epsilon)$.

1. $\hat{H} := 0$;
2. **if** $D = \emptyset$ **then return** 0;
3. **if** $p > 0$ **then**
4.   **if** $q > 0$ **then**
5.   **for each** $B = \text{bucket}(D, A_1, \epsilon)$ **do** $\hat{H} := \hat{H} + \text{Cover\_Estim}(B, \langle A_2, \ldots, A_p \rangle, \epsilon, q-1)$;
6.   **return** $\hat{H}$.
7.   **else return** $\min\{\text{diam}_X(D)/\epsilon, |\pi_X(D)|\}$.
8. **else return** 1.

Fig. 2. Procedure for estimating the size of a covering.

attributes $X' = \langle A_{q+1}, \ldots, A_p \rangle$. Note that q is a user-chosen parameter that determines how many attributes should be processed using the bucket refinement procedure. If $q = p$, the recursion stops when all attributes have been processed. In this case, a recursive call with $X = \emptyset$ is made, indicating that the non-empty bucket B under consideration does not need any further refinement and thus will contribute a count of 1 to the estimate (line 8). Also note that if $q = 0$ then no recursion takes place and the naive upper bound $\min\{\text{diam}_X(D)/\epsilon, |\pi_X(D)|\}$ is returned for the complete attribute set X (line 7).

It remains to detail the sub-procedure $\text{bucket}(D, A_i, \epsilon)$ which, given a database D, attribute A_i and accuracy value ϵ, constructs a partition $D_1 \cup \cdots \cup D_k$ of the data corresponding to a histogram $H(D, A_i, \epsilon)$. Assuming that the sorted attribute values for A_i can be cast as numerical values, which is often the case in practice, we can leverage the presence of the Width_bucket function in SQL. This function takes as input an attribute A_i, the minimal and maximal value of the active domain of A_i in D, and a desired number of buckets. The result consists of pairs (t, i) where $t \in D$ and i is the unique bucket number to which t belongs. From this, the number of non-empty buckets and a histogram can easily be computed. More specifically, $\text{bucket}(D, A_i, \epsilon)$ can be implemented by means of the following SQL expression:

```
SELECT Width_bucket(A_i, min, max, min{⌈diam_{A_i}(D)/ε⌉, |π_{A_i}(D)|}) AS bucket
FROM R  GROUP BY bucket ORDER BY bucket
```

where we use the estimate $\min\{\lceil \text{diam}_{A_i}(D)/\epsilon \rceil, |\pi_{A_i}(D)| \}$ for an upper bound on the number of buckets. Furthermore, it should come as no surprise that the recursive procedure Cover_Estim can be implemented entirely in SQL, provided of course that we know the attributes in X up front (otherwise a recursive SQL query is needed). Indeed, the SQL implementation of Cover_Estim consists of nested variants of the SQL query given above, where the nesting level is determined by the number of attributes in X. Observe that Cover_Estim not only computes size bounds but returns actual coverings.

Example 2. Recall the `Lineitem` database D given in Example 1. For the extended price (EP) attribute the diameter of D is simply given by $\mathsf{diam}_{\mathsf{EP}}(D) = \max_{t \in D} t[\mathsf{EP}] - \min_{t \in D} t[\mathsf{EP}] = 52559.92$. Observe that $|\pi_{\mathsf{EP}}(D)| = 10$, hence at most 10 buckets are needed to exactly cover D on attribute EP. Taking our $\epsilon = 8770$, the quantity $\min\{\lceil \frac{52559.92}{8770} \rceil, 10\} = 6$ is an upper bound on the number of buckets needed.

We now illustrate the Cover_Estim procedure. Firstly consider the evaluation of Cover_Estim $(D, \mathsf{EP}, 0, 8770)$. In this case, all attributes are processed by the naive method ($q = 0$) and therefore the procedure outputs the upper bound of 6 buckets. Next, we set $q = 1$ so that the procedure Cover_Estim $(D, \mathsf{EP}, 1, 8770)$ now uses the non-naive method by evaluating the above SQL query. We obtain a covering $C = \{t_1, t_4, t_6, t_7, t_{10}\}$ of D of size 5 buckets as follows: $B_1 = \{t_7, t_2\}, B_2 = \{t_6, t_8\}, B_3 = \{t_4, t_5\}, B_4 = \{t_{10}\}$ and $B_5 = \{t_1, t_3, t_9\}$. Tuples are sorted in each bucket. It is readily verified that C is an (EP, 8770)-covering of D. From this small example we can already see that the non-naive method improves on the trivial upper bound on the number of buckets (five buckets rather than six). Furthermore we also note that C is an ($\{\mathsf{EP}, \mathsf{QT}\}, 8770$)-covering of D. This is so because after the Cover_Estim procedure processes the EP attribute, the buckets do not require any further refinement. Indeed, for each bucket B_i, for $i \in \{1, \ldots, 5\}$, we already have that $\mathsf{diam}_{\mathsf{QT}}(B_i) \leqslant \epsilon$. □

Remarks. (1) We described a very specific method for estimating $N(D, X, \epsilon)$. However, any other method (*e.g.*, based on k-means clustering or other kinds of histograms) can be easily plugged into our query approximation system. (2) It is important to observe that the cost of estimating $N(D, X, \epsilon)$ is a one-time cost and can be done when the DBMS is idle. (3) Clearly, the order in which the attributes in X are fed to Cover_Estim(D, X, ϵ) directly impacts the estimate of $N(D, X, \epsilon)$. Further investigation is required to determine heuristics to select the best order.

5   Error Propagation

The second challenge that we have to address is the efficient estimation of $\mathsf{d}_Z(Q(C), Q(D))$ for (X, ϵ)-coverings (C, d) of (D, d). That is, we need to estimate the error on the query result due to the use of coverings rather than the original database. Since our aim is to speed-up the query evaluation of Q, one cannot rely on computing $Q(D)$, $Q(C)$ and $\mathsf{d}_Z(Q(D), Q(C))$, as this requires evaluating the queries. The estimation procedure for $\mathsf{d}_Z(Q(D), Q(C))$ should thus be *independent* of (D, d) and thus also of the chosen covering (C, d). This bears the question whether the knowledge of the query Q and the fact that there is an (X, ϵ)-covering of the data is sufficient to obtain an accuracy bound on the query result. We answer this question affirmatively, provided that we slightly relax the query Q into an approximate query $\widetilde{Q}$, as will be explained shortly.

The overall strategy to estimate $\mathsf{d}_Z(Q(D), Q(C))$ then consists of showing under which conditions an (X, ϵ)-covering (C, d) of (D, d) can be transformed in a (Z, ϵ')-covering $\widetilde{Q}(C)$ of $Q(D)$. Given this, we can then estimate $\mathsf{d}_Z(Q(D), Q(C))$

using $\mathsf{d}_Z(Q(D), \widetilde{Q}(C)) = \epsilon'$. In particular, we show how ϵ' can be expressed in terms of ϵ, hereby alleviating the need for evaluating any query in the estimation process.

▷ *Query Relaxation.* Let us first explain why query relaxations are needed. Suppose that $Q = \sigma_{A=a}(R)$ and let (D, d) be a metric database. Then $Q(D)$ contains all tuples $t \in D$ with $t[A] = a$. Take any (A, ϵ)-covering (C, d) of (D, d). Then, unless C contains tuples t of D with $t[A] = a$, we have that $Q(C) = \emptyset$ and thus $Q(C)$ is not a covering of $Q(D)$. In other words, we cannot guarantee that any (A, ϵ)-covering (C, d) of (D, d) suffices to approximate $Q(D)$. A similar situation arises when considering $Q = \sigma_{A=B}(R)$.

To remedy this situation, we not only approximate the data but also consider *relaxations* of queries. More specifically, we compare $Q = \sigma_{A=a}(R)$ on (D, d) with its relaxation $\widetilde{Q} = \sigma_{\mathsf{d}(A,a)\leqslant\eta}(R)$ on (C, d), for some value η. The semantics of $\widetilde{Q}$ is as expected: $\sigma_{\mathsf{d}(A,a)\leqslant\eta}(C)$ selects all tuples t in C for which $\mathsf{d}_A(t[A], a) \leqslant \eta$. When considering (A, ϵ)-coverings (C, d) of (D, d) with $\epsilon \leqslant \eta$, we then have that $\mathsf{d}_A(Q(D), \widetilde{Q}(C)) \leqslant \epsilon$. Similarly, one can verify that when $Q = \sigma_{A=B}(R)$ is relaxed to $\widetilde{Q} = \sigma_{\mathsf{d}(A,B)\leqslant\eta}(R)$, then $\mathsf{d}_{A,B}(Q(D), \widetilde{Q}(C)) \leqslant \epsilon$ for any $(\{A, B\}, \epsilon)$-covering (C, d) of (D, d) with $\epsilon \leqslant \frac{\eta}{2}$. Note that for a selection condition $A = B$ to make sense, attributes A and B must have the same domain and distance function. We denote $\mathsf{d}_A = \mathsf{d}_B$ by d. Hence, $\sigma_{\mathsf{d}(A,B)\leqslant\eta}(C)$ selects all tuples t in C such that $\mathsf{d}(t[A], t[B]) \leqslant \eta$. Now, given a tuple t in D such that $t[A] = t[B]$ and given an $(\{A, B\}, \epsilon)$-covering (C, d) of (D, d) with $\epsilon \leqslant \frac{\eta}{2}$, we have that there exists a tuple $t' \in C$ such that $\mathsf{d}_{A,B}(t, t') \leqslant \epsilon$. Indeed, observe that $\mathsf{d}(t'[A], t'[B]) \leqslant \mathsf{d}(t[A], t'[A]) + \mathsf{d}(t[B], t'[B]) = 2\epsilon \leqslant \eta$. Hence, $t' \in \widetilde{Q}(C)$ and $\widetilde{Q}(C)$ is a covering of $Q(D)$.

For a CQ query Q we denote by $\widetilde{Q}_\eta$ the query obtained by replacing any selection predicate in Q by its relaxed version, *i.e.*, all occurrences of $\sigma_{A=a}$ and $\sigma_{A=B}$ in Q are replaced by $\sigma_{\mathsf{d}(A,a)\leqslant\eta}$ and $\sigma_{\mathsf{d}(A,B)\leqslant\eta}$, respectively.

Example 3. Recall again our `Lineitem` database D from Example 1. Consider the constant selection query Q which selects all tuples $t \in D$ with $t[\mathsf{EP}] = 18102.24$ and projects on the EP attribute. We reuse our covering from Example 2, *i.e.*, $C = \{t_1, t_4, t_6, t_7, t_{10}\}$. The relaxed query $\widetilde{Q}$ selects all tuples $t \in C$ with $t[\mathsf{EP}] \in [18102.24 - 8770, 18102.24 + 8770]$ and also projects on the EP attribute. Evaluating both queries, we have $Q(D) = \{t_8\}$ and $\widetilde{Q}(C) = \{t_4, t_6\}$, from which we obtain

$$\mathsf{d}_{\mathsf{EP}}(Q(D), \widetilde{Q}(C)) = 4238.88 \leqslant 8770.$$

Hence, $\tilde{Q}(C)$ is $(\mathsf{EP}, 8770)$-approximation of $Q(D)$.   □

▷ *Propagation Algorithm.* We next provide an algorithm, Error_Prop, that given a CQ query Q, a set X of attributes in $\mathcal{R}$, an accuracy value ϵ, and relaxation parameter η for $\widetilde{Q}_\eta$, returns:

(a) a set of attributes $\mathsf{prop}(Q, X)$ in the result schema of Q; and
(b) an error bound $\mathsf{err}(Q, \epsilon)$,

Procedure Error_Prop (Q,X,ϵ,η)
1. **switch**
2.   **case** $Q=R(A_1,...,A_k)$
3.   **return** $\mathrm{prop}(Q,X):=X\cap\{A_1,...,A_k\}$; and $\mathrm{err}(Q,\epsilon):=\epsilon$;
4.   **case** $Q=\rho(Q'(A_1,...,A_k))$ for some renaming $\rho=(A_1\mapsto B_1,...,A_k\mapsto B_k)$
5.   **return** $\mathrm{prop}(Q,X):=\rho(\mathrm{prop}(Q',X))$; and $\mathrm{err}(Q,\epsilon):=\mathrm{err}(Q',\epsilon)$;
6.   /* Where $\mathrm{prop}(Q',X)$ and $\mathrm{err}(Q',\epsilon)$ are the result of Error_Prop(Q',X,ϵ,η) */
7.   **case** $Q=\sigma_{A=a}(Q')$
8.   **return** $\mathrm{prop}(Q,X):=\mathrm{prop}(Q',X)$ if $A\in\mathrm{prop}(Q',X)$ and $\mathrm{prop}(Q,X):=\emptyset$ other-
9.   wise; and $\mathrm{err}(Q,\epsilon):=\mathrm{err}(Q',\epsilon)$ if $\epsilon\leqslant\eta$ and $\mathrm{err}(Q,\epsilon):=+\infty$ otherwise;
10. **case** $Q=\sigma_{A=B}(Q')$
11.   **return** $\mathrm{prop}(Q,X):=\mathrm{prop}(Q',X)$ if $A,B\in\mathrm{prop}(Q',X)$ and $\mathrm{prop}(Q,X):=\emptyset$ other-
12.   wise; and $\mathrm{err}(Q,\epsilon):=\mathrm{err}(Q',\epsilon)$ if $\epsilon\leqslant\eta/2$ and $\mathrm{err}(Q,\epsilon):=+\infty$ otherwise;
13. **case** $Q=\pi_Y(Q')$
14.   **return** $\mathrm{prop}(Q,X):=\mathrm{prop}(Q',X)\cap Y$; and $\mathrm{err}(Q,\epsilon):=\mathrm{err}(Q',\epsilon)$;
15. **case** $Q=Q_1\times Q_2$
16.   **return** $\mathrm{prop}(Q,X):=\mathrm{prop}(Q_1,X)\cup\mathrm{prop}(Q_2,X)$; and $\mathrm{err}(Q,\epsilon):=\max\{\mathrm{err}(Q_1,\epsilon),$
17. $\mathrm{err}(Q_2,\epsilon)\}$.
18. /* Here, $\mathrm{prop}(Q_i,X)$ and $\mathrm{err}(Q_i,\epsilon)$ are given by Error_Prop(Q_i,X,ϵ,η), for $i=1,2$.*/

Fig. 3. Error propagation algorithm.

such that for *any* (X,ϵ)-covering (C,d) of (D,d) it is guaranteed that

$$\mathsf{d}_Z(Q(D),\widetilde{Q}_\eta(C))\leqslant\mathrm{err}(Q,\epsilon)$$

for any non-empty $Z\subseteq\mathrm{prop}(Q,X)$. The algorithm Error_Prop works inductively on the structure of the query Q and is described in Fig. 3. Its correctness can be readily verified but this is omitted due to space limitations.

▷ *Query Column Sets and Error Guarantees.* One can see from the description of Error_Prop(Q,X,ϵ,η) in Fig. 3 that certain conditions on coverings need to hold when using them to approximate $Q(D)$. That is, when $\mathrm{prop}(Q,X)$ is empty, insufficiently many attributes are covered to approximate Q. Observe that $\mathrm{prop}(Q,X)$ is empty when X does not contain (i) any attribute in the relations occurring in Q (line 3); (ii) an attribute that appears in a selection condition (lines 8, 11); or (iii) any of the projected attributes in Q (line 14). Given a set of attributes Z in the result schema of Q, one can compute for each relation R_i in $\mathcal{R}$ the minimal set of attributes X_i such that for $X=X_1\cup\cdots\cup X_n$, $Z\subseteq\mathrm{prop}(Q,X)$. In other words, the X_i's are the attributes that are required to be covered in R_i in order to approximate Q on Z. We denote by $\mathrm{qcs}(Q,Z)$ the set of pairs (R_i,X_i). In analogy with the BlinkDB system, we also refer to $\mathrm{qcs}(Q,Z)$ as the *query column set of Q relative to Z*. The query column sets can be computed by starting from Z and by reversely applying the different cases in Error_Prop(Q,X,ϵ,η) for $\mathrm{prop}(Q,X)$. We omit the details due to space limitations.

Furthermore, even when the query column set $X = \mathsf{qcs}(Q, Z)$ is covered it may be that $\mathsf{err}(Q, X) = +\infty$ and thus no approximation is achieved. This happens when ϵ is too large compared to the chosen relaxation parameter η and when Q contains selection conditions (lines 9, 12). In particular, from $\mathsf{Error\_Prop}(Q, X, \epsilon, \eta)$ we obtain the following error guarantees:

$$\text{If} \left\{ \begin{array}{l} \text{(a) } Q \text{ does not contain selection conditions} \\ \qquad\qquad\qquad\qquad \text{or} \\ \text{(b) } Q \text{ only contains constant selection conditions and } \epsilon \leqslant \eta \\ \qquad\qquad\qquad\qquad \text{or} \\ \text{(c) } Q \text{ contains an equality selection conditions and } \epsilon \leqslant \eta/2 \end{array} \right\} \implies \mathsf{err}(Q, \epsilon) = \epsilon.$$

Otherwise, we have an unbounded $(+\infty)$ error. Of course, this also implies that unbounded errors can be avoided altogether by considering relaxations $\widetilde{Q}_\eta$ that depend on ϵ, i.e., by letting $\eta = 0$ in case (a); $\eta = \epsilon$ in case (b); and $\eta = 2\epsilon$ in case (c). In the following, we always assume that these relaxations are used when approximating Q and simply denote the relaxation by $\widetilde{Q}$.

Example 4. Consider query Q' obtained from modifying query Q given in Example 3 by projecting on two attributes $\{\mathsf{EP}, \mathsf{QT}\}$ instead of only projecting on EP. Suppose that we want to approximate Q' on these attributes, i.e., $Z = \{\mathsf{EP}, \mathsf{QT}\}$ with user-defined threshold $\eta = 8770$. To identify which (X, ϵ)-coverings of D can be used to approximate Q', we must have that $Z \subseteq \mathsf{prop}(Q', X)$. Evaluating the procedure $\mathsf{Error\_Prop}$ tells us that any (X, ϵ)-covering of D such that $\{\mathsf{EP}, \mathsf{QT}\} \subseteq X$ will do. This follows directly from the selection (line 8) and projection (line 14) rule in the procedure. Consider the $(\{\mathsf{EP}, \mathsf{QT}\}, 8770)$-covering C from Example 2. It is readily verified that for $\epsilon = 8770$, we get an error bound $\mathsf{err}(Q', \epsilon) = \epsilon = 8770$. Hence, for any $Z \subseteq \{\mathsf{EP}, \mathsf{QT}\}$ we guarantee that $\mathsf{d}_Z(Q'(D), \widetilde{Q'}_\eta(C)) \leqslant 8770$, i.e., we can approximate Q' within the user-defined threshold η on attributes Z. □

6   Valid Covering Selection

We now have all ingredients at hand to describe our approach for solving the valid covering selection problem. Let $\mathcal{Q} = \{Q_1, \ldots, Q_\ell\}$ be the query workload consisting of ℓ CQ queries. For each query Q_i the user specifies its importance p_i, the set of attributes Z_i in Q_i's result schema to be approximated, and desired error bound δ_i. Furthermore, a space budget $\mathbb{B}$ is given. Our approach works in four steps:

1. We collect the query column sets $\mathsf{qcs}(Q_i, Z_i)$, for $i \in \{1, \ldots, \ell\}$. Recall that in order to approximate $Q_i(D)$ on attributes Z_i, one minimally needs to cover all attributes in $\mathsf{qcs}(Q_i, Z_i)$.
2. Next, we inspect the DBMS catalog that, by using the size estimation method described in Sect. 4, is now extended with quadruples $(R_j, X_j, \epsilon_j, N_j)$, indicating that there is an (X_j, ϵ_j)-covering of size N_j of the instance I_j of R_j.

Note that there may be multiple coverings on each relation. Denote by $\mathsf{cov}(D)$ the collection of all such quadruples in the catalog. Clearly, considering all possible coverings on $\mathcal{R}$ would lead to exponentially many coverings in $\mathsf{cov}(D)$. Instead, we assume that a set of candidate covering attributes is provided based on those that actually were needed in the past or simply by inspecting the query column sets of the workload queries.

3. We then solve a mixed integer linear program (MILP). Part of the solution of this program are variables x_{ij} that, when set to 1, indicate that the ith covering in $\mathsf{cov}(D)$ is used to approximate Q_j.

4. We materialise all coverings $(C_1, \mathsf{d}), \ldots, (C_\ell, \mathsf{d})$ identified in the previous step, hereby avoiding replicating the same covering on a relation. Just like in BlinkDB, the materialisation step is a one-time cost and if the query workload $\mathcal{Q}$ is representative for the past, present and future workload, the stored coverings can be used for any future incoming queries as well. For now, to obtain an approximation for queries in $\mathcal{Q}$ we evaluate $\widetilde{Q_j}(C_j)$ on the stored coverings. Recall that $\widetilde{Q_j}$ is the relaxation of Q_j by setting η to the appropriate value 0, ϵ_j, or $2\epsilon_j$, where ϵ_j is the accuracy of covering (C_j, d_j) (See Sect. 5).

The MILP ensures that (i) all coverings fit into the available space budget; and (ii) the best possible accuracies of these coverings are selected for approximating the workload queries. Observe that, assuming that $\mathsf{cov}(D)$ is available, we only need to evaluate the relaxed queries on coverings. No other query evaluation or access to the data is needed. This implies, among other things, that the size of the MILP is independent of the size of D and that solving it is a cost that is negligible. We verify this in the experimental section. It remains to detail the mixed integer linear program.

▷ *MILP Formulation.* Part of the MILP consists of a simple set covering problem: for each $Q_i \in \mathcal{Q}$ find $(R_1, X_1, \epsilon_1, N_1), \ldots, (R_n, X_n, \epsilon_n, N_n)$ in $\mathsf{cov}(D)$ that cover $\mathsf{qcs}(Q_i, Z_i)$. More specifically, if $\mathsf{qcs}(Q_i, Z_i) = \{(R_1, Y_1), \ldots, (R_n, Y_n)\}$ then we must have that $Y_i \subseteq X_i$ for $i \in \{1, \ldots, n\}$. We encode this set cover problem in the MILP in the standard way. Let $I = \{1, \ldots, |\mathsf{cov}(D)|\}$ and $J = \{1, \ldots, \ell\}$. For each $(i, j) \in I \times J$ and relation name $R \in \mathcal{R}$, we introduce a constant c_{ij}^R and boolean variable x_{ij}^R. Here, $c_{ij}^R = 1$ if the ith covering in $\mathsf{cov}(D)$ contains the attributes in $\mathsf{qcs}(Q_j, Z_j)$ corresponding to R; and $c_{ij}^R = 0$ otherwise. Furthermore, $x_{ij}^R = 1$ is to indicate that this covering on R is used to approximate Q_j on the Z_j attributes and $x_{ij}^R = 0$ indicates the opposite. To ensure that $\mathsf{qcs}(Q_j, X_j)$ is fully covered we thus require that

$$\sum_{i \in I} c_{ij}^R x_{ij}^R \geqslant 1 \qquad \text{(for each } j \in J, R \in \mathcal{R} \text{ such that } (R, X) \in \mathsf{qcs}(Q_j, Z_j) \\ \text{for some non-empty set } X \text{ of attributes.)}$$

In addition, all coverings in $\mathsf{cov}(D)$ that are used for approximating queries in $\mathcal{Q}$ need to be stored and must fit within the available space budget $\mathbb{B}$. For each $i \in I$ we therefore introduce a variable x_i that will be set to 1 if any of the x_{ij}^R's for $j \in J$ is 1; and x_i is set to 0 otherwise. In other words, $x_i = \max\{x_{ij}^R \mid j \in J, R \in \mathcal{R}\}$ and indicates which coverings in $\mathsf{cov}(D)$ are being used. We therefore require

$$x_{ij}^R \leqslant x_i \quad \text{(for each } i \in I, j \in J, R \in \mathcal{R}\text{), and } x_i \leqslant \sum_{j \in J, R \in \mathcal{R}} x_{ij}^R \text{ (for each } i \in I\text{).}$$

The space budget constraint is simply given by $\sum_{i \in I} x_i N_i \leqslant \mathbb{B}$, where N_i is the size of the ith covering in $\mathsf{cov}(D)$. It remains now to relate the selected coverings (i.e., those with $x_{ij}^R = 1$) to the error bound on the corresponding query Q_j. Since this error is given by $\mathsf{err}(Q_j, \epsilon) = \epsilon$ we need to determine the maximum ϵ used in the coverings for Q_j. For this purpose we introduce variables y_j, for $j \in J$, and require that

$$x_{ij}^R c_{ij}^R \epsilon_i \leqslant y_j \quad \text{(for each } i \in I, j \in J, R \in \mathcal{R}\text{)}$$

where ϵ_i is the accuracy value of the ith covering in $\mathsf{cov}(D)$. Finally, the objective function of the MILP is

$$\textbf{minimise}: \quad \max_{j \in J} p_j |\delta_j - y_j|,$$

where δ_j is the user-specified accuracy threshold. It is easily verified that the objective function can be transformed into a linear constraint, i.e., without using max and absolute value $|\cdot|$.

Example 5. We next illustrate the interaction of the space budget and accuracy threshold in the valid covering selection method. Recall query Q' from Example 4 and the $(\{\mathsf{EP}, \mathsf{QT}\}, 8770)$-covering, here denoted by C_1, from Example 2. Let C_2 be another covering of our Lineitem database D consisting of three buckets, i.e., $C_2 = \{t_4, t_7, t_{10}\}$. It is readily verified that C_2 is a $(\{\mathsf{EP}, \mathsf{QT}\}, 17520)$-covering. Then we have $\mathsf{cov}(D)$ with two coverings $\{C_1, C_2\}$ of size 5 and 3 respectively. We know that $\mathsf{qcs}(Q', Z) = \{\mathsf{EP}, \mathsf{QT}\}$. Let $p_1 = 1$, $\delta_1 = 5500$ and $\mathbb{B} = 4$.

The MILP can now be formulated with the required parameters as above. In this case the program has a simple task: to decide which of the two coverings should be used to approximate Q'. Based on the constraints, an optimal solution would set x_{21} to 1, i.e., covering C_2 is chosen since only C_2 fits into $\mathbb{B}$. Let us consider another scenario where more space is available and we increase our space budget, i.e., $\mathbb{B} = 7$. Now, we see that any of the two coverings can fit into the available budget. Note that the user desired error bound remains unchanged. Again, by looking at the objective function of the MILP, it is easy to see that an optimal solution would set x_{11} to 1, i.e., covering C_1 is chosen because the propagated error for covering C_1, ($\epsilon_1 = 8770$), is more close to the desired error bound, δ_1 than the propagated error of C_2, ($\epsilon_2 = 17520$).   □

7   Experimental Evaluation

In this section, we evaluate the performance of the procedure Cover_Estim for estimating the size and computing the coverings of the data, and the accuracy of our solution to the valid selection problem on individual queries and on queries in some workload.

▷ *Evaluation Setting.* Our experiments were run on a GNU/Linux machine with Intel(R) Xeon(R) CPU 2.90 GHZ (16 Cores) and 32 GB memory. We use PostgreSQL as the underlying database system. All experiments were repeated five times and averages are reported. We used two datasets: (i) the TPC-H benchmark data[1] (scale factor 1) consisting of 9 million tuples (1 GB); and (ii) the Big Data benchmark[2] (scale factor 1) consisting of tables uservisits and rankings of 8 million (1.28 GB) and 155 million (25.4 GB) tuples, respectively. For schema details on the datasets, consult the links in the footnote. On the TPC-H data, our query workload Q consists of variants of Q_1, Q_3, Q_6, Q_{13}, Q_{19} of the TPC-H queries. For the Big Data benchmark data, we use a variant of their scan and aggregate query. In all of our experiments we consider coverings on attributes that have a sorted domain and use distance functions as described in Sect. 2.

▷ *Covering Computations.* We first experimentally validate the efficiency of the procedure Cover_Estim, as described in Sect. 4, then illustrate how different datasets can be compressed by means of coverings, and finally investigate the impact of the parameter q on the quality and efficiency of the bounds returned by Cover_Estim.

Figure 4 shows the time to compute the size of coverings and the time to materialise them on the Big Data benchmark dataset. More specifically, we varied (i) the sizes of the input tables uservisits (1000 k to 10000 k) tuples and rankings (1000 k to 10000 k) tuples; and (ii) the sets X of attributes to be covered. On the left, we report the times for individual attributes *duration* and *adrevenue* in uservisits, and attributes *avgduration* and *pagerank* in rankings; on the right we consider the combined attribute sets {*duration, adrevenue*} and {*avgduration, pagerank*}. We fixed ϵ to be 0.0001. Not surprisingly, estimating the size of coverings requires considerably less time than materialising them. Indeed, while the size estimation typically takes a couple of seconds, the materialisation takes tens of seconds. This verifies our claim that extending the DBMS's catalog with quantitative information on coverings is feasible, especially since this is a one-time cost and can be computed when the system is idle. We further observe that the running times strongly depend on the set X of attributes. In particular, the running time increases when X consists of more attributes. This is not unexpected since a larger X results in a larger (X, ϵ)-covering. A similar behaviour is observed when varying ϵ, *i.e.*, the smaller the ϵ, the more time it takes to bound the size of the coverings. Experiments on the TPC-H data gave analogous results (not reported).

We next considered the compressibility of the datasets. Figures 5(a) and (b) show the size of the resulting covering on the two tables in the Big Data benchmark data set for varying values of ϵ and for the attribute sets considered earlier. We fixed the size of both tables to 10 million tuples. Similarly, Fig. 5(c) shows the size

---

[1] http://www.tpc.org/tpch/.
[2] https://amplab.cs.berkeley.edu/benchmark/.

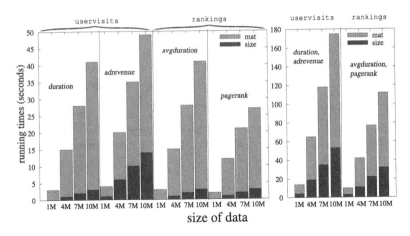

size of data

Fig. 4. Efficiency of Cover_Estim for computing size and materialisation of coverings.

of coverings on the `lineitem` table in the TPC-H data set for the following sets of attributes: {*lextendedprice*}, {*lextendedprice, lquantity*} and {*lextendedprice, lquantity, llinestatus*}. One can see that the datasets compress rather well: for reasonable values of ϵ the size of the corresponding covering provides a considerable reduction compared to the size of the original data. In other words, even for a small space budget $\mathbb{B}$ one can find accurate coverings of the data. As before, the more attributes are used in the covering, the larger the covering.

Finally, Fig. 5(d) shows the impact of the parameter q in Cover_Estim. Recall from Sect. 4 that q indicates how many attributes are processed by the recursive bucketisation process, and consequently, how many attributes are treated by the naive upper bound. Figure 5(d) reports the effect on the `lineitem` table and attribute set {*lextendedprice, lquantity, ldiscount*}. We let $q = 3$ (most fine-grained upper bound), $q = 2$ (last attribute is treated by naive upper bound), and $q = 1$ (last two attributes are estimated by naive upper bound). Not surprisingly, the quality of the size estimate degrades with decreasing q. On the other hand, the running times for Cover_Estim decrease when more attributes are treated by the naive upper bound. Indeed, our experiments (not reported due to space limitations) show that the size estimation for $q = 2$ takes half the time when compared to $q = 3$.

▷ *Query Approximation and Valid Covering Selection.* Our next set of experiments concerns the use of coverings to approximate query results. We first consider individual queries and compare the theoretical upper bound with the actual error made by our query approximation method. Next, we consider a query workload and investigate our solution to the valid cover selection problem.

Figure 6 shows the comparison of the actual error $\mathsf{d}_Z(Q(D), \widetilde{Q}(C))$ with the theoretical upper bound $\mathsf{err}(Q, \epsilon)$ given in Sect. 5, for varying sizes of coverings C of D. More specifically, we express the size $|C|$ of C as a percentage of the size $|D|$ of D and report the error $\mathsf{err}(Q, \epsilon)$, where ϵ is the accuracy associated with the covering C. The computation of $\mathsf{d}_Z(Q(D), \widetilde{Q}(C))$ is done by evaluating

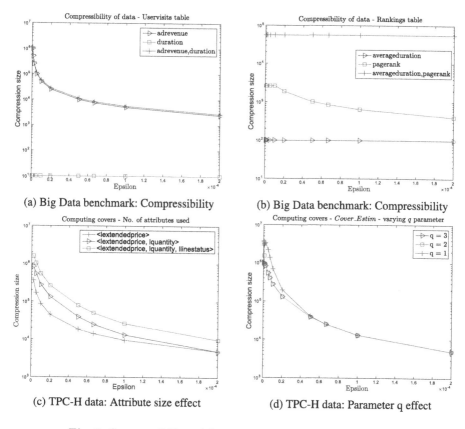

Fig. 5. Compressibility of datasets and impact of parameter q.

$Q(D)$ and $\widetilde{Q}(C)$ and by computing the distance between them. Due to space limitations we only report two settings. In Fig. 6(a) we consider the lineitem table of the TPC-H data and a constant selection query Q on the *lextendedprice* attribute. Coverings are on this attribute only. In Fig. 6(b) we consider a join query Q (*i.e.,* cartesian product followed by equality selection) involving both tables in the Big data benchmark data. Coverings are on attribute sets {*duration, adrevenue*} on uservisits and {*avgduration, pagerank*} on rankings.

These experiments show that our approach actually provides better actual accuracy bounds on the query results than is anticipated by the theoretical upper bound. In particular, we note that for highly compressible attributes, such as *lextendedprice* in lineitem we can get an actual error of 0 using only a small fraction (13.2 %) of the original dataset. For this particular setting, Fig. 7(a) verifies that answering queries on coverings takes less time than when using the original data and, more importantly, that the error made by the approximation is within reasonable bounds. In particular, Fig. 7(a) shows the actual error for the constant selection query on the TPC-H data for various sizes of coverings and the time it takes to answer its relaxation on the coverings. For example,

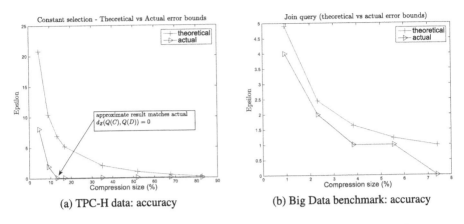

Fig. 6. Comparison of actual and theoretical accuracy of query results.

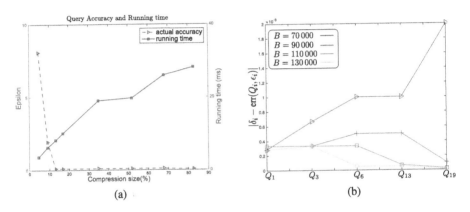

Fig. 7. (a) Accuracy vs space trade-off for single query on TPC-H data; (b) effect of space budget $\mathbb{B}$ on accuracy $|\delta_i - \text{err}(Q_i, \epsilon_i)|$ of queries in workload for the coverings selected by the MILP.

evaluating the relaxation using 13.2 % of the data takes 1/4 of the time needed to evaluate the query on the original data without loss of accuracy.

Finally, we consider a query workload $\mathcal{Q}$ of the 5 TPC-H queries mentioned earlier. We extended the DBMS catalog with 50 different coverings (some of them over the same sets of attributes). We have arbitrarily chosen the weights and desired accuracy thresholds for each of the queries. Figure 7(b) shows the errors, $|\delta_i - \text{err}(Q_i, \epsilon_i)|$, made on each of the queries by using the coverings as identified by the MILP given in Sect. 6, and this for varying space budgets $\mathbb{B}$ (70 0000 to 130 000 tuples). As expected, increasing $\mathbb{B}$ results in a better approximation of the queries. Furthermore, the increase in accuracy when increasing the budget is more noticeable for queries with high importance (*e.g.*, query Q_{19}). Solving the MILP took a few milliseconds, which is negligible in the overall query approximation process.

8   Related Work and Conclusions

Approximate query answering (AQA) in relational databases has been the subject of extensive research. We refer to the recent survey [5] for more details. Most research has focused on AQA systems that make use of concise data structures called synopses built from the database. The synopses techniques can be divided into two broad categories: non-sampling based, and sampling based. Examples of non-sampling based synopses are wavelets [6], histograms [7,8], and kernels [9]. Sampling-based methods are the key components of AQA systems as described in [1,10,11], among others. Most of these works, however, consider simple aggregate queries. A notable exception is [7] where set-valued conjunctive queries are approximated by means of a rewriting in terms of a compact histogram representation of the data. The result of this rewriting is a histogram that is an approximation of the query result. Although close in spirit to our use of coverings, [7] does not provide accuracy guarantees and cannot be easily generalised to non-histogram synopses. By contrast, our notion of covering is more general and we do provide guarantees. Furthermore, [7] considers single queries only and does not impose an upper bound on the available space. Finally, we recall that the valid covering selection problem is inspired by the sampling-based BlinkDB system [1], as mentioned in the Introduction.

▷ *Conclusions.* We have presented a formal approach for space bounded query approximation and experimentally validated it. Much more needs to be done, however: Can we enrich coverings so that they become samples that can be used to approximate aggregate queries? How to incorporate other error measures? Are there special classes of queries for which better (more compact and accurate) coverings can be computed? Can our query approximation be integrated in indexing methods or be part of the DBMS query optimiser? How can a large number of coverings be efficiently stored and accessed? Can our approach benefit from moving to other platforms, such as Apache Hive? These are just a number questions that need to be addressed.

References

1. Agarwal, S., Mozafari, B., Panda, A., Milner, H., Madden, S., Stoica, I.: BlinkDB: queries with bounded errors and bounded response times on very large data. In: Proceedings of ECCS, pp. 29–42 (2013)
2. Garey, M.R., Johnson, D.S.: Computers and Intractability: A Guide to the Theory of NP-Completeness. W. H. Freeman & Co., New York (1979)
3. Gonzalez, T.F.: Clustering to minimize the maximum intercluster distance. Theor. Comput. Sci. **38**, 293–306 (1985)
4. Samet, H.: Foundations of Multidimensional and Metric Data Structures. Morgan Kaufmann Publishers Inc., San Francisco (2005)
5. Cormode, G., Garofalakis, M., Haas, P.J., Jermaine, C.: Synopses for massive data: samples, histograms, wavelets, sketches. Found. Trends Databases 4(1–3), 1–294 (2012)

6. Chakrabarti, K., Garofalakis, M.N., Rastogi, R., Shim, K.: Approximate query processing using wavelets. In: Proceedings of VLDB, pp. 111–122 (2000)
7. Ioannidis, Y.E., Poosala, V.: Histogram-based approximation of set-valued query-answers. In: Proceedings of VLDB, pp. 174–185 (1999)
8. Poosala, V., Ganti, V.: Fast approximate answers to aggregate queries on a data cube. In: Proceedings of SSDBM, pp. 24–33 (1999)
9. Gunopulos, D., Kollios, G., Tsotras, V.J., Domeniconi, C.: Approximating multi-dimensional aggregate range queries over real attributes. In: Proceedings of SIGMOD, pp. 463–474 (2000)
10. Chaudhuri, S., Das, G., Narasayya, V.: Optimized stratified sampling for approximate query processing. ACM TODS **32**(2), 1–50 (2007)
11. Gibbons, P.B., Poosala, V., Acharya, S., Bartal, Y., Matias, Y., Muthukrishnan, S., Ramaswamy, S., Suel, T.: Aqua: system and techniques for approximate query answering. Bell Labs Technical report (1998)

Hybrid Web Service Discovery
Based on Fuzzy Condorcet Aggregation

Hadjila Fethallah[✉], Belabed Amine, and Halfaoui Amel

Computer Science Department Informatics Research Laboratory,
Université de Abou Bekr Belkaid Temcen, Tlemcen, Algeria
{f_hadjila,belabed.amine,
a_halfaoui}@mail.univ-tlemcen.dz

Abstract. The web service discovery is a major issue in service oriented computing. We distinguish several semantic service matchmakers that cover multiple matching criteria, these approaches may use crisp logic-based matching, token based similarity measures, and eventually machine learning that combines many individual rankings into a global matching list. This latter category aims to boost the discovery performance by using various matching algorithms, but it introduces additional difficulties, such as the weighting of the matching algorithms components, and the compensation between the matching criteria(inputs, outputs, pre-conditions, effects....). The purpose of this paper is to handle the aforementioned difficulties by introducing a majority vote based approach. This technique fuses five individual rankings (four textual similarity measures and a pure logic matching algorithm) into a global ranking. The different scores are aggregated according to the condorcet principle. More specifically we use a fuzzy dominance relationship, to compare the services, and thus we infer the condorcet order of the final ranking. We have tested our approach, on the OWLSTC benchmark, and the preliminary results are very encouraging.

Keywords: Web service discovery · Service matching · Fuzzy dominance · Rank aggregation · Theory of social choice

1 Introduction

The web service technology is considered as an ideal scheme for realizing the interoperability and the integration of distributed applications. To fulfill this aim, the community has developed many standard, such as SOAP, UDDI, WSDL, [7] and BPEL [20]. Due to the rapid increase in the number of services over the internet, service matchmaking has been an active area of research during the last years. One of the major concerns, is to assess and rank services that fulfill (partially/or globally) a given request. More specifically, we need an efficient and effective approach for retrieving the top-k services according to several matching criteria.

In what follows, we present an example that clearly shows the different difficulties involved in this issue.

Let us consider a user request that consists of a set of input concepts Pin1, Pin2... and output concepts Pout1, Pout2...., (for the sake of simplicity we neglect the other parameters such as preconditions, effects, paths...). To fulfill this request, we can use

© Springer International Publishing Switzerland 2015
T. Morzy et al. (Eds.): ADBIS 2015, LNCS 9282, pp. 415–427, 2015.
DOI: 10.1007/978-3-319-23135-8_28

several matchmaking algorithms or similarity functions termed f1...fn, each function is applied on set of parameters such as inputs or outputs.

Let RP be the request' parameters set, i.e. RP = {Pin1,Pin2,... Pout1,Pout2....}. Each matchmaking function f_j matches the advertised services against the request (i.e. the set of parameters RP). For each service S, if more than one parameter matches a requested parameter, the most appropriate match is retained. More formally we match the pair < RP, S > as follows:

$$\forall \, p_i \in RP, \; match(p_i, S) = MAX_{l=1}^{|S|} \, f_j(p_i, \, p'_l) \tag{1}$$

Such that p'$_l$ is an input/output parameter belonging to S.

Each cell of the Table 1, indicates a matching score denoted: match(p$_i$,S).

For more simplification, we also suppose, that all the services have a single input Pin and a single output Pout, the same assumption is considered for the request (see Table 1). In order to design an effective discovery system, we have to select a consistent matchmaking algorithm.

In practice, there is no ideal approach that always gives the right degree of Match. In fact, we discern different types of semantic matchmakers: we have logic approaches, that leverage reasoning techniques such as subsumption test, non-logic approaches that use information retrieval techniques, and datamining to assess the similarity between the request and the services, and hybrid approaches that combine the previous categories in order to enhance the retrieval system performance. This class, may also apply machine learning, so as to make the final decision.

To leverage the advantages of the proposed matching techniques, and thus boost the performances, we suggest the use of a set of matching algorithms. We refer to this suggestion as Sug1. Hence, the proposed approach belongs to the hybrid class.

The second key component of the discovery system consists in choosing the aggregating mechanism of the partial matching scores, a partial matching score measures the closeness degree between a single parameter of the request and a single parameter of the advertised service. To compute the global score (that handles all the parameters), we adopt a majority vote approach, and more specifically we compute the ranking according to the Condorcet principle.

To this end, we propose a fuzzy relationship to compare the services. This latter is a fuzzified version of the pareto dominance relation. We recall that, the condorcet winner

Table 1. Partial matching scores of services

Services	Parameter	f1	f2	f3
X	Pin	0.88	0.82	0.89
	Pout	0.84	0.85	0.79
Y	Pin	0.85	0.84	0.87
	Pout	0.83	0.78	0.80
W	Pin	0.75	0.66	0.79
	Pout	0.82	0.67	0.71
Z	Pin	0.70	0.61	0.80
	Pout	0.76	0.70	0.74

of a set of candidate objects, is the one that beats or ties with every other candidate in a pair-wise comparison. In general, we notice that, several existing aggregating schemes are not suitable for service matchmaking, the major reasons are:

(1) First, these techniques allow the compensation between parameters and thus, they entail an information loss. More specifically, the fact of averaging the partial scores (before comparison), may cause an erroneous ranking.
(2) Second, these techniques cannot pick up the tradeoffs given by the advertised services, for example if the user focuses on services having medium output scores and higher input scores, then the averaging process is not proper, because it smoothes the final scores.

Thus, we propose in this paper, the use of all partial matching scores during the comparison and without averaging them. After that, we sum the comparison results, to get a final decision. This suggestion is referred to as Sug2.

Our aim is to propose a hybrid matching algorithm, which takes into account the aforementioned suggestions. More specifically, we adopt a fuzzy version of the dominance relationship, to compare the services.

The example revealed in Table 1, presents a user request and four advertised services. These services are evaluated according to m different matchmaking functions f1, f2,f3...., We also observe, that the matching degrees between two particular parameters (for instance the service output and the request output) are non-deterministic and seem to be stochastic, For instance, according to f1, W dominates Z, however, according to f3, Z is better than W. To deal with this kind of ambiguity, we introduce the fuzzy dominance score which measures the extent to which a service S1 dominates another service S2. This fuzzy score is used for comparing and ranking the services.

The main ideas of our proposition are given below:

1. We propose a Condorcet based ranking of the advertised services, that takes into account all the matching parameters (inputs, outputs,....) and all the matching algorithms (f1,f2,....).
2. To implement the fuzzy version of the Condorcet approach, we propose a fuzzy relation, so as to compare the services, and consequently we can compute the Condorcet winner of an answer set.

Our choice is mainly motivated by May's theorem, which states that in the case of a two candidate election, "majority voting is the only method that is anonymous (equal consideration of all voters), neutral (equal consideration of the candidates), and monotonic (more support for a candidate does not jeopardize its election)" [19]. The rest of this paper is organized as follows: the Sect. 2 presents the related work, the third section introduces the developed approach, the fourth section shows the results and finally we present in the fifth section our conclusions.

2   State of the Art

We focus in this section on two areas: the web service discovery as well as the data fusion problem. For each of them we present the existing approaches.

Semantic Web Service Discovery: We discern 03 types of semantic approaches. The first class is based on pure logic matchmaking, and more specifically these algorithms use subsumption test, or consistence test to establish the relatedness between a request and a service.

In OWLMX [14], the authors propose 04 discrete scores for OWLS web services, The major drawback of this class is the high rate of false positives and false negatives, and the exponential complexity [2].

For this reason the research community has developed another kind of algorithms which are based on non-logic approaches such as, graph matching, datamining, optimization, probabilistic matching, information retrieval mechanisms… [11, 21, 23].

In [5] the authors use Probabilistic Latent Semantic Analysis (PLSA), [12] and Latent Dirichlet Allocation (LDA) [4] to extract latent factors from semantic service descriptions and retrieve services in latent factor space. Each service is represented as a probability distribution over latent factors. A latent factor represents a group of concepts and more formally it is a probability distribution over semantic concepts. In URBE [22], the researchers use different service descriptors to assess the similarity between the request and the services. They leverage a path based similarity for measuring the closeness between concepts. In addition to that, they use mixed integer programming to find the optimal matching.

The third category [13, 15] combines the previous types, in order to improve the performances. More specifically, they aggregate the scores given by the individual matching algorithms, so as to provide a global score. The aggregating scheme can be static such as [15] or dynamic (or adaptive) such as [16, 24]. The static case, means that the individual scores have a fixed priority (or the scores' ranking is static), for instance in OWLMX [14], or OWLMX2 [15] the logic score "exact" is always ranked before the textual similarity scores. In contrast, to the aforementioned systems, the adaptive approaches leverage machine learning, to boost the performances. For instance OWLMX3 [16], uses SVM to decide whether the advertised service is proper to the request or not. In [24] the authors propose three metrics for retrieving and sorting web services, each of them uses a set of matching algorithms (logic, textual similarities..), so as to provide a combined ranking.

Data Fusion: The rank aggregation (or data fusion) aims to build a global ranking from a set of ranked lists of objects (or services/documents). These lists are given by different search engines. Even if the problem seems to be simple, it is worth mentioning that, the retrieval of the optimal combined ranking is NP-hard [8] under certain constraints.

Several aggregating techniques have sprung up [9, 10, 18], in particular we notice four categories of data fusion techniques [1]. The hierarchy is based on two criteria:

(1)   The presence/absence of the relevance scores
(2)   The presence/absence of machine learning

If the scores are given by the individual ranking algorithms, the combined method leverages them to build the novel order, for instance the work presented in [1] follows this policy. Since almost no search engine provides the ranking scores, it is possible to convert local ranks into relevance scores. In CombSUM [10], the combined relevance score of a document is the sum of the scores affected by each input ranking. Likewise the CombANZ (CombMNZ) systems [10], compute the final score of a document, in similar fashion with CombSUM, except that we divide (multiply) the score by the number of rankings in which the document appears. According to [17], the CombMNZ approach provides the highest search quality.

If the scores are not present (for rank aggregation), we use only the order of the input ranking [1].We call these approaches order based methods or positional methods. The first positional method is the Borda-fuse model [1]. For each document, it affects as score the summation of its rank (position) in each list.

The Condorcet-fuse approach [18], leverages a majoritarian voting model. More specifically, a document d1 is ranked before another document d2 in the fused list, if d1 is ranked before d2 more times than d2 is ranked before d1. The outranking approach [9], leverages the majoritarian model by introducing several kinds of thresholds.

3 Web Service Retrieval and Ranking

3.1 Problem Statement

Before introducing the developed contribution, we begin by formalizing the concept of top K dominating web services under multiple matching functions.

First of all we suppose that, each service (or request) is represented as a vector containing the inputs/outputs parameters. A matching degree between a parameter of the request RP = $\{p_1 \ldots p_d,\}$ and a service S, under the matching function f_j is referred to as: $match_j(p_i,S)$.

Since we have d requested parameters, we obtain d matching scores, each of them belongs to [0,1]. Each function f_j provides a partially ordered set of services, these elements are labeled with a vector of matching degrees having d dimensions. This vector is called matching instance. It is denoted V_{qj}. q represents the service identifier and j represents the matching function identifier.

Simply speaking, the partially ranked list of f_j is represented as follows:
$PRL_j = <(S_1,V_{1j}), (S_2,V_{2j}),\ldots (S_{|base|},V_{|base|j}) >$, each $V_{qj} \in [0,1]*[0,1]*\ldots[0,1]$.

For instance, if the user's request has a single input and a single output, then, $V_{qj} \in [0,1]*[0,1]$.

Our purpose is to construct a fused (combined) list FL from a set of partially ranked lists PRL1 ….PRLm, (given by m matching functions), such that the FL's Top-K elements, have the best precision rate and the best recall rate. We also notice that, the implemented fusion technique, must take into account the suggestions Sug1 and Sug 2.

In order to satisfy the requirement Sug2, we use a fuzzy version of the dominance relationship, we adapt the function proposed by [3] to our context. We notice that, the (fuzzy) dominance function, compares (simultaneously) in a pairwise manner all the vector components of the matching instances, hence, the contribution largely fulfills the second suggestion Sug2.

In what follows we present the pareto dominance and its fuzzified version.

3.2 Pareto Dominance

Let us consider two d-dimensional vectors u and v, we say that u dominates v, i.e. u > v, if and only if u is at least as good as v in all dimensions and (strictly) better than v in at least one dimension, i.e., $\forall i \in [1, d]$, $u_i \geq v_i \wedge \exists j \in [1, d]$, $u_j > v_j$.

Form the definition, we observe that, the Pareto dominance is not always decisive in comparing objects, in fact, it does not permit the differentiation between vectors with a large variance, i.e., vectors which are very good in some dimensions and mediocre in other ones (e.g., (0.95, 0.1) and (0.80, 0)) and good vectors, i.e., points that are (in general) adequate in all dimensions (e.g., (0.8, 1) and (1, 0.7)). To explain this fact, let us consider u = (u1, u2) = (0.95, 0.2) and v = (v1, v2) = (0.85, 0.9). According to the Pareto principle, we have neither u > v nor v > u, i.e., the instances u and v are incomparable. However, we can say that v is more appropriate than u since $v_2 = 0.9$ is bigger than $u_2 = 0.2$, while $v_1 = 0.85$ is very close to $u_1 = 0.95$.

To take advantage of this observation, we propose a fuzzy version of the Pareto dominance relationship to express the dominance degree between two matching instances, i.e. we assess, the extent to which a matching instance vector (more or less) dominates another one.

3.3 Fuzzy Dominance

Given two d-dimensional vectors u and v, the fuzzy dominance function adapted from [3] computes the extent to which u dominates v:

$$FD(u, v) = (1/d) \sum_{i=1}^{d} EFD(u_i, v_i) \tag{2}$$

Where the elementary fuzzy dominance EFD(x,y) is defined as follows:

$$EFD(x, y) = \begin{cases} 0 & \text{if } (x - y) \leq \varepsilon \\ (x - y - \varepsilon) & \text{Otherwise} \end{cases} \tag{3}$$

Where, $\varepsilon \in [0,1]$, this parameter is chosen by an expert or empirically. ε allows the control of the dominance score between x and y. In this version, we neglect the additional parameter λ proposed by [3], because the experiments do not show a high importance of the parameter λ on the result's quality. EFD constitutes a membership function of the fuzzy relation "x dominate y". It expresses the strength of the relation" x is more or less greater than y". The dominance function given in formula 2, is simply an average value of the different elementary fuzzy dominance degrees.

3.4 Service Comparison

Since we have various relevance scores for each service (i.e., the matching scores given by the different matching functions), we should design a comparison operator that takes into account all of them (i.e.,we accomplish the first requirement Sug1). In what follows, we elucidate our idea by considering a simple scenario. Let us consider two services S_1 and S_2, such that each of them is characterized by a set of matching instances derived from the m matching functions f_1 f_2 ....f_m. First, we compute the extent to which S1 is better than S_2, by adopting the following formula:

$$\text{Dominance} - \text{degree}(S_1, S_2) = (1/m) \sum\nolimits_{j=1}^{m} (1/m) \sum\nolimits_{k=1}^{m} FD(V_{1j}, V_{2k}) \quad (4)$$

Where V_{1j} is the J th matching instance of S_1
Where V_{2k} is the K th matching instance of S_2
Informally, this function computes the average dominance score between the instances of S_1 and the instances of S_2.

Second, we compute the extent to which S_2 is better than S_1, by adopting the same formula:

$$\text{Dominance} - \text{degree}(S_2, S_1) = (1/m) \sum\nolimits_{j=1}^{m} (1/m) \sum\nolimits_{k=1}^{m} FD(V_{2j}, V_{1k})$$

Third, we rank the services according to the obtained scores:
If Dominance-degree(S_1,S_2) > Dominance-degree(S_2,S_1) then S_1 is ranked before S_2, and this means that S_1 is better than S_2.
If Dominance-degree(S_1,S_2) < Dominance-degree(S_2,S_1) then S_2 is ranked before S_1, and this means that S_2 is better than S_1
If Dominance-degree(S_1,S_2) = Dominance-degree(S_2,S_1) then S_1 and S_2 have same rank, and this means that S_2 and S_1 are equivalents ($S_1 \equiv S_2$).
To clarify the comparison mechanism, we consider the following example: For the sake of simplicity, we handle only two parameters: one input and one output and 03 matching functions f_1 f_2 f_3 (i.e.03 matching instances per service), we also set ε to 0.

S_1: {(1,0.6),(0.4,0.5),(0.7, 0.3)}, S_2:{(0.7,0.7),(0.5,0.5),(0.6,0.7)}
Dominance-degree(S_1,S_2)=

1/3.[1/3.[(FD((1,0.6),(0.7,0.7)) + FD((1,0.6), (0.5,0.5)) + FD((1,0.6), (0.6,0.7))] +
1/3.[(FD((0.4,0.5),(0.7,0.7)) + FD((0.4,0.5), (0.5,0.5)) + FD((0.4,0.5), (0.6,0.7))] +
1/3.[(FD((0.7,0.3),(0.7,0.7)) + FD((0.7,0.3), (0.5,0.5)) + FD((0.7,0.3), (0.6,0.7))]] =
0.0888.

In a similar manner: Dominance-degree (S_2,S_1) = 0.1222. Since the second score is bigger than the first score, we infer that S2 is better than S1 in the aggregated list.

3.5 Fuzzy Ranking Algorithm

Our proposition, referred to as FCAA (Fuzzy Condorcet Aggregating Algorithm), computes the top-k Web services according to the Condorcet ranking principle. The main ide a is to build a graph, such that, the nodes represent the services, and the edges represent the relation "is better than". If two services are equivalents then, we put two edges S → S' and S'→ S.

The major problem of the Condorcet based ranking, is the presence of cycles. Several solutions are proposed to this issue, for instance, we may simply, view cycles as ties. In [18] the local ordering of objects within a cycle is only of secondary importance, while their ordering with respect to the rest of the candidates is a prime priority. In this contribution, we rank the services within a cycle, by using a simple heuristic. This latter favors the services having the smallest variance, in other words if we have for instance two equivalent services S and S' such that S has two matching instances (a1,b1), (a2,b2) and S' has two matching instances (a'1,b'1), (a'2,b'2), then

Mean(S) = ((a1 + a2)/2, (b1 + b2)/2) we denote this mean by (a*,b*)
Mean(S') = ((a'1 + a'2)/2, (b'1 + b'2)/2) we denote this mean by (a'*,b'*)
Variance(S) = Variance1(S) + variance2(S), where:
Variance1(S) = $1/2.[(a1-a*)^2 + (a2-a*)^2]$
Variance2(S) = $1/2.[(b1-b*)^2 + (b2-b*)^2]$

In a similar manner, we define Variance(S'), and hence S is better than S' if and only if Variance(S) < Variance(S'). The idea behind this heuristic is that we want to privilege the services that have consistent scores in all the matching algorithms. This policy will ensure more chances to retain reliable services. To infer the aggregated ranking, we simply search a hamiltonian traversal of the constructed graph. This task can be done in a polynomial time [18]. In what follows, we present the pseudo code of the proposed ranking algorithm:

Algorithm : FuzzyCondorcetAggregatingAlgorithm

Inputs: a set of input source lists RPL1,.... RPLm

Output: Top-K(Global-Ranking)

1   **For each** input list RPLj **Do**

2 | RPLj'= prune (RPLj)

3   **End**

4   Initialize C= (RPL1')∪(RPL2')∪ ...∪(RPLm').

5   CondorcetGraph=(C,null) // the edges are not initialized.

6   **For each** service S_i belonging to C **Do**

7 | **For each** service S_j belonging to C and different from S_i **Do**

8 | **If** (S_i is better than S_j) **then** add the edge (i,j) in CondorcetGraph **End**

9 | **If** (S_j is better than S_i) **then** add the edge (j,i) in CondorcetGraph **End**

10 | **If** ($S_i \equiv S_j$) **then** add the edges (i,j), and (j,i) in CondorcetGraph **End**

11 | **End**

12 **End**

13 Global- Ranking = Hamiltonian-Path(CondorcetGraph).

14 Result=Top-K(Global- Ranking)

15 **Return** Result

The explanation of the algorithm is given below:

(1) (Lines 1,2,3), we remove all the services where the matching scores (for inputs and outputs) are 0 in all the matching algorithms. This phase will alleviate the time complexity of the ordering. We discard all the services that have a matching instance = (0,0....0), in all the matching functions.
(2) (Line 4), we fuse the pruned source lists in the set C.
(3) (Line 5) the graph is created by considering the elements of C as nodes, and the edges are inserted according to the dominance degree, initially the edges are empty.
(4) (Lines 6..12), we build the Condorcet graph by leveraging the fuzzy dominance relationship.
(5) (Lines 13), we construct the hamiltonian path of the Condorcet graph (see Fig. 1), this task is done by counting the out-degree of each service(or node), the root of the hamiltonian graph is the node having an out degree = |C|-1, the other nodes of the hamiltonian path are computed with an almost similar manner.
(6) (Lines 14..15), we extract the Top-K dominating services from the hamiltonian path, for instance, in Fig. 1 Top-2(Global- Ranking) = <S19,S4>

We notice that Fig. 1 shows only the edges connecting two consecutive levels, or nodes belonging to the same level. In our work, we use conjointly five matching functions: four of them are similarity measures, more specifically we adopt: the loss-of-information measure (IL), the extended Jaccard similarity (EJ), cosine similarity (Cosine), and Jensen-Shannon information divergence based similarity (JS). The last matching function (Logic) is a pure logic-based reasoning. This latter uses the subsumption test, for making the decision, it gives fives scores: Exact, Plugin, Subsume, SubsumedBy, and Fail. All these algorithms are presented in detail in [14]. These similarity measures are preferred in this work, because they represent the most promising techniques in information retrieval [14].

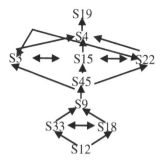

Fig. 1. Global ranking

4 Experimental Study

We conduct a set of experiments to assess the effectiveness and the efficiency of the proposal, the used test collection OWLTC V2.2, (http://www-ags.dfki.uni-sb.de/ ~klusch/owls-mx/) involve real-world Web service descriptions, extracted principally from public IBM UDDI registries. The data set contains: (a) 1007 service descriptions, (b) 29 sample requests, and (c) a manually identified relevance set for each request. We use 19 requests for assessing the approach quality (precision and recall).

In order to compare the efficiency, we measure, the mean execution time of the different matching algorithms. All the algorithms were implemented in Java, and the experiments were conducted on a Core I3 GHz machine with 4 GB of RAM, running Window7.

In this, study we have implemented five matching functions and one fusion algorithm, which are respectively: Extended Jaccard similarity measure (EJ), Information Loss similarity measure (IL), jenson shanon similarity measure (JS), cosine similarity measure (cosine), logic matching (Logic), and Fuzzy Condorcet Aggregating Algorithm (FCAA).

We also compare our aggregating algorithm with the results given by the Borda fuse model, we assume that the Borda's experiments use a set of source lists having 50 elements. To choose the best value of ε, we have conducted several simulations of the Fuzzy Condorcet Aggregating Algorithm, the best values belong to [0, 0.09], in what follows we take ε = 0.05.

As demonstrated in Fig. 2, FCAA presents a little overhead, in comparison with the 05 matching algorithms. This cost is mainly due to the comparison task (see lines 8, 9,10), The fast algorithm is the logic approach, this latter is based on an ontological encoding technique [6] for reducing the execution time of the subsume test. The rest of algorithms have almost similar execution times. We also observe that fact of pruning the input ranked lists (see step 2), enables the reduction of the computational cost, because the entire data set is replaced with a small subset C.

Fig. 2. Average execution time

We also notice that, the time complexity of the dominance-degree (formula 4) is $O(d.m^2)$. Because we need to perform m^2 elementary comparisons and each of them requires d actions. To rank the services we must execute the dominance degree twice for each pair (S,S') (we neglect for the moment the equivalence case), and since we have $|C|$ services to rank, then each service is compared with $|C|-1$ elements. Finally, to build the Condorcet graph and extract the Hamiltonian path, we need $O(|C|^2.d.m^2)$ steps. It obvious that this polynomial complexity enables a high efficiency for large data sets.

The Tables 2 and 3, show the precision and the recall associated to the TOP 10 up to TOP 60 answers. In general, the results indicate that FCAA is more effective than the other algorithms, in terms of both criteria. Furthermore the FCAA superiority is independent from the list size (K).

Since the Borda approach is very sensitive to the service ranks in the source lists, its average recall is very weak in comparison with FCAA (especially for high values of k). Likewise the precision superiority of the proposed approach is more apparent for k = 20 and k = 30. These experiments confirm the effectiveness of the Condorcet voting model.

The Table 4 provides the details that concern the R-precision. This criterion represents a special execution scenario, in which the recall and the precision are equal.

According to the results, FCAA is better than the individual matching functions and the Borda fuse model, we also notice that the information loss gives an almost similar performance.

The only case where FCAA will give a poor performance, is described below:

Table 2. Average precision for 19 requests

Function	Top 10	Top 20	Top 30	Top 40	Top 50	Top 60
EJ	0,81	0,64	0,58	0,51	0,42	0,36
IL	0,81	0,64	0,58	0,5	0,42	0,36
JS	0,8	0,65	0,57	0,49	0,41	0,36
LOGIC	0,73	0,53	0,48	0,42	0,37	0,31
COSINE	0,81	0,64	0,57	0,48	0,4	0,36
BORDA	0,83	0,67	0,58	0,5	0,42	0,38
FCAA	**0,83**	**0,69**	**0,59**	**0,5**	**0,42**	**0,37**

Table 3. Average recall for 19 requests

Function	Top 10	Top 20	Top 30	Top 40	Top 50	Top 60
EJ	0,33	0,59	0,73	0,79	0,83	0,85
IL	0,33	0,59	0,74	0,79	0,84	0,86
JS	0,33	0,59	0,72	0,79	0,83	0,86
LOGIC	0,3	0,46	0,6	0,66	0,72	0,69
COSINE	0,33	0,59	0,72	0,78	0,82	0,86
BORDA	0,39	0,58	0,73	0,81	0,84	0,9
FCAA	**0,4**	**0,61**	**0,77**	**0,84**	**0,87**	**0,9**

Table 4. Average R-precision for 19 requests

Function	R-precision
EJ	0,697
IL	0,704
JS	0,696
LOGIC	0,562
COSINE	0,693
BORDA	0,70
FCAA	**0,709**

– When the majority of the matching functions commits an error about a good service denoted S* and they ranked it beyond the first K elements, then the FCAA will not correct the mistake because it is based on the majority vote principle.
– In this situation FCAA is not able to outperform the good matching algorithms.

We notice that these cases are very rare, because the majority of the similarity measures are very effective in the Top 30 services (see Tables 2 and 3).

5   Conclusion

In this paper, we have proposed an aggregating algorithm based on Condorcet principle for retrieving and ranking web services. Our proposition takes into account several parameters as well as various matching functions during the ranking process. Our contribution leverages five matching functions: four similarity measures and one logic matching approach. The experimental results show that the proposed algorithm is quite effective and efficient.

There are several directions for future work. First, we can learn additional parameters such as ε and the number of input lists that participate in the election process, second we can add additional matching functions, especially edge based similarities, or even kernel based similarities. These directions may ensure more chances to pick up the most relevant services.

References

1. Aslam, J.A., Montague, M.H.: Models for metasearch. In: SIGIR, pp. 275–284 (2001)
2. Baader, F., Sattler, U.: An overview of tableau algorithms for description logics. Stud. Logica **69**, 5–40 (2001)
3. Benouaret, K.: Advanced techniques for web service query optimization. Ph.D. thesis in Computer Science. Université Claude Bernard Lyon1 (2012)
4. Blei, D.M., Ng, A.Y., Jordan, M.I.: Latent dirichlet allocation. J. Mach. Learn. Res. **3**, 993–1022 (2003)
5. Cassar, G., Barnaghi, P., Moessner, K.: Probabilistic matchmaking methods for automated service discovery. IEEE Trans. Serv. Comput. J. **7**(4), 1 (2013)

6. Caseau, Y., Habib, M., Nourine, L., Raynaud, O.: Encoding of multiple inheritance hierarchies and partial orders. Comput. Intell. **15**, 50–62 (1999)
7. Curbera, F., Duftler, F., Khalaf, R., Nagy, W., Mukhi, N., Weerawarana, S.: Unraveling. the web services web: an introduction to SOAP, WSDL, and UDDI. IEEE Internet Comput. **6** (2), 86–93 (2002)
8. Dwork, C., Kumar, R., Naor, M., Sivakumar, D.: Rank aggregation methods for the web. In: Proceedings of the ACM International Conference on World Wide Web (WWW), pp. 613–622 (2001)
9. Farah, M., Vanderpooten, D.: An outranking approach for rank aggregation in information retrieval. In: SIGIR, pp. 591–598 (2007)
10. Fox, E.A., Shaw, J.A.: Combination of multiple searches. In: 2nd TREC, NIST, pp. 243–252 (1993)
11. Hadjila, F., Chikh, A., Belabed, A.: Automated discovery of web services: an interface matching approach based on similarity measure. In: Proceedings of the 1st International Conference on Intelligent Semantic Web-Services and Applications, ISWSA 2010, pp. 13:1–13:4. ACM, New York (2010)
12. Hofmann, T.: Probabilistic latent semantic analysis. In: Proceedings of Uncertainty in Artificial Intelligence, UAI99, pp. 289–296 (1999)
13. Klusch, M., Kapahnke, P.: Semantic web service selection with SAWSDL-MX. In: CEUR Proceedings of 2nd International Workshop on Service Matchmaking and Resource Retrieval in the Semantic Web (SMR2), Karlsruhe, Germany. CEUR 416 (2008)
14. Klusch, M., Fries, B., Sycara, K.: Automated semantic web service discovery with OWLS-MX. In: Proceedings of 5th International Conference on Autonomous Agents and Multi-Agent Systems (AAMAS), Hakodate, Japan. ACM Press (2006)
15. Klusch, M., Fries, B., Sycara, K.: OWLS-MX: a hybrid semantic web service matchmaker for OWL-S services. Web Semant. **7**(2), 121–133 (2009)
16. Klusch, M., Kapahnke, P.: OWLS-MX3: an adaptive hybrid semantic service matchmaker for OWL-S. In: Proceedings of 3rd International Workshop on Semantic Matchmaking and Resource Retrieval (SMR2) at ISWC, Washington, USA (2009)
17. Lee, J.-H.: Analyses of multiple evidence combination. In: SIGIR, pp. 267–276 (1997)
18. Montague, M.H., Aslam, J.A.: Condorcet fusion for improved retrieval. In: ACM CIKM, pp. 538–548 (2002)
19. Moulin, H.: Axioms of Cooperative Decision Making. Cambridge University Press, Cambridge (1988)
20. OASIS. Web services business process execution language, April 2007. http://docs.oasis-open.org/wsbpel/2.0/wsbpel-v2.0.pdf
21. Platzer, C., Rosenberg, F., Dustdar, S.: Web service clustering using multidimensional angles as proximity measures. ACM Trans. Internet Technol. **9**(3), 1–26 (2009)
22. Plebani, P., Pernici, B.: URBE: web service retrieval based on similarity evaluation. IEEE Trans. Knowl. Data Eng. **21**(11), 1629–1642 (2009)
23. Segev, A., Toch, E.: Context-based matching and ranking of web services for composition. IEEE Trans. Serv. Comput. **99**(PrePrints), 210–222 (2009)
24. Skoutas, D., Sacharidis, D., Simitsis, A., Kantere, V., Sellis, T.: Ranking and clustering web services using multi-criteria dominance relationships. IEEE Trans. Serv. Comput. J. **3**(3), 163–177 (2010)

Confidentiality and Trust

Confidentiality Preserving Evaluation
of Open Relational Queries

Joachim Biskup$^{(\boxtimes)}$, Martin Bring, and Michael Bulinski

Fakultät für Informatik, Technische Universität Dortmund, Dortmund, Germany
{joachim.biskup,martin.bring,michael.bulinski}@cs.tu-dortmund.de

Abstract. Relational database systems may serve to evaluate an open query, returning a relation with all those tuples that satisfy the properties expressed in the query, complemented with an often implicit statement about the completeness of the result. Known inference control procedures for enforcing a confidentiality policy have to inspect by theorem-proving explicit completeness sentences expressed in first-order logic. Unfortunately, experiments indicate that standard theorem provers are not efficient enough to deal with completeness sentences of a larger size. We describe and evaluate approaches to overcome the performance issues by suitably transforming an original completeness sentence and by optimizing the number of prover calls with a completeness sentence involved.

Keywords: Active domain · Apriori knowledge · Binary search · Closed-world assumption · Combined lying and refusal · Confidentiality policy · Constant symbol · Completeness sentence · Dictionary · Enumeration of a domain · Free variable · Inference control · Open query · Theorem-proving

1 Introduction

Inference control for information systems in general and relational databases in particular is a mechanism to confine the information content and thus the usability of data made accessible to a client to whom some piece(s) of information should be kept confidential, see, e.g., [2,6]. Thus inference control aims at protecting *information* rather than just the underlying *data*, as achieved by traditional access control or simple encryption. Though protection of information is a crucial requirement for many applications, the actual enforcement is facing great challenges arising from conceptual and computational problems.

In this work, we focus on the problems arising from controlling *open queries* to a *relational database*, as managed by well-known products of a DBMS complying with the SQL-standard. Basically, given a database schema and corresponding relation instances (sets of tuples) an open query requests to return an answer relation (set of tuples) that contains exactly those tuples that both fit the format

This work has been partially supported by the Deutsche Forschungsgemeinschaft (German Research Council) under grant no. BI 311/12-2.

© Springer International Publishing Switzerland 2015

T. Morzy et al. (Eds.): ADBIS 2015, LNCS 9282, pp. 431–445, 2015.
DOI: 10.1007/978-3-319-23135-8_29

and satisfy the properties expressed in the query. Notably, all tuples fitting the format but not satisfying the properties are not explicitly returned. Rather, the issuer of the query and receiver of the answer is assumed to apply a *closed-world assumption*, which says that each format-fitting tuple not contained in the answer does not satisfy the requested properties. Under the assumption that infinite type extensions (domains) are declared by the schema, there are infinitely many such "negative tuples". Accordingly, controlling an open query necessarily has to identify and, as far as requested by a confidentiality policy, to confine the information supplied by all these tuples.

We will treat the problems raised in a formal approach to relational databases based on first-order logic [1,8], where, e.g., tuples are seen as ground atoms, queries are expressed by formulas which may contain free variables, semantic constraints and other a priori knowledge are specified by sentences, and potential secrets, i.e., elements of a confidentiality policy, are declared by sentences as well.

Example 1. Consider a database db with a relation ill relating patients to illnesses as present for example in hospitals: $db := \{ill(Smith, cancer), ill(Miller, flu), ill(Miller, rheumatism)\}$. Let $psec$ be the confidentiality policy stating that the information of person $Smith$ suffering from $cancer$ should be kept confidential: $psec := \{ill(Smith, cancer)\}$. Additionally the requestor is assumed to have the knowledge $prior$ that person $Smith$ or person $Miller$ is actually suffering from $cancer$: $prior := \{ill(Smith, cancer) \vee ill(Miller, cancer)\}$. Assuming the requestor is interested in the illnesses of person $Miller$, he submits the following open query with a free variable x to the information system: $ill(Miller, x)$.

The answer relation of this query would then consist of the two tuples $ill(Miller, flu)$ and $ill(Miller, rheumatism)$. Applying the closed world assumption yields, among others, the "negative tuple" $\neg ill(Miller, cancer)$ which enables the requestor to infer the confidential information of person $Smith$ suffering from $cancer$, as formally captured by the following entailment:

$$\begin{aligned} \big(ill(Smith, cancer) \vee ill(Miller, cancer)\big) \wedge \\ \neg ill(Miller, cancer) \quad \models \quad ill(Smith, cancer) \end{aligned}$$

Elaborated as part of a specific approach to inference control called Controlled Interaction Execution, see [3,4] for an introduction, Biskup/Bonatti [5] proposed and verified control procedures for open relational queries within a dedicated logic-oriented relational model dealing with different settings of a reaction on detecting harmful information, including *refusal*, *lying* and the *combination* thereof. As already indicated above, suitably *representing* and *handling* the pertinent closed-world assumptions are a most crucial aspect.

Representation is enabled by expressing the information content of the infinitely many "*negative tuples*" by a single *completeness sentence* in first-order logic. Conceptually, handling of completeness sentences is managed in two ways, either in advance by determining a suitable bound for the set of tuples to be explicitly inspected for inclusion into the controlled answer or repeatedly while inspecting tuples for inclusion one after another until the pertinent completeness sentence for the answer becomes true, i.e., all remaining tuples are guaranteed to not satisfying the pertinent properties. For both ways, not only the set of "*positive tuples*" but also the corresponding completeness sentence are explicitly returned to the requestor, and memorized by the control system.

Example 2. As seen in Example 1, the information $\neg ill(Miller, cancer)$ has to be restricted in order to avoid an information flow violating the confidentiality policy. This may lead to the following result of an explicit finite answer relation together with a suitable completeness sentence:

$$\{ill(Miller, flu),\ ill(Miller, rheumatism)\}$$
$$\forall x\ \big[(x \neq flu \wedge x \neq rheumatism \wedge x \neq cancer) \Rightarrow \neg ill(Miller, x)\big]$$

Algorithmic handling of completeness sentences, however, turned out to be a major obstacle to achieve *efficient* and *scalable* controlled query evaluation. To confine information, possible inferences revealing confidential information have to be detected by the control system by employing a *theorem prover*. The difficulties in handling completeness sentences arise in the internal treatment of completeness sentences by theorem provers. In this report we present a detailed description of these difficulties together with some approaches to overcome them. More specifically, focussing on the combination of refusal and lying, we will

- summarize the basic control procedures for open queries (Sect. 2);
- outline an implementation of those control procedures (Sect. 3);
- enhance theorem-proving with a completeness sentence (Sect. 4);
- optimize the number of pertinent prover calls (Sect. 5).

2   Basic Control Procedures for Open Queries

Regarding open queries there are three basic control procedures corresponding to the three possible reaction types, namely refusal, lying and the combination thereof, see [5] for details. The basic strategy of all of these procedures is the simulation of an open query by a *sequence* of closed queries. These *closed queries* result from substituting the free variables by domain elements in all possible ways. The order of inspecting tuples corresponding to the closed queries is determined by a *fixed enumeration sequence* of the underlying *domains* of the free variables contained in the open query to be evaluated. As mentioned in the introductory section, a completeness sentence, formulated as a closed query as well, is employed to detect a suitable bound for terminating the simulation sequence.

Unfortunately, *fixing* the domain enumerations will be *costly* in general, since it requires to explicitly consider a large number of syntactically possible tuples only a small fraction of which might turn out to be actually relevant for a concrete application. However, the fixing is crucially *necessary* for proving that the control procedures satisfy the formal notion of confidentiality preservation used for Controlled Interaction Execution. In more intuitive terms, this notion guarantees that a user always receives answers to his queries such that, from his point of view, for each element of the confidentiality policy the answers could be explained to result from relation instances in which the policy element is not valid. In other words, for the user such possible "harmless" instances are *indistinguishable* from the hidden actual instances (in which the policy element might be valid or not).

In the basic control procedure of the *refusal* approach the tuples are treated stepwise one after another. In each step the current tuple and a completeness sentence – stating that all further tuples will belong to the set of "negative tuples" – are inspected concerning their compliance with the confidentiality policy. If the *current tuple* does not violate confidential information, it is added to the set of "positive tuples" or the set of "negative tuples", respectively, according to its containment in the underlying database instance; otherwise, the current tuple is added to a third collection of "refused answers". Analogously, if the *completeness sentence* does not violate any confidential information, it is returned to the requestor according to its validity with respect to the database instance. If the completeness sentence is harmless and true in the underlying database instance then the enumeration is stopped, since all remaining tuples are covered by this completeness sentence. In case of a harmful completeness sentence, it is refused and the enumeration is also stopped, because information about further tuples would be harmful as well.

Whereas the refusal approach obtains the final completeness sentence as a result of inspecting the tuples in a sequential manner, the lying and combined approach determine a suitable bound for considering tuples in advance. However, for lying this bound is only tentative and in need for a later adjustment. For lack of space, we omit all further explanations for the *lying* approach.

Regarding the control procedure of the (so-called alternative[1]) *combined* approach, in the following we outline the essentials of the four-step forward-backward algorithm to determine the final completeness sentence in a first phase; formal details of the algorithm and justifications by means of providing formal arguments in the proofs can be found in [5].

The *first* step determines a position k of the enumeration sequence as the least position that makes the corresponding completeness sentence true in the underlying database instance (but may still violate confidential information). In a *second* forward searching step this position is increased until the corresponding completeness sentence does not violate confidential information anymore yielding a position m. In order to obtain the last "positive tuple" (which might be a lie), by a *third* backward searching step beginning at position m, the position k is then modified to a position k^* satisfying the condition that the last "positive tuple", represented by k^*, together with the completeness sentence belonging to m do not violate confidential information. The choice of k^* may arise the necessity (namely, if $k^* < k$) of adjusting the completeness sentence belonging to m by a *fourth* forward search step starting at k^* and yielding the position m^*. This adjustment results in the final completeness sentence which is specified

---

[1] see Sect. 5.2 of [5].

by the position m^* (possibly also expressing a lie). Notably, the positions k^* and m^* are mutually optimal, i.e., k^* is maximal with respect to m^*, and m^* is minimal with respect to k^*, and they are determined in an instance-independent manner, i.e., without further considering the actual relation instances. The final completeness sentence is returned to the requestor, and the last "positive tuple" is added to the set of "positive tuples".

In a second phase, all tuples in the ranges of $[1,\ k^* - 1]$ and $[k^* + 1,\ m^*]$ are then treated as closed queries and appropriately added to the sets of "positive tuples" and "negative tuples" or discarded as refused, respectively.

As needed for Controlled Interaction Execution in general, the outcome of each step might depend on the results of previous steps, and thus the overall control procedure is *inherently sequential*, offering only limited options of employing parallel computing.

3 An Implementation within the CIE-System

The control procedures for open relational queries outlined in Sect. 2 have been implemented as part of the CIE-System described in [3]. The CIE-System serves as a frontend for the underlying database, which is managed by an Oracle DBMS. To ensure the preservation of confidential information, the view of the client must not contain any confidential information, neither directly nor indirectly in form of possible implications. This requirement is guaranteed by the use of a *theorem prover*. The CIE-System uses the Prover9 [9], alternatively also the E-Prover[2], and provides an interface complying with the TPTP format [10].

At declaration time, the *administrator* has to declare a *dictionary* for each attribute occurring in a relation. A dictionary contains sufficiently many elements of the corresponding domain in a *fixed* order. Since domains are infinite, a dictionary cannot cover all domain elements; accordingly, the CIE-System returns an error message if a dictionary will turn out to be too small.

For a given open query, the *enumeration sequence* for its free variables (which defines the order in which the substituted tuples get inspected) is determined in two steps. First, each variable is *individually* assigned an enumeration sequence: if the variable occurs only in one predicate (relation) and always for the same attribute, then the enumeration sequence is simply obtained by the dictionary for that attribute; otherwise, if the variable occurs multiple times for different attributes, then the dictionaries for these attributes are merged using the zipper method. Second, if there is only *one* free variable, the enumeration sequence determined so far is taken; otherwise, if there are *at least two* free variables, the *joint* enumeration sequence for them is obtained by diagonalization, such that each element of the enumeration sequence contains exactly one constant for each of the free variables. The diagonalization has been implemented by inductively using the Cantor pairing function π and its inverses.

---

[2] http://www.eprover.org.

4   Theorem-Proving with Completeness Sentences

During the evaluation of an open query, a theorem prover is applied to ensure that no confidential information is revealed by the query answer. In the process of theorem-proving, a special challenge arises from the treatment of completeness sentences the formal structure of which is explained in the following.

Let $\Phi(x)$ be a safe open relational query with free variables x and $c_1, \ldots, c_n, c_{n+1}, \ldots$ the fixed enumeration sequence of the infinite combined domains of x computed as described in Sect. 3. An answer of an open query generally consists of a positive and a negative part. The *positive part* is given by a set of tuples $\Phi(c_{i_1}), \ldots, \Phi(c_{i_k})$ where the free variables of the open query are substituted by the elements $c_{i_1}, \ldots, c_{i_k}$ of the enumeration sequence. The *negative part* is covered by a completeness sentence stating that all further substitutions of the open query lead to "negative tuples":

$$Compl(\Phi(x), \{c_{i_1}, \ldots, c_{i_k}\}) \equiv \forall x \left[\, x = c_{i_1} \vee \cdots \vee x = c_{i_k} \vee \neg\Phi(x)\,\right] \quad (1)$$

While computing the answer of an open query, each enumeration step $n \in \mathbf{N}$ has to deal with a completeness sentence of the following kind:

$$Compl(\Phi(x), n) \equiv \forall x \left[\, x = c_1 \vee \cdots \vee x = c_n \vee \neg\Phi(x)\,\right] \quad (2)$$

It states the following: all further substitutions, which are different from each of the substitutions already considered before, will lead to "negative tuples". Notably, in both sentences a disjunct of the form $x = c_i$ actually consists of a *conjunction*, namely $x_1 = c_{i,1} \wedge \ldots \wedge x_w = c_{i,w}$, where $x = (x_1, \ldots, x_w)$ and $c_i = (c_{i,1}, \ldots, c_{i,w})$, relating each of the free variables of x to the corresponding constant of the enumeration sequence element c_i.

The fundamental problem of theorem-proving in the presence of a completeness sentence is the performance: it drastically degrades when (i) there is more than one free variable and (ii) the size of the completeness sentence, i.e., the number of excluded elements of the enumeration sequence, increases. One cause of this performance issue is the *conversion* of the completeness sentence to *conjunctive normal form* which is done internally by the theorem provers.

More specifically, the above presented completeness sentence basically consists of one big disjunction the operands of which are conjunctions of equality formulas. Thus, the original completeness sentence is a formula in *disjunctive normal form*, resulting in an exponential blow up during the conversion to conjunctive normal form. This observation particularly applies for completeness sentences covering formulas with more than one free variable. In case of just one free variable, for each excluded element of the enumeration sequence there is only one equality subformula needed, which can directly be added as operand of the overall disjunction. To treat the performance issue two workarounds have been developed which are described below.

4.1   Employing the Active Domain of a Completeness Sentence

The first workaround is based on the idea of an *active domain* of a completeness sentence w.r.t. the constants of the excluded combinations and an explicit listing of all "negative tuples" which are covered by the completeness sentence and consist solely of elements of the domain of the completeness sentence.

Let $Compl(\Phi(\boldsymbol{x}), n)$ be a completeness sentence of the form presented in equation (2) where each $\boldsymbol{c_i}$ ("excluded combination") is a vector consisting of w constants denoted as $(c_{i,1}, \ldots, c_{i,w})$. The active domain of $Compl(\Phi(\boldsymbol{x}), n)$ w.r.t. the constants of excluded combinations is then defined as

$$active\_domain\_cs := \{c_{1,1}, \ \ldots \ , c_{1,w}, \quad \ldots\ldots \quad , c_{n,1}, \ \ldots \ , c_{n,w}\}. \quad (3)$$

For describing this active domain of the completeness sentence the additional predicate ad is introduced together with the following formulas:

$$ad(c_{1,1}), \ \ldots \ , ad(c_{1,w}), \quad \ldots\ldots \quad , ad(c_{n,1}), \ \ldots \ , ad(c_{n,w}) \quad (4)$$

$$\forall z[\neg ad(z) \lor z = c_{1,1} \lor \ldots \lor z = c_{1,w} \lor \ldots \lor z = c_{n,1} \lor \ldots \lor z = c_{n,w}] \quad (5)$$

The formulas in (4) define which constants comprise the active domain of the completeness sentence and the formula (5) states the completeness of the domain by expressing that all constants which differ from the constants covered by the formulas in (4) are not part of the active domain.

The original completeness sentence states which substitutions of the free variables of the open query lead to "negative tuples" and therefore being different from the "excluded combinations". The transformation approach splits this set of "negative tuples" described by the original completeness sentence into two parts.

The first part deals with those "negative tuples" that can be constructed by substituting the free variables of the open query only by using constants of the active domain of the completeness sentence. Since the completeness sentence is a finite formula, and thus the set of constants of the active domain of the completeness sentence is finite as well, there are only finitely many "negative tuples" of this form. The idea of the transformation consists of an explicit listing of these "negative tuples", which are represented by the following formulas where w is the number of free variables of the open query and $\boldsymbol{c_i}$ with $i \in \{1, \ldots, n\}$ are the "excluded combinations" (see Eq. (2)):

$\neg\Phi(\boldsymbol{c})$, for each vector $\boldsymbol{c}$ of length w over $active\_domain\_cs$

satisfying for all $i \in \{1, \ldots, n\}$: $\boldsymbol{c} \neq \boldsymbol{c_i}$ ($\boldsymbol{c}$ is "not excluded")   (6)

The second part of "negative tuples" takes into account all further possible substitutions of the free variables of the open query containing at least one constant which is not contained in the active domain of the original completeness sentence. The representation of this second part is based on the following observation: whenever a substitution contains a constant outside of the active domain of the original completeness sentence, then this substitution cannot coincide with an "excluded combination". For, otherwise, this constant would be part of the active domain of the original completeness sentence as well. This observation leads to the following formulas for representing the second part of "negative tuples" where the vector $\boldsymbol{x}$ of free variables of the open query has the form $\boldsymbol{x} = (x_1, \ldots, x_w)$:

$$\forall\boldsymbol{x}\,[ad(x_1) \lor \neg\Phi(\boldsymbol{x})] \qquad \ldots \qquad \forall\boldsymbol{x}\,[ad(x_w) \lor \neg\Phi(\boldsymbol{x})] \quad (7)$$

These formulas state the following: whenever one variable is substituted by a constant that is not part of the active domain of the original completeness sentence then the corresponding formula $\Phi(x)$ is false.

Finally, the formulas (4), (5), (6) and (7) are added to the prover's input in order to replace the original completeness sentence.

Example 3. Let $ill(x, y)$ be a query with free variables x and y and $Compl(ill(x, y), 2) \equiv \forall x, y\,[\,(x = Smith \wedge y = cancer) \vee (x = Miller \wedge y = flu) \vee \neg ill(x, y)\,]$ a completeness sentence. The transformation of $Compl(ill(x, y), 2)$ results in the following formulas (i) for the active domain of the completeness sentence, (ii) describing the explicit "negative tuples", and (iii) referring to "negative tuples" with at least one element outside the active domain, respectively:

$$ad(Smith),\ ad(cancer),\ ad(Miller),\ ad(flu)$$
$$\forall z[\neg ad(z) \vee z = Smith \vee z = cancer \vee z = Miller \vee z = flu]$$

$\neg ill(Smith, Smith)$	$\neg ill(Smith, Miller)$	$\neg ill(Smith, flu)$
$\neg ill(cancer, Smith)$	$\neg ill(cancer, cancer)$	$\neg ill(cancer, Miller)$
$\neg ill(cancer, flu)$	$\neg ill(Miller, Smith)$	$\neg ill(Miller, cancer)$
$\neg ill(Miller, Miller)$	$\neg ill(flu, Smith)$	$\neg ill(flu, cancer)$
$\neg ill(flu, Miller)$	$\neg ill(flu, flu)$	

$$\forall x, y\,[ad(x) \vee \neg ill(x, y)] \qquad \forall x, y\,[ad(y) \vee \neg ill(x, y)]$$

The effect of the described workaround is the avoidance of the costly conversion of the original completeness sentence in conjunctive normal form resulting in a huge improvement in performance. The obvious drawback is the drastically increasing size of the input of the theorem prover by defining the active domain of the original completeness sentence together with explicitly listing lots of "negative tuples" which can be constructed by active domain constants.

4.2   Introducing New Constants

A second workaround is based on the idea of reducing the general case of completeness sentences corresponding to open queries with arbitrary many free variables to completeness sentences similar to the ones corresponding to queries with only one free variable. Therefore, for each excluded combination c_i of the completeness sentence, a *new constant* c_{combi_i} is introduced. To relate the new constants to its origin, the following mapping functions are introduced as well:

$$f_{map_{x_1}}(c_{combi_i}) = c_{i_1} \qquad \cdots \qquad f_{map_{x_w}}(c_{combi_i}) = c_{i_w} \qquad (8)$$

Furthermore, a new predicate *combi* is introduced with the following intended meaning: the predicate is true for exactly the excluded combinations of the original completeness sentence which are denoted by the new constants c_{combi_i}. This is achieved by the following formulas:

$$combi(c_{combi_1}) \wedge \ldots \wedge combi(c_{combi_n}) \tag{9}$$

$$\forall z \, [z = c_{combi_1} \vee \ldots \vee z = c_{combi_n} \vee \neg combi(z)] \tag{10}$$

The reduction of a completeness sentence with multiple variables to a completeness sentence with only one variable is realized by formula (10). To relate the new completeness sentence regarding the new predicate $combi$ back to the original completeness sentence with the formula Φ, the following formula is used:

$$\forall \boldsymbol{x} \, [\exists z \, [combi(z) \wedge f_{map_{x_1}}(z) = x_1 \wedge \ldots \wedge f_{map_{x_w}}(z) = x_w] \vee \neg \Phi(\boldsymbol{x})] \tag{11}$$

Example 4. Let $ill(x, y)$ be a query with free variables x and y and $Compl(ill(x, y), 2) \equiv \forall x, y \, [(x = Smith \wedge y = cancer) \vee (x = Miller \wedge y = flu) \vee \neg ill(x, y)]$ a completeness sentence. The transformation of $Compl(ill(x, y), 2)$ results in the following formulas for (i) the new constants c_{combi_1} and c_{combi_2} with the mapping functions, (ii) the new predicate $combi$ and the new completeness sentence, and (iii) the relationship between new and original completeness sentence, respectively:

$$f_{map_x}(c_{combi_1}) = Smith \qquad f_{map_y}(c_{combi_1}) = cancer$$
$$f_{map_x}(c_{combi_2}) = Miller \qquad f_{map_y}(c_{combi_2}) = flu$$

$$combi(c_{combi_1}), combi(c_{combi_2})$$
$$\forall z \, [z = c_{combi_1} \vee z = c_{combi_2} \vee \neg combi(z)]$$

$$\forall x, y \, [\exists z \, [combi(z) \wedge f_{map_x}(z) = x \wedge f_{map_y}(z) = y] \vee \neg ill(x, y)]$$

The hope of this transformation has been that theorem provers will profit from it. Unfortunately, however, we show below that the theorem provers suffer in finding suitable substitutions for the variables $\boldsymbol{x}$ and z for formula (11) offering in practice no substantial benefit.

4.3   Runtime Evaluation

We have experimentally compared the two ideas for a transformation of the theorem prover's input regarding their runtime. The comparison is based on input instances which comprise exactly *one* completeness sentence with varying size belonging to queries with two free variables. Additionally, formulas for the unique name assumption of the used constants are added to the prover's input.

The task for the theorem prover then consists of proving or refuting the implication of one specific tuple by the completeness sentence. The constants of the tuple are different from the ones used in the completeness sentence. The results of the runtime evaluation are presented in Table 1. They are based on the Prover9 on an Intel Core 2 Duo system using one core at 2.4 GHz.

Table 1. Runtime evaluation of theorem-proving with completeness sentences.

Size ("exclusions")	Without transformation	Active domain transformation	Reduction transformation
50	5 s	0 s	0 s
75	4 min 30 s	0 s	0 s
85	23 min 25 s	0 s	2 s
100	> 5 hrs	0 s	5 s
500	n.a	0 s	44 min 7 s
1,000	n.a	1 s	> 5 hrs
2,000	n.a	6 s	n.a
5,000	n.a	1 min 3 s	n.a
25,000	n.a	42 min 36 s	n.a
50,000	n.a	> 5 hrs	n.a

Handling of completeness sentences without a transformation yields the worst run times, exceeding the limit for a practical employment in an interactive system already for 100 excluded combinations. Fortunately, the *active domain transformation* essentially enhances the performance. However, at sizes of 2,000 and above this transformation becomes impractical as well. The *reduction transformation* outperforms the handling without a transformation as well, but cannot keep up with the active domain transformation.

These results refer to only *one* theorem-proving process, whereas the overall control procedures in general require *multiple, partly sequential* proving tasks. As a rule of thumb, for the lying and combined approaches the overall computing time can be estimated as the runtime of one single theorem-proving process multiplied with the number of "excluded combinations".

4.4 Extension of the Theorem Prover

None of the transformations described so far solves the problem of efficient handling of completeness sentences in theorem-proving tasks completely. A further idea – not been realized yet – is extending a theorem prover by implementing a *built-in* treatment of completeness sentences in an efficient manner.

5 Optimizing the Number of Prover Calls

Aiming at reducing the number of calls to the theorem prover specifically for the combined approach, we developed two related optimizations for the implementation described in Sect. 3. In accordance with the outline of the basic control procedure presented in Sect. 2, this implementation searches for an appropriate completeness sentence and inspects the closed queries generated by substitutions, respectively, in an essentially linear way. Inspired by the benefits of binary searching, we replaced the linear behavior by a processing that tries to cover dynamically determined larger parts of the pertinent problem space.

5.1 Applying the Divide-and-Conquer Heuristic

Regarding searching for an appropriate *completeness sentence*, we have to deal with the 4-step forward-backward search for the positions k, m, k^*, m^* in the underlying enumeration of the domain. Basically, position k is directly calculated as the maximum of all positions of domain elements that lead to a (positive) tuple in the answer relation. In contrast, each of the remaining positions m, k^* and m^* is determined by repeated probing, originally performed linearly with increases or decreases, respectively, of 1, starting with k, m and k^*, respectively.

Now, to speed up step 2 of searching for m, starting at position k with an increase of 1, we first iteratively double the increase until we have found the respective completeness sentence being harmless, possibly for a still non-optimal position. To find the optimal position m, we then start a binary search for the range between the last and the immediately preceding position. The correctness of the optimized search is justified by the fact that for any positions $i < j$ the completeness sentence for j is a weakening of the completeness sentence for i.

For step 3 of searching for k^*, which has to be in the range $[0, m]$, we cannot rely on a corresponding property that would justify a similar optimized searching.

For speeding up step 4 of searching for m^*, which is guaranteed to be either equal to m or in the range $[k^*, k - 1]$, we can proceed either as in step 2 or directly by a binary search in the full range.

Regarding the inspection of the *closed queries generated by substitutions*, we have to evaluate all tuples generated by a substitution with a position in the sequence $\langle 1, \ldots, k^* - 1, k^* + 1, \ldots, m^* \rangle$ in a controlled way by using the original *combined censor for closed queries*. For each such tuple submitted as a query, this censor checks by a *first* call to the theorem prover whether adding the correct answer to the current knowledge would be harmless: if this is the case, the correct answer is returned and the current knowledge is updated accordingly; otherwise, the negated query is checked by a *second* call to the theorem prover and then treated either by returning the negated query as a lie or requiring a refusal. Essentially depending on the results of preceding substitutions as represented in the current knowledge, and thus keeping track of them, conceptually we have to proceed strictly in the given enumeration sequence. Clearly, any optimized procedure returning the same result is equally acceptable.

Our optimization tentatively bundles several adjacent substitutions, say with positions $i, i+1, \ldots, j-1, j$. If there is actually only one position left, i.e., $i = j$, then the original combined censor for closed queries is used for the respective single tuple. Otherwise, if $i < j$, by calling the theorem prover *only once* it is checked whether adding the *conjunction of the correct answers* to the resulting tuples to the current knowledge would be harmless: if this is the case, all these correct answers together are used to update the current knowledge; otherwise, the positions are divided into a left part (of smaller positions) and a right part (of larger positions) and then first the *left lower* part and subsequently the *right upper* part are treated recursively. The recursion is initialized for the range $[1, k^* - 1]$. Subsequently, the remaining range $[k^* + 1, m^*]$ is inspected linearly with the original combined censor for closed queries since, by the choice of k^* and m^*, in that range there is no single correct "positive" tuple that is harmless nor is the conjunction of all "negative" tuples harmless.

Example 5. To simplify the discussion, we consider (the front of length 8 of) an abstract dictionary with positions $1, 2, \ldots, 8$ to be inspected in principle. We assume that only the constant c_3 at position 3 requires a distortion, while for all other positions the correct answer can be returned. The standard linear probing according to the given enumeration sequence of the dictionary performs as follows regarding the number of prover calls: position 3 leads to 2 calls, while all other positions need only 1 call, summing up to 9 calls in total. The optimized approach conceptually traverses an inspection tree in a depth-first manner pruning a subtree once at its top node its range has been proved to not requiring any distortion. Again, position 3 leads to 2 prover calls and the sibling position in the dictionary to only 1 call, while each range captured by an inner node of the pruned subtree also needs only 1 call, in this example summing up to 8 calls in total. Figure 1 visualizes the overall situation.

Depending on the number and the distribution of positions requiring a distortion, the effectiveness of the optimization might vary considerably. More, specifically, let *P(rover)C(all)S(tandard)* and *P(rover)C(all)O(ptimized)* be the functions that count the number of calls of the theorem prover for a range r using the standard and the optimized approach, respectively. Then we observe:

$$length(r) \leq PCS(r) \leq 2 \cdot length(r) \text{ and}$$
$$PCO(r) = 1 + PCO(left(r)) + PCO(right(r))$$

with the following special cases:

- for a range r_{dist_0} that requires no distortion at all,
  $PCS(r_{dist_0}) = length(r)$ and
  $PCO(r_{dist_0}) = 1$;
- for a range r_{dist_1} that requires exactly 1 distortion,
  $PCS(r_{dist_1}) = 2 + (length(r_{dist_1}) - 1)$ and
  $PCO(r_{dist_1}) = 1 + 2 \cdot log_2(length(r_{dist_1})) + 1$;
- for a range $r_{dist_{all}}$ that requires a distortion at all positions,
  $PCS(r_{dist_{all}}) = 2 \cdot length(r_{dist_{all}})$ and
  $PCO(r_{dist_{all}}) = (length(r_{dist_{all}}) - 1) + 2 \cdot length(r_{dist_{all}})$.

So we want to figure out when the optimized approach actually outperforms the standard approach. Facing the difficulties of a general analytical evaluation, we experimentally explored the threshold up to which the optimized approach is expected to perform better than the standard approach, in terms of the number d of positions requiring a distortion, for each d taking the average over all $\binom{d}{l}$ selections of d many positions out of a range of length l, when fixing l. We performed such an exploration for all lengths l up to 500 and found that the

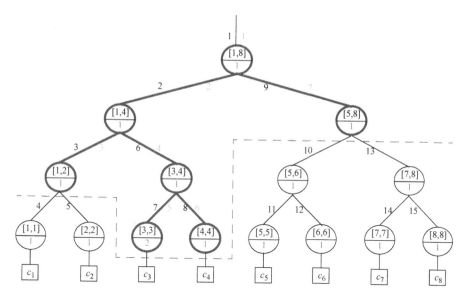

Fig. 1. The inspection tree of an abstract dictionary, constituted by the leaves, showing (1) the complete depth-first traversal as left edge annotation, (2) a pruned depth-first traversal as right edge annotation assuming that only position 3 requires a distortion, (3) the inspected ranges as upper node marks, and (4) the required number of prover calls as lower node marks.

threshold is approximately $l/6$, i.e., the overhead of the optimization resulting from forming an inspection tree is expected to be profitable as long as there are relatively few distortion positions for a sufficiently large range.

5.2   Experimental Runtime Results

As an example of our achievements when also using the active domain transformation described in Sect. 4.1, we summarize our findings for five experiments dealing with a somehow special situation:

- The fixed underlying database relation ill_S over attributes *Patient*, *Diagnostic* and *Symptom* has only 8 tuples.
- The *queries* Φ_i have either one free variable x for the attribute *Patient* or two free variables x and y for the attributes *Patient* and *Symptom*, respectively:
  $\Phi_1 \equiv ill_S(x, cancer, fever)$, $\Phi_2 \equiv ill_S(x, cancer, fever)$,
  $\Phi_3 \equiv ill_S(x, pneumonia, y)$, $\Phi_4 \equiv ill_S(x, virus, y)$, and
  $\Phi_5 \equiv ill_S(x, prolapse, y)$.
- Each of the confidentiality policies has only *one potential secret* Ψ_i:
  $\Psi_1 \equiv ill_S(Smith, cancer, fever)$ at position 2 for x,
  $\Psi_2 \equiv ill_S(Mann, cancer, fever)$ at position 100 for x,
  $\Psi_3 \equiv ill_S(Morris, pneumonia, dyspnea)$ at position 128 for (y, x),
  $\Psi_4 \equiv \neg ill_S(Beier, virus, stomach)$ at position 462 for (y, x), and
  $\Psi_5 \equiv \neg ill_S(Bach, prolapse, lumbago)$ at position 801 for (y, x).

ill_S	Patient	Diagnostic	Symptom	Pos1	Pos2
	Smith	cancer	anemia	2	5
	Smith	cancer	fever	2	3
	Jones	cancer	fever	7	28
	Bloch	cancer	fever	30	465
	Mann	cancer	fever	100	5050
	Derp	cancer	fever	101	5151
	Morris	pneumonia	dyspnea	8	128
	Gray	burnout	fatigue	13	716
		...			
$\neg ill_S$	Beier	virus	stomach	27	462
		...			
	Bach	proplapse	lumbago	21	801
		...			

Patient Dict		Diagnostic Dict		Symptom Dict	
Entry	Pos	Entry	Pos	Entry	Pos
Smith	2	cancer	1	fever	1
Jones	7	pneumonia	3	anemia	2
Morris	8	virus	7	stomach	4
Gray	13	proplapse	10	dyspnea	9
Bach	21	burnout	18	lumbago	20
Beier	27			fatique	26
Bloch	30				
Mann	100				
Derp	101				

Fig. 2. Left: All "positive tuples" and two "negative tuples" of the relation ill_S and their positions *Pos1* and *Pos2* for the free variable x and the pair (y,x) of free variables, respectively. **Right:** The positions of the relevant entries within their dictionaries.

The tuples of the database relation and also two further "negative tuples" are shown in the left part of Fig. 2, together with their positions in the enumeration sequences needed for evaluating the queries, whereas the right part specifies the (relevant part of the) declared domains of the attributes. Figure 3 shows the values of the positions computed in phase 1 and the means of the measured runtimes of 10 repetitions (with small variances not shown).

Dealing with only one potential secret, the measured improvements of the optimization crucially depend on the range of candidate tuples to be inspected. In general, however, the number and the distribution of positions requiring a distortion are also important. Even in our simple case, the only obvious difference between the first and the second experiment, namely the positions 2 and 100 of the potential secrets Ψ_1 and Ψ_2, respectively, causes an essential decrease of the runtimes for phase 2, presumably since the early lie for Ψ_1 makes the work for the theorem prover harder. Each of the last two experiments exhibits an extreme advantage of the divide-and-conquer heuristic, since the negative potential secret requires to already lie for k^* in phase 1 such that phase 2 needs only one prover call and runs without further distortions.

Ex	Positions				Phase 1										Phase 2		Total	
					k		m		k^*		m^*		total					
	k	m	k^*	m^*	PCS	PCO	PCS	PCO	PCS	PCO	PCS	PCO	PCS	PCO	PCS	PCO	PCS	PCO
1	101	101	101	101	0.1	0.1	2.0	2.0	2.1	2.1	2.0	2.0	6.1	6.1	51.9	11.9	58	18
2	101	101	101	101	0.1	0.1	2.0	2.0	2.1	2.1	2.0	2.0	6.1	6.1	51.9	2.9	58	9
3	128	128	127	127	0.1	0.1	0.2	0.2	0.2	0.2	0.2	0.2	0.8	0.8	24.2	11.2	25	12
4	0	462	462	462	< 0.1	< 0.1	220	10.0	1.5	1.5	1.5	1.5	220	14	935	2	950	16
5	0	801	801	801	< 0.1	< 0.1	600	42.0	4.5	4.5	4.5	4.5	609	50.5	n.a.	5.5	n.a	56

Fig. 3. For the different experiments, (i) the positions found in phase 1 and (ii) the runtimes in seconds for phase 1 (for computing the positions and in total), for phase 2 and for the total answer generation, using the standard and the optimized approach.

6 Conclusions

Based on a theoretical design presented in [5], we discussed an *implementation* within the CIE-System [3] for inference control of *open relational queries* under the closed-world assumption. We mainly focussed on the *complexity issues* resulting from two needs: (i) calling a *theorem prover* [9,10] for deciding implication problems with a *completeness sentence* involved to detect potential violations of the confidentiality policy – a functionality not covered by the expressiveness of the underlying SQL-based DBMS – and (ii) inspecting many candidate tuples according to query-/instance-independent *fixed domain enumerations* – to provably guarantee enforcement of a formal notion of confidentiality [5] in the spirit of [7]. The latter need implies to deal with a large number of prover calls with completeness sentences having a large number of conjunctive disjuncts.

We explored two optimizations: *transformations* of a single completeness sentence occurring as an input of a *single* prover call and a divide-and-conquer *heuristic* to reduce the *number* of such calls. Selected experiments confirmed a substantial improvement regarding runtimes. Nevertheless, so far our implementation can handle only relatively small problem instances. This insight suggests two challenging *research topics*: extending current theorem provers with dedicated rules for relational completeness sentences and inventing an inference-proof design of controlled open query evaluation without fixed domain enumerations.

References

1. Abiteboul, S., Hull, R., Vianu, V.: Foundations of Databases. Addison-Wesley, Reading (1995)
2. Biskup, J.: Inference control. In: van Tilborg, H.C.A., Jajodia, S. (eds.) Encyclopedia of Cryptography and Security, pp. 600–605. Springer, Heidelberg (2011)
3. Biskup, J.: Inference-usability confinement by maintaining inference-proof views of an information system. Int. J. Comput. Sci. Eng. **7**(1), 17–37 (2012)
4. Biskup, J.: Logic-oriented confidentiality policies for controlled interaction execution. In: Madaan, A., Kikuchi, S., Bhalla, S. (eds.) DNIS 2013. LNCS, vol. 7813, pp. 1–22. Springer, Heidelberg (2013)
5. Biskup, J., Bonatti, P.A.: Controlled query evaluation with open queries for a decidable relational submodel. Ann. Math. Artif. Intell. **50**(1–2), 39–77 (2007)
6. Farkas, C., Jajodia, S.: The inference problem: a survey. SIGKDD Explor. **4**(2), 6–11 (2002)
7. Halpern, J.Y., O'Neill, K.R.: Secrecy in multiagent systems. ACM Trans. Inf. Syst. Secur. **12**(1), 5.1–5.47 (2008)
8. Levesque, H.J., Lakemeyer, G.: The Logic of Knowledge Bases. MIT Press, Cambridge (2000)
9. McCune, W.: Prover9 and Mace4 (2005–2010). http://www.cs.unm.edu/mccune/prover9/
10. Sutcliffe, G.: The TPTP problem library and associated infrastructure: the FOF and CNF parts, v3.5.0. J. Autom. Reason. **43**(4), 337–362 (2009)

A General Trust Management Framework for Provider Selection in Cloud Environment

Fatima Zohra Filali and Belabbas Yagoubi[(✉)]

Department of Computer Science,
Oran 1 University - Ahmed Ben Bella, Oran, Algeria
{filalifz7,byagoubi}@gmail.com

Abstract. Trust has been a predominant issue in adopting cloud service among consumers. It has been a critical concern for business application and sensitive information. In a previous work [1], we proposed a trust computing model for Cloud Computing which we validated by simulation with well-known trust model. This paper proposes a framework for trust management in Cloud Computing environment, based on the proposed computing model. It describes the implementation of the general trust framework. The proposed framework identifies the metrics of performance to select the most suitable provider for performing service transaction and integrate it into trust rating process while filtering biased opinions.

Keywords: Cloud computing · Security · Trust · Performance · CertainLogic · Opinion · Service selection · Trust management

1 Introduction

For the last years, Cloud services have grown to become an essential paradigm for both industry and academia, by allowing Cloud users to rent computing, network, and storage resources. In that way, users pay for their use of services without apprehensions about maintenance, management or cost.

In spite of all importance of Cloud Computing, most of the organizations are not making a trend of Cloud Computing, and its evolution has raised many concerns and was encountered by various obstacles. Security is one of the most crucial problems for this model, and the risks accompanying the deployment of services and applications are more important with the architecture of Cloud environment [2].

Besides, Cloud provider's interaction with users is not achieved completely, and trust must be granted in this relationship. Otherwise, many organizations are not willing to adopt Cloud Computing, without strong evidence of trustworthiness; thus establishing trust between these entities becomes a very effective way to secure the Cloud.

In [3] authors observed that organizations have limited knowledge about Cloud Computing. In a survey of different organizations, they reported that

© Springer International Publishing Switzerland 2015
T. Morzy et al. (Eds.): ADBIS 2015, LNCS 9282, pp. 446–457, 2015.
DOI: 10.1007/978-3-319-23135-8_30

40 % do not know about Cloud's component (such as SLA, SaaS, PaaS, IaaS), 20 % are concerned about security problems, 40 % are unaware of the services provided, and 20 % do not trust Services providers. Therefore, we can remark that is necessary to establish trust between Cloud provider and Cloud users, since certain data is confidential by its nature, and handing it over to providers without having a very high level of trust beforehand is definitely unacceptable for users.

The rest of this paper is organized as follows. The Sect. 2 overviews the issues and challenges in service selection and trust computing. Our proposed framework with design details is presented in Sect. 3. Implementation and evaluation are presented in Sect. 4. Finally the conclusion and future perspectives are discussed in Sect. 5.

2   Challenges in Trust Management and Service Selection

Cloud service selection is among the most challenging issues in Cloud Computing security [4–6]. In fact, selecting the best service is equivalent to selecting the most trusted Cloud service. Thus, many studies focus on adopting trust as a solution for cloud service selection. With that, many challenges arise, such as: How a user can use a trustworthy service? On which metrics the trust can be established? What is the metrics that define the performance offered by the provider?

In the next section, we present several dimensions for classification of majors trust issues around Cloud systems. This classification served as a basis for designing the proposed framework.

2.1   Assurance

The trust represents an important concern for any system, especially in Cloud Computing [7–9]. To guarantee trust, most of the organizations negotiate a Service Level Agreement (SLA) between providers and users [10,11]. However, no guarantee is granted by the SLA, it only provides an assurance in case of data damage. Many research have been made to assess the trust problem in the Cloud computing. However, a robust trust management system needs to be developed to enhance the security in providing trust and making the cloud more reliable.

2.2   Ratings and Feedbacks Trustworthiness

Trust systems represent a significant trend in decision making of Cloud service. The basic principle is to let users rate the used services, for example at the end of a transaction, and use the aggregated ratings about a given service to derive a trust value. This value can be used to support other users in deciding whether to interact with that service. A side effect that arises is provision of various dishonest rating, resulting in incorrect ratings of the services.

2.3   Trust Sources

Another issue in trust management is the sources of the trust information. Trust systems can use both implicit and explicit information for decision-making. These evidences can be collected from different sources:

- Users, the evidences from direct interactions with the service.
- Cloud Service Provider, the information provided by registration.
- Recommendation from other sources and feedback given by other consumers.
- QoS values and performance of the service from the SLA agreement or monitoring services.

2.4   Trust Degree

Many researches have been made to model the trust and calculate the trustworthiness of a service provider. Some models are based on probabilistic approaches [12,13], others are based on Bayesian representation [14–16], some others use fuzzy logic [17–19] or subjective logic [20–22].

These approaches are not suitable for the modelling of our solution. First, the probabilistic approaches represent uncertainty of the evidences but the probabilities are assumed to be known, which, is most likely difficult in a Cloud environment. The approaches based on Bayesian probabilities suppose the use of the probability density function, which bring to complex mathematical distributions and hard interpretations. The approaches based on fuzzy logic models represent a different sort of certainty, more oriented to linguistic uncertainty or fuzziness. Finally, the approaches based on subjective logic are more appropriates, but the parameters of belief, disbelief and certainty are dependent on each other.

Moreover, all these models quantify the trust as a probability value based on direct trust or recommended trust. However, they ignore the objective factors for the provided service (QoS). Hence, these models cannot assess the accuracy of the trust value made by itself. In [23] the author proposed a model for the assessment of propositional logic terms under uncertainty. The model has been proved to be compliant with the standard probabilistic evaluation of propositional logic terms and with subjective logic, which provides the justification for the mathematical validity of the model. The proposed approach is more expressive than the standard probabilistic approach, and although it is as expressive as subjective logic. It provides simpler representation since it is based on independent parameters and provides a more intuitive and more expressive graphical representation. Furthermore, it has been shown that the parameters for assessing opinions in Certain-Logic can be derived using multiple approaches and sources. Finally, they have shown the applicability and the benefits of the model in a use case. They have evaluated the trustworthiness of their system in Cloud Computing scenario.

Hence, we can remark that in the most proposed works, trust has been computed using user feedback in some formalism. Most of these methods are based on rating and neglect the fact that the provided performance of the service takes an important part to win user confidence. Hence, a model that integrate these two aspects would make the trust system more reliable.

2.5   Other Issues

Furthermore, a better trust management system for cloud services should take into consideration the following requirements:

1. The trust should cover several factors of QoS among a bunch of services offering similar features. Users need to know not only the purpose of the service but also the qualities of the service.
2. The trust should be computed based on user preferences since different users may be interested in different characteristics, and a service adapted to a particular user may be not suitable for other users.
3. It is necessary to combine both subjective aspect (users feedback) and objective dimension (QoS performance) to evaluate the trust. Trust is a subjective notion that predicts future action of an entity based on past actions. Thus, the ratings of users are an important factor to sharing knowledge about direct experiences in using Cloud services.

3   Proposed Framework

In this section, we describe the proposed framework for trust management and service provider selection within cloud computing. We define the system architecture, and the trust relationships between the involved parties.

3.1   Framework Architecture

In this part, we describe the general system architecture for the proposed framework of trust management.

Figure 1 depicts the main components of the trust selection system, which consists of three different layers, namely Service Requester Layer, Service Manager Layer and Service Provider Layer.

The Service Manager Layer. This level represents the system core. It includes the trust management service where a service requester can give trust feedbacks to a particular service, and where a service provider can register to the service. The General Trust Calculator constitutes the essential part of this layer. It is responsible for the service provider selection based on the model presented below. The General Trust Monitor contains two monitors: performance and trust responsible for supervising the execution of the application. The General Trust Broker is in charge of exchanging resources between users and providers. It selects suitable providers able to deliver the required service.

The Service Requester Layer. This level consists of different service requesters who consume services in the service layer. For example, a new organization that has limited funding can use cloud services (e.g., hosting their services in Amazon S3). Service requesters can give trust feedbacks of a particular cloud service by invoking the trust system.

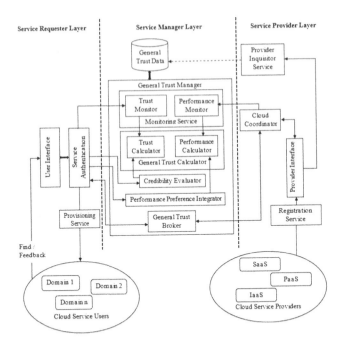

Fig. 1. Framework architecture

The Service Provider Layer contains cloud services, IaaS (Infrastructure as a Service), PaaS (Platform as a Service), and SaaS (Software as a Service).

3.2   General Trust Framework Components

In this section, we present the descriptions of the main framework's components architecture and detailed formulas for computing the trustworthiness of the providers and the selection process.

The proposed framework is composed of the following components:

User Services. The user will be given a set of services for registration to the framework, finding the suitable service and providing the framework with the feedback after completing a transaction with the selected service.

User Interface (UI) : a web interface provided to the users for interacting with the platform.

Authentication Service (AS) : each user must be authenticated to accessing the required service and rating the services

Provisioning Service (PS): The interaction of the user with the platform.

Framework Core Services. The basis of the framework is the General Trust Manager (GTM), which is responsible for performing monitoring, evaluating and calculating the trust.

Performance Monitor (PM): is responsible for collecting the QoS from the delivered service.

Trust Monitor (TM): manages the trust values from the user and feedbacks.

General Trust Calculator (GTC): The GTC is composed of two calculator : Trust and Performance. The TC is based on the CertainTrust [24]. The PC integrates a performance value to provide a more accurate evaluation of our model. For the final selection of the reliable service, both the trust and performance values are used. The proposed model ha been validated in a previous work [1]

Performance Calculator (PC): In [25] the authors proposed a trust model based on QoS for Cloud Computing, based on four attributes: availability, reliability, data integrity and Turnaround Efficiency. However, computing performance based on these four attributes is insufficient to achieve a valid model. It must rely on standardized and approved measures in the context of Cloud Computing. In order to form a performance model, attributes defined by Service Measurement Index (SMI) are used. Cloud Service Measurement Index Consortium (CSMIC) [26] proposes a framework based on common characteristics of cloud services. The purpose of this consortium is to express each of QoS attributes given in the framework and offer a methodology for computing a relative index for comparing different cloud services. CSMIC has designed the Service Measurement Index (SMI), which consists of a set of Key Performance Indicators (KPI) that aids to standardize the measurement of services.

The attributes (x_i) represent the performance factors of our model, which include power, cost, response time, efficiency, transparency, interoperability, reliability, availability, security.

Each of these features is included in a set of Key Performance Indicators (KPIs), which describe the data to be gathered for measurement.

The performance value for Cloud Service (x) is computed by a utility function used with the described objective attributes.

$$P(x) = \sum_{i=0}^{9} w_i * x_i$$

w_i represent a weight for each attribute with $\Sigma w_i = 1$

Trust Calculator (TC): Trust is a term that describes how much one believes in another. Evaluating trust for service can help users to predict its future behavior [12]. For service selection, the trust value is denoted as T(x).

In our opinion, the three main features that a reliable service must offer besides the performance values are the time, the cost and the overall satisfaction of the service.

Thus, to compute the final value of the trust, these three factors: time, cost, satisfaction that are presented in the SMI framework [26] have been used. For each transaction, the trust value is computed as a combination of these factors.

$$T(x) = \begin{cases} 0.5 & initial \\ \alpha \frac{\sum_{k=1}^{n} (f(t_k)*f(c_k)*f(s_k))}{n} \end{cases}$$

k : kth use of the service,
α represent the adjustment factor, and is calculated by :

$$\alpha = \sqrt{\frac{n}{T+1}}$$

n : Number of satisfactory service use,
T : Total use of the service.
$f(t_k)$: Attenuation factor,

$$f(t_k) = \frac{e^{-(t_0 - t_{k-1})}}{T}$$

$f(c_k)$: Cost factor [26]

$$f(c_k) = \frac{c_k}{cpu^a * net^b * vm^c * capacity^d}$$

with $a + b + c + d = 1$
$f(s_k)$: Satisfaction factor,

$$f(s_k) = \begin{cases} 1 & if\ user\ satisfied \\ \frac{\Sigma\ Satisfaction\ criteria}{Total\ criteria} & else \end{cases}$$

Credibility Evaluator (CE): component responsible for filtering biased opinions. It represents the euclidean distance between the user feedback and the expectation of others users feedbacks for a service. If the distance is higher than a threshold the opinion is biased and the feedback average are used instead.

Performance Preferences Integrator (PPI): component responsible for getting the preferences given by the user in the interface to integrate it into the calculation process of the performance value.

General Trust Broker (GTB): responsible for selecting the most reliable provider, based on the general trust value calculated by the General Trust Calculator component.

Provider Services

Provider Inquisitor Service (PIS): proceeds to a look up for the different provider values. It supplies the provider a limited access to the stored information and statistics about the provided services, via the web interface.

Provider Authentication (PA): authenticate the provider to get access to the stored information about the trust framework.

Provider Interface (PI): The provider will be given an interface to performing registration and accessing statistical information of the different services.

Registration Service (RS): responsible for indexing the list of the offered services by the providers.

Cloud Coordinator (CC): manages the different providers to access to the framework. It gets the information from providers for monitoring. It is also responsible for coordinating the broker of the framework with the provided services.

4 Performance Analysis

In this section, we evaluate the effectiveness of the proposed framework. We analyse the framework reliability in the experiments presented below.

4.1 Framework Implementation

Experiments were conducted on an Intel Core i7 2.2 GHz processor, 8.0 GB RAM running under Windows 8 Ultimate. The proposed framework prototype was built from scratch using Java Enterprise Edition in Eclipse.

The service selection and trust computation contain several modules running at the provider end. These components were implemented using java. A web interface is provided to both provider and user, which uses Apache web server for deploying the web system. MySQL at the backend database to store the trust values, selection data and other results. The web interface is developed with the JSP and servlets. The framework components are also developed with java. It includes the following modules: GTC (TC and PC), GTB, CC, PPI, CE.

4.2 Metrics

In addition to developing and implement the framework, we also conduct a comprehensive performance analysis using various metric values to estimate the accuracy of the trust as discussed here.

Efficiency represents the number of trusted service provided per the total number of the service.

Average Trust represents the ratio of the sum of the trust values to the number of trust values.

Trust Bias represents the filtered value of user feedbacks.

4.3 Experiments

This section analyses the performance of the proposed trust system model in terms of the discussed metrics.

Performance. The proposed framework is tested with 20 services (10 trusted services and 10 untrusted services). We perform 200 requests users in 10 cycles of time. Then, we counted the number of user requests allocated to trusted services, by using the proposed trust management framework, and without using trust management (selection based on the performance of the service). The results are presented in the Fig. 2 below.

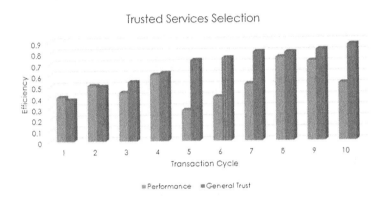

Fig. 2. Trusted service selection

In the experiment, we note that the proposed trust framework performs a higher efficiency compared to performance selection. For the first cycle of time, the proposed framework takes some time to stabilize due to initialization of the first transactions. In addition, the performance selection chooses the service by its performance value and does not consider the trustworthiness of the provider compared to the proposed framework

Feedback Filtering. In this experience, the average trust value is computed for the proposed model, by integrating the bias factor for filtering feedback and without using the bias factor. We generate a set of random requests for different datasets of users. In each round, the average trust values of the General Trust with a credibility evaluator are compared to the average trust values without a credibility evaluator. The Fig. 3 gives comparison between these values.

In the results, we can observe that even if the number of users increases, the general trust value computed with the bias value produces regular results. Thus, the selection of the more suited provider will increase for the proposed framework with filtering the biased opinion even if the number of the user decrease compared to a solution without filtering.

User feedback

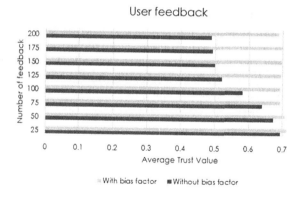

Fig. 3. Filtering user feedbacks

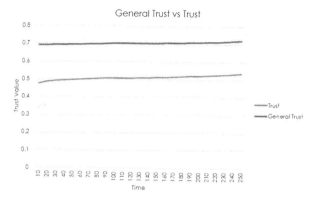

Fig. 4. Average trust selection

Trust Selection. In the last experiment, we compared the proposed framework model to a solution based on trust (without including the performance in the selection process). We generate a set of user requests over periods. Then, we computed the average trust value for both approaches. The results are shown in the Fig. 4.

From the result, we can notice the difference in the average values of trust between the proposed solution and the approach based only on trust. Since, the General trust take into account the opinion of the performance to compute the average value. It gives a more reliable result, in comparison to the trust provided by the feedbacks. Hence, it increases the number of trusted selected services.

5 Conclusion

This paper proposed a prototype framework for trust management. It is based on a trust computing model that achieves a high trust accuracy.

The proposed trust model is based on the point that trustworthiness can be evaluated by both objective and subjective metrics. Our method has also proposed the user satisfaction estimation in the credibility evaluator. Even more, we take into account the user preference in the overall selection process.

We plan to extend our work by conducting further experimentation under several attacks and malicious users, to integrate a threat model into the proposed framework.

References

1. Filali, F.Z., Yagoubi, B.: Global trust: a trust model for cloud service selection. Int. J. Comput. Netw. Inf. Secur. **7**, 41–50 (2015)
2. Puthal, D., Sahoo, B., Mishra, S., Swain, S.: Cloud computing features, issues and challenges: a big picture. In: Computational Intelligence (2015)
3. Ahmad, S., Ahmad, B., Saqib, S.M., Khattak, R.M.: Trust model: clouds provider and clouds user. Int. J. Adv. Sci. Technol. **44**, 69–80 (2012)
4. Fernandes, D.A.B., Soares, L.F.B., Gomes, J.V.P., Freire, M.M., Inácio, P.R.M.: Security issues in cloud environments: a survey. Int. J. Inf. Sec. **13**(2), 113–170 (2014)
5. Noor, T.H., Sheng, Q.Z., Zeadally, S., Yu, J.: Trust management of services in cloud environments: obstacles and solutions. ACM Comput. Surv. **46**(1), 12 (2013)
6. Pearson, S., Benameur, A.: Privacy, security and trust issues arising from cloud computing. In: Proceedings of Second International Conference on Cloud Computing, CloudCom 2010, 30 November - 3 December 2010, pp. 693–702, Indianapolis, Indiana, USA (2010)
7. Subashini, S., Kavitha, V.: A survey on security issues in service delivery models of cloud computing. J. Netw. Comput. Appl. **34**(1), 1–11 (2011)
8. Sherchan, W., Nepal, S., Paris, C.: A survey of trust in social networks. ACM Comput. Surv. **45**(4), 47 (2013)
9. Han, G., Jiang, J., Shu, L., Niu, J., Chao, H.: Management and applications of trust in wireless sensor networks: a survey. J. Comput. Syst. Sci. **80**(3), 602–617 (2014)
10. Wu, L., Garg, S.K., Buyya, R.: Sla-based resource allocation for software as a service provider (saas) in cloud computing environments. In: 11th IEEE/ACM International Symposium on Cluster, Cloud and Grid Computing (CCGrid), pp. 195–204. IEEE (2011)
11. Kouki, Y., Ledoux, T.: Csla: a language for improving cloud sla management. In: International Conference on Cloud Computing and Services Science, CLOSER 2012, pp. 586–591 (2012)
12. Muller, T., Schweitzer, P.: On beta models with trust chains. In: Fernández-Gago, C., Martinelli, F., Pearson, S., Agudo, I. (eds.) Trust Management VII. IFIP AICT, vol. 401, pp. 49–65. Springer, Heidelberg (2013)
13. van Deursen, T., Koster, P., Petkovic, M.: Hedaquin: a reputation-based health data quality indicator. Electr. Notes Theor. Comput. Sci. **197**(2), 159–167 (2008)
14. Teacy, W.T.L., Luck, M., Rogers, A., Jennings, N.R.: An efficient and versatile approach to trust and reputation using hierarchical Bayesian modelling. Artif. Intell. **193**, 149–185 (2012)

15. Tavakolifard, M., Knapskog, S.J.: A probabilistic reputation algorithm for decentralized multi-agent environments. Electr. Notes Theor. Comput. Sci. **244**, 139–149 (2009)
16. Whitby, A., JÃsang, A., Indulska, J.: Filtering out unfair ratings in Bayesian reputation systems. In: The Third International Joint Conference on Autonomous Agenst Systems (2004)
17. Iltaf, N., Ghafoor, A.: A fuzzy based credibility evaluation of recommended trust in pervasive computing environment. In: 10th IEEE Consumer Communications and Networking Conference, CCNC 2013, Las Vegas, NV, USA, 11–14 January 2013, pp. 617–620 (2013)
18. Song, S., Hwang, K., Kwok, Y.: Risk-resilient heuristics and genetic algorithms for security-assured grid job scheduling. IEEE Trans. Comput. **55**(6), 703–719 (2006)
19. Bharadwaj, K.K., Al-Shamri, M.Y.H.: Fuzzy computational models for trust and reputation systems. Electron. Commer. Res. Appl. **8**(1), 37–47 (2009)
20. Jøsang, A., Hayward, R., Pope, S.: Trust network analysis with subjective logic. In: Twenty-Nineth Australasian Computer Science Conference on Computer Science 2006, (ACSC2006), Hobart, Tasmania, Australia, 16–19 January 2006, pp. 85–94 (2006)
21. Wang, Y., Singh, M.P.: Trust representation and aggregation in a distributed agent system. In: Proceedings of the 21st National Conference on Artificial Intelligence - Volume 2. AAAI 2006, pp. 1425–1430. AAAI Press, 2006
22. Jøsang, A.: Subjective logic, University of Oslo, Technical report (2013)
23. Ries, S., Habib, S.M., Mühlhäuser, M., Varadharajan, V.: CertainLogic: a logic for modeling trust and uncertainty. In: McCune, J.M., Balacheff, B., Perrig, A., Sadeghi, A.-R., Sasse, A., Beres, Y. (eds.) Trust 2011. LNCS, vol. 6740, pp. 254–261. Springer, Heidelberg (2011)
24. Ries, S.: Extending Bayesian trust models regarding context-dependence and user friendly representation. In: Proceedings of the 2009 ACM Symposium on Applied Computing (SAC), Honolulu, Hawaii, USA, 9–12 March 2009, pp. 1294–1301 (2009)
25. Manuel, P.: A trust model of cloud computing based on quality of service. Ann. Oper. Res., 1–12 (2013)
26. Garg, S.K., Versteeg, S., Buyya, R.: A framework for ranking of cloud computing services. Future Gener. Comp. Syst. **29**(4), 1012–1023 (2013)

Sybil Tolerance and Probabilistic Databases to Compute Web Services Trust

Zohra Saoud[1] (✉), Noura Faci[1], Zakaria Maamar[2], and Djamal Benslimane[1]

[1] Claude Bernard Lyon 1 University, Lyon, France
zohra.saoud@univ-lyon1.fr
[2] Zayed University, Dubai, UAE

Abstract. This paper discusses how Sybil attacks can undermine trust management systems and how to respond to these attacks using advanced techniques such as credibility and probabilistic databases. In such attacks end-users have purposely different identities and hence, can provide inconsistent ratings over the same Web Services. Many existing approaches rely on arbitrary choices to filter out Sybil users and reduce their attack capabilities. However this turns out inefficient. Our approach relies on non-Sybil credible users who provide consistent ratings over Web services and hence, can be trusted. To establish these ratings and debunk Sybil users techniques such as fuzzy-clustering, graph search, and probabilistic databases are adopted. A series of experiments are carried out to demonstrate robustness of our trust approach in presence of Sybil attacks.

Keywords: Trust · Credibility · Sybil · Fuzzy clustering · Web service

1 Introduction

There is a large consensus in the R&D community about the role of trust in Web services (WS)s selection [27,28]. Most existing Trust Management Systems (TMS)s rely on end-users' interactions with WSs to compute trust. To this end ratings, tags, and even narrative reviews out of these interactions are used. However TMSs are vulnerable to different attacks for instance, biased feedback and Sybil, which could undermine their efficiency. Biased feedback refers to ratings that either promote or demote falsely a WS's non-functional properties. And Sybil refers to end-users who have purposely different identities and hence, can provide inconsistent ratings over the same WSs. To deal with biased feedback approaches, such as Cloud Armor [17] and RateWeb [14], consider *user's credibility* when computing trust.

Credibility-based trust approaches assume that end-users are either experts or untrustworthy. When these end-users do not agree on a certain WS rating the majority opinion helps reach a consensus. Those who have close ratings to the majority opinion are more credible than those who have distant ratings. However these approaches overlook end-users who are simultaneously experts and

© Springer International Publishing Switzerland 2015
T. Morzy et al. (Eds.): ADBIS 2015, LNCS 9282, pp. 458–471, 2015.
DOI: 10.1007/978-3-319-23135-8_31

trustworthy. In a previous work we refer to such end-users as *strict experts* who usually do not have specific interest in aligning themselves with the majority [18]. We use fuzzy-clustering technique to reduce the gap between *strict experts'* ratings and the current majority's opinion so that a consensus is reached.

In this paper we put additional efforts into WSs trust by examining Sybil attacks. Recently, social network-based trust approaches (e.g., [29] and [30]) have drawn the attention of the R&D community to tackle these attacks. Their main assumption is that Sybils tend to be connected to a limited number of non-Sybils in the network. A trust graph that represents end-users' identities (nodes) and trust-relations between end-users (edges) is built and labeled with capacities (i.e., a certain amount of ratings that can be broadcast by end-users through edges). Using graph search techniques these approaches aim to decrease the impact of Sybil attackers on the quality of the computed trust value. However they assign capacities to nodes randomly, which does not help establish a concise picture of the role of Sybil in either promoting or demoting WSs.

To address the random capacity assignment we identify a maximum number of non-Sybil credible users who will be able to provide accurate ratings. Our approach to deal with Sybil advocates for problem-specific knowledge graph search with the user credibility as a heuristic. This credibility would guide the way the ratings are collected from the most non-Sybil credible users. We rely on our fuzzy credibility model reported in [18] to compute this credibility. A non-Sybil user should collect a maximum number of ratings from credible peers in his neighborhood graph. The proposed approach uses then selected-users' credibility to model their ratings in terms of probabilistic databases and compute WS trust as a query evaluation.

The remainder of this paper is organized as follows. Section 2 discusses some work on trust computing in presence of Sybils. Section 3 describes the proposed trust approach. Section 4 gives details on our credibility model and defines the both capacity distribution mechanism and user selection strategy used. Section 5 depicts how WS trust is computed. Section 6 analyzes experimental results obtained. Finally, concluding remarks and future work are reported in Sect. 7.

2   Related Work

This section presents how some existing works tackle Sybil attacks and how trust is computed in the context of social networks.

Sybil works. Sybil attacks are widely reported in the literature. Some solutions advocate for early detection of Sybil by analyzing user profiles and relationships between users in social networks. Other solutions focus on how to decrease the impact of Sybil attacks on the quality of the computed trust value.

In [7] Danezis et Mittal propose SybilInfer, an algorithm that labels nodes in a social network as either honest user or Sybil user. The authors define a probabilistic model of honest networks that SybilInfer uses to infer potential regions of dishonest nodes. Similarly Mislove et al. propose Ostra, an algorithm that prevents undesirable communication between honest and dishonest nodes by using trust relationships such as social links [15]. Both Ostra and SybilInfer assume a global knowledge of the social network such as structure and node profiles. Ostra does not provide guarantees to properly discard Sybil nodes. And Sybil-Infer determines the probability via sampling that, unfortunately, concurrently reduces uncertainty and introduces additional uncertainty.

In [10] Hota et al. present two algorithms to detect Sybil nodes: multi-path routing and verification. The first searches for any common segment in a path between a node called verifier and the group of nodes suspected as Sybil. The second confirms the status of this group by randomly selecting few nodes and polling them. These nodes must then reply to this polling within a constrained time frame. If an entity has more than one identity it will fail to reply within this time.

In [24] Tran and al. propose SumUp, an online content rating system. SumUp emphasizes on users accounts in social networks to respond to Sybil attacks. By assigning capacities on the social links and collecting ratings in an approximate max-flow fashion, SumUp can successfully decreases the number of ratings broadcast by the attacker through attack edges with high probability. SumUp also relies on users' feedback to further reduce the number of ratings provided by the attacker if this latter still behaves inappropriately.

In [26] Viswanath et al. propose CANAL, a system that uses landmark routing and credit payments over large networks to efficiently tackle Sybil attacks. Canal can be easily integrated into existing Sybil tolerance schemes and can deploy solutions such as SumUp and Bazar over real systems.

To wrap up this first part of related work the random choice selection (e.g., blind breadth-first search) in the aforementioned algorithms challenges the number and quality of the ratings considered while computing WS trust. To this end we advocate for heuristic search to guide the way the ratings are collected from the most appropriate users. This should lead to better quality of the computed WS's trust value.

Trust Management Works. In existing probabilistic trust management approaches (e.g., [23] and [31]) users rely on direct interactions with services or ratings received from other users. False feedback/ratings are handled through a suitable filtering mechanism. In the following we describe three relevant probabilistic approaches.

TRAVOS is a trust model used in open Agent systems [23]. An agent trusts a peer based on previous direct interactions. Interactions' outcomes represent ratings to express success or failure. The obtained binary ratings are then used to form the probability-density function that models the probability of a successful interaction with an agent. If there is a lack of direct experiences the model uses other agents' experiences to compute the trust value. The model determines the credibility of agents to filter ratings provided by agents that are inaccurate due to their limited knowledge or malicious behaviors.

PowerTrust is a trust system for P2P networks [31]. Initially nodes rate individual interactions and compute local trust values using a *Bayesian* learning technique [3]. These local trust values are then used to evaluate a global trust value. This value is updated periodically using the Look-ahead Random Walk (LRW) algorithm [13]. Along with a distributed ranking module, LRW identifies the nodes that assess the reputation of providers.

To wrap up this second part of related work the aforementioned probabilistic approaches overlook the ratings interpretation in the presence of uncertainty, i.e. each rating is true at some extent and false at another extent. We thus propose to model user ratings as a probabilistic database, interpret ratings in terms of possible words and compute Web service trust as a query evaluation over probabilistic database.

3   Approach to Tackle Sybil Attacks

Our trust approach relies on user ratings to compute WS trust. Users register in a social network with an identity and hence have the right to rate WSs they interact with. A user u_i can also establish trust links with other users (u_k). A trust link represents the u_i's belief that u_k is not a Sybil and will evaluate honestly the WS. In [5] Cheng and Friedman demonstrate a trust function that using only the structure of a social network does not allow to distinguish between non-Sybil users and Sybil users. There is a need to know at least one non-Sybil node, known as source to identify a trust graph of users. The source operates as a rating collector. Max-flow algorithms have been applied to the trust graph to limit the number of ratings that Sybil users can propagate on some resource (e.g., [24]). However a user in the trust graph does not necessarily provide accurate ratings.

We thus propose an approach that relies on both user's trustworthiness in the network and credibility to build the labeled trust graph. Formally, the trust graph G is a tuple $< S, N, E, Cr^N, C^E >$ where S represents the source, $N = \{N_1, N_2, ...\}$ is a set of nodes that refer to user identities, $E = \{E^{(N_1, N_2)}, ..., E^{(N_i, N_j)}, ...\}$ is a set of edges (i.e., trust links) between nodes, $Cr^N : N \rightarrow [0, 1]$ assigns a credibility value to each N_i, and $C^E : E \rightarrow \mathbb{N}$ assigns a capacity value to each $E^{(N_i, N_j)}$.

Figure 1 illustrates our four-step trust approach to debunk Sybil users and promote non-Sybil credible users.

– **Social Network Pruning.** The attack capability of a Sybil user refers to the overall capacity of edges connected to this user. So smaller the value of the incoming edges is fewer the number of biased ratings that the source will collect is. In this step the in-coming edges per node in the social network should not exceed some predefined threshold e_{in} as suggested in [24]. Network pruning would prevent the case where a Sybil has several incoming edges from non-Sybil nodes and will consequently have a high attack capability. It also eliminates redundant paths in the trust graph and consequently speeds up the rating collection process.
– **Trust Graph Building:** Upon attack-capability decrease of Sybil users this step aims to increase rating capability of non-Sybil credible users. First the users in the pruned social network provide ratings to the credibility model

described in Sect. 4.1. This model computes the user i's credibility (Cr_i) by using a fuzzy clustering technique. Then capacities are assigned to edges through a ticket distribution mechanism that starts from the source acting as the rating collector (cf. Sect. 4.2).

- **User Selection:** This step collects the maximum number of users within the trust graph. For this purpose, we adapts the max-flow algorithm proposed by Ford and Fulkerson [8] through heuristic search to guide user selection towards non-Sybil credible users. More details on this step are given in Sect. 4.2.
- **Database-based Trust Evaluation:** This step assesses WS trust using the selected users' ratings. The ratings are stored in a probabilistic database along with their credibility values. Ratings inconsistencies lead to disagreement amongst end-users' opinions. Troffaes shows that probabilities can address this disagreement [25]. End-user's credibility helps tackle uncertainty over ratings. Therefore we associate credibility with probabilities. WS trust is assessed using a query evaluation on the probabilistic ratings database as per Sect. 5.

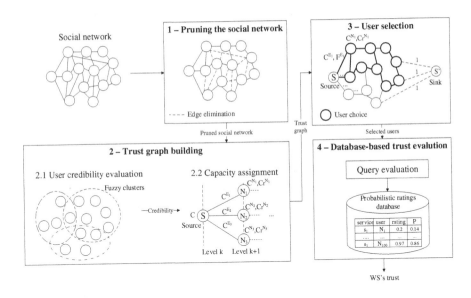

Fig. 1. Approach overview

4 Credibility-based User Selection

This section first discusses the appropriateness of fuzzy clustering for establishing users' credibility. It also describes how credibility is used to build the labeled trust graph G and the proposed user selection strategy.

4.1 Credibility Model

Credibility has two components [2]: expertise and trustworthiness. Our model target *strict users* who are experts and trustworthy. These users stick to their ratings regardless of the majority for reasons listed in [20] including their veracity and objectivity, and accuracy of ratings. Several studies in social psychology (e.g., [12] and [21]) evaluate the impact of source credibility on belief and attitude changes. These studies demonstrate that credible sources are persuasive and can affect existing beliefs (e.g., ratings) and attitudes more than non-credible sources. Therefore, *strict users* can "push" the majority to question (even review) their ratings.

The credibility model aims at reducing the gap between *strict users'* ratings and the current majority's rating so that a consensus is reached. Since *strict users* can be in several groups they can affect groups' beliefs in different manners (e.g., strongly and weakly). *Strong* and *weak* membership terms are fuzzy because they are not well-defined and/or their semantics are dependent on domains and/or user preferences. To deal with uncertainty in group membership and derive overlapping groups the credibility model uses fuzzy clustering. Consensus clustering algorithms like $\mathcal{K}$-means and fuzzy $\mathcal{C}$-means generate robust clusters, detect "unusual" ones, and handle noise and outliers [16]. Existing credibility-based trust approaches such as [14] and [17] rely on $\mathcal{K}$-means to compute the Majority ($\mathcal{M}_K$) consensus as a centroid of the most populated cluster.

We use Bezdek's Algorithm discussed in [1] to reduce the gap between *strict users'* (u_i) ratings and $\mathcal{M}_K$ consensus. Each u_i provides a set of ratings ($\mathcal{X}_i$) on a set of common WSs. The algorithm takes as input $\mathcal{ME}=\{\mathcal{ME}_{i,j}\}$ a membership matrix where $\mathcal{ME}_{i,j}$ represents the membership degree of $\mathcal{X}_{i=1,n}$ in the Cluster $\mathcal{C}_j$ and generates as output a number of clusters ($Nb_{cluster}$) with fuzzy boundaries. A new *Majority Cluster* ($\mathcal{C}_{Maj}$) needs to be identified taking into account that *each user's rating has a degree of membership per cluster*. We use three strategies to decide on $\mathcal{C}_{Maj}$ that rely on qualitative values of membership degree in a fuzzy cluster: weak, moderate, and strong. The experimental results we obtain in [18] show that the strong strategy gives the most accurate results. This strategy selects the cluster with the highest membership degree of ratings as $\mathcal{C}_{Maj}$. The next step is computing the credibility of users in this cluster. Equation 1 identifies u_i's credibility as a distance from a rating to the majority opinion represented by the centroid of $\mathcal{C}_{Maj}$. This credibility is computed using the normalized euclidean distance $\|*\|_{\mathcal{N}}$ as the similarity measure:

$$\mathcal{CR}_i^j = 1 - \left\| \mathcal{X}_i - centroid(\mathcal{C}_{Maj}^{strong\_strategy}) \right\|_{\mathcal{N}} \tag{1}$$

Now credibility is assessed for each user, we explain in the following subsections how we use it for assigning capacities to the edges of the trust graph and then for user selection.

4.2   User Selection

In order to bootstrap non-Sybil credible users' rating capability we develop a novel credential mechanism. This mechanism distributes rating tickets (or capacity) along the edges in the trust graph, using user credibility. The level of a node is defined by letting the source to be at level 0, while a node at level k has at least one parent at level $k - 1$. Node N_i receives C tickets through edges from parent nodes ($Parent_i$) at upper levels, consumes one ticket and re-distributes the remaining tickets to child nodes ($Child_i$) at lower levels. We associate N_i with a capacity C^{N_i} as follows:

$$C^{N_i} = \sum_{\forall N_k \in Parent_i} C^{E^{(N_k, N_i)}} - 1 \tag{2}$$

We propose a ticket distribution strategy where a node N_k re-distributes C^{N_k} through its outgoing edges according to Eq. 3. For flow conservation purpose in the trust graph we round up or down the capacity values.

$$C^{E^{(N_k, N_i)}} = round(\frac{C^{N_k} * Cr_i}{\sum_{\forall N_j \in Child_k} Cr_j}) \tag{3}$$

Upon trust graph building the next step is to select the higher number of non-Sybil credible users. We formalize this selection constraint as a maximum-flow problem. This problem refers to maximize the flow paths from the source S to the sink S' in the trust graph. For applicability purpose of maximum-flow algorithm a sink is added to the trust graph and linked to the different leaf nodes. The edges between the sink and these leaf nodes are labeled with 1. We are inspired by Ford-Fulkerson algorithm based on three functions. The first function $F : E \rightarrow \mathbb{N}$ assigns to each edge ($E^{(N_i, N_j)}$) with some capacity ($C^{E^{(N_i, N_j)}}$) a flow value ($F(E^{(N_i, N_j)}) \leq C^{E^{(N_i, N_j)}}$). $F(E^{(N_i, N_j)})$ represents the flow that could be passed through $E^{(N_i, N_j)}$. The second function $R_F : E \rightarrow \mathbb{N}$ deals with residual capacity. R_F represents the remaining flow that could be passed through $E^{(N_i, N_j)}$ (i.e., $R_F(E^{(N_i, N_j)}) = C^{E^{(N_i, N_j)}} - F^{E^{(N_i, N_j)}}$). The third function ($G_F$) builds the residual graph that represents the graph's state when some flow is passed through one of the edges in this graph. Formally, G_F is the tuple $< S, N, E_F, Cr^N, C^E >$, where $E_F = \{E^{(N_i, N_j)} \in E \backslash R_F(E^{(N_i, N_j)}) \geq 0\}$.

When $R_F(E^{(N_i, N_j)})$ equals to 0 $E^{(N_i, N_j)}$ is considered as saturated so no more flow could be passed through it. To find an augmenting path, there exists path-search methods such as depth-first-search (DFS) used in [24]. However these search methods rely only on the structure of the graph. We thus propose Algorithm 1 that relies on greedy best-first search ($findGreedyBFSPath$) with user credibility as heuristic. Algorithm 1 returns the maximum flow (F) and the set of users (U) kept for rating. It aims to reach a global optimum (i.e., a path from the source to the sink that contains the maximal number of non-Sybil credible users) by finding a locally optimal choice. At each iteration Algorithm 1 starts from the source S and expands the flow path at the next level with the most credible node N_i with the lowest value of the heuristic function $h(N_i) = 1/Cr^{N_i}$ and a non-null residual capacity $R_F(E^{(S, N_i)}) > 0$.

Algorithm 1. User selection algorithm

Require: $G = <S, N, E, Cr^N, C^E>$ and S'
Ensure: F, U
   for each $E^{(N_i, N_j)} \in E$ **do**
     $F(E^{(N_i, N_j)}) \leftarrow 0$
   end for
   $U \leftarrow \emptyset$
   repeat
     $path=$findGreedyBFSPath(G_F)
     for each $E^{(N_i, N_j)} \in path$ **do**
       $F \leftarrow F + 1$
       $U \leftarrow U \cup \{N_i, N_j\}$
     end for
   until $path = NULL$
   return F, U

---

5 Probabilistic Trust Assessment

In this section, we describe our proposed probabilistic database-based trust model. It consolidates end-users' ratings taking into account end-user credibility. We first discuss how our probabilistic database is structured using a tuple-independent model and then how trust is assessed as a query evaluation.

5.1 Our Probabilistic-Data Model

Our trust approach designs a probabilistic database $ProbDB$ in order to assess trust. Formally, $DataBase$ $ProbDB=(S, T, prob)$ is a triple consisting of a database $Schema$ (S), a finite set of $Tuples$ (T), and a function $prob$ that assigns a probability value to each tuple $t \in T$. S defines $Probabilistic$ $Relations$ $ProbR$ represented as $ProbR(A_1,\ldots, A_m, p)$ where $A_1,\ldots, A_m$ denote a finite set of Attributes and p denotes the probability value attached to t in a relation instance of $ProbR$. The value $p = prob(t)$ represents the confidence that the tuple exists in the database. The $Semantics$ (Sem) of $ProbDB$ is defined through the *possible worlds model* [6]. In [4] Cavallo and Pittarelli define $Sem(ProbDB)$ as a discrete probability space over a finite number (n) of database instances. They refer to the various alternative states of $ProbDB$ as "possible worlds" (pwd_k). $ProbDB$ with n tuples will include 2^n possible worlds. Formally, $Sem(ProbDB)=(PWD, P)$ where $PWD = \{pwd_1, \ldots, pwd_n\}$ is the set of possible worlds and $P : PWD \rightarrow [0, 1]$ such that $\sum_{j=1, n} P_j = 1$ is the probability associated to the existence of each possible world.

Let us consider the following tuple t: `u`$_i$ `has correctly observed that` WS_j `satisfies his requests`. The probability $(prob(t))$ means the extent to which this observation is true. When $prob(t)$ is equal to 1 (resp. 0) t is valid (resp. is not) in all cases. We model this probability by the user credibility CR_i. To design $ProbDB$ we first pre-process a traditional relational database (DB) that contains on top of collected ratings additional information on service providers and evaluation periods. To obtain $\mathcal{P}rob\mathcal{DB}$ we extract from $\mathcal{DB}$ relevant views for trust assessment and add extra details such as credibility values obtained by the credibility model in Sect. 4 to these views. Thus, $\mathcal{DB}$ is built upon an extended schema compared to $\mathcal{P}rob\mathcal{DB}$. Different data models exist to handle uncertainty in databases (e.g., [9] and [19]) according to whether the uncertainty is related to tuples or attributes of the database. We use the independent tuple-level uncertainty-model like the one in [6] where $\mathcal{P}rob\mathcal{DB}$ is an ordinary relational database where each tuple is associated with a probability of being true regardless of any other tuple.

For illustration purposes let assume a database that contains one probabilistic relation $\mathcal{P}rob\mathcal{R}(service, end\text{-}user, rating, p)$ where $service, end\text{-}user, rating$ denote service's identifier, end-user's name, and satisfaction degree of end-user in this service (Fig. 2a). $\mathcal{P}rob\mathcal{R}$ consists of three tuples t_1, t_2, and t_3 with probabilities 0.12, 0.84, and 0.88, respectively. These latter correspond to credibility values computed using our credibility model on a random dataset.

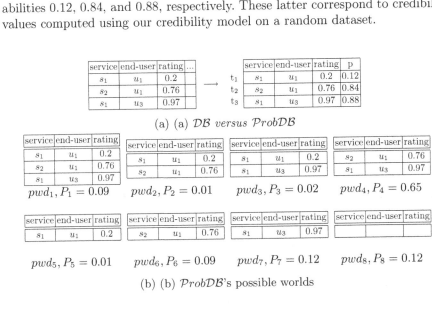

(a) (a) $\mathcal{DB}$ versus $\mathcal{P}rob\mathcal{DB}$

(b) (b) $\mathcal{P}rob\mathcal{DB}$'s possible worlds

Fig. 2. Probabilistic database illustration

Figure 2b shows the possible worlds pwd_k for $\mathcal{P}rob\mathcal{DB}$ and their associated probabilities ($\mathcal{P}_k$). Each pwd_k contains a subset of the tuples present in $\mathcal{P}rob\mathcal{DB}$. $\mathcal{P}_k$ is calculated using the independence assumption (multiply together the existence probabilities of tuples present in pwd_k and non-existence probabilities of tuples not present in pwd_k). For example, $\mathcal{P}_2$ for $pwd_2 = \{t_1, t_2\}$ is computed as $0.12*0.84*(1-0.88) = 0.01$.

We note that $\mathcal{P}rob\mathcal{R}$ contains tuples linked to end-users who provide ratings for different services. These end-users can be constant (i.e., always credible or not) or inconsistent (i.e., swing from credible to uncredible and *vice versa*) in their evaluations. Indeed some end-users are more credible than others and provide correct ratings, while others are less credible and do the opposite. Let consider two tuples t_1 and t_2 related to u_1. If t_1 is false, then it is false because u_1 is wrong. t_2 is likely to be false, too. Thus, if one tuple is false, the probability that the other tuple is false increases as well. Therefore, the proposed probabilistic data-model does not comply with the independent tuple model (e.g., [22]); each tuple is associated with a probability that needs to be independent from the rest of tuples.

In order to achieve the independence rule in the model, We normalize $ProbDB$ ($ProbDB^N$) into two tuple-independent probabilistic relations $PEER$ and $ProbR_1$. $PEER$ stores all end-users and their respective credibility values. Since $PEER$ should often be updated we consider it as a view instead of a table. u_i is credible about WS_j if his ratings are consistent. Equation 4 assesses u_i's credibility (CR_i) over the ratings he provided in the past.

$$CR_i = \prod_j CR_i^j \qquad (4)$$

From $ProbR$ we compute CR_1 as $0.12 * 0.84 = 0.1$. As u_3 provides only one rating, CR_3 remains the same in $PEER$. $ProbR_1$ stores all tuples that now are independent subject to the end-user credibility (Fig. 3).

$PEER$

end-user	p
u_1	0.1
u_3	0.88

$ProbR_1$

service	end-user	rating	p
s_1	u_1	0.2	0.12
s_2	u_1	0.76	0.84
s_1	u_3	0.97	0.88

Fig. 3. $ProbDB$ normalization

5.2 Trust Assessment as a Query Evaluation

To establish WS's trust from $ProbDB^N$ we develop specific queries. An end-user trusts WS_j if it has successfully satisfied a large number of end-users' requests. We establish WS's trust by aggregating end-users' ratings into one probabilistic value. This can be expressed using a SQL query *SELECT AVG* to obtain the *rating* average value from $ProbR_1$. Intuitively, applying this query on pwd_k means that end-users in pwd_k **jointly** observe that WS_j satisfies their requests with probability P_k. Let $\mathcal{F}_{AVG(rating)}(\sigma_{service=WS_1})$ be the following SQL query:

SELECT AVG(*rating*)   **FROM** $ProbDB^N$
WHERE *service* $= WS_1$;

$ProbDB^N$ is interpreted as $2^5 = 32\ pwd_k$. Figure 4a shows pwd_1's content. Figure 4b shows that $\mathcal{F}_{AVG(rating)}(\sigma_{service=WS_1})$'s evaluation returns four possible answers for trust value 0.585, 0.2, 0.97 and empty set ordered by existence probability. In [11] Jayram et al. represents $\mathcal{F}_{AVG()}(\sigma)$'s result over probabilistic databases as a weighted average of possible answers for the trust value.

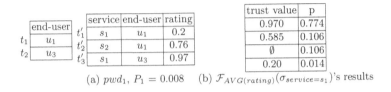

(a) pwd_1, $P_1 = 0.008$ (b) $\mathcal{F}_{AVG(rating)}(\sigma_{service=s_1})$'s results

Fig. 4. Query evaluation on $\mathcal{P}rob\mathcal{DB}^{\mathcal{N}}$

Despite the simplicity of possible worlds semantics it raises some challenging computational concerns even for simple query operations like in [6]. Many studies have shown that the query evaluation problem is ♯P-hard. Several algorithms (e.g., [6] and [11]) are provided to handle complex queries over massive data streams.

6   System Development

We implemented a JAVA trust assessment system using Eclipse IDE and PostgreSQL to store users' ratings on WSs. The experiments' objective is to challenge the robustness of our approach in the presence of Sybils by injecting invalid ratings for the same user. Robustness is an important quality attribute when it comes to executing critical applications.

6.1   Parameter Setting

Nowadays there is a serious lack of publicly available real datasets on WSs' ratings. To address this limitation we looked for a dataset that could encompass similar information suitable for WSs evaluation and could also represent users' trust relationships. The social rating network dataset Epinions[1] seems to be a good dataset for our work. Epinions.com is a well-known knowledge sharing and review site. Users need to register for free so they can submit their personal reviews (e.g., narrative reviews and integer ratings from 1 to 5) on various items such as products, companies, and movies. Every Epinions user maintains a list of trust relationships with other users. The dataset used in our experiments was crawled from Epinions.com in November and December 2013. This latter contains 664, 824 ratings from 49, 290 users on 139, 738 items. We consider an item as a WS and normalize the items' rating values. We also use the open-source library Apache Mahout for the fuzzy $\mathcal{C}$-means algorithm (Sect. 4) and Jung library for max-flow algorithm. To prune the social network we fix the incoming edges threshold e_{in} to 3 based on the experiments done in [24] that show that more than 80 % of ratings are collected when $e_{in} \geq 3$.

In our experiments we alter a variable ratio of existing users' ratings in the dataset to make them act like Sybils and preserve ratings of the remaining

---

[1] http://www.epinions.com.

users. Since Sybils provide inconsistent ratings their satisfaction is reversed. This permits to disturb the trust model by making users act as Sybils and then non-Sybils and so on.

6.2 Experiments

We carried out several experiments that compute trust values based on two parameters: (i) ratio of altered users in the dataset; and (ii) choice of a trust model $\mathcal{M}_i$. Three models are selected to assess trust as per Sect. 5. Database tuples' probabilities are assessed using the fuzzy credibility model described in Sect. 4. The models differ in terms of using or not a Sybil-defense mechanism and user selection mechanism. $\mathcal{M}_1$ does not use a Sybil-defense mechanism and involves all users in trust assessment. Both $\mathcal{M}_2$ and $\mathcal{M}_3$ use a Sybil-defense mechanism. The former uses DFS search to select users and the latter use the heuristic user selection mechanism defined in Sect. 4.2. Trust is assessed as a query evaluated on the probabilistic database $\mathcal{P}rob\mathcal{DB}^{\mathcal{N}}$. The experiment analyzes the performance of these models in achieving realistic trust values when altering the ratio of Sybils. Figure 5 shows that both $\mathcal{M}_2$ and $\mathcal{M}_3$ provide better accurate trust values (i.e., closer to those obtained before rating alterations) than $\mathcal{M}_1$. This shows the importance of Sybil-defense in improving the trust model robustness. However it can be seen that trust results given by $\mathcal{M}_2$ oscillate significantly compared to those given by $\mathcal{M}_3$. This is due to the arbitrary choice of users by operating a DFS search compared to our selection mechanism that relies almost on credible users.

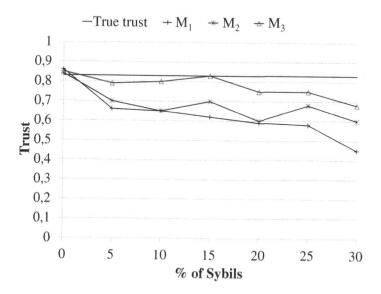

Fig. 5. Quality of trust

7 Conclusion

In this paper, we proposed an approach that addresses the attacks of Sybil users on Web services with focus on their trust computation. These users provide inconsistent ratings over the same Web services since they use different identities. To debunk Sybil users the approach relies on credible users who provide consistent ratings and hence, can be trusted. Both users, Sybil and non-Sybil, are part of a graph that is searched using a rating-ticket distribution mechanism and an adapted max-flow algorithm, both based on user's credibility. The objective is to select the highest number of non-Sybil credible users and decrease the attack capability of Sybil users. In term of future work we would like to investigate further credit payment models to improve the quality of trust value in the presence of Sybils.

References

1. Bezdek, J.: Pattern Recognition with Fuzzy Objective Function Algorithms. Kluwer Academic Publishers, Norwell (1981)
2. Bordens, K., Horowitz, I.: Social Psychology. Psychology Press, Mahwah (2001)
3. Buchegger, S., Boudec, J.Y.L.: A robust reputation system for peer-to-peer and mobile ad-hoc networks. In: P2P Econ, Cambridge, USA (2004)
4. Cavallo, R., Pittarelli, M.: The theory of probabilistic databases. In: Very Large Data Bases Conferences, Brighton, England (1987)
5. Cheng, A., Friedman, E.: Sybilproof reputation mechanisms. In: Proceedings of the 2005 ACM SIGCOMM Workshop on Economics of Peer-to-peer Systems. ACM, New York, NY, USA (2005)
6. Dalvi, N., Suciu, D.: Efficient query evaluation on probabilistic databases. VLDB J. **16**(4), 523–544 (2007)
7. Danezis, G., Mittal, P.: Sybilinfer: detecting sybil nodes using social networks. Technical report (2009)
8. Ford, L.R., Fulkerson, D.R.: A simple algorithm for finding maximal network flows and an application to the hitchcock problem. Can. J. Math. **9**(2), 210–218 (1957)
9. Fuhr, N., Rölleke, T.: A probabilistic relational algebra for the integration of information retrieval and database systems. ACM Tran. Inf. Syst. (TOIS) **15**(1), 32–66 (1997)
10. Hota, C., Srikanth, M., Yla-Jaaski, A., Lindqvist, J., Kristiina, K.: Safeguarding against sybil attacks via social networks and multipath routing. In: 2007 Second International Conference on Communications and Networking in China, CHINA-COM 2007 (2007)
11. Jayram, T.S., Kale, S., Vee, E.: Efficient aggregation algorithms for probabilistic data. In: Annual ACM-SIAM Symposium on Discrete Algorithms, New Orleans, USA (2007)
12. Lesko, W.: Readings in Social Psychology: General, Classic and Contemporary Selections. Allyn & Bacon, Boston (1997)
13. Mihail, M., Saberi, P.A.: Random walks with lookahead in power law random graphs. Internet Math. **1**(1), 147–152 (2007)
14. Malik, Z., Bouguettaya, A.: Rateweb: reputation assessment for trust establishment among web services. Very Large Data Bases (VLDB) J **18**(4), 885–911 (2009)

15. Mislove, A., Viswanath, B., Gummadi, K., Druschel, P.: You are who you know: inferring user profiles in online social networks. In: Proceedings of the Third ACM International Conference on Web Search and Data Mining. WSDM 2010 (2010)
16. Nguyen, N., Caruana, R.: Consensus clusterings. In: International Conference on Data Mining, Omaha, USA (2007)
17. Noor, T., Sheng, Q., Ngu, A., Alfazi, A., Law, J.: Cloud armor: a platform for credibility-based trust management of cloud services. In: The ACM Conference on Information and Knowledge Management (CIKM) (2013)
18. Saoud, Z., Faci, N., Maamar, Z., Benslimane, D.: A fuzzy clustering-based credibility model for trust assessment in a service-oriented architecture. In: International Conference on Enabling Technologies: Infrastructure for Collaborative Enterprises (WETICE), Parma, Italy (2014)
19. Sarma, A., Benjelloun, O., Halevy, A., Widom, J.: Working models for uncertain data. In: International Conference on Data Engineering (ICDE), Atlanta, USA (2006)
20. Schum, D., Morris, J.: Assessing the competence and credibility of human sources of intelligence evidence: contributions from law and probability. Law Probab. Risk 6(1), 247–274 (2007)
21. Sternthal, B., Phillips, L., Dholakia, R.: The persuasive effect of source credibility: a situational analysis. Public Opin. Q. 42(3), 285–314 (1978)
22. Suciu, D.: Probabilistic databases. SIGACT News 39(2), 17–43 (2008)
23. Teacy, W.T., Patel, J., Jennings, N.R., Luck, M.: Travos: trust and reputation in the context of inaccurate information sources. Auton. Agent. Multi-Agent Syst. 12(2), 239–256 (2006)
24. Tran, N., Min, B., Li, J., Subramanian, L.: Sybil-resilient online content voting. In: Proceedings of the 6th USENIX Symposium on Networked Systems Design and Implementation, Berkeley, CA, USA (2009)
25. Troffaes, M.: Generalizing the conjunction rule for aggregating conflicting expert opinions. Int. J. Intell. Syst. 21(3), 229–259 (2006)
26. Viswanath, B., Mondal, M., Gummadi, K., Mislove, A., Post, A.: Canal: scaling social network-based Sybil tolerance schemes. In: Proceedings of the 7th European Conference on Computer Systems (EuroSys 2012), Bern, Switzerland (2012)
27. Vu, L.-H., Hauswirth, M., Aberer, K.: QoS-based service selection and ranking with trust and reputation management. In: Meersman, R., Tari, Z. (eds.) OTM 2005. LNCS, vol. 3760, pp. 466–483. Springer, Heidelberg (2005)
28. Wang, Y., Singh, M.: Formal trust model for multiagent systems. In: Proceedings of the International Joint Conference on Artifical Intelligence, Hyderabad, India (2007)
29. Yu, H., Gibbons, P.B., Kaminsky, M., Xiao, F.: Sybillimit: A near-optimal social network defense against sybil attacks. In: Proceedings of the 2008 IEEE Symposium on Security and Privacy. IEEE Computer Society, Washington, DC, USA (2008)
30. Yu, H., Kaminsky, M., Gibbons, P.B., Flaxman, A.: Sybilguard: defending against sybil attacks via social networks. SIGCOMM Comput. Commun. Rev. 36, 267–278 (2006)
31. Zhou, R., Hwang, K.: Powertrust: a robust and scalable reputation system for trusted peer-to-peer computing. IEEE Trans. Parallel Distrib. Syst. 18(4), 460–473 (2007)

Erratum to: ForCE: Is Estimation of Data Completeness Through Time Series Forecasts Feasible?

Gregor Endler[(✉)], Philipp Baumgärtel, Andreas M. Wahl, and Richard Lenz

Computer Science 6 (Data Management),
Friedrich-Alexander-Universität Erlangen-Nürnberg, Erlangen, Germany
{gregor.endler,philipp.baumgaertel,andreas.wahl,
richard.lenz}@fau.de
https://www6.cs.fau.de

Erratum to:
Chapter 'ForCE: Is Estimation of Data Completeness
Through Time Series Forecasts Feasible?' in:
T. Morzy et al. (Eds.):
Advances in Databases and Information Systems, LNCS,
DOI: 10.1007/978-3-319-23135-8_18

The authors corrected errors in the figures appearing in Sect. 3.2 and the Appendix and adjusted the text referring to the figures.

The updated original online version for this chapter can be found at
DOI: 10.1007/978-3-319-23135-8_18

© Springer International Publishing Switzerland 2017
T. Morzy et al. (Eds.): ADBIS 2015, LNCS 9282, p. E1, 2015.
DOI: 10.1007/978-3-319-23135-8_32

Author Index

Printed in the United States
By Bookmasters